Back to the Roots and Back to the Future

Towards a New Synthesis amongst Taxonomic, Ecological and Biogeographical Approaches in Carabidology

Proceedings of the XIII European Carabidologists Meeting, Blagoevgrad, August 20-24, 2007

Back to the Roots and Back to the Future

Towards a New Synthesis amongst Taxonomic, Ecological and Biogeographical Approaches in Carabidology

Proceedings of the XIII European Carabidologists Meeting, Blagoevgrad, August 20-24, 2007

Edited by

L. Penev, T. Erwin & T. Assmann

PENSOFT

Sofia–Moscow

2008

Back to the Roots and Back to the Future

Towards a New Synthesis amongst Taxonomic, Ecological and Biogeographical Approaches in Carabidology

Proceedings of the XIII European Carabidologists Meeting, Blagoevgrad, August 20-24, 2007

Editors:
L. Penev, T. Erwin & T. Assmann

First published 2008
ISBN 978-954-642-325-2 (HB)
ISBN 978-954-642-424-2 (e-book)

Pensoft *Series Faunistica* No 75
ISSN 1312-0174

Pensoft Publishers
Geo Milev Str. 13a, Sofia 1111, Bulgaria
Fax: +359-2-870-42-82
info@pensoft.net
www.pensoft.net

Printed in Bulgaria, June 2008

This volume is dedicated to
Professor George E. Ball

Prof. George E. Ball, Emeritus Curator of the Strickland Entomological Museum and Emeritus Professor of Entomolgy at the University of Alberta, Edmonton, Canada enjoying a beer, as he listens to Achille Casale explain why his study of Neotropical *Calleida* is so difficult (color morphs, pronotum morphs, getting geographic representation, etc.). George, the ultimate Mentor with patience with older distinguished colleages, as well as with the students participating in the XIII Carabidologists Meeting.

Contents

Preface

Carabidae, the ground beetles, is one of the most diverse taxa on Earth. Nearly 40,000 species are already taxonomically known and some hundreds of new species are described each year from the Palaearctic realm alone. Moreover, ground beetles are distributed in nearly all terrestrial habitats and on all continents (perhaps with the exception of present day Antarctica, although fossil Trechini are known from the interior of that Continent). Therefore, it is a real challenge to detect, describe, and categorize the taxa of this hyperdiverse family. The systematic and taxonomic work on this group is highly important because they are needed for organizing our basic knowledge of nature, as well as for other disciplines of carabidology, as this beetle family is a preferred study subject of ecologists and evolutionary biologists both in basic and applied research.

The 13th European Carabidologists' Meeting held from the 20th to 24th of August 2007 in Blagoevgrad (Bulgaria) was the first meeting in the series of these scientific gatherings which focused explicitly on the ecology, biogeography, and evolution, *as well as* on taxonomic aspects of Carabidae. It was attended by almost 90 participants from 20 countries (see photo and list of participants).

The Meeting was honored in an outstanding way by the active participation of Professor Dr. George E. Ball, senior carabidologist and one of the most influential coleopterists worldwide. His opening speech was characterized by his usual wit and indeed his broad and perceptive view on the development of American carabidology during the last half of the 20th century (plus the first decade of the 21th century) (Ball, this volume). To honor Professor Ball's life work and his influence on carabidology, all of the contributors and editors dedicate this proceedings volume to him with a heartfelt **thank you** for his life time of carabidological endeavors.

The contributions included in this proceedings volume cover a broad spectrum of research on carabids ranging from taxonomy and systematics to genetics, morphology, Pleistocene faunal reconstructions, ecology, evolutionary biology, methodological approaches, ethology, faunistics, and conservation biology. All manuscripts were peer-reviewed before acceptance and underwent scrutiny in the Q & A session following their oral presentations at the Meeting. The rapid publication was possible due to excellent cooperation among authors, editors, and referees. The latter are acknowledged for their valuable contributions in improving the manuscripts.

In the opening presentation of the Meeting, Professor Ball emphasized the need for systematic work on ground beetles, not only for the American realm, but also for most regions worldwide. The contribution of Borislav V. Guéorguiev and Roman Lohaj on the microphthalmic Pterostichini, genus *Speluncarius*, is in this tradition and combines morphology-based taxonomy and phylogeny. An overview of the genera of Western Hemisphere Pseudomorphini is presented by Terry L. Erwin and Christy J. Geraci. Both contributions are not restricted to pure taxonomy's thinking and give also notes on distribution, habitat selection, and – very important for other carabidologists – offer identification keys.

Professor Ball called on our community several times to take into account larval characters and morphology into consideration for present day carabid taxonomy. The following contributions comply with this challenge: Kirill V. Makarov worked on larval chaetotaxy in the genus *Rhysodes* to examine the systematic position of Rhysodidae. Artjem A. Zaitsev studied the thoracic endoskeleton of larvae belonging to a series of tribes for comparison. A synthesis on morphology-based alpha taxonomy and molecular methods is suggested by Assmann and co-authors.

Morphological approaches are presented by Anita Giglio and co-authors to compare antennal morphometry and sensilla distribution patterns in three *Siagona* species. Although sibling species were studied, surprising differences were detected. Wilfried Paarmann and Dietrich Mossakowski, each together with their co-authors, were able to demonstrate that the different elytral colors of *Poecilus lepidus* are controlled by a single gene with several alleles. The color forms of the phenotypes are caused by a multilayer system of electron dense and less dense layers in the exocuticle.

Different approaches to reconstruct past distributions and population history of carabids were used by Evgeniy Zinovyev and Nordfried Kamer and his co-authors. The first mentioned author used fossils from more than 100 deposits originated during the Late Pleistocene to Holocene. He is able to reconstruct the past carabid communities and species' range shifts during the last thousands of years. DNA sequences are used by Kamer et al. to reconstruct the phylogeography of ground beetles mostly from the Baltic Sea coast. Marked genetic differences support the assumption of multiple glacial refuges for *Carabus clatratus*.

Three conservation biological approaches deal with endangered carabids and their habitats. Achille Casale and Enrico Busato describe the decline of *Carabus clatratus* in Italy. The authors have every reason to suppose that the decline is caused by an alien, very invasive species, the red swamp crayfish *Procambrus clarkii*. The contribution of Erik Arndt introduced the Habitats Directive of the European Union. Habitat types and some protected species require the designation of special areas and conservation programs. Ground beetles can also be used as indicator groups in the monitoring program of protected habitats in the directive's framework. Sándor Bérces and co-authors summarize all available data on the current distribution of *Carabus hungaricus* in Hungary. The species is listed in the Habitats Directive of the European Union and it is characteristic for the Pannonian sand steppe habitats, the most vulnerable of the dolomitic grasslands in Hungary.

Since the first European Carabidologists' Meeting in Wijster (Wageningen, The Netherlands), 1969, all the following ones have contained some ecological contributions. This was also true for the meeting in Blagoevgrad. The spectrum is broad and ranges from methodological and local aspects to community ecology.

Herein, three contributions focus on methodological approaches in carabid ecology: Evan D. Esch and co-authors use correlations of the probability of capture in pitfall traps and the parameters body size and temperature to estimate a correction factor. The use of this factor increases the correlation between pitfall trap samples and known or estimated abundance of carabids in two study sites. Anika Timm and co-authors compared pitfall trap catches with the results from hand picking and sifting in Mediterranean habitats of Israel. Claudia Drees and co-authors propose a method of direct observation of nocturnal beetles under red-light conditions in order to gain insight into the biology and the behavior of ground beetle species. The results from both a small ground beetle and a larger one illustrate the easy applicability of the method.

Forest ecology is the main topic of two contributions from Canada and Poland: Joshua M. Jacobs and co-authors describe the influences of succession and harvest intensity on ground beetles in a boreal mixed wooded forest in Alberta, Canada. By the way, J. Jacobs and Evan Esch (see above) are "academic grandchildren" of Professor Ball because Ball was the academic supervisor of John Spence, who is now supervising both Jacobs and Esch. Jaroslaw Sklodowski used capture-recapture rates to describe the locomotory movement of forest carabids in a clear-cut.

Community ecology is the topic of four contributions: Yurii I. Chernov and Olga L. Makarova describe the species-poor beetle fauna of the arctic tundra and polar deserts. The present day fauna is composed of relatively young migratory elements containing large portions of macropterous species and by continental elements which comprise mainly wingless or wing-dimorphic species. Pavel Saska used metal enclosures with nested pitfall traps to study impacts of weed communities on carabid assemblages. His results indicate that composition of weed populations influences the structure of ground beetle assemblage. A subproject of GLOBENET, the largest international carabid project, is presented by Lyubomir Penev and co-authors. They describe patterns of urbanization for Sofia, Bulgaria and its environs. Axel Schwerk and Jan Szyszko studied patterns of succession of post-industrial habitats in central Poland. With ongoing succession , species characterized by small size are replaced by large bodied species, so that Mean Individual Biomass (MIB) increases as succession progresses.

Inessa Kh. Sharova developed a new version of larval Carabidae life forms, based on numerous literature sources. The general trend seems to be the diversification of zoophags, which reflects the expansion into all main terrestrial habitats and numerous microhabitats. Another tendency of larval morphology is the transformation from zoophagy through mixophyto-, to phytophagy.

Andrey V. Matalin used a meta-analysis to describe the evolution and gonad development of biennial life cycles in ground beetles of the western Palaearctic. His data show that the proportion of ground beetles with biennial life cycles in different habitats within the temperate zone is rather high and biennial development is not unusual. Andrey V. Matalin and Kirill V. Makarov determined life cycles of Pogonini species. Most species are monovoltine with spring or spring-summer reproduction and adult hibernation. However, one species is an autumn-breeding iteroparous species, its life cycle being monovoltine with obligate larval hibernation and obligate adult aestivation parapause.

A review of defensive strategies against predators is presented for carabid beetles by Tullia Zetto Brandmayr and co-authors. The authors consider unpalatability, gregariousness and aposematism as important strategies of ground beetles applying to defense against predators. Moreover they tested possible aposematism via warning signals in the laboratory. The highly similar cuticle molecular composition between two species, the bombardier beetle *Brachinus sclopeta* and the platynine, *Anchomenus dorsalis*, is hypothesized as a mechanism similar to that involved in color similarity, i.e., effective in reducing the predation risk by non-visual predators of ground beetles.

The organizing committee for the 13th European Carabidologists' Meeting consisted of: Dr. Lyubomir Penev, Chairman, (Central Laboratory for General Ecology, BAS), Dr. Nesho Chipev (Central Laboratory for General Ecology, BAS), Dr. Borislav Gueorguiev (National Museum of Natural History, BAS), Dr. Ivailo Dedov (Central Laboratory for General Ecology, BAS), Dr. Vlada Peneva (Central Laboratory for General Ecology, BAS), Dr. Gergana Vasileva (Central Laboratory for General Ecology, BAS), Dr. Vera Antonova (Central Laboratory for General Ecology, BAS), Ivailo Stoyanov (Central Laboratory for General Ecology, BAS), Teodor Georgiev (Pensoft Publishers).

We thank also the American University in Bulgaria for providing the venue of the Meeting, as well as the team of Pensoft Publishers for their help in organising the Meeting and efforts to publish this volume.

Lyubomir Penev
Terry Erwin
Thorsten Assmann

List of the Proceedings of the European Carabidologists Meetings

Den Boer, P.J. (Ed.) 1971. Dispersal and dispersal power of carabid beetles. Miscellaneous Papers Landbouwhogeschool Wageningen, No. 8. (European Meeting no.1)

Den Boer, P.J., Thiele, H-U. & Weber, F. (Eds) 1979. On the evolution of behaviour in carabid beetles. Miscellaneous Papers, Agricultural University Wageningen, No. 18. (European Meeting no.3)

Erwin, T.L., Ball, G.E., Whitehead, D.E. & Halpern, A.L. (Eds) 1979. Carabid beetles: their evolution, natural history and classification. Junk, The Hague. (International Symposium no.1)

Brandmayr, P., den Boer, P.J. & Weber, F. (Eds) 1983. Ecology of carabids: the synthesis of field study and laboratory experiment. Report of the 4th Meeting of European Carabidologists. PUDOC, Wageningen

Den Boer, P.J., Grüm, L. & Szyszko, J. (Eds) 1986. Feeding behaviour and accessibility of food for carabid beetles. Warsaw Agricultural University Press, Warsaw (European Meeting no. 5)

Den Boer, P.J., Luff, M.L., Mossakowski, D. & Weber, F. (Eds) 1986. Carabid beetles. Their adaptations and dynamics. G. Fischer Verlag, Stuttgart-New York. (symposium at the 17th Int. Entomological Conference)

Den Boer, P.J., Lövei, G.L., Stork, N.E. & Sunderland, K.D. (Eds) 1987. Proceedings of the 6th European Carabidologists' Meeting. Acta Phytopathologica et Entomologica Hungarica 22: 1-458.

Stork, N.E. (Ed.) 1990. The role of ground beetles in ecological and environmental research. Intercept, Andover, U.K. (European Meeting no.7)

Desender, K., Dufrene, M., Loreau, M., Luff, M.L. & Maelfait, J-P. (Eds) 1994. Carabid beetles: ecology and evolution. Series Entomologica, vol. 51. Kluwer Academic Publishers, Dordrecht. (European meeting no.8 & International Meeting no.2)

Niemelä, J. (Ed.) 1996. Annales Zoologici Fennici (International Meeting no.3)

Brandmayr, P., Lövei, G.L., Zetto Brandmayr, T., Casale, A. & Vigna Taglianti, A. (Eds) 2000. Natural History and Applied Ecology of Carabid Beetles. Proceedings of the IX European Carabidologists' Meeting (26-31 July 1998, Camigliatello, Cosenza, Italy). Pensoft Publishers, Sofia–Moscow.

Szyszko, J., Den Boer, P. J. & Bauer, Th. (Eds) 2001. How to protect or what we know about carabid beetles. Proceedings of the X European Carabidologists' Meeting, Tuczno, Poland. Warsaw Agricultural University Press, Warsaw.

Lövei, G.L. & Toft, S. (Eds) 2005. European carabidology 2003. Proceedings of the 11th European carabidologists meeting, Århus, 21-24 July 2003 – DIAS report plant production 114, Århus, Denmark.

Serrano J., Gomez-Zurita, J. & Ruiz, C. (Eds) 2005. Ground beetles as a key group for biodiversity conservation in Europe. Proceedings volume of the XII European Carabidologsts Meeting, Murcia, September 19-22 2005. Nausícaä, Murcia.

Serrano, J., Koivula Matti & Lövei, G. (Eds) 2006. Proceedings of the XII Carabidologists' Meeting. Entomologica Fennica 17(3).

Photo by Vera Antonova

1. R. Babko
2. V. Chikatunov
3. M. Kirichenko
4. S. Golovatch
5. A. Turin
6. L. Golovatch
7. M. Babajko
8. J. Fermin-Sanchez
9. L. Jelaska
10. S. Jelaska
11. Z. Elek
12. T. Magura
13. B. Tóthmérész
14. G. Pozsgai
15. B. Noll
16. C. Drees
17. A. Materin
18. M. Kleinwächter
19. A. Timm
20. S. Bérces
21. J. Skłodowski
22. T. Assmann
23. S. Vujčić-Karlo
24. D. Prins
25. A. Schwerk
26. A. Brigić
27. A. Casale
28. J. Bohac
29. C. Bergeron
30. T. Zetto Brandmayr
31. D. Paarmann
32. S. Tejero-Garcia
33. L. Penev
34. A. Taboada
35. D. Mossakowski
36. W. Paarmann
37. J. Jacobs
38. P. Brandmayr
39. E. Esch
40. P. Saska
41. J. Niemelä
42. G. Lövei
43. E. Arndt
44. E. Dauffy-Richard
45. L. Khobrakova
46. R. Vermeulen
47. M. Koivula
48. T. Erwin
49. A. Matalin
50. H. Turin
51. H. Dhuyvetter
52. R. Pizzolotto
53. G. Ball
54. O. Makarova
55. O. Shelef
56. W. Dekoninck
57. M. Gerisch
58. B. Guéorguiev
59. E. Zinovyev
60. R. Kostova
61. A. Gasith
62. S. Venn
63. P.K. Poulsen

List of Participants of the XIII European Carabidologists Meetings

Veronica Agostinelli – Technische Universität Berlin, GERMANY, *veronicaago@yahoo.it*

Carmelo Andujar – Department of Zoology, Faculty of Veterinary, Campus de Espinardo, 30071 Murcia, SPAIN, *candujar@um.es*

Vera Antonova – Central Laboratory for General Ecology, Yuri Gagarin Street 2, 1113 Sofia, BULGARIA, *vera.antonova@yahoo.com*

Erik Arndt – Anhalt University of Applied Sciences, Department LOEL, Strenzfelder Allee 28, D-06406 Bernburg, GERMANY, *earndt@loel.hs-anhalt.de*

Thorsten Assmann – Institute of Ecology and Environmental Chemistry, University of Lüneburg, Scharnhorststr. 1, D-21314 Lüneburg, GERMANY, *assmann@uni-lueneburg.de*

Morana Babajko – Public institution for management of protected areas in the county of Zadar, B.Vranjana 11, 23000 Zadar, CROATIA, *morana.babajko@gmail.com*

Roman Babko – Department of Ethology and Sociobiology of Insects, Schmalhausen Institute of Zoology, NAS of Ukraine, B. Khmielnicky str. 15, Kiev-30, 01601 UKRAINE

George E. Ball – Department of Biological Sciences, University of Alberta, Edmonton, Alberta, T6G 2E9, CANADA, *gball@ualbera.ca*

Arvīds Barševskis – Institute of Systematic Biology, Daugavpils University, Vienības Str. 13, Daugavpils, LV-5401, LATVIA, *arvids.barsevskis@biology.lv*

Sándor Bérces – Duna-Ipoly National Park Directorate, Hűvösvölgyi út 52., Budapest, 1021 HUNGARY, *bercess@gmail.com*

Colin Bergeron – 442 Earth Sciences Building, University of Alberta, Edmonton, AB, T6G 2E3, CANADA, *cb1@ualberta.ca*

Jaroslav Bohac – University of South Bohemia, Faculty of Agriculture, Studentska 13, 370 05 Ceske Budejovice, CZECH REPUBLIC *ardaboh@seznam.cz*

Pietro Brandmayr – Dipartimento di Ecologia, Università della Calabria, 87036 Arcavacata di Rende, (CS), ITALY, *brandmayr@alice.it*

Tullia Zetto Brandmayr – Dipartimento di Ecologia, Università della Calabria, 87036 Arcavacata di Rende, (CS), ITALY, *tuzetto@unical.it*

Andreja Brigić – Department of Zoology, Faculty of Science, University of Zagreb, Rooseveltov trg 6, 10000 Zagreb, CROATIA, *andrejab@zg.biol.pmf.hr*

Andris Bukejs – Institute of Systematic Biology, Daugavpils University, Vienības Str. 13, Daugavpils, LV-5401, LATVIA, *carabidae@inbox.lv*

Achille Casale – Università di Sassari, Dipartimento di Zoologia e Genetica evoluzionistica, Via Muroni 25, 07100 Sassari, ITALY, *a_casale@libreto.it*

Vladimir Chikatunov – Department of Zoology, Tel-Aviv University, Tel-Aviv 69978, ISRAEL

Emmanuelle A. Dauffy-Richard – Cemagref – Agricultural and Environmental Engineering Research, BIOFOR Team – Sustainable management and biodiversity of forest ecosystems, Domaine des Barres, F-45290 Nogent-sur-Vernisson, FRANCE, *emmanuelle.richard@cemagref.fr*

Ivailo Dedov – Central Laboratory for General Ecology, Yuri Gagarin Street 2, 1113 Sofia, BULGARIA, *dedov@ecolab.bas.bg*

Wouter Dekoninck – Royal Belgian Institute of Natural Sciences (RBINS), Departement of Entomology, Vautierstraat 29, 1000 Brussel, BELGIUM, *wouter.dekoninck@natuurwetenschappen.be*

Hilde Dhuyvetter – Entomology Department, Royal Belgian Institute of Natural Sciences, Vautierstreet 29, Brussel, BELGIUM, *Hilde.Dhuyvetter@naturalsiences.be*
Claudia Drees – Institute of Ecology and Ecological Chemistry, University of Lüneburg, D-21314 Lüneburg, GERMANY, *cdrees@uni-lueneburg.de*
Zoltán Elek – Szent István University, Faculty of Veterinary Sciences, Zoological Institute, Department of Ecology, H-1077 Budapest, Rottenbiller str. 50., HUNGARY, *elek.zoltan@aotk.szie.hu*
Terry L. Erwin – Department of Entomology, MRC 187, National Museum of Natural History, Smithsonian Institution, P.O. Box 37012, Washington, DC, 20013-7012 USA, *erwint@si.edu*
Evan D. Esch – University of Alberta, Department of Renewable Resources, Edmonton, Alberta, ESB 2-36 CANADA, *eesch@ualberta.ca*
Avital Gasith – Zoology Department, Faculty of Life Sciences, Tel-Aviv University, Tel-Aviv 69978, ISRAEL, *avitalg@tauex.tau.ac.il*
Michael Gerisch – UFZ – Helmholtz-Centre for Environmental Research, Permoser Str. 15, 04318 Leipzig, GERMANY, *michael.gerisch@ufz.de*
Lyuba Golovatch – Accompanying person with Sergei Golovatch
Sergei Golovatch – Severtsov Institute of Ecology and Evolution, Russian Academy of Sciences, Leninsky pr., 33, Moscow, 11907 RUSSIA, *sgolovatch@yandex.ru*
Borislav V. Guéorguiev – Natural Museum of Natural History, 1 Blvd. Tzar Osvoboditel, 1000 Sofia, BULGARIA, *bobivg@yahoo.com*
Joshua M. Jacobs – 442 Earth Sciences Building, Department of Renewable Resources, University of Alberta, T6G 2E3 CANADA, *email@joshuajacobs.org*
Lucija Šerić Jelaska – Department of Zoology, Faculty of Science, University of Zagreb, Rooseveltov trg 6, HR-10000 Zagreb, CROATIA, *slucija@zg.biol.pmf.hr*
Sven Jelaska – Accompanying person
Larisa Khobrakova – Institute of General and Experimental Biology, Siberian Branch of the Russian Academy of Sciences, Sakhjanovoj street 6, Ulan-Ude 670047, RUSSIA, *khobrakova77@mail.ru*
Marina Kirichenko – Department of Ethology and Sociobiology of Insects, Schmalhausen Institute of Zoology, NAS of Ukraine, B. Khmielnicky str. 15, Kiev-30, 01601 UKRAINE, *marina_kiri@yahoo.com*
Meike Kleinwächter – Institute for Geoecology, Technical University Braunschweig, D-38092 Braunschweig, GERMANY, *meike.kleinwaechter@tu-bs.de*
Matti J. Koivula – Finnish Museum of Natural History, University of Helsinki, P.O. Box 26, FI-00014, Helsinki, FINLAND, *mjkoivula@gmail.com*
Rumyana Kostova – Faculty of Biology, University of Sofia, Dragan Tsankov Str. 8, 1164 Sofia, BULGARIA, *rkostova@biofac.uni-sofia.bg*
Tal Levanony – Department of Zoology, Tel-Aviv University, Tel-Aviv 69978, ISRAEL, *Levanony@post.tau.ac.il*
Gabor L. Lövei – University of Aarhus, Faculty of Agricultural Sciences, Department of Integrated Pest Management, Flakkebjerg Research Centre, DK-4200 Slagelse, DENMARK, *gabor.lovei@agrsci.dk*
Tibor Magura – Hortobágy National Park Directorate, H-4002 Debrecen, POB. 216, HUNGARY, *magura@www.hnp.hu*

Olga L. Makarova – Severtsov Institute of Ecology and Evolution, Russian Academy of Sciences, Leninsky pr., 33, Moscow 11907, RUSSIA

Andrey V. Matalin – Department of Zoology, Moscow State Pedagogical University, Kibalchicha str. 6, Build. 5, Moscow 129164, RUSSIA, *a_matalin@tochka.ru*

Andrea Matern – Institute of Ecology and Environmental Chemistry, University of Lüneburg, Scharnhorststr. 1, D-21314 Lüneburg, GERMANY, *andrea.matern@enett.de*

Dietrich Mossakowski – Institute for Ecology & Evolutionary Biology, University of Bremen, P.O.Box 330440, D- 28334 Bremen, GERMANY, *dmossa@uni-bremen.de*

Jari Niemelä – University of Helsinki, P.O. Box 65 (Viikinkaari 1), FI-00014 FINLAND, *jari.niemela@helsinki.fi*

Britta Noll – Institute of Ecology and Environmental Chemistry, University of Lüneburg, Scharnhorststr. 1, D-21335 Lüneburg, GERMANY, *britta.noll@gmx.de*

Doris Paarmann – Accompanying person with Wilfried Paarmann

Wilfried Paarmann – HAWK, Fakultät Ressourcenmanagement, Büsgenweg 1A, D 37077 Göttingen, GERMANY, *w-paar@t-online.de*

Ainārs Pankjāns – Institute of Systematic Biology, Daugavpils University, Vienības Str. 13, Daugavpils, LV-5401, LATVIA

Lyubomir Penev – Central Laboratory for General Ecology, Yuri Gagarin Street 2, 1113 Sofia, BULGARIA, *info@pensoft.net*

Roberto Pizzolotto – Universita della Calabria, Dip. Ecologia, via P. Bucci 4b, Rende (CS) 87036, ITALY, *rotopito@unical.it*

Gabor Pozsgai – Macaulay Institute, Craigiebuckler Aberdeen, AB15 8QH UNITED KINGDOM, *g.pozsgai@macaulay.ac.uk*

Debbie Prins – Willem Beijerinck Biologisch Station, Drenthe, THE NETHERLANDS

Poul Kry Poulsen – Benloseparken 19, 1th, DK-4100 Ringsted, DENMARK, *kry@privat.dk*

Zsolt Sághy – Novochem Trading and Service Co. Ltd., pf. 13, Györ 9011, HUNGARY, *saghy.zsolt@chello.hu*

Jose-Fermin Sanchez – Department of Zoology, Faculty of Veterinary, Campus de Espinardo. 30071 Murcia, SPAIN, *jfermin@um.es*

Pavel Saska – Crop Research Institute, Drnovska 507, Praha 6, Ruzyne 161 06 CZECH REPUBLIC, *saska@vurv.cz*

Axel Schwerk – Warsaw Agricultural University, Laboratory of Evaluation and Assessment of Natural Resources, Nowoursynowska Street 166, 02-787 Warsaw, POLAND, *aschwerk@yahoo.de*

Jose Serrano – Department of Zoology, Faculty of Veterinary, Campus de Espinardo, 30071 Murcia, SPAIN, *jserrano@um.es*

Grace P. Servat – Accompanying person with Terry Erwin

Oren Shelef – Mitrani Department of Desert Ecology, Ben-Gurion University of the Negev, Sede-Boqer Campus, 84990 Midreshet Ben-Gurion, ISRAEL, *milioren@yahoo.com*

Inna Shtirberg – Department of Zoology, Tel Aviv University, Tel Aviv 69978, ISRAEL, *innashti@post.tau.ac.il*

Jarosław Skłodowski – Warsaw Agricultural University, Nowoursynowska 159, 02-776 Warszawa, POLAND, *Jarek.Sklodowski@wl.sggw.pl*

John R. Spence – Department of Renewable Resources, 4-42 ESB, University of Alberta, Edmonton AB, T6G 2E3 CANADA

Ivailo Stoyanov – Central Laboratory for General Ecology, Yuri Gagarin Street 2, 1113 Sofia, BULGARIA

Angela Taboada – Area of Zoology, Department of Biodiversity and Environmental Management, University of León, Campus de Vegazana s/n, E-24071 León, SPAIN, *angela.taboada@unileon.es*

Sergio Tejero-Garcia – Department of Biodiversity and Environmental Management, University of León, Campus de Vegazana s/n, E-24071 León, SPAIN, *sgart@unileon.es*

Anika Timm – Institute of Ecology and Environmental Chemistry, University of Lüneburg, Scharnhorststr. 1, D-21335 Lüneburg, GERMANY, *anika.timm@arcor.de*

Annelies Turin – Accompanying person with Hans Turin

Hans Turin – Esdoorndreef 29, 6871 LK, Renkum, THE NETHERLANDS, *h.turin@hccnet.nl*

Bela Tóthmérész – Department of Ecology, University of Debrecen, H-4010 Debrecen, POB. 71, HUNGARY, *tothmerb@delfin.klte.hu*

Uldis Valainis – Institute of Systematic Biology, Daugavpils University, Vienības Str. 13 – 229, Daugavpils, LV-5401 LATVIA, *uldis.valainis@biology.lv*

Stephen Venn – University of Helsinki, P.O. Box 65 (Viikinkaari 1), FI-00014 FINLAND, *stephen.venn@helsinki.fi*

Rikjan Vermeulen – Willem Beijerinck Biologisch Station, Drenthe, THE NETHERLANDS, *stichting.wbbs@freeler.nl*

Snejana Vujčić-Karlo – Natural History Department, National Museum of Zadar, Medulićeva 2, 23000 Zadar, CROATIA, *snjezana_karlo@hotmail.com*

Evgeniy Zinovyev – Institute of Plant and Animals Ecology, Urals Branch of the Russia Academy of Sciences, Ekaterinburg, RUSSIA, *zin@etel.ru*

The European Union created the Natura 2000 network to counteract the loss of biodiversity and natural [illegible]. Natura 2000 forms the main legal framework to protect nature [illegible] EU level. It is the legal basis for:

- [illegible] network of protect[illegible]
- provisions [illegible] [identi]fication and designation of [illegible]
- [illegible] promote connectivity between [illegible]

Natura 2000 [illegible] Habitats and Birds directives [illegible] [conser]vation of wild living [illegible] concerned with the Habitats Directive (Directive [illegible])

[illegible] to demonstrate the importance of the EU Natura 2000 network [illegible] [cara]bidology. The subject should be [illegible] "new" EU Member States. Following questions are [illegible] [distr]ibution? What is the aim of the Habitats Directive? Why [illegible] Directive of any interest for carabidologists? Is there an intersection [illegible] ground beetles and protected habitat types listed in Annex I of Habitats Directive?

AIMS AND CONTENTS OF THE HABITATS DIRECTIVE

The main aims of the Directive 92/43/EEC are:

- to [illegible] biodiversity through the conservation of natural habitats and [illegible] in the European territory of the Member [States]
- [illegible] restore natural habitats and species of community interest [illegible] conservation sta[tus]
- [illegible] of economic, social and cultural requirements and regional [illegible]

[illegible] Habitats Directive should lead to a [illegible] network [illegible] conservation under the title "Natura 2000". This [illegible] is comp[osed] [illegible] sites of community-interest [illegible]. These sites [illegible] by the [illegible] Member States as [illegible] of [illegible] Member States [illegible] special areas of conservation (SAC) [illegible] area of SACs [illegible] of the [illegible] than 20% [illegible]

Further informa[tion] [illegible] actually available [illegible] [con]servation/index [illegible]

L. Penev, T. Erwin & T. Assmann (Eds) 2008
Back to the Roots and Back to the Future. Towards a New Synthesis amongst Taxonomic, Ecological and Biogeographical Approaches in Carabidology
Proceedings of the XIII European Carabidologists Meeting, Blagoevgrad, August 20-24, 2007, pp. 23-38.

Sofia–Moscow

Twentieth Century carabidology in the Nearctic region

George E. Ball

Department of Biological Sciences, University of Alberta, Edmonton, Alberta, Canada T6G 2E9. E-mail address: gball@ualberta.ca

INTRODUCTION

In extending to me the opportunity to open the meetings that served as the basis for these Proceedings, the organizers probably hoped for an address that would have considerable breadth taxonomically and otherwise. But I have chosen to focus on that part of the world that I inhabit, and in which I first developed my interest in carabid beetles. Since my own working life has spanned mainly the last half of the 20th Century, it seems appropriate to review that period (plus the first decade of the 21st Century) by tracing the history of carabidology for North America north of México, or temperate North America, emphasizing taxonomic aspects. Thus restricted, one could query the appropriateness of this subject for presentation to a principally European audience. But, as shown and emphasized below, the contribution of European carabidologists was vital to the development of North American carabidology during the latter half of the 20th Century. My presentation, then, may be seen as an expression of appreciation and acknowledgement of a debt to Western European carabidology.

Carabidology is the study of Carabidae in the broadest taxonomic sense of that term. Many taxonomically conservative authors exclude from the Carabidae (ground beetles) the Cicindelidae (tiger beetles) and Rhysodidae (wrinkled bark beetles). Some taxonomically radical authors include virtually only members of the Tribe Carabini in the family, assigning other caraboids to as many as 56 families. I prefer to include ground beetles, tiger beetles and wrinkled bark beetles in a single family, the Carabidae. Nonetheless, the ground beetles, tiger beetles, and wrinkled bark beetles have each had a rather different and virtually independent, taxonomic development. I elect to deal here only with the ground beetles.

I use the term "Nearctic" as shorthand for "North America north of México". The area of attention is in reality only part of the Nearctic Region, which extends south of the United States into Middle American México. Development of knowledge of the ground beetles of México is markedly different from that of their more northern counterparts, and is thus a different, though related, story (Ball & Shpeley, 2000: 366-371) from the one that I will tell here.

My review begins with consideration of where we have come from in the past half century, the focus being on progress that has been made in understanding the ground beetle fauna of North America. This review will lead to a consideration of the future – the way forward – which will include comments about advancing knowledge of ground beetles generally, not only of the North American fauna. This account is reminiscent and personal, and accordingly, I hope I may be forgiven if, by accident or short sightedness, I overlook points or people that are intrinsically important to the topic.

THE PAST: WHERE WE HAVE COME FROM

Three rather ill-defined sequential periods are recognized in the past history of Nearctic carabidology: quiescent; re-vitalization; and the thrust forward.

The quiescent period

I present here the perspective of a university student, beginning formal studies in the late 1940's and early 1950's. Taxonomic work on insects seemed to be on a plateau of relative inactivity, probably because of proximity to a period that was racked by two world wars, with a depression in between. Further, even before those disasters, North America lacked a strong phalanx of amateur entomologists. For Coleoptera, the situation, paradoxically, was exacerbated by good cataloguing, giving an impression of finality – that is, that most species were known, and there was little more for the taxonomically inclined to do except build collections for their own edification. In addition to these constraints, another had delayed work on ground beetles: at the species level, this group was perceived generally to be so difficult that it was wise for a dedicated coleopterist-in-training to look elsewhere for a beetle group to study. This last perception was the result primarily of the extensive revisionary work of Colonel Thomas L. Casey (1913-1924), whose keys and descriptions were difficult to interpret, and who described numerous species that were based on few specimens, many from single localities.

Many coleopterists regarded Casey's work with suspicion, but did not openly challenge it. Philip J. Darlington, Jr. (1938) was the first to establish just how inadequate Casey's efforts were with respect to ground beetles, showing for the Patrobini that many of the characters used by Casey were individually or geographically variable, leading to abundant synonymy. More generally, Darlington showed that the North American ground beetles were not as well known as the catalogues seemed to suggest, and thus the group was suitable for intense investigation.

Especially for taxonomic tyros, institutional collections were of limited access, most curators being reluctant to loan material for study. (One distinguished retired curator advised his successor that he would roll over in his grave if his much beloved cychrines were ever loaned! Evidently, he believed that nothing more was to be learned about that material on which he had worked). For my Master's work, I received part of a determined collection of *Helluomorphoides* declared by the curator to be correctly determined. Preliminary study showed that half of those specimens were misidentified. Unfortunately, loan of types was virtually unheard of, and depending upon curatorial predilection, they might or might not be made available even to a visiting student, as a fellow student of mine discovered shortly before departing on a trip to study museum material.

The literature was scattered, and older publications were not easily obtained. General works that included keys to carabid genera were not readily available. The only publications with broad taxonomic treatment of carabids were limited to states or provinces (Blatchley, 1910; Chagnon, 1940), and they were out of print and difficult to obtain as working copies in the days before photocopiers. A compilation of keys to the beetle genera, including carabids (Bradley, 1930) was also out of print. These works were present in libraries well-stocked in entomology, but were available there to students only on a short term basis (a week or two at a time). These limitations required long hours in libraries to hand-copy text, or writing to authors for reprints, or ordering publications from dealers in entomological publications, when one had succeeded in saving the funds necessary to make such purchases.

As discouraging as these circumstances were, during my years as a graduate student (1949-1950, University of Alabama; 1950-1954, Cornell University) three other like-minded graduate students were developing theses about carabids, each in a different institution: Clarence Benschoter (University of Minnesota); Ross T. Bell (University of Illinois); and Thomas C. Barr (Vanderbilt University). All produced and published taxonomic theses, and Barr, Bell, and I developed our academic careers, based in large part on continuing study of carabids.

Taxonomic catalogues give one an idea of the magnitude of taxa, and this information is valuable to a beginning graduate student in deciding if a group is too large or too small to form the basis for a thesis. Ready access was essential to the still in-print Catalogue of Coleoptera of North America north of Mexico (Leng, 1920, and five supplements, variously Leng & Mutchler, 1927 and 1933, and Blackwelder 1939, and Blackwelder & Blackwelder, 1948), which together listed names of 2342 ground beetle species and subspecies, arranged in 220 genera and 42 tribes. The arrangement of taxonomic names in the text represented a workable classification, dating from the late 19th Century (LeConte & Horn, 1883), but one that required substantial modification to meet the standards already set by European carabidologists (Csiki, 1927-1933; Jeannel, 1941-1942; see below). For study material, we had access to our respective institutional collections, borrowed specimens from those other institutions whose curators were willing to make loans, and our respective field-collected specimens.

We received inspiration and encouragement from various like-minded experienced seniors, and their publications provided excellent carabidological models. Of particular importance to me were three publications of P. J. Darlington, Jr. (1938, 1943, and 1950; see Ball, 1985 for a detailed appraisal of Darlington's contributions) and of J. Manson Valentine (1935, 1936), a southern gentleman of means, and a former curator of Coleoptera at the United States National Museum. These papers featured the use of male genitalia in species recognition, thus being unique at the time within the Nearctic carabidological literature. Those publications included also figures of other diagnostic features, range maps, and evolutionary and biogeographical considerations, pointing the way to future developments in the field. Also of substantial value, though geographically restricted, was the ground beetle part of a treatment of the Coleoptera of northwestern North America (Hatch, 1953) that eventually extended to five volumes.

The re-vitalization of North American Carabidology

If wartime and economic disturbances influenced adversely development of systematic entomology in general, and carabidology in particular, the reverse could have been important in its re-vitalization. In fact, peacetime and improved economic conditions had implications for science, in general. Public funding of basic biological research was undertaken both in the United States and Canada. This funding, principally by means of competitive grants through the National Science Foundation in the United States, and in Canada, the National Research Council (later the Natural Sciences and Engineering Research Council of Canada), made possible development or expansion of university science faculties and especially their graduate programs. Thus, the foundation was laid for a flowering of scientific endeavor. A small but, for this audience, most significant part of this development was the re-vitalization of carabidology.

An important element in the re-vitalization was the liberalization of institutions and curators regarding their collections. Loans of specimens began to be made freely, and types became more readily accessible.

Re-vitalization had four major catalysts that were made evident in publications. Two were at root philosophical. One of these forces was named "The New Systematics", highlighted in, for example, Mayr (1942). The second philosophical catalyst appeared as Hennig's (1966) influential book "Phylogenetic Systematics". A third influence was Arnett's (1960; obituary, Gerberg, 1999) broad treatment of North American beetle genera. The fourth was strictly carabidological, as embodied in a series of publications treating the ground beetles of Canada and Alaska (Lindroth, 1961-1969; obituary and appraisal, Ball, 1981).

The New Systematics challenged the fundamental approach of previous taxonomic work, opening the way to question the validity of each and every previously described species—a powerful stimulant to the young, ill-informed and naïve. In retrospect, however, the overall effect of the New Systematics because of its undue emphasis on population-level investigation, had negative implications for taxonomy (Wheeler, 1995: 48-50).

Hennig's book provided the background and methods of phylogenetic analysis, emphasizing its importance in postulating relationships among taxa, in evolution, and in developing classifications reflecting those relationships. Himself at heart and in practice a highly skilled and productive insect taxonomist, Hennig intended that the methods he developed would be of substantial value in taxonomy. His ideas were adopted readily by elements of the North American taxonomic community, and were especially influential among carabidologists.

Ross Arnett's publication provided keys to the Nearctic beetle genera. This made possible identification to at least that level of all ground beetles in the area covered. Further, it provided the opportunity to do some re-classification, bringing in ideas on the subject developed in Europe – principally in France, by René Jeannel (1941, 1942), arguably the foremost systematic carabidologist of his time.

Arnett's contribution extended beyond a volume for identification of adult beetles. He instituted "The Coleopterists Bulletin" a journal now in its sixty second year, and he was instrumental in beginning "The Coleopterists Society". Also he published various other books and journals, all of which had a positive effect on carabidology.

Carl H. Lindroth was a distinguished Swedish entomologist. He contributed his superb skills in taxonomy and ecology to conducting exemplary field and museum work on the northern ground beetle fauna, through his initial work on the Newfoundland fauna, and with subsequent encouragement and cooperation from Darlington at the Museum of Comparative Zoology, and George P. Holland and Williamson J. Brown at the Canadian National Collection of Insects (Division of Entomology, Agriculture Canada, Ottawa). The resulting "Ground beetles of Canada and Alaska" (Lindroth 1961-1969) was of such value and high quality as to be inspirational, thereby setting a standard and an example to be emulated. Lindroth established for North Americans the routine use of microsculpture and male genitalia in recognizing and diagnosing ground beetle species. He simplified the understanding of the Canadian-Alaskan ground beetle fauna by synonymization of many of the Casey names, and by enlightened recognition and description of many species unrecognized previously.

His publication on invasive species (Lindroth, 1957) was a major synthetic effort in that field, and demonstrated for North American carabidologists the presence of an important faunal element – the trans-Atlantic connection. He studied also the older trans-Pacific connection, between northeastern Siberia and western Alaska, by way of the Pleistocene-emergent (now submerged) Beringian land bridge (Lindroth, 1979).

Behind these highly influential scientific productions was an alert, highly perceptive, charming, self-effacing personality, laced with a whimsical sense of humor, and with a passion for ground beetles. He offered advice and encouragement to those with interests similar to his. Some 28 years after Carl's death, when I think of him, as I often do, I experience a wave of nostalgia washing over me, as I recall the grand times we enjoyed together on both sides of the Atlantic Ocean, in our respective homes, the field and museums, and at scientific meetings. I like to think that Carl's spirit is somewhere across the River Styx, contentedly smoking a pipe, and searching for fine ghostly carabids along the streams, in the marshes, and in the grasslands of the Elysian fields.

The Thrust Forward

Flowing from the background outlined above, the way forward was marked by essentially uncoordinated individual efforts in the field and museum. The individuals involved (professors, graduate students, and professional taxonomists) were in different institutions — universities and museums (state, provincial and federal); or were amateurs, working from their own homes. The systematic work done was primarily revisionary, treating smaller tribes, moderately large genera, or close-knit, difficult small species clusters. In an effort to link the North American fauna to other geographical assemblages, some revisions extended to those other regions, and for some groups, those other regions dominated the revisions. The common goal was straightforward: to make known the ground beetle species and their relationships.

Much of the work was related to classification including phylogenetic analyses, some of these being quite simple, others, highly complex. Bell (1964, 1967, 1983), for example, provided useful insights on carabid classification through his studies of comparative morphology. Based on pioneering work of European colleagues (Schuler, 1963; Deuve, 1993; Serrano, 1981), features of the female genital tract (Liebherr & Will, 1998), and karyotypic features (Maddison, 1985) were employed in taxonomy and reconstructing phylogeny. Kane et al. (1993) used gel electrophoresis to determine genetic divergence and gene flow among Appalachian species of Trechini. Some recent systematic efforts have focused on higher carabid taxa, using principally molecular features (Maddison et al., 1999).

Collective efforts, international and national, during the thrust forward, played a role as well. These included a symposium, during the 1976 International Congress of Entomology (Erwin et al., 1979), a symposium honoring the memory of Maximilien de Chaudoir (Whitehead, 1983), a memorial volume dedicated to P.J. Darlington, Jr. (Ball, 1985), a symposium (not published) honoring the memory of Henry Walter Bates (1992, Annual Meeting of the Entomological Society of America), a symposium, during the XX International Congress of Entomology (Ball et al., 1998) and, more recently, an 80th birthday celebration for the putatively oldest living North American member of the carabidological community (2006, Carnegie Museum of Natural History, Pittsburg, Pennsylvania, the proceedings yet to be published). These joint efforts manifested and promoted common interests, personal friendship and mutual respect among members of the carabidological community.

In passing, I note that because of the possibility of making accurate species identifications, and stimulated by the publications of den Boer and colleagues (summarized in the Proceedings of the Meetings of the European Carabidologists, cited by Ball et al., 1998: 16) in western Europe, ecological studies began to flourish in North America. In these studies, carabid communities were used to measure changes in various environmental parameters such as forest harvesting; agriculture; and fire in grasslands and forests. Cave communities were studied, particularly in the Appalachian Mountains of eastern North America. The geographical spread of invasive species and their effects on the native carabid fauna was also studied. Autecological observations made about life history, habitats, and habits of carabids were compiled and summarized in a remarkable book by Larochelle and Lariviere (2003).

Similarly, a substantial amount of work on Quaternary-age ground beetle fossils, principally in Canadian and Alaskan localities, was accomplished by a few individuals, under the influence and inspiration of G. Russell Coope (University of Birmingham, U.K.) and Carl Lindroth. Much of this work was summarized by Elias (1994).

THE PRESENT: WHERE WE ARE

Table 1 summarizes the numerical aspects of taxonomic changes in the North American ground beetle fauna from 1947 to the present. The number of known species has increased at about a rate of 11 per year. Interestingly, 74 new species have been discovered within the past 10 years, and it is not clear that an asymptote is yet approached closely. In other words it is highly probable that many more North American species remain to be discovered. (see Erwin & Geraci, this volume)

Number of genera recognized has decreased, indicating a generally conservative approach to recognition of higher taxa, just as Lindroth advocated. Nonetheless, many previously described (especially by Casey) genera linger in use as subgenera. Tribal numbers have remained nearly constant, but change has occurred. Of seven monogeneric tribes recorded in the Leng Catalogue, six included slightly aberrant members of the groups to which they have been assigned now: Zacotini, Micratopini, Anillini, Nomiini, Agrini, and Egini. One of the seven tribes, Trachypachini, was changed to family rank. Eight tribes were newly recognized in North America: Pelophilini, Clivinini, Trechini, Loxandrini, Perigonini, Pentagonicini, Cyclosomini, and Zuphiini.

Based on phylogenetic considerations, there has been a marked increase in number of higher taxonomic ranks from two (tribe and subfamily) to five (subtribe, tribe, supertribe, subfamily, and division), and a concomitant increase in number of subfamily taxa (from two to 13). This more complex classification indicates major progress in understanding relationships of ground beetles (Erwin, 1985). For ordinary purposes, however, the only suprageneric rank used is that of tribe.

Carabid taxa are now very well catalogued from a Nearctic perspective (Bousquet & Larochelle, 1993); from a Western Hemisphere perspective (Erwin, 2007); and from

Table 1. Numbers of ground beetle taxa (Carabidae, excl. Cicindelinae and Rhysodini) recognized in 1948 (Leng Catalogue, 1920-1948) and 2007 (Erwin, 2007).

Taxa	1947	2007
Spp. + Subspp.	2342	2975
Genera	220	189
Tribes	42	43
Subfamilies	2	13

a world perspective (Lorenz, 2005). A key to tribes and genera is available (Ball & Bousquet, 2001: 36-61), modeled after and brought up to date from the ground beetle part of the Arnett (1960-1962) volume.

During the past 60 years, virtually every tribe has received at least some study. Thirty-five tribes are reasonably well known: that is, it is possible to provide accurate identification of adults, without much difficulty. But much remains to be learned about them, taxonomically as well as ecologically. Thirteen tribes, in my opinion, require sustained, serious study, in particular clivinines, trechines, bembidiines, and within the Harpalini, the bradycellines and selenophorines. Also, as demonstrated by Erwin & Geraci in this volume, the North American pseudomorphines are very inadequately known.

Readily useable treatments of regional faunas of various extents are now available. See, for example: Beetles of the Pacific Northwest (Hatch, 1953); ground beetles, for Canada and Alaska (Lindroth, 1961-1969), for Atlantic Northeast (Downie and Arnett, 1996), South Carolina (Ciegler, 2000) and Connecticut (Krinsky & Oliver, 2001).

On the negative side, virtually all Nearctic literature about ground beetle taxonomy is restricted to adults. Immature stages have not received much attention, but at least Thompson (1979) and the classic publication by van Emden (1942) are available and permit at least tribal identification. Additionally, comparatively recent Western European treatments of carabid larvae are available (see Arndt (1991), and Luff (1993)), which will provide at least useful background information to anyone planning to study Nearctic carabid larvae.

Progress has been made also in increasing interest in ground beetles: some 62 members of The Coleopterists Society have declared an interest in ground beetles, though active taxonomists are appreciably fewer than that.

THE FUTURE: NEEDS AND DIRECTIONS

The Nearctic Region

Following the lead of the cicindelophiles, I think we may look forward to production of more regional handbooks about ground beetles, printed or electronic-based. More than a quarter of a century ago, Arnett advocated such regional works as a means to eventually cover the continent, though he had in mind more inclusive publications that would treat all of the beetles of a more or less natural region rather than those of single states, or for single families. Encouragingly, there is an upward trend in number of regional treatments of ground beetles (Table 2). Perhaps this will be the path leading to completion of revision of the North American ground beetle fauna.

But an adequate understanding of ground beetles for North America must be underlain by a substantial amount of taxonomic work, beginning with the groups noted above that have yet to be revised. A synthetic publication based on these separate synthetic treatments would appropriately follow. This could well be "cybertaxonomic", meaning that it would be produced using electronic techniques now available or soon to become

Table 2. Number of ground beetle revisions and of regional publications treating North American ground beetles (Carabidae, excl. Cicindelinae) and Rhysodini) from 1946 to 2005.

Periods	Revisions	Regional publications
1946- 1965	23	2
1966- 1985	67	1
1986- 2005	53	5

available (Wheeler, 2007: 12). Such an electronic publication is envisioned to provide a multi-entry key for identification of larvae and adults of all known North American species. Such a publication will be profusely illustrated, principally with habitus figures. It would contain natural history information for each species, and maps showing species ranges. One of the basic building blocks may be the proposed multi-volume "treatise" by Erwin (2007) that will provide a workable classification and basic information about each taxon, from division to species. What could be viewed as an experimental model is the plan by Will (2006, pers. com.) to produce an expanded electronic version (e-Carl) of Lindroth's treatment of the Canadian-Alaskan ground beetles.

To produce this synthetic treatment, we must have relatively current Nearctic-wide revisions of all tribes, and much more information about larvae and natural history than is available currently. Maps of species ranges, supplemented with the ecological information, will provide the base for a biogeographical analysis, that will include Pleistocene age fossils and that will take account of Late Cretaceous-Tertiary events. Following these taxonomic and biogeographical treatments, a phylogenetic analysis may be produced that will address carabid evolution in the Nearctic Region.

Obtaining the basic information to achieve these goals will require the sustained effort of many individuals, or perhaps the effort of one or two who are dedicated, ambitious, and highly skilled. The relatively easy, though mind-numbing part of the task is data entry. The much harder but more pleasurable and more intellectually stimulating part will be undertaken in the field and museum: collection, curation, dissection, and morphological analysis. These operations are time-consuming, and mostly without short cuts.

At present, I am not overwhelmed by a sense of optimism that such a goal will be achieved any time soon, at least not during my lifetime. Although a goodly number of coleopterists have expressed interest in ground beetles, taxonomic work on the North American fauna has taken something of a downturn (Table 2). Few of the important revisions that we need are being undertaken. I know of no one who is doing serious work on larvae. Community ecologists are using the beetles in applied entomology to indicate changes in climate or as environmental indicators in forestry and agriculture. Meanwhile, little effort is being directed to elucidating way of life of the beetles themselves – studies that as by-products could provide badly needed information about immature stages.

For the immediate future, I expect to see taxonomic treatments based primarily on molecular features by specialists whose perceptions and activities are mainly conducted in laboratories, and devoted to testing what is already known, or to phylogenetic analysis virtually divorced from taxonomy and not seeking out taxa yet to be discovered. These studies will be useful at some stages but are likely to be disruptive to achieving a timely, overall taxonomic synthesis of the North American ground beetle fauna (Wheeler, 2007: 16).

Carabidology, world-wide

Although this has been a focused regional review, it would be inappropriate, if not unwise, to ignore carabidology in the rest of the world because understanding of the Nearctic ground beetles has been stimulated or increased profoundly by the work of carabidologists elsewhere, and that benefit is likely to continue. Beyond the gains to be made locally, possibly the Nearctic experience may be of benefit to carabidologists in other zoogeographic regions, and that would be rewarding. With that in mind, I offer a few general comments about carabidology in those regions.

The Neotropical, Palaearctic, and Australian regions are being served by resident carabidologists, plus others—not as many as one might hope for, but sufficient to make sustained progress. Recently published catalogues are available for the carabid fauna of each of those regions.

Useful taxonomic publications are available for the fauna of the Western European part of the Palaearctic Region. Many of these are nation-restricted, and to my knowledge there is no region-wide generic synopsis. About the same is true for the Australian Region, except that the New Zealand Sub-Region has a succinct, well illustrated generic synopsis, including a classification and keys (Larochelle & Larivière, 2007). A similar publication is available for the Neotropical Region (Reichardt, 1977), but it is out of date. The Afrotropical and Oriental Regions are without widely recognized resident carabidologists, and their carabid faunas are without recent catalogues or generic synopses.

Finally, I venture to suggest that catalogues and broadly based taxonomic publications, such as generic synopses, are important as bases for initial understanding of carabid faunas and for undertaking efficiently more detailed study. Preparing such publications is difficult and challenging. I hope that residents will take up the challenge in each region, but failing that, someone, anyone, who will do the job. It would be a fine thing if carabidology were to move forward on a worldwide front.

SUMMARY AND CONCLUSIONS

1. This historical review charts progress in development of knowledge of Nearctic ground beetles (Carabidae, excluding tiger beetles and wrinkled bark beetles), through three periods: quiescent period; revitalization; and the thrust forward.
2. The quiescent period reflected a general depression in taxonomic work on beetles, resulting from a prolonged time of political and social unrest, occasioned by two world wars and in between an economic depression.

3. Ground beetle study was further inhibited, on one hand, by the excellent Leng catalogue that gave the impression of relatively complete knowledge of the group, and on the other, by the extensive publications by Thomas L. Casey that were difficult to use, and gave the impression of the near impossibility of recognition and identification of ground beetle species.
4. Limited revisionary work by P. J. Darlington, Jr. showed that the problem of identification of ground beetle species was primarily an artifact of a species concept that was faulty. That revelation indicated that the ground beetle fauna was not that well known, and that species were indeed identifiable, and thus a suitable group for taxonomic study, even by beginning coleopterists.
5. Revitalization was catalyzed by four events: the "New Systematics"; development of phylogenetic systematics, by Willi Hennig; publication of "The ground beetles of Canada and Alaska", by Carl Lindroth; and publication of "The beetles of the United States", by Ross Arnett.
6. During The Thrust Forward, the number of carabidologists increased, and as a result of their revisionary work which was based on a more detailed understanding of standard morphological features, as well as on features not used previously or extensively in North America, the number of known species increased and some new genera were described. In addition, some previously recognized genera were combined, resulting in an overall decrease in genera. Number of tribes remained about the same, but some rearrangements occurred: six of the monogeneric tribes recognized in the Leng catalogue were combined with related groups, and eight groups were newly recognized and separated from the tribes in which they had been included. A marked increase in number of suprageneric classificatory ranks and in number of subfamilies reflected a more detailed understanding of putative phylogenetic patterns.
7. Some recent systematic efforts have focused on higher carabid taxa, using principally molecular features.
8. In addition to standard taxonomic publications (descriptions and revisions), volumes treating politically demarcated (regional or state) ground beetle assemblages were published, intended to aid and encourage identification and study of carabids.
9. In spite of the substantial progress that has been made, much remains to be done in the way of revisionary work, and study of life stages (particularly larvae) other than the adult.
11. The proposed goal to strive toward is production of a general synthesis of knowledge of the Nearctic carabid fauna. Such a work could be a printed multi-volume publication, or more likely, it will be in electronic form, including keys, habitus illustrations, and maps of species ranges.
12. Future progress might begin with taxonomic treatments of the groups that have not been revised recently: most clivinine genera, trechines, bembidiines, pseudomorphines, and bradycelline and selenophorine Harpalini. Extensive work on larvae is required. Having achieved such, the groundwork will have been laid for a synthetic publication about the Nearctic carabid fauna.

13. Progress in gaining knowledge of the carabid faunas of the rest of the world would be accelerated if for each zoogeographic region there were a catalogue and a generic synopsis, including a classification and keys (Larochelle & Larivière, 2007:15).

ACKNOWLEDGEMENTS

I am pleased to thank the Organizing Committee of the XIII European Carabidologists Meeting, particularly Lyubomir Penev, for the invitation to attend, and to offer the opening address. I appreciate also the superb hospitality that he and Vlada Penev extended during our time together in Sofia and Blagoevgrad.

Useful personally solicited reviews were received from Yves Bousquet (Canadian National Collection of Insects, Agriculture and Agri-Food Canada, Ottawa, Ontario), Terry L. Erwin, Department of Entomology, Smithsonian Institution, Washington, D.C.) Bruce S. Heming (Department of Biological Sciences, University of Alberta, Edmonton, Alberta), and John R. Spence (Department of Natural Resources, University of Alberta, Edmonton, Alberta). I am grateful for their promptly submitted comments on a previous draft, incorporation of which markedly improved the text. Further improvement was provided by Thorsten Assmann (Institute of Ecology and Environmental Chemistry, University of Lüneburg, Lüneburg, Germany), who reviewed the manuscript at the request of the editors.

REFERENCES

Arndt, E. (1991). Familie Carabidae, pp. 45-141. – In: Die Käfer Mitteleuropas, Larven 1. (Klausnitzer, B., ed.). Goecke and Evers, Krefeld. 273 pp.

Arnett, R.H. (1960-1962). The beetles of the United States. The Catholic University of America Press, Washington, D.C. xi+ 1112 pp.

Ball, G.E. (1981). Carl H. Lindroth: contributions of a Swedish naturalist to systematics and biogeography in North America. Entomologica Scandinavica. Supplement 15: 17-32.

Ball, G.E. (Ed.) (1985). Taxonomy, phylogeny and zoogeography of beetles and ants: a volume dedicated to the memory of Philip Jackson Darlington, Jr. (1904-1983). Dr. W. Junk, Publishers, Dordrecht/ Boston/ Lancaster. XIV+514 pp.

Ball, G.E. & Bousquet, Y. (2001). [Chapter] 10. CARABIDAE (Latreille, 1810), pp. 32-132. – In: American Beetles, Volume 1 (Thomas, M.C. & Arnett, R.H. Jr., eds). CRC Press, Boca Raton. xiii+331 pp.

Ball, G.E., Casale, A. & Vigna Taglianti, A. (Eds) (1998). Phylogeny and classification of Caraboidea (Coleoptera: Adephaga). Atti, Museo Regionale di Scienze Naturali, Torino, Italy. 543 pp.

Ball, G.E. & Shpeley, D. (2000). [Chapter] 19. Carabidae, pp. 363-399. – In: Biodiversidad taxonomía y biogeografía de artropodos de México: hacia una syntesis de su conocimiento, Volumen II. (Llorente, B.J., Gonzaléz, S.E., y Papavero, N., eds). Universidad Nacional Autónoma de México, México, D.F. XVI + 676 pp.

Bell, R.T. (1964). Does *Gehringia* belong to the Isochaeta? The Coleopterists Bulletin, 18: 59-61.

Bell, R.T. (1967). Coxal cavities and the classification of the Adephaga (Coleoptera). Annals of the Entomological Society of America, 60: 101-107.

Bell, R.T. (1983). What is *Trachypachus*? (Coleoptera: Trachypachidae), pp. 590-596. – In: The Baron Maximilien de Chaudoir (1816-1881): a symposium to honor the memory of a great Coleopterist during the centennial of his death. (D.R. Whitehead, ed.). The Coleopterists Bulletin, 36 (1982): 459-609.

Blackwelder, R.E. (1939). Fourth supplement 1933 to 1938 to the Leng Catalog of Coleoptera of America north of Mexico. Sherman, Mount Vernon [New York]. 146 pp.

Blackwelder, R.E. & Blackwelder, R.M. (1948). Fifth supplement 1939-1947 (inclusive) to the Leng Catalog of Coleoptera of America north of Mexico. Sherman, Mount Vernon [New York]. 87 pp.

Blatchley, W.S. (1910). An illustrated descriptive catalogue of the Coleoptera or beetles (exclusive of the Rhynchophora) known to occur in Indiana – with bibliography and descriptions of new species. The Nature Publishing Co., Indianapolis. 1386 pp.

Bousquet, Y & Larochelle, A. (1993). Catalogue of the Geadephaga (Col. Trachypachidae, Rhysodidae, Carabidae, incl. Cicindelini) of America north of Mexico. Entomological Society of Canada, Memoir No. 167. 395 pp.

Bradley, J.C. (1930). A manual of the genera of beetles of America north of Mexico. Daw, Illston and Company, Ithaca, New York. vii+360 pp.

Casey, T. L. (1913). Studies in the Cicindelidae and Carabidae of America. Memoirs on the Coleoptera, vol. IV, pp. 1-192. The New Era Printing Company, Lancaster, Pa. 355 pp.

Casey, T. L. (1914). *Ibid.,* vol. V, 305 pp.

Casey, T. L. (1918). *Ibid.,* vol. VIII, 427 pp.

Casey, T. L. (1920). Some observations on the Carabidae, including a new subfamily. *Ibid.*, vol. IX, pp. 25-299.

Casey, T. L. (1924). Additions to the known Coleoptera of North America. *Ibid*, vol. XI, 347 pp.

Chagnon, G. (1940). Contribution a l'Étude des Coléoptères de la Province de Québec. Les Presses de l'Université Laval. 385 pp.

Ciegler, J. (2000). Ground beetles and wrinkled bark beetles of South Carolina (Coleoptera: Geadephaga: Carabidae and Rhysodoidae). South Carolina and Forestry Research System, Clemson University, Clemson, South Carolina. VI+149 pp

Csiki, E. (1927-1933). Carabidae. – In: Coleopterorum Catalogus (Junk & Schenkling, eds), Berlin.

Darlington, P.J., Jr. (1938). The American Patrobini (Coleoptera: Carabidae). Entomologica Americana, 18 (New Series), 4: 135-183.

Darlington, P.J., Jr. (1943). Carabidae of mountains and islands: data on the evolution of isolated faunas and atrophy of wings. Ecological Monographs, 13: 37-61, 8 figures.

Darlington, P.J., Jr. (1950). Paussid beetles. Transactions of the American Entomological Society, 76: 47-142, 207 figs, 3 maps.

Deuve, T. (1993). L'abdomen et les genitalia des femelles de Coléoptères Adephaga. Mémoires du Muséum National d'Histoire Naturelle, 155: 1-184.

Downie, N.M. & Arnett, R.H., Jr. (1996). The Beetles of Northeastern North America. Volume 1. Introduction, Suborders Archostemata, Adephaga, and Polyphaga thru [*sic*!] Superfamily Cantharoidea. The Sandhill Crane Press, Publisher,. Gainesville, Florida. xiv + 880 pp.

Elias, S.A. (1994). Quaternary insects and their environments. Smithsonian Institution Press, Washington/ London. xiii + 284 pp.

Emden, F.I. van. (1942). A key to the genera of larval Carabidae. Transactions of the Royal Entomological Society of London, 92: 1-99, 100 figs.

Erwin, T.L. (1985). The taxon pulse: a general pattern of lineage radiation and extinction among carabid beetles, pp. 437-493. – In: Taxonomy, phylogeny and zoogeography of beetles and ants: a volume dedicated to the memory of Philip Jackson Darlington, Jr. (1904-1983) (Ball, G.E., ed.) Dr. W. Junk, Publishers, Dordrecht/ Boston/ Lancaster XIV + 514 pp.

Erwin, T.L. (2007). A treatise on the Western Hemisphere Carabidae (Coleoptera). Their classification, distributions and ways of life. Volume 1. Trachypachidae, Carabidae—Nebriiformes. Pensoft, Sofia-Moscow. 323 pp. + 22 color plates.

Erwin, T.L., Ball, G.E., Whitehead, D.R. & Halpern, A.L. (Eds) (1979). Carabid beetles: their evolution, natural history, and classification. Dr W. Junk bv Publishers, The Hague, The Netherlands. X+644 pp.

Gerberg, E.J. (1999). Obituary, Ross Harold Arnett, Jr., 1919-1999. Florida Entomologist, 82: 644-645.

Hatch, M.H. (1953). The beetles of the Pacific Northwest. Part 1: Introduction and Adephaga. University of Washington Press, Seattle, Washington. vii + 340 pp., 37 plates, 2 text figs.

Hennig, W. (1966). Phylogenetic systematics. University of Illinois Press, Urbana/ Chicago/ London. 263 pp.

Jeannel, R. (1941). Coléoptères carabiques, première partie. Faune de France, 39: 1 -571, figs 1-213. Paris.

Jeannel, R. (1942). Coléoptères carabiques, deuxième partie. *Ibid.,* 40: 573-1173, figs 214-368. Paris.

Kane, T.C., Barr, T.C. & Badarraca, W.J. (1993). Cave begtle genetics: geology and gene flow. Heredity, 68: 277-286.

Krinsky, W.L. & Oliver, M.K. (2001). Ground beetles of Connecticut (Coleoptera: Carabidae, excl. Cicindelini). Bulletin 117, State Geological and Natural History Survey, Connecticut. a-d+308 pp.

Larochelle, A. & Larivière, M.-C. (2003). A natural history of the ground beetles (Coleoptera: Carabidae) of America north of Mexico. Pensoft, Sofia-Moscow. 583 pp.

Larochelle, A. & Larivière, M.-C. (2007). Carabidae (Insecta: Coleoptera): synopsis of supraspecific taxa. Fauna of New Zealand, No. 60: 1-188.

LeConte, J.L. & Horn G.H. (1883). Classification of the Coleoptera of North America. Smithsonian Miscellaneous Collections, 507, xxxviii+567 pp.

Leng, C.W. (1920). Catalogue of the Coleoptera of America, north of Mexico. Sherman, Mount Vernon, [New York]. x + 470 pp.

Leng, C.W. & Mutchler, A.J. (1927). Supplement 1919-1924(inclusive) to catalogue of the Coleoptera of America, north of Mexico. Sherman, Mount Vernon, [New York]. 78 pp.

Leng, C.W. & Mutchler, A.J. (1933). Second and third supplements 1925-1932(inclusive) to catalogue of the Coleoptera of America, north of Mexico. Sherman, Mount Vernon, [New York]. 112 pp.

Liebherr, J.K. & Will, K.W. (1998). Inferring phylogenetic relationships within the Carabidae (Insecta, Coleoptera) from characters of the female reproductive tract, pp. 107-170. – In: Phylogeny and classification of Caraboidea (Coleoptera: Adephaga)

(Ball G.E., Casale, A. & Vigna-Taglianti, A., eds). Atti, Museo Regionali di Scienze Naturali, Torino. 543 pp.

Lindroth, C.H. (1957). The faunal connections between Europe and North America. John Wiley and Sons, New York, Almqvist and Wiskell, Stockholm. 344 pp.

Lindroth, C.H. (1961). The ground-beetles (Carabidae, excl. Cicindelinae) of Canada and Alaska. Part 2. Opuscula Entomologica Supplementum No. 20. Pp. 1-200.

Lindroth, C.H. (1963). *Ibid.* Part 3. *Ibid*, No. 24. Pp. 201-408.

Lindroth, C.H. (1966). *Ibid.* Part 4. *Ibid.*, No. 29. Pp. 409-648.

Lindroth, C.H. (1968). *Ibid.* Part 5. *Ibid.*, No. 33. Pp. 649-944.

Lindroth, C.H. (1969a). *Ibid.* Part 6. *Ibid.*, No. 34. Pp. 945-1192.

Lindroth, C.H. (1969b). *Ibid.* Part 1. *Ibid.*, No. 35. Pp. I-XLVIII.

Lindroth, C.H. (1979). [Chapter] 2.36. The importance of Beringia as reflected in the present fauna, pp. 349-367. – In: Carabid beetles: their evolution, natural history, and classification. (Erwin, T.L., Ball, G.E., Whitehead, D.R. & Halpern, A.L., eds). Dr W. Junk bv Publishers, The Hague, The Netherlands. x + 644 pp.

Lorenz, W. (2005). Systematic list of extant ground beetles of the world (Insecta Coleoptera 'Geodephaga': Trachypachidae and Carabidae and Paussinae, Cicindelinae, Rhysodinae). Published by author, Tutzing, Germany. 530 pp.

Luff, M.L. (1993). The Carabidae (Coleoptera) larvae of Fennoscandia and Denmark. Fauna Entomologica Scandinavica, 27: 1-186.

Maddison, D.R. (1985). Chromosomal diversity and evolution in the ground beetle genus *Bembidion* and related taxa (Coleoptera: Carabidae: Trechitae). Genetica, 66: 93-114.

Maddison, D.R., Baker, M.D. & Ober, K.A. (1999). Phylogeny of carabid beetles as inferred from 18S ribosomal DNA (Coleoptera: Carabidae). Systematic Entomology, 24: 103-138.

Mayr, E. (1942). Systematics and the origin of species from the view point of a zoologist. Columbia University Press, New York. xiv + 334 pp.

Reichardt, H. (1977). A synopsis of the genera of Neotropical Carabidae (Insecta: Coleoptera). Quaestiones Entomologicae, 13(4): 346-493.

Schuler, L. (1963). Les organs genitaux chez les Pterostichidae de France. Les tribus Anchomenini et Sphodrini. Le cas des Patrobidae. Bulletin de la Societe entomologique de France, 68: 13-26.

Serrano, J. (1981). Chromosome numbers and karyotypic evolution of Caraboidea. Genetica, 55: 51-60.

Thompson, R.G. (1979). Larvae of North American Carabidae with a key to the tribes, pp. 209-291. – In: Carabid beetles: their evolution, natural history, and classification. (Erwin, T.L., Ball, G.E., Whitehead, D.R. & Halpern, A.L., eds). Dr W. Junk bv Publishers, The Hague, The Netherlands. x + 644 pp.

Valentine, J.M. (1935). Speciation in *Steniridia* a group of cychrine beetles. Journal of the Elisha Mitchell Science Society, 51: 341-375, pls. 65-73 (1 map).

Valentine, J.M. (1936). Raciation in *Steniridia andrewsii* Harris, a supplement to speciation in *Steniridia*. *Ibid.*, 52: 223-234, pl. 17.

Wheeler, Q.D. (1995). The "Old Systematics: classification and phylogeny, 31-62. – In: Biology, phylogeny, and classification of Coleoptera, Papers celebrating the 80th birthday of Roy A. Crowson (Pakaluk, J. & Slipinski, S.A., eds). Muzeum I Institut Zoologii PAN, Warszawa. xii + 558 pp.

Wheeler, Q.D. (2007). Invertebrate systematics or spineless taxonomy, pp. 11-18. – In: Linnaeus tercentenary: progress in invertebrate taxonomy (Zhang, Z.-Q. & Shear, W.A., eds). Zootaxa, 1668: 1-766.

Whitehead, D.R. (Ed.). (1983). The Baron Maximilien de Chaudoir (1816-1881): a symposium to honor the memory of a great coleopterist during the centennial of his death Coleopterists Bulletin 36 (1982): 459-609.

Taxonomy, Morphology and Biogeography

L. Penev, T. Erwin & T. Assmann (Eds) 2008
BACK TO THE ROOTS AND BACK TO THE FUTURE. TOWARDS A NEW SYNTHESIS AMONGST TAXONOMIC, ECOLOGICAL AND BIOGEOGRAPHICAL APPROACHES IN CARABIDOLOGY
Proceedings of the XIII European Carabidologists Meeting, Blagoevgrad, August 20-24, 2007, pp. 41-76.

From Latreille to DNA systematics – towards a modern synthesis for carabidology

Thorsten Assmann[1], Joern Buse[1], Claudia Drees[1], Jan Habel[2], Werner Härdtle[1], Andrea Matern[1], Goddert von Oheimb[1], Andreas Schuldt[1] & David W. Wrase[3]

[1] Institute of Ecology and Environmental Chemistry, Leuphana University of Lüneburg, Scharnhorststr. 1, D-21314 Lüneburg, Germany. E-mail: assmann@uni.leuphana.de, joernbuse@gmx.de, cdrees@uni.leuphana.de, whaerdtle@uni.leuphana.de, matern@uni.leuphana.de, vonoheimb@uni.leuphana.de, schulte@uni.leuphana.de
[2] Musée national d'histoire naturelle, Section Zoologie des Invertébrés, L-2160 Luxembourg, and University of Trier, Biogeographie, D-54296 Trier, Germany. E-mail: janchristianhabel@gmx.de
[3] Dunckerstr. 78, D-10437 Berlin, Germany. E-mail: carterus@gmx.de

SUMMARY

The aim of this contribution is a compilation of the present-day status of carabid taxonomy and systematics with a special focus on the Palaearctic. We give a short review on morphology-based alpha taxonomy (MORAT), morphometry, karyotypes, and molecular systematics. We believe that MORAT has to be subdivided in two periods, a classical and a modern one. The latter is mainly characterized by the excessive use of genital characters for classification and ranking of taxa. Important molecular marker systems for carabid taxonomy and systematics are allozymes, mitochondrial DNA sequences, microsatellite DNA and nuclear DNA sequences. We discuss the use of molecular methods to solve taxonomic problems at the species level and plead for a combination of molecular techniques, morphometrics and morphology-based taxonomy. Following challenges of present-day taxonomy and systematics are discussed: (1) stability of nomenclature, (2) tremendous amount of undescribed taxa, even from the Palaearctic, (3) the role of different aedeagus shapes to delimit species, and (4) incongruence of the results from MORAT, morphometrics, and molecular techniques.

Keywords: Morphology-based alpha taxonomy, MORAT, molecular taxonomy, morphometry, karyotypes, allozymes, mtDNA, nuclear DNA, microsatellites, lock-and-key hypothesis, undescribed species, stability of nomenclature

INTRODUCTION

The genera *Carabus* and *Cicindela* have been established by von Linné (1758) in his 10th edition of the 'Systema Naturae'. The publication of this work was the birth date of scientific nomenclature in zoology. About half a century later, Latreille (1810) established the family Carabidae and hence started carabidology in its narrower sense. Since the beginning of the 19th century, interest in ground beetles has grown rapidly. Nowadays ground beetles are regarded as a very large animal group, from which almost 40,000 species have already been described (Lorenz, 2005a).

Carabid beetles are not only an object of taxonomy and systematics, but also of ecology, evolutionary research, physiology, biogeography and many other disciplines of basic biology. Some facets of the biology of this fascinating animal group are already known to both scientists and many amateurs interested in natural history, e.g. ectoparasitic life forms, incorporation of toxins, mimicry, explosive mechanism of bombardier beetles, adaptation to anoxybiotic conditions etc. (e.g. Conradi-Larsen & Somme, 1973a, 1973b; Duman et al., 2004; Eisner et al., 2001; Erwin, 1967; Lindroth, 1971; Neuwinger, 2004; Saska & Honek, 2004; Schildknecht et al., 1968; Thiele, 1977; Turin et al., 2003). Apart from these interesting aspects of life history, ground beetles are important subjects especially in ecology and evolutionary research. The number of papers dealing with these topics seems to be innumerable. Moreover ground beetles are also an important subject of applied biology: They are antagonists of some pest species and their significance for conservation biological approaches is increasing (e.g. species are listed in the U.S. Endangered Species Act and in the Annexes II and IV of the European Habitats Directive, NATURA 2000).

Taxonomy and systematics are two of the most important bases for all these biological disciplines. Nevertheless, an increasing debate concerns the future of taxonomy, stimulated by modern, especially molecular, techniques on the one hand and a changing interest of the scientific community on the other hand. Therefore the time seems to be ripe both for a compilation of the present-day status and a perspective for the future orientation of carabid taxonomy. Owing to our areas of research this contribution focuses mainly on the Palaearctic carabid fauna and taxa of genera and species groups. This is true especially because molecular and morphometric techniques are used mainly on Palaearctic and Nearctic carabid beetles. The usefulness of these techniques for ground beetles still has to be proven for other biogeographic realms. We divide our contribution into two main sections, a short review (Section "Carabid systematics and taxonomy – a short overview") and more subjectively coloured chapters on current challenges of taxonomy (Sections "Current taxonomy" and "How to implement contemporary systematics and taxonomy on carabids – plea for a modern synthesis").

CARABID SYSTEMATICS AND TAXONOMY – A SHORT OVERVIEW

The history of ground beetle taxonomy and systematics has been already reviewed by leading carabidologists (Ball, 1979a, 1979b, 2008; Ball et al., 1998; Lindroth, 1979). These overviews focus mainly on morphology-based alpha taxonomy (MORAT). One of the best introductions to MORAT, not only for ground beetles, is given by Darlington (1971). Therefore our account of this period has to be short (Section "Morphology-based alpha taxonomy (MORAT)"). Nevertheless, we believe that MORAT has to be split in two periods, a classical and a modern one. In the last decades our knowledge on carabid systematics, taxonomy and phylogeny increased enormously due to new, especially molecular methods. A synopsis of these approaches is still missing and we attempt an overview here (Section "Alternative approaches").

Morphology-based alpha taxonomy (MORAT)

Classical morphology-based alpha taxonomy (classical MORAT)

From the middle of the 18th century until the turn of the 19th to 20th century the pioneers of carabid systematics described many of the widely distributed ground beetles. We cannot present a list of these scientists because of its enormous length. These pioneers of carabidology were able to recognize many species, although the quality of the optical devices was sometimes poor.

Unfortunately some taxa were wrongly synonymized despite the fact that they had been identified and described by some of these early carabidologists as "good" species, for example *Brachinus hebraicus* Reiche & Saulcy, 1855. At the turn of the 19th to 20th century this taxon was already treated as a junior synonym of *B. exhalans* Rossi, 1792 (Reitter, 1919). A detailed inspection of specimens from the Middle East revealed differences in the male genitalia between both taxa. The proportion of the differentiation is comparable to sympatric *Brachinus* sibling species of the same genus. As a consequence, the species status of *B. hebraicus* has to be re-established.

This example demonstrates the advantages and disadvantages of a long-lasting history of taxonomical studies: the better knowledge of the fauna inhabiting the given regions on the one hand and the necessity of a detective's keen perception to "dig up ancient pseudo-synonyms" on the other hand. The careful re-examination of old types is an important task for the analysis of the West-Palaearctic realm.

At the end of the 19th and at the beginning of the 20th century, the foundation for the systematics of higher taxa of ground beetles was developed. Especially Ganglbauer contributed to this milestone in an outstanding way. His work "Die Käfer von Mitteleuropa" (Ganglbauer, 1891) summed up the development of the different systems used during the latter half of the 19th century and presented a system which was the basis for many later carabidologists and consequently for the modern morphology based systematics (e.g. Jeannel and Lindroth) (Ball, 1979a: 84).

Moreover classical MORAT was characterized by the development of important identification keys for large regions. Many of them are still used, or they are at least the basis for many newer keys (e.g. for the Palaearctic realm: Apfelbeck, 1904; Chaudoir, 1876; de la Fuente, 1927; Reitter, 1900; Reitter, 1919).

Modern morphology-based alpha taxonomy (modern MORAT)

At the end of the 19th century Escherich (1892, 1893, 1894) studied the attributes of "genital appendages" of beetles and their potential role in the prevention of cross-specific hybridizations in detail. His work stimulated the young Holdhaus (1912) who wrote the first revision among beetles based on both the external and internal structure of the aedeagus to discriminate sibling species from each other. In contrast to many later revisions he studied also the female genitalia (which are generally more weakly sclerotized and therefore not comparably useful for taxonomic revisions).

Holdhaus' revision on the Palaearctic *Microlestes* species revealed the "discovery" of the genitalia as a tool for carabid taxonomy and systematics (Lindroth, 1979). The careful examination of the aedeagus, frequently also of the internal structures (endophallus) to discriminate species, especially sibling species, from each other is a standard method in carabid taxonomy. The frequent use of these characters is revealed in the monographs of Jeannel, Lindroth, Darlington, Ball, and many other authors at first sight.

The cryptic identification characters of the aedeagus require the study of types in a more compelling way than in classical MORAT. Consequently, the study of holotypes and paratypes has been developed as an important sign for the quality of many monographs and descriptions.

An excellent overview on the present day status of modern MORAT is given by Ball et al. (1998). After their review further important work on the taxonomy and systematics of genera and higher taxa and for geographical regions has been published (Andújar & Serrano, 2001; Baehr, 1998; Ball & Shpeley, 2005; Barr, 2004; Belousov, 1998; Casale, 1998; Facchini, 2003; Kirschenhofer, 1999, 2004; Ledoux & Roux, 2005; Liebherr & Schmidt, 2004; Liebherr & Zimmermann, 2000; Moravec, 2002; Moret, 2005; Naviaux, 1998; Naviaux & Pinratana, 2004; Obydov, 2002; Ortuno & Toribio, 2005; Sciaky & Facchini, 2005). The list of all contributions had to be restricted to a synopsis, as it would go beyond the scope of our contribution.

Breuning (1932ff) revised the genus *Carabus* in a monumental work of about 1600 pages. Although he reduced the numerous names below the species level, his infrasubspecific nomenclature is inconsistent with the International Code of Zoological Nomenclature. However, the discussion of "different" hierarchical subspecies levels is still ongoing (Deuve, 2004).

Despite the fact that catalogues have a long-lasting tradition in carabidology (Winkler, 1924-1927), the last decades stand out for their publication of important lists: Lorenz (1998a, 1998b) published a systematic list of extant ground beetles of the world (a second edition is already published; Lorenz, 2005a, 2005b) and a team of authors published the carabid section of the "Catalogue of Palaearctic Coleoptera" under the editorial work of Löbl & Smetana (2003).

During the last two decades a remarkable and very positive trend to study unusual characters can be recognized, e.g. muscles, digestive system, defensive glands, egg shell, wing, and endophallus (Ball et al., 1998; Deuve, 2005; Eisner et al., 2001; Hurka, 1999; Hurka & Ruzickova, 1999). In general a large set of characters is used in contemporary studies of carabid MORAT.

Alternative approaches

Morphometry

Morphometry or morphometrics is a quantitative way of addressing the shape comparisons of organisms by using at least few, in most cases many characters. Morphometric approaches started in the first half of the last century. The scientific aim was mainly to understand evolutionary processes. The Evolutionary Synthesis of Dobzhansky, Mayr, Rensch and Simpson conveyed the ideas of population genetics to other fields of biology. These authors argued persuasively that mutation, recombination, selection and gene flow operating within species ("microevolution" in the sense of Dobzhansky) account for the origin of new species and for the long-term effects of evolution ("macroevolution" in the sense of Dobzhansky) (Futuyma, 2005).

The European exponent of the Evolutionary Synthesis was Bernhard Rensch who emphasized, in his publications (and lectures; Friedrich Weber, personal communication), that carabids, especially those of the genus *Carabus*, are favourable subjects for studying processes of genetic variation below the species level. With a special focus on microevolutionary processes some papers dealing with the morphometric differentiation of conspecific *Carabus* populations were published until the middle of the 20th century (Boettger, 1921; Krumbiegel, 1932, 1936a, 1936b, 1936c; Rensch, 1943, 1948, 1950, 1954, 1958; Zarapkin, 1934).

In the second half of the 20th century multivariate approaches and many exoskeleton characters have been used. The main insight resulting from these studies is the extraordinary amount of morphometric differentiation between conspecific populations in the genus *Carabus*. This is also true if habitats of the studied populations are only few kilometres apart (e.g. Gries et al., 1973; Mossakowski, 1971; Mossakowski et al., 1986; Mossakowski & Weber, 1976; Terlutter, 1991). Winglessness, low vagility and stenotopy seem to be prerequisites for a strong differentiation at the population level.

Very recently a new morphometric approach, landscape morphometry (Zelditch et al., 2004), showed significant shape variation of the beetles' body among sites and significant positive correlations between morphological and geographical distances between populations (Alibert et al., 2001).

Two results of these studies are important: (1) Some species of the genus *Carabus* show an extraordinary differentiation at the population level. (2) Selection as well as random genetic drift seem to be the driving forces of the differentiations below the species level.

Liebherr (1986) and Saskawa & Kubota (2007b) used morphometric data as a tool for the identification of taxa at the species level, other authors applied morphometry successfully to clarify systematic problems below the species level. Examples for the latter approaches are:

(1) Lindroth (1968) showed that the Icelandic population of *Carabus problematicus* represents an independent subspecies. We believe that Lindroth's description of the subspecies *C. p. islandicus* after the morphometric analyses of North European populations should be a shining example for *Carabus* systematics: His approach was to prove the independence of the taxon not by (subjective) eye inspection (as it is done by most so-called *Carabus* specialists) but by measurements which are objective. Just this objectivity lacks most of the other assessments/evaluations of subspecies in the subgenus *Carabus*.

This and other contributions on refugia of the glacial period in northern Europe (including Iceland) (Lindroth, 1968, 1969, 1970) were intensively discussed. Most scientists rejected the hypothesis of northern refugia developed by Lindroth. Genetic approaches which are helpful in the analyses of glacial refugia were not available during Lindroth's active scientific period. More recent results from other animals give further evidences for glacial refugia in northern Europe. If species such as the rock ptarmigan (Holder et al. 1999) were able to survive the last glacial period on Iceland, we have to postulate complex tundra ecosystems for these periods. In the light of the new findings Lindroth's hypotheses regarding refugia in northern Europe seem to be more probable than ever.

(2) Terlutter (1991) demonstrated that current subspecies taxonomy of *Carabus auronitens* is only partly supported by morphometry (using discriminant analysis).

(3) Bonadona (1973) used morphometry to reduce the inflation of described subspecies of *Carabus solieri*.

(4) Assmann & Schnauder (1998) revealed excessive gene flow in a broad hybrid zone between *Carabus violaceus* s.str. and *C. v. purpurascens* in north-western Germany despite the fact that both taxa show clear differences in the external shape of the aedeagus. Moreover they demonstrated that some characters were not helpful to separate the subspecies from each other, despite the fact that these characters are frequently used in identification keys.

Morphometric approaches are helpful for systematics and taxonomy in many carabid groups, but a general problem is not resolved: We do not know to which extent environmental factors act on the characters of the exoskeleton (see Maynard-Smith, 1998 for a detailed discussion). Moreover ordinary phenotypic characters can not provide any information on the number of variable gene loci and the frequencies of alleles, because we cannot count how many genes contribute to phenotypically uniform traits. To understand the genetic basis of quantitative variations cross-breeding experiments are necessary (cf. Section "Different genitalia = different species! State of the art or traditional paradigm ?").

Karyotypes

Since the 60s of the last century karyotypes have been determined for hundreds of ground beetle species (Nettmann, 1986; Serrano & Galián, 1998). While some genera show uniform chromosome numbers (e.g. *Carabus* and *Campalita*: Kudoh et al., 1970; Weber, 1966), some sibling species are differentiated regarding the number of chromosomes (e.g. *Pterostichus nigrita* group, Koch & Thiele, 1980). Thus, karyotypology can be a helpful tool for systematics and taxonomy, at least in some ground beetle groups (Serrano & Galián, 1998; they also give a compilation of chromosome formulas for all studied carabids).

Multiple sex chromosome systems (MSCS)* occur in some Cicindelinae species, which have between two and four X chromosomes (X_nY in males and X_nX_n in females) (Smith & Edgar, 1954). Not only this system is very peculiar, but also the finding that in *Cicindela hybrida* the sex chromosomes do not form chiasmata (Giers, 1977). Some sibling species of tiger beetles show different numbers of sex chromosomes (e.g. *Cicindela campestris* and *C. maroccana*: X_3Y and X_4Y in males, X_3X_3 and X_4X_4 in females; Serrano, 1980a). However, it is more remarkable that MSCS occur only in the Cicindelinae tribes Cicindelini and Collyrini, but they are not recorded in the basal lineages of Megacephalini and Manticorini. Therefore MSCS seem to go back to a single common ancestor (of Collyrini and Cicindelini) (Proenca et al., 1999a; Proenca et al. 1999b). Despite the fact that there are some exceptions for the MSCS in Cicindelini (e.g. the genus *Odontocheila* has an X0 chromosome system and the sibling species *Cicindela* (*Cylindera*) *germanica* and *C.* (*C.*) *paludosa* have simple sex chromosome systems), the chromosomes offer good possibilities for phylogenetic and taxonomic interferences in tiger beetles (Pearson & Vogler, 2001).

At lower taxonomic levels, especially below the species level, there is no evidence that the karyotype is an analytical tool, despite the fact that differences between conspecific *Carabus* populations were found (Mossakowski & Weber, 1972; Weber, 1967, 1968). Serrano & Galián (1998) concluded rightly however, that the use of more refined techniques, especially by combining cytological and molecular approaches, have the potential to discover relationships of phylogenetic, systematic and taxonomic importance (cf. Galián et al., 1999, 2002; 2007; Galián & Vogler, 2003; Martinez-Navarro et al., 2004; Proenca & Galián, 2003; Sanchez-Gea et al., 2004; Sanchez-Gea et al., 2000).

Molecular systematics

Without any question, molecular systematics (MOSY) has a great influence on present-day taxonomy. It has its origin in molecular population genetics, which started when Lewontin & Hubby (1966) and Harris (1966) demonstrated that native **allozymes**** can

* Multiple sex chromosomes are also evolved in some other ground beetle taxa (e.g. *Scarites*, *Brachinus*, *Ceroglossus*; Galián et al., 1990, 1996; Serrano, 1980b; Serrano & Yadav, 1984).

** Allozymes are electrophoretically distinguishable forms of an enzyme (respectively an isozyme) that are encoded by different alleles of the same gene.

be visualized by staining techniques after electrophoresis. In the following years up to now, many allozyme studies dealt with population and evolutionary genetics, systematics, taxonomy, and conservation genetics.

Allozyme differentiation of ground beetles was studied at first in North America in cave inhabiting ground beetles (*Rhadine*, which is a microphthalmic Platynini genus, Avise & Selander, 1972, already a short time after the first work on allozymes was published; *Neaphaenops tellkampfii* and other trechine beetles; Kane, 1982; Kane et al., 1992; 1990; Turanchik & Kane, 1979). Many other ground beetles from America, Asia and Europe followed (e.g. Assmann & Weber, 1997; Desender et al., 2000, 2002, 2005a; Dhuyvetter et al., 2004; Dhuyvetter et al., 2007; Gaublomme et al., 2002; Liebherr, 1986; Pavlicek & Nevo, 1996; Terlutter 1990).

Especially unwinged ground beetles show strongly differentiated populations. In general this result is congruent with the findings from morphometry. Some ground beetles belong to the insect species with the most differentiated populations ever studied (e.g. *Carabus variolosus*; Matern et al., 2008a, 2008b).

Allozymes are a useful tool for the identification of sibling ground beetle species: Individuals of the four Central European species of the *Amara communis* group, the three Central European *Bembidionetolitzkya* species of the *fasciolatum-ascendens* group and some *Pogonus* species related to *P. chalceus* are easy to separate by few allozyme loci. In all of these cases hybrids could not be detected (Hurka, Manderbach & Assmann, unpubl. results; Desender, personal communication). We believe that possibilities and easy use of allozyme techniques are not exhausted for the discrimination of critically discussed sibling species. This conclusion is in general also stressed by Allendorf & Luikart (2007: 55): "Protein electrophoresis is the quickest and best initial method for detecting cryptic species in a sample of individuals from an unknown taxonomic group. Individuals from different species will generally be fixed for different alleles at some loci. The absence of heterozygotes at these loci would suggest the presence of two reproductively isolated, genetically divergent groups …".

Allozymes are an important tool for many approaches at and below the species level. Above the species level the genetic distance values strongly increase and the resolution for phylogenetic and taxonomic challenges decreases (Tables 1 and 2).

The greatest possible information at the molecular level is offered by **DNA sequencing**. Mostly conservative genes of the mitochondrion are sequenced, in the last years also nuclear genes. A good overview on the DNA techniques which are related to systematics and taxonomy is given by Hillis et al. (1996). DNA sequence data allow numerous analyses at all taxonomic levels. But there is a drawback in practical approach: Due to the expenditure of work and money mostly only one or few genes are sequenced, even if more than thousand base pairs (bp) are sequenced. Additionally, it is sometimes very difficult to identify heterozygote specimens, a drawback that seems not that important in taxonomy, but especially in evolutionary and population genetics.

By using sequence data some working groups contributed substantially to our understanding of the molecular evolution and of the phylogeny of taxa (Barraclough & Vogler, 2002; Bruckner & Mossakowski, 2006; Cardoso & Vogler, 2005; Düring et al., 2000, 2006; Martinez-Navarro et al., 2005; Mossakowski, 2005; Pearson & Vogler, 2001; Prüser & Mossakowski, 1998; Sanchez-Gea et al., 2004; Sota et al., 2001; Sota & Vogler, 2003). For instance, the phylogenetic relationship of the carabids within the Adephaga and Coleoptera is much better understood owing to molecular data (Hunt et al., 2007; Maddison et al., 1998; Maddison et al., 1999). Moreover, sequence data enable a combining of phylogeography and systematics, which leads to an understanding of the history of well known to poorly known taxa. The *Calathus* species from the Macaronesian Islands in the northern Atlantic Ocean are one example. By using mitochondrial cytochrome oxidase I and II sequence data, hypotheses of monophyly are tested for this highly radiated island fauna. Data suggest that the Canary Islands have been colonized three times and Madeira twice. At least four of these colonization events are of continental origin, but it is possible that one clade from Madeira archipelago may be monophyletic with a Canarian clade (Emerson et al., 1999, 2000).

RAPDs and AFLPs (and other fingerprinting methods) are mainly used for the identification of breeding structures within populations or landscape genetics of ground beetles (Sander et al., 2006a; 2006b). Other sequences, such as Ubiquitin-single-strand conformation polymorphism are only rarely used (Sedlmair et al., 2000) mainly due to their very conservative character. However, these markers may be helpful to resolve basal splits of higher taxa of carabids.

Microsatellite DNA is studied not only in landscape genetics (Brouat et al., 2003, 2004; Keller & Largiader, 2003b; Keller et al., 2004; Drees et al., 2008), but also in order to reconstruct phylogeography and population history, in some cases also the relationships of different subspecies (Garnier et al., 2004; Rasplus et al., 2000, 2001).

All in all the use of molecular methods significantly enlarged our knowledge of carabid evolution and systematics. Many aspects important for general aspects of evolution are clarified by these approaches and revolutionize our understanding of evolutionary processes: Here we only want to name the ongoing speciation under sympatric conditions in *Pogonus chalceus* (Dhuyvetter et al., 2007) and the detection of horizontal gene flow across species boundaries (e.g. Brouat et al., 2006; During et al., 2006; Sota & Vogler, 2003; Streiff et al., 2005).

During the last decades molecular techniques developed rapidly and we believe that at least two tendencies seem to be probable for the near future: (1) DNA sequencing becomes faster and cheaper which will result in an increasing use of this technique for taxonomic issues. (2) Better insights into breeding structures and evolutionary processes are possible by the development of new marker systems (e.g. Allendorf & Luikart, 2007).

Table 1. Strengths and weaknesses of different molecular marker systems (modified after Scribner, 2006; Allendorf & Luikart, 2007).

Marker system/ method	Strengths	Weaknesses	Literature (general and examples from ground beetles)
1) Allozymes	• Large number of loci can be studied • Relatively easy, quick and low-cost • Codominant expression of alleles and relatively constant number of subunits reveals genetic basis and heterozygotes • Adaptable to "classical" population genetics (e.g. Hardy-Weinberg equilibrium) • Data sets from different laboratories can be combined	• Only a specific set of genes can be studied (water-soluble enzymes) • Only genetic changes affecting the amino acid sequence can be detected (no silent substitutions) • Individuals or tissues to be analysed must be stored under proper refrigeration (not available from museum material)	Assmann & Günther, (2000); Assmann & Janssen (1999); Assmann & Weber (1997); Desender et al. (2000, 2005a, 2005b); Desender & Serrano (1999); Dhuyvetter et al. (2007); Matern et al. (2008a); Reimann et al. (2002); Vogler et al. (1993)
2) Mitochondrial DNA (mtDNA)	• Maternal inheritance • No recombination	• limited information on migration and exchange between populations	
Restriction fragment length polymorphism (RFLP)	• Detection of single mutations • > 400 different restriction enzymes • Available from pinned museum material	• Expensive (especially if many restriction enzymes are used) • Weak relationship to gene loci (exception: RFLP after amplification of a given DNA sequence)	Takami & Suzuki (2005)
Sequencing after Polymerase chain reaction (PCR)	• Best information on the studied genes • "universal" primers for "conservative" genes	• Development of primers for variable genes is expensive	Clarke et al. (2001); Matern et al. (2008b); Sota et al. (2001); Sota & Vogler (2001); Zhang et al. (2005, 2006); Zhang & Sota (2007)
3) Nuclear DNA	• Vast number of gene loci • Heterozygotes • Recombination • Available also from pinned museum material		

PCR, microsatellites	• Rapid evolving sequences • "Phylogeny" of alleles at least partly possible • Many alleles • Sequencer not necessary	• Expensive development of primers • Sub-bands or stutter bands in the gel (result of "slippage" during PCR amplification) make it difficult to detect some heterozygotes	Brouat et al. (2002, 2003, 2004) ; Dhuyvetter et al. (2007); Drees et al. (2008); Garnier et al. (2002, 2004); Keller et al. (2005); Keller & Largiader (2002, 2003a, 2003b); Keller et al. (2004); Rasplus et al. (2000, 2001)
PCR and sequencing	• Best information on the studied genes • "Universal" primers for "conservative" genes available • Exons and introns • Enormous variability • Sequences for comparisons and outgroups available	• Expensive development of primers for genes which are not conservative • High costs for sequencing • Time consuming (especially for larger numbers of specimens) • Heterozygotes somesometimes difficult to detect	Brückner & Mossakowski (2006); Cardoso & Vogler (2005); Contreras-Diaz et al. (2007); Düring et al. (2006); Kim et al. (2000); Ribera et al. (2006); Sasakawa & Kubota (2007); Sota & Ishikawa (2004); Sota et al. (2005); Zhang et al. (2005, 2006); Zhang & Sota (2007)
Minisatellites	• Highly polymorphic • Useful for parentage analysis and individual identification	• Problems to identify alleles of a given locus	Not yet used in carabids
Randomly amplified polymorphic DNA (RAPDs)	• Polymorphic markers for species if no sequence information exists	• Difficult to achieve reproducible results • Some journals (e.g. Molecular Ecology) do not accept studies using RAPDs	Clarke et al. (2001); Sander et al. (2006a)
Amplification of DNA fragments produced by cleaving genomic DNA (AFLPs)	• Polymorphic markers for species if no sequence information exists • Fast, less laboratory intensive • More reproducible than RAPDs	• Sometimes difficult to identify alleles of a given locus	Sander et al. (2006b)

Table 2. Usefulness of different alternative methods for systematics, taxonomy, and related fields of phylogeny and evolutionary genetics of carabids. ***: very useful, **: useful, *: sometimes useful, -: not useful.

Marker system	Breeding structure and within population variability	Differentiation between populations and taxa below species level	Taxa of species level	Taxa of genus level	Higher taxa (e.g. tribes, subfamilies)
Morphometry	-	***	**	-	-
Karyotypes	-	-	***	*	*
Allozymes	**	***	***	*	-
mtDNA, RFLPs	*	***	***	**	*
mtDNA, sequences of base pairs	*	***	***	***	***
Nuclear DNA, microsatellites	**	***	*	-	-
Nuclear DNA, sequences	*	***	***	***	***
Fingerprinting technique 1: minisatellites	***	**	-	-	-
Fingerprinting technique 2: RAPDs	*	*	*	-	-
Fingerprinting technique 3: AFLPs	***	**	*	-	-

CURRENT TAXONOMY

Linking proven and new approaches

In the past 250 years since Linné or about 200 years since Latreille numerous species and genera have been described and amount to presently more than 35,000 species (Lorenz, 2005b). Numerous taxonomic and faunistical publications (including identification keys) are available, especially from the western part of the Palaearctic region (Table 3). From other parts of the world there are many excellent contributions (e.g. Larochelle & Larivière, 2001; Moret, 2005). An updated compendium on the forthcomings for the Nearctic region is given by Ball (this volume). Further monographs are in preparation (e.g. identification keys for the ground beetle genera of South Africa, Schüle & Lorenz, personal communication; Greece, Arndt, personal communication).

There is no serious alternative to MORAT. Molecular techniques can contribute substantially to many facets of biology, and of course to taxonomy, too. But the exclusive use of DNA barcoding as a sole "taxonomic" approach as suggested by some authors (Tautz et al., 2003) will not be able to replace taxonomy and systematics in its wider and classical sense (Valdecasas et al., 2008; Wheeler et al., 2004). The overall goal of taxonomy is to name taxa for other biological disciplines. Identification keys are an important service, as they are essential for many other scientists, the "clients" who are interested in ecology, biochemistry, biogeography, ethology, physiology, etc. For these "users" of taxonomy it seems not practicable to sequence DNA of large series of individuals (e.g. from pitfall traps) for identification. Moreover DNA barcoding by itself cannot solve the problem of species and genus naming and biologists favouring DNA barcoding use names from the 'traditional' taxonomy.

As we have already shown, alternative approaches to MORAT can contribute substantially to taxonomy, phylogeny, evolutionary biology and other systematics related fields in carabidology. It is therefore consequential to combine MORAT and the new approaches, especially molecular techniques. In general the number of publications with different methodological approaches, especially a combination of MORAT and alternative approaches is increasing (e.g. Casale et al., 1998; Prüser & Mossakowski, 1998).

In some outstanding works molecular methods complement classical alpha-taxonomy, especially for resolving the uncertain species status. Both, the (re-) establishment and the synonymisation of previously described species are known. Two examples can illustrate this:

Table 3. Examples of important faunistical and taxonomic publications (including identification keys from family to subspecies level) on carabids in the western parts of the Palaearctic region.

Region	Literature
Scandinavia	Lindroth (1985f)
Britain, Ireland	Anderson et al. (2000); Lindroth (1974); Luff (1998)
France	Bonadona (1971); Forel & Leplat (1995, 2001, 2003, 2005); Jeannel (1941f)
Central Europe (including Austria, Belgium, Czech Republic, Denmark, Germany, Netherlands, Slovak Republic, Switzerland, Netherlands)	Bangsholt (1983); Desender (1986); Hurka (1996); Marggi (1992); Marggi & Luka (2001); Müller-Motzfeld (2004); Turin (2000)
Poland	Burakowski et al. (1973f)
Russia and adjacent countries	Kryzhanovskij et al. (1995), Retezár (2008)
Iberian Peninsula	Serrano (2003)
Italian Peninsula	Brandmayr et al. (2005)
Balkan Peninsula (including Albania, Bulgaria, Serbia, Slovenia, Greece)	Drovenik & Peks (1999); Ćurčić et al. (2007), Guérguiev & Guérguiev (1995); Guérguiev et al. (1997); Hieke & Wrase (1988); Neculiseanu & Matalin (2000)
Northwestern Africa (Morocco, Algeria, Tunisia)	Antoine (1955ff); Kocher (1963)
North-eastern Africa (Egypt, Libya)	Alfieri (1976); Hosni et al. (2003)
Turkey	Casale & Vigna-Taglianti (1999)
Middle East	Kleinfeld & Rapuzzi (2004); Nussbaum (1987)

1 Woodcock et al. (2007) discovered an exceptionally high pairwise sequence divergence for cytochrome b and cytochrome oxidase subunit I (mtDNA) between *Cicindela* (*Cylindera*) *lunalonga* and the other members of the *Cicindela terricola* group (mean of 6.36 %). They conclude that based on strict monophyly, pairwise sequence divergence, and the lack of hybrids, *C. lunalonga* is a distinct species.
2 To determine the relationship of the species of the *Nebria gregaria* group, Clarke et al. (2001) conducted phylogenetic analyses on nucleotide sequence data obtained from five regions of the mitochondrial DNA (in total 1835 bp were sequenced). The results revealed very low genetic divergence (less than 1 % bp divergence). These results and random amplified polymorphic DNA fingerprinting analyses suggest that the described taxa may be postglacial in origin and that they might represent local variations of a single species.*

We believe that it is no coincidence that both studies deal with North American taxa. The use of molecular methods to solve taxonomic problems at the species level is still poorly developed in Europe.

The work of some authors from all over the world illustrate that molecular data can provide substantially to taxonomy above the species level (Barraclough & Vogler, 2002; Hunt et al., 2007; Maddison et al., 1998). The genus *Carabus* was intensively studied at the molecular level by European and Japanese working groups (During et al., 2006; Mossakowski, 2005; Osawa et al., 2003; Prüser & Mossakowski, 1998). Our knowledge on the molecular evolution and the classification above the species level increased enormously during the last two decades. However, some, especially molecular based changes at the genus group level are criticized due to statistical weakness (Mossakowski, 2005). Therefore, molecular techniques are not a guarantee to solve long-lasting problems in systematics or taxonomy (e.g. the subgenera and phylogeny within the genus *Carabus*), but it would be a mistake to ignore the benefits molecular techniques can offer.

Challenges

The occurrence of overlooked cryptic species, the establishment of evolutionarily significant units (ESUs), management units (MUs), and – more recently – conservation units (CUs) (for an overview see Allendorf & Luikart, 2007) and in general the barely understood taxonomic variety has been described as the taxonomic impediment (New, 1984), and more recently as the taxonomic challenge (Samways, 2005).

Without question, there is a general discussion on and within taxonomy. Sometimes taxonomists are reproached of poor intellectual content in their work (but see Lipscomb et al., 2003; Valdecasas et al., 2008), or the general right of the existence of taxonomy is questioned (but see Wheeler et al., 2004). Despite the fact that most criticism does not

* We have to emphasize that the establishing of species cannot be done exclusively by determining the DNA basepair substitutions. But the DNA methods can contribute their share especially if other criteria related to the Biological Species Concept do not work (e.g. in cases of strict allopatry).

do justice to taxonomy and systematics, we plead for a synthesis of different methods (see Section "Linking proven and new approaches").

Stability of nomenclature

Some criticism seems to contain a core of truth, e.g. the unnecessary change of names. Although the International Code of Zoological Nomenclature (1999) intends to stabilize the system of nomenclature, especially the names of genera and species, occasionally names must be changed due to scientific progress (e.g. Dominguez & Wheeler, 1997). The main criticism is directed to really superfluous changes of names:

1 If a name is well established, it should have priority over any changes in nomenclature. An example might be the attempt to change the frequently used species name of *Carabus lineatus* Dejean, 1826 into *C. basilicus* Chevrolat, 1837 (cf. Bousquet et al., 2003; Lorenz, 1998b). Another example is *Abax ater* (Villers, 1789) and *Abax parallelepipedus* (Piller & Mitterpacher, 1783). Taking into account that taxonomy is not done for itself but should be an important service for other biological disciplines these changes in nomenclature are confusing and not understood by most scientists.

2 Many non-taxonomists criticize changes in the rank of subspecies and species (and of other taxonomic levels) without any new results and any necessity (Isaac et al., 2005; Isaac et al., 2004; Wiens, 2007). Such changes are regularly done in many carabid catalogues without any discussion of reasons. A discourse on the underlying reasons would be an important prerequisite for a true scientific basis of naming. In general it is possible to make the reasons for changes in taxonomic ranks available for the scientific public (e.g. Wrase, 2005).

3 In some cases changes in the rank of taxa without a scientific discussion led to great confusions in other fields of biology. An example is *Carabus variolosus* which is listed in the Annex of the European Union Habitats and Species Directive as a species of Community interest. Numerous catalogues and identification keys list the taxon *nodulosus* as a subspecies, others as a species. Meanwhile there is a still ongoing discussion if this taxon has to be protected under the EU Directive or not (Müller-Kroehling, 2006).

4 "Private" taxonomy in the genus *Carabus?* A view into some catalogues from the last years (Brezina, 1999; Bousquet et al., 2003; Deuve, 1994, 2004) reveals conspicuously diverging opinions in terms of naming, systematics and rank of taxa above and below the species level of *Carabus*. Moreover, in *Carabus* taxonomy the tendency to name subspecies of different levels is still widely distributed, despite the fact that it is not supported by the International Code for Zoological Nomenclature (e.g. Deuve, 2004). – This "subtaxonomy at the subspecies level" has the potential to damage taxonomy's good reputation, because many (if not most) scientists cannot understand why naming of a given species has to change several times within few years (without any discussion of the reasons). Moreover the uncertainty in taxonomy interferes with efficient development of applied fields such as conservation biology (see above).

5 Names at the genus level (and partly also of subfamily and family rank) are used inconsistently by some carabidologists. Jeannel (1941f) initiated the uplevelling of taxa which are ranked as subgenera by other authors (e.g. of the genera *Bembidion* and *Carabus*). Some "francophone" and "romanophone" authors followed him, but most other carabiodologists (including those from North America) follow Lindroth and downlevel these taxa. The consequences of this unfortunate or even vexating "double taxonomy" are numerous synonyms, unavailable names (depending on the system used), and a broad confusion in nomenclature from subgenus to family level.

A huge amount of undescribed taxa

Another just accusation is the enormous amount of undescribed (and therefore unknown) species despite 250 years of research in taxonomy. This reproach seems to fit for carabid taxonomy as well. Following the estimations and assumptions of Baehr and some colleagues (Baehr, 2005; Stork, 1988), the probable number of carabid species exceeds the number of described species by two or more powers of ten. Therefore, many if not most carabid species of the tropics are unknown. There is good evidence that numerous carabid species even in the Palaearctic realm are still undiscovered and unknown.

Only few new species have been described in the last decade from the western Palaearctic, especially Europe.* But for the entire Palaearctic region, the total number of new taxa described per year is still clearly increasing (Fig. 1). Most new species and genera originate from southern China and other parts of southeastern Asia. The Catalogue of Palaearctic Coleoptera lists about 3200 species and subspecies from China (Löbl & Smetana, 2003). From 1995 to 1999 about 200 to 400 species and subspecies from the entire Palaearctic realm were described per year, half of which from China alone. Interpreting the results from a macroecological analysis in terms of Palaearctic carabid diversity very carefully and conservatively, Schuldt et al (in prep.) expect at least two to three times more carabid species in China than described so far.

Sometimes records or discoveries from the Chinese mountains come as a surprise for biogeographical, faunistical and phylogenetical aspects of carabidology. An example might be the discovery of the first known representative of the tribe Metriini in the Palaearctic (*Sinometrius turnai*, Wrase & Schmidt, 2006). The tribe was presumed to be monogeneric and restricted to western Nearctic North America (Ball, 2001). The southeastern Asian biodiversity hotspots (for other taxa see Tang et al., 2006) seem to accommodate a rich and still poorly explored carabid fauna.

* Mainly anophthalmic and microphthalmic cave or deeper soil horizons dwelling carabids are exceptions, e.g. the genera *Tinautius*, *Dalyat*, and *Galiciotyphlotes* which are described in the last years from the Iberian Peninsula (Assmann, 1999; Mateu, 1997, 2001). The small sized endogeic Anillini seem still to be incompletely studied on a taxonomic level as it is revealed by numerous recently discovered species (e.g. Serrano & Aguiar, 2002; Serrano & Aguiar, 2004, 2006; Serrano et al., 2003; Zaballos, 2005; Zaballos & Ruiz-Tapiador, 1997). Large and beautiful species are only occasionally overlooked, e.g. *Cephalota* (*Taenidia*) *dulcinea* (Lopez et al., 2006).

The recent exploration of the eastern and southeastern parts of Asia in the last decades triggered more taxonomic descriptions from the Palaearctic than ever before (Fig. 2). Therefore we cannot recognize a decrease in the taxonomic activity of carabidologists. The progress of taxonomy depends not only on the engagement of professional entomologists,

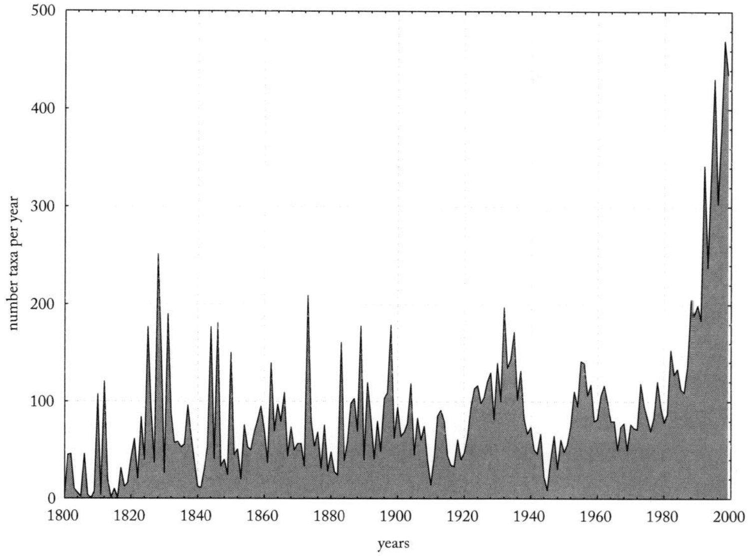

Fig. 1. Number of carabid taxa described from the Palaearctic per year (1800-1999) (generated after the Catalogue of Palaearctic Coleoptera; Löbl & Smetana 2003).

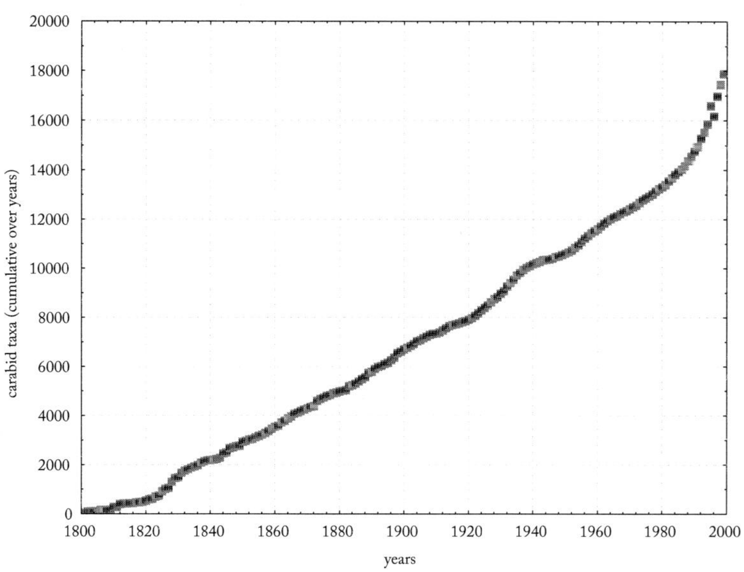

Fig. 2. Cumulative number of carabid taxa described from the Palaearctic (1800-1999) (generated after the Catalogue of Palaearctic Coleoptera; Löbl & Smetana 2003).

but it reflects also the contribution of enthusiastic collectors and amateur carabidologists, whose contributions to our knowledge about ground beetles were substantial and should not be underestimated. In the last decades clearly more than half of all authors who described species from the Palaearctic region are amateur entomologists (revealed by authors' addresses of about 200 randomly chosen reprints screened by us).

Different genitalia = different species!
State of the art or traditional paradigm?

Since the pioneer work of Escherich and Holdhaus (see section "Morphometry") differences in the male genitalia have been used to discriminate species from each other. About two decades ago the lock-and-key hypothesis was introduced by Eberhard (1985; cited after Shapiro & Porter, 1989) to explain species-specific genital morphology in terms of mechanical reproductive isolation.

Perhaps best evidence for this hypothesis comes from Japanese *Carabus* species. A section of this outstanding work should give an insight:

- Sota & Kubota (1998) demonstrated experimentally that differences in genital characters impose a direct cost of interspecific copulation in the two closely related species *C.* (*Ohomopterus*) *maiyasanus* and *C.* (*O.*) *iwawakianus*, that share a hybrid zone. Females experiencing heterospecific mating often die due to rupture of their vaginal membranes. Males of one of the two species, but not of the other one, often had broken copulatory pieces following interspecific copulations. "Because of female mortality and low fertilization rate, the estimated fitness cost of interspecific mating was very large in terms of the reduction in the number of offspring hatching larvae for both sexes and both species. Thus, genital lock-and-key appears to exert significant selection against hybridization in the hybrid zone of these carabid beetles" (Sota & Kubota, 1998: 1507).
- Biometric analyses on experimental F-1 and backcross offspring of *Ohomopterus* revealed that inheritance of genital morphology is polygenic and a relatively small number of loci is responsible for species differences in genital morphology. Overall, the genetic basis of male and female genitalia of *Ohomopterus* species seems to be fairly simple, enabling these traits to respond quickly to selection pressures and to diverge rapidly (Sasabe et al., 2007).
- Additional results on the insemination, species-specific shapes and lengths of genitalia support the lock-and-key-hypothesis for Japanese *Carabus* species (Takami, 2003, 2007; Usami et al., 2006).

These results suggest ranking taxa with different genitalia at least as different species due to assumed barriers of gene flow (which is important if Biological Species Concept is accepted for carabids, cf. Claridge et al., 1997).

However, despite significant differences in terms of shape of the genitalia, the two *Carabus* (*Megodontus*) *violaceus* subspecies, *C. v. violaceus* and *C. v. purpurascens*, form broad hybrid zones without evidence of selection pressure against hybrids or cross-breeding

(Assmann & Schnauder, 1998; Eisenacher et al., in prep.). The differences in external aedeagus shape were the reason for an uncertain rank of the taxa (e.g. Arndt & Trautner, 2004). Endophalli of both subspecies are very similar (Muergues & Ledoux, 1966). The lack of considerable differences in the copulatory pieces of the aedeagus and congruence of the shape of vaginal appendix and *Bursa copulatrix* in both taxa distinguish them from species of the genus *Ohomopterus*. In contrast to *C. violaceus*, the external shape of the aedeagus does not differ significantly for *Ohomopterus* species.

From these results it has to be concluded that the apex is not a functional part of the aedeagus of *Carabus* species (Eisenacher et al., in prep.) and differences in the external shape of this organ should not be used as sole characters to delimit taxa at the species rank. Surprisingly, this conclusion results in conservative rankings for some *Carabus* taxa. This is not only true for *C. violaceus* s.l., but also for *C. variolosus* s.str. and *C. v. nodulosus*, which are ranked as subspecies by Breuning (1926), who knew about the divergences of the aedeagus apex of both forms. Deviating from Breuning's opinion some authors ranked the taxa as separate species (see Müller-Kroehling, 2006 for a detailed discussion). Clear differences are revealed exclusively by the apex of the aedeagus, which seems to have no function following the conclusions from the above mentioned work. Consequently both forms have to be ranked as subspecies Matern et al. (2008b).

DNA sequence data support this conclusion and stress our plea for the necessity to link molecular and morphology-based approaches. The mean genetic divergence of the COI gene (mtDNA) is about 1.2 % for the allopatric subspecies of *C. variolosus nodulosus* (Matern et al., 2008b). The differentiation is larger than those of *Nebria* taxa which have been synonymized by Clarke et al. (2001), but clearly much lower than the high divergence of *Cicindela lunalonga* and other species of the *C. terricola* group (6.3 %). Surprisingly, the subspecies of *Carabus violaceus* exhibit a comparably high mean genetic divergence at the same gene (6.4 %), despite their broad hybrid zones. Low divergence rates for mitochondrial DNA have been proposed for carabids and cicindelids (Barraclough & Vogler, 2002; Prüser & Mossakowski, 1998). Hence, the split between the two last mentioned pairs of taxa occurred perhaps in the Pliocene, but it is unlikely that it is more recent. In contrast the *Carabus variolosus* subspecies might have split up during quaternary glacial cycles.

The mtDNA divergence cannot be used to define exact thresholds for species and subspecies rank as every other character unsuitable for proving gene flow, but these genetic data offer new/additional criteria which are independent from morphology and help to open the window to the history of taxa and populations. This phylogenetic approach supports especially systematics (and by that alpha taxonomy) at and below the species level (Habel & Assmann, 2008).

Incongruence from different approaches

Species and subspecies delimitation is complicated if morphological variation is continuous or poorly subdivided, but for taxonomy it is necessary to separate and name phylogenetic entities to capture this variation. Molecular and genetic based approaches can be more useful than those based exclusively on morphology. Especially DNA techniques offer the potential to test wether groups in fact represent historically divided, discrete entities. In any case the results from different approaches have to be combined for common taxonomy. We would like to quote and interpret some results:

- Cardoso & Vogler (2005) sequenced a discontinuous segment of 1899 bp of mtDNA for about a hundred specimens of the *Cicindela hybrida* complex from localities across Europe. They found four major clades corresponding to geographical groups from central Iberia, Ukraine, Central Europe, and a band from the Atlantic Iberian coast to northern Europe. Within the latter group specimens belonging to *transversalis* and *hybrida* s.str. form three and two entities, respectively, which are supported by high bootstrap values. This result indicates a paraphyletic origin of the named taxa.
- Differentiation of populations of the known subspecies of *Carabus solieri* has been studied by morphometry and at the molecular level (Garnier et al., 2004, 2005; Rasplus et al., 2001). The results confirmed the hybrid origin of a group of populations which are listed by some authors as a valid subspecies. Moreover many subspecies systematics of *Carabus solieri* are incongruent with the findings of morphological and genetic approaches (e.g. Deuve, 2004).
- *Carabus auronitens* is the most widely distributed species of the subgenus *Chrysocarabus*. Numerous subspecies are described. A huge bulk of populations has been screened for allozyme differentiation. Substantial inconsistencies between the molecular divergence and current subspecies taxonomy have been demonstrated (Assmann & Weber, 1997; Reimann et al., 2002). Neither morphometry nor molecular data support any of the current morphology-based subspecies taxonomies of the given species.

Especially subspecies, mostly from the genus *Carabus*, and only rarely species show these conspicuous divergences between current taxonomical rankings and evidences from morphometric and molecular analyses. To avoid these problems in future, we recommend the use of objective comprehensible methods to delimit subspecies. Morphometry and molecular methods are outstanding tools for this type of taxonomy (nevertheless classical descriptions with type specimens are still necessary). Descriptions of subspecies exclusively based on eye-inspection of external morphology have to stop.

HOW TO IMPLEMENT CONTEMPORARY SYSTEMATICS AND TAXONOMY ON CARABIDS – PLEA FOR A MODERN SYNTHESIS

There are some publications on how to practise MORAT in general and especially in carabidology (e.g. Baehr, 2005; Ball, 2008; Winston, 1988). The necessity to proceed intensively on MORAT is clearly given by the fact that only a small portion of ground beetles is known (and described) so far.

The enormous quantity of yet undescribed species demands an efficient work from taxonomists which might be fulfilled not only by new descriptions but by incorporating the new taxa into existing identification keys (Baehr, 2005). In many cases new keys have to be worked out. – Keys are also an important element in George Ball's (this volume) vision on future carabidology. He also emphasizes the development of new keys for the genus level as it is done for North America. Similar synopses are lacking from the Palaearctic and most other parts of the world, but for the western Palaearctic numerous keys and faunistic works are available (Table 3). Taxonomy is done not for itself, as keys are an important service for other biologists. The same holds true for overall synopses on the biology, ecology, and evolutionary biology of ground beetles from larger regions. These works model themselves on Lindroth's famous "Die fennoskandischen Carabidae" (Lindroth, 1945, 1949). Now there is an increasing number of that type of work for other regions (Erwin, 2007; Larochelle & Larivière, 2001, 2003).

We would ignore scientific progress if we work exclusively on MORAT. Present-day taxonomy needs pluralistic approaches and alternative methods, especially morphometry and molecular methods to solve uncertain ranks. Taxonomy incorporated new methods in the past (such as the study of genitalia as diagnostic features) and has to continue doing it, because the ignorance of huge datasets would be a nonscientific action.

Just the combination of morphology, morphometry and molecular based information will develop a modern taxonomy which has the potential to manage the scientific basis for other studies on ground beetles which are an important model group for organismic biology.

ACKNOWLEDGEMENTS

We would like to thank Dr. Martin Baehr, Munich, Prof. Dr. Alfried Vogler, London, and Prof. em. Dr. Friedrich Weber, Münster, for stimulating discussions.

LITERATURE

Alfieri, A. (1976). The Coleoptera of Egypt. – Mémoires de la Société Entomologique d'Égypte 5: 1-361.

Alibert, P., Moureau, B., Dommergues, J.L. & David, B. (2001). Differentiation at a microgeographical scale within two species of ground beetle, *Carabus auronitens* and *C. nemoralis* (Coleoptera, Carabidae): a geometrical morphometric approach. – Zoologica Scripta 30: 299-311.

Allendorf, F. W. & Luikart, G. (2007). Conservation and genetics of populations. – Blackwell, Malden, Oxford, Victoria.

Anderson, R., McFerran, D. & Cameron, A. (2000). The ground beetles of northern Ireland (Coleoptera: Carabidae). – Ulster Museum, Belfast.

Andújar, A. & Serrano, J. (2001). Revisión y filogenia de los *Zabrus* Clairville, 1806 de la Península Ibérica (Coleóptera, Carabidae). – Monografias S.E.A. 5: 1-90.

Antoine, M. (1955ff). Coléoptères carabiques du Maroc. – Memoires de la Société des Sciences Naturelles et Physiques du Maroc, Zoologie 1ff: 1-692.

Apfelbeck, V. (1904). Die Käferfauna der Balkanhalbinsel, mit Berücksichtigung Klein-Asiens und der Insel Kreta. – R. Friedländer und Sohn, Berlin.

Arndt, E. & Trautner, J. (2004). 4. Unterfamilie: Carabinae, 4. Tribus: Carabini. – In: Die Käfer Mitteleuropas. (Müller-Motzfeld, G., editor). Spektrum, Heidelberg, Berlin, p. 30-60.

Assmann, T. (1999). A new anophthalmic genus of Perigonini from the Iberian Peninsula (Insecta, Coleoptera, Carabidae). – Spixiana 22: 237-244.

Assmann, T. & Günther, J. (2000). Relict populations in ancient woodlands: genetic differentiation, variability, and power of dispersal of *Carabus glabratus* (Coleoptera, Carabidae) in north-western Germany. – In: Natural history and applied ecology of carabid beetles (Brandmayr, P., Lövei, G., Brandmayr, T. Z., Casale, A. & Vigna Taglianti, A., eds). Pensoft, Sofia and Moscow, p. 197-206.

Assmann, T. & Janssen, J. (1999). The effects of habitat changes on the endangered ground beetle *Carabus nitens* (Coleoptera: Carabidae). – Journal of Insect Conservation 3: 107-116.

Assmann, T. & Schnauder, C. (1998). Morphometrische Untersuchungen an einer Kontaktzone zwischen *Carabus* (*Megodontus*) *violaceus* und *purpurascens* (Coleoptera, Carabidae) in Südwest-Niedersachsen. – Osnabrücker Naturwissenschaftliche Mitteilungen 24: 111-138.

Assmann, T. & Weber, F. (1997). On the allozyme differentiation of *Carabus punctatoauratus* Germar (Coleoptera, Carabidae). – Journal of Zoological Systematics & Evolutionary Research 35: 33-43.

Avise, J.C. & Selander, R.K. (1972). Evolutionary genetics of cave-dwelling fishes of the genus *Astyanax*. – Evolution 26: 1-19.

Brezina, B. (1999). World Catalogue of the Genus *Carabus* L. – Pensoft, Sofia.

Baehr, M. (1998). A preliminary survey of the classification of the Psydrinae (Coleoptera: Carabidae). – In: Phylogeny and classification of Caraboidea (Coleoptera: Adephaga) (Ball, G.E., Casale, A. & Vigna-Taglianti, A., eds). Museo Regionale di Scienze Naturali, Torino, p. 359-368.

Baehr, M. (2005). Sollen wir noch Arten beschreiben? Und wenn ja, wie? – Entomologische Nachrichten und Berichte 49: 91-95.

Ball, G.E. (1979a). Conspectus of carabid classification: history, holomorphologhy, and higher taxa. – In: Carabid beetles: their evolution, natural history, and classification (Erwin, T.L., Ball, G.E. & Whitehead, D.R., eds). Dr W Junk, The Hague, Boston, London, p. 63-111.

Ball, G.E. (1979b). Introduction – three leaders. – In: Carabid beetles: their evolution, natural history, and classification. Proceedings of the First International Symposium of Carabidology (Erwin, T.L., Ball, G.E. & Whitehead, D.R., eds). Dr. W. Junk The Hague, Boston, London, p. 1-5.

Ball, G.E. (2001). Carabidae Latreille, 1810. – In: American beetles, volume 1: Archostemata, Myxophaga, Adephaga, Polyphaga: Staphyliniformia (Arnett, R.S. & Thomas, M.C., eds). CRC Press, Boca Raton, London, New York, Washington, D.C, p. 32-132.

Ball, G.E. (2008). Twentieth century carabidology in the Nearctic region. – In: Back to the Roots and Back to the Future. Towards a New Synthesis amongst taxonomical, ecological and biogeographical approaches in carabidology (Penev, L., Erwin, T. & Assmann, T., eds). Pensoft, Sofia-Moscow, p. 23-38.

Ball, G.E., Casale, A. & Vigna-Taglianti, A. (1998). Introduction. – In: Phylogeny and classification of Caraboidea (Coleoptera: Adephaga) (Ball, G.E., Casale, A. & Vigna-Taglianti, A., eds). Museo Regionale di Scienze Naturali, Torino, p. 15-52.

Ball, G.E. & Shpeley, D. (2005). Taxonomic review of the tribe Melaenini (Coleoptera: Carabidae), with observation on morphological, ecological and chorological evolution. – Zootaxa 1099:1-120.

Bangsholt, F. (1983). Sandspringernes og lobebillernes udbredelse og forekomst i Danmark ca. 1830-1981. – Scandinavian Science Press, Kobenhavn.

Barr, C. (2004). A classification and checklist of the genus *Psudanophthalmus* Jeannel (Coleoptera: Carabidae: Trechinae). – Virginia Museum of Natural History, Virginia.

Barraclough, T.G. & Vogler, A.P. (2002). Recent diversification rates in North American tiger beetles estimated from a dated mtDNA phylogenetic tree. – Molecular Biology and Evolution 19: 1706-1716.

Belousov, I.A. (1998). Le complex générique de *Nannotrechus* Winkler du Caucase et de la Crimée (Coleoptera, Carabidae, Trechini). – Pensoft Publishers, Sofia-Moscow-St. Petersburg.

Boettger, C.R. (1921). *Carabus morbillosus* Fabr. und *Otala tigri* Gerv., eine Studie gegenseitiger Anpassung. – Abhandlungen der Senckenbergischen Naturforschenden Gesellschaft 37: 321-329.

Bonadona, P. (1971). Catalogue des Coléoptères Carabiques de France. – Nouvelle Revue d'Entomologie Suppl.: 1-177.

Bonadona, P. (1973). Nouvelle contributiuon a la connaissance des races francaises de *Chrysocarabus solieri* (Col., Carabidae). – Annales de la Société entomologique de France 9: 759-812.

Bousquet, Y., Brezina, B., Davies, A., Farkac, J. & Smetana, J. (2003). Tribe Carabini Latreille, 1802. – In: Catalogue of Palaearctic Coleoptera Vol. 1: Archostemata, Myxophaga, Adephaga (I. Löbl & Smetana, A., eds). Apollo Books, Stenstrup, p. 118-201.

Brandmayr, P., Zetto, T., Pizzolotto, R., Casale, A. & Vigna-Taglianti, A. (2005). I Coleotteri Carabidi per la valutazione ambientale e la conservazione della biodiversità. Agenzia per la protezione dell'ambiente per i servizi tecnici. – Manuali e Linee Guida, Rome.

Breuning, S. (1926). Über *Carabus variolosus* Fabr. – Koleopterologische Rundschau 12: 19-25.

Breuning, S. (1932). Monographie der Gattung *Carabus* L. – Reitter, Troppau.

Brouat, C., Chevallier, H., Meusnier, S., Noblecourt, T. & Rasplus, J.-Y. (2004). Specialization and habitat: spatial and environmental effects on abundance and genetic diversity of forest generalist and specialist *Carabus* species. – Molecular Ecology 13: 1815-1826.

Brouat, C., Meusnier, S., Veyrier, R. & Streiff, R. (2006). Haldane's rule in *Carabus*: interspecific mating between *Carabus punctatoauratus* and *Carabus splendens* using experimental tests and molecular markers. – Entomologia Experimentalis et Applicata 120: 189-194.

Brouat, C., Mondor, G., Audiot, P., Sennedot, F., Lesobre, L. & Rasplus, J.-Y. (2002). Isolation and characterization of microsatellite loci in the ground beetle *Carabus nemoralis* (Coleoptera, Carabidae). – Molecular Ecology Notes 2: 119-121.

Brouat, C., Sennedot, F., Audiot, P., Leblois, R. & Rasplus, J.-Y. (2003). Fine-scale genetic structure of two carabid species with contrasted levels of habitat specialization. – Molecular Ecology 12: 1731-1745.

Brückner, M. & Mossakowski, D. (2006). Phylogeny of the genus *Percus* (Coleoptera: Carabidae) – nuclear genes and the basal splits. – Entomologica Fennica 17: 195-199.

Burakowski, B., Mroczkowski, M. & Stefanska, J. (1973f). Coleoptera – Carabidae. – Panstwowe Wydawnictwo Naukowe, Warszawa.
Cardoso, A. & Vogler, A.P. (2005). DNA taxonomy, phylogeny and Pleistocene diversification of the *Cicindela hybrida* species group (Coleoptera: Cicindelidae). – Molecular Ecology 14: 3531-3546.
Casale, A. (1998). Phylogeny and biogeography of Calleidina (Coleoptera: Carabidae: Lebiini): a preliminary survey. – In: Phylogeny and classification of Caraboidea (Coleoptera: Adephaga) (Ball, G.E., Casale, A. & Vigna-Taglianti, A., eds). Museo Regionale die Scienze Naturali, Torino, p. 381-428.
Casale, A., Prüser, F., Arndt, E. & Mossakowski, D. (1998). Phylogenetic relationship in the subgenus *Platycarabus* Morawitz, 1886 (Coleoptera: Carabidae: Carabini). – In: Phylogeny and classification of Caraboidea (Coeloptera: Adephaga) (Ball, G.E., Casale, A. & Vigna-Taglianti, A., eds). Museo Regionale di Scienze Naturali, Torino, p. 429-448.
Casale, A. & Vigna-Taglianti, A. (1999). Caraboid beetles (excl. Cicindelidae) of Anatolia, and their biogeographical significance (Coleoptera, Caraboidea). – Biogeographia 20: 277-406.
Chaudoir, M. d. (1876). Monographie des chléniens. – Annali del Museo Civico di Storia Naturale di Genova 8: 3-315.
Claridge, M.F., Dawah, H.A. & Wilson, M.R. (1997). Species: The units of biodiversity. – Chapman and Hall, London
Clarke, T.E., Levin, D.B., Kavanaugh, D.H. & Reimchen, T.E. (2001). Rapid evolution in the *Nebria gregaria* group (Coleoptera: Carabidae) and the paleogeography of the Queen Charlotte Islands. – Evolution 55: 1408-1418.
Conradi-Larsen, E.-M. & Somme, L. (1973a). Anaerobiosis in the overwintering beetle *Pelophila borealis*. – Nature 245:388-390.
Conradi-Larsen, E.-M. & Somme, L. (1973b). The overwintering of *Pelophila borealis* Payk. II. Aerobic and Anaerobic Metabolism. – Norsk Entomologisk Tidsskrift 20:325-332.
Contreras-Diaz, H.G., Moya, O., Oromi, P. & Juan, C. (2007). Evolution and diversification of the forest and hypogean ground-beetle genus *Trechus* in the Canary Islands. – Molecular Phylogenetics and Evolution 42: 687-699.
Ćurčić, S.B., Brajković, M.M. & Ćurčić, B.P.M. (2007). The carabids of Serbia. – Institute of Zoology, Faculty of Biology, University of Belgrade, Belgrade.
Darlington, P.J. (1971). Modern taxonomy, reality, and usefulness. – Systematic Biology 20: 341-365.
de la Fuente, J.M. (1927). Tablas analíticas para la clasificación de los Coleópteros de la Península Ibérica. I. Cicindelidae. II. Carabidae. – Imprenta Altés, Barcelona.
Desender, K. (1986). Distribution and ecology of carabid beetles in Belgium (Coleoptera, Carabidae). – Studiedocumenten, Bruxelles, 26-29.
Desender, K., Gaublomme, E., Dhuyvetter, H. & Verdyck, P. (2005a). Interaction between regional forest history, ecology and conservation genetics of *Carabus problematicus* in Flanders (Belgium). – DIAS report 114: 73-87.
Desender, K. & Serrano, J. (1999). A genetic comparison of Atlantic and Mediterranean populations of a saltmarsh beetle. – Belgian Journal of Zoology 129: 83-94.
Desender, K., Small, E., Gaublomme, E. & Verdyck, P. (2005b). Rural-urban gradients and the population genetic structure of woodland ground beetles. – Conservation Genetics 6: 51-62.
Desender, K., Verdyck, P., Gaublomme, E., Dhuyvetter, H. & Rasplus, J.Y. (2002). Extreme genetic differentiation and isolation by non-distance in *Carabus auronitens* in relation to forest historical ecology in Flanders (Belgium). – In: How to protect or what we know

about Carabid Beetles (Szyszko, J., den Boer, P.J. & Bauer, T., eds). Warsaw Agricultural University Press, Warsaw, p. 227-235.

Desender, K., Serrano, J. & Verdyck, P. (2000). Genetic diversity and wing polymorphism in the saltmarsh beetle *Pogonus chalceus*: an Atlantic-Mediterranean comparison. – In: Natural history and applied ecology of carabid beetles (Brandmayr, P., Lövei, G., Zetto Brandmayr, T. Casale, A. & Vigna Taglianti, A., eds). Pensoft, Sofia-Moscow, p. 35-44.

Deuve, T. (1994). Une classification du genre *Carabus*. – Sciences Nat, Venette.

Deuve, T. (2004). Illustrated Catalogue of the Genus *Carabus* of the World (Coleoptera: Carabidae). – Pensoft, Sofia-Moscow.

Deuve, T. (2005). Les Gehringiidae: famille relistuelle de Coléoptères Adephaga, alticole dans les Montagnes Rocheuses et déserticole dans le Sinai. – Annales de la Société entomologique de France 41: 69-80.

Dhuyvetter, H., Gaublomme, E. & Desender, K. (2004). Genetic differentiation and local adaptation in the salt-marsh beetle Pogonus chalceus: a comparison between allozyme and microsatellite loci. Molecular Ecology 13: 1065-1074.

Dhuyvetter, H., Hendrickx, F., Gaublomme, E. & Desender, K. (2007). Differentiation between two salt marsh beetle ecotypes: Evidence for ongoing speciation. – Evolution 61: 184-193.

Dominguez, E. & Wheeler, Q.D. (1997). Taxonomic stability is ignorance. – Cladistics: the International Journal of the Willi Hennig Society 13: 367-372.

Drees, C., Matern, A., Rasplus, J.-Y., Assmann, T. & Weber, F. (2008). Microsatellites and allozymes as the genetic memory of habitat fragmentation and defragmentation in populations of the ground beetle *Carabus auronitens* (Col. Carabidae). - Journal of Biogeography (in press). DOI: 10.111/j.1365-2699.2008.01949.x.

Drovenik, B. & Peks, H. (1999). Catalogus faunae. Carabiden der Balkanländer. Coleoptera Carabidae. Coleoptera – Schwanfelder Coleopterologische Mitteilungen Sonderheft 1: 1-123.

Duman, J.G., Bennett, V., Sformo, T., Hochstrasser, T. & Barnes, B.M. (2004). Antifreeze proteins in Alaskan insects and spiders. – Journal of Insect Physiology 50: 259-266.

Düring, A., Brückner, M. & Mossakowski, D. (2006). Different behaviour of mitochondrial and nuclear markers: introgression and the evolutionary history of *Chrysocarabus* (Coleoptera: Carabidae). – Entomologica Fennica 17:200-206.

Düring, A., Brückner, M., Zimmermann, M., Bobenhausen, I., Eicke, N. & Mossakowski, D. (2000). Geographic distribution of two highly different mtDNA haplotypes in *Carabus* (*Chrysocarabus*) *splendens* (Olivier). – In: Natural history and applied ecology of Carabid Beetles (Brandmayer, P., Lövei, G., Zetto Brandmayer, T., Casale, A. & Vigna Taglianti, A., eds). Pensoft, Sofia-Moscow, p. 5-9.

Eisner, T., Aneshansley, D.J., Yack, J. Attygalle, A.B. & Eisner, M. (2001). Spray mechanism of crepidogastrine bombardier beetles (Carabidae; Crepidogastrini). – Chemoecology 11: 209-219.

Emerson, B.C., Oromi, P. & Hewitt, G.M. (1999). MtDNA phylogeography and recent intra-island diversification among Canary Island *Calathus* beetles. – Molecular Phylogenetics and Evolution 13:149-158.

Emerson, B.C., Oromi, P. & Hewitt, G.M. (2000). Interpreting colonization of the *Calathus* (Coleoptera: Carabidae) on the Canary Islands and Madeira through the application of the parametric bootstrap. – Evolution 54: 2081-2090.

Erwin, T.L. (2007). A treatise on the Western Hemisphere Carabidae (Coleoptera). Their classification, distribution s and ways of life. Volume 1. Trachypachidae, Carabidae-Nebriiformes. – Pensoft, Sofia-Moscow.

Erwin, T.L. (1967). Bombardier beetles (Coleoptera, Carabidae) of North America: Part II. Biology and behavior of *Brachinus pallidus* Erwin in California. – Coleopterists Bulletin 21: 41-55.
Escherich, K. (1892). Die biologische Bedeutung der "Genitalanhänge" der Insekten – Ein Beitrag zur Bastardfrage. – Verhandlungen der zoologisch-botanischen Gesellschaft Wien 42: 225-240.
Escherich, K. (1893). Vorläufige Erwiderung auf Verhoeff's Kritik über meine Arbeit "die biologische Bedeutung der Genitalanhänge der Insekten". – Entomologische Nachrichten 19: 129-133.
Escherich, K. (1894). Anatomische Studien über das männliche Genitalsystem der Coleopteren. – Zeitschrift für wissenschaftliche Zoologie 57: 620-641.
Facchini, S. (2003). Notes on *Axinotoma*, with description of two new species from Africa (Coleoptera: Carabidae, Harpalinae). – Koleopterologische Rundschau / Coleopterological Revue 73: 19-30.
Forel, J. & Leplat, J. (1995). Les Carabes de France. – Science Nat, Venette.
Forel, J. & Leplat, J. (2001). Faune des Carabiques de France 1. – Magellanes, Andrésy.
Forel, J. & Leplat, J. (2003). Faune des Carabiques de France XI. – Megellanes, Andrésy.
Forel, J. & Leplat, J. (2005). Faune des Carabiques de France X. – Magellanes, Andrésy.
Futuyma, D.J. (2005). Evolution. – Sinauer Associates, Sunderland, Massachusetts (U.S.A.).
Galián, J., de la Rua, P., Serrano, J., Juan, C. & Hewitt, G.M. (1999). Phylogenetic relationships in West Mediterranean Scaritina (Coleoptera: Carabidae) inferred from mitochondrial COI sequences and karyotype analysis. – Journal of Zoological Systematics and Evolutionary Research 37: 85-92.
Galián, J., Hogan, J.E. & Vogler, A.P. (2002). The origin of multiple sex chromosomes in tiger beetles. – Molecular Biology and Evolution 19: 1792-1796.
Galián, J., Ortiz, A.S. & Serrano, J. (1990). Cytogenetics and cytotaxonomy of 7 Iberian species of *Brachinus* Weber (Coleoptera, Carabidae). – Cytobios 63: 185-192.
Galián, J., Proenca, S.J.R. & Vogler, A.P. (2007). Evolutionary dynamics of autosomal-heterosomal rearrangements in a multiple-X chromosome system of tiger beetles (Cicindelidae). – BMC Evolutionary Biology 7: 158.
Galián, J., Prüser, F., de la Rua, P., Serrano, J. & Mossakowski, D. (1996). Cytological and molecular differences in the *Ceroglossus chilensis* species complex (Coleoptera: Carabidae). – Annales Zoologici Fennici 33: 23-30.
Galián, J. & Vogler, A.P. (2003). Evolutionary dynamics of a satellite DNA in the tiger beetle species pair *Cicindela campestris* and *C. maroccana*. – Genome 46: 213-223.
Ganglbauer, L. (1891). Die Käfer von Mitteleuropa. Die Käfer der österreichisch-ungarischen Monarchie, Deutschlands, der Schweiz sowie des französischen und italienischen Alpengebietes. Erster Band. Familienreihe Caraboidea. – C. Gerold's Sohn, Vienna.
Garnier, S., Alibert, P., Audiot, P., Prieur, B. & Rasplus, J.Y. (2004). Isolation by distance and sharp discontinuities in gene frequencies: implications for the phylogeography of an alpine insect species, *Carabus solieri*. – Molecular Ecology 13: 1883-1897.
Garnier, S., Brouat, C., Mondor-Genson, G., Prieur, B., Sennedot, F. & Rasplus, J.Y. (2002). Microsatellite DNA markers for two endemic ground beetles: *Carabus punctatoauratus* und *C. solieri*. – Molecular Ecology Notes 2: 572-574.
Garnier, S., Magniez-Jannin, F., Rasplus, J.Y. & Alibert, P. (2005). When morphometry meets genetics: inferring the phylogeography of *Carabus solieri* using Fourier analyses of pronotum and male genitalia. – Journal of Evolutionary Biology 18: 269-280.

Gaublomme, E., Desender, K., Verdyck, P., Dhuyvetter, H. & Rasplus, J.Y. (2002). Non-destructive sampling for genetic studies on *Carabus auronitens* and *Carabus problematicus*: a study based on allozymes and microsatellites. – In: How to protect or what we know about carabid beetles (Szyszko, J., den Boer, P.J. & Bauer, T., eds). Warsaw Agricultural University Press, Warsaw, p. 337-344.

Giers, E. (1977). Die Nicht-Homologen-Assoziation multipler Geschlechtschromosomen in der Spermatogenese von *Cicindela hybrida* (Coleoptera). University of Münster. PhD Thesis, Münster.

Gries, B., Mossakowski, D. & Weber, F. (1973). Coleoptera Westfalica: Familia Carabidae Genera *Cychrus*, *Carabus* und *Calosoma*. – Abhandlungen aus dem Westfälischen Museum für Naturkunde 35: 3-80.

Guérguiev, V. B. & Guérguiev, B.V. (1995). Catalogue of the ground-beetles of Bulgaria (Coleoptera: Carabidae). – Pensoft, Sofia-Moscow.

Guérguiev, V.B., Sakalian, V.P. & Guérguiev, B.V. (1997). Biogeography of the endemic Balkan ground-beetles (Coleoptera: Carabidae) in Bulgaria. – Pensoft, Sofia-Moscow.

Habel, J.C. & Assmann, T. (Eds) (2008). Survival on changing climate – phylogeography and conservation biology of relict species. – Springer, Heidelberg (in press).

Harris, H. (1966). Enzyme polymorphism in man. – Proceedings of the Royal Society B-Biological Sciences 164: 298-310.

Hieke, F. & Wrase, D.W. (1988). Faunistik der Laufkäfer Bulgariens (Coleoptera, Carabidae). – Deutsche Entomologische Zeitschrift 35: 1-171.

Hillis, D.M., Moritz, C. & Mable, B.K. (Eds) (1996). Molecular systematics. – Sinauer, Sunderland.

Holder, K., Montgomerie, R. & Friesen, V.L. (1999). A test of the glacial refugium hypothesis using patterns of mitochondrial and nuclear DNA sequence variation in rock ptarmigan (*Lagopus mutus*). – Evolution 53: 1936-1950.

Holdhaus, K. (1912). Monographie der paläarktischen Arten der Coleopterengattung *Microlestes*. – Denkschrift der Kaiserlichen Akademie der Wissenschaften in Wien, mathematisch-naturwissenschaftliche Klasse 88: 477-540.

Hosni, M.T., Al-Sharkawy, A.Z., Oshaibah, A.A., Hassan, M.I. & El-Gharbawy, A.A. (2003). Classification of subfamily Harpalinae (Family: Carabidae – Order: Coleoptera) in Egypt. II – Tribes Harpalini, Masoreini and Amarini. – Egyptian Journal of Zoology 40: 549-573.

Hunt, T., Bergsten, J., Levkanicova, Z., Papadopoulou, A., John, O.S., Wild, R., Hammond, P.M., Ahrens, D., Balke, M., Caterino, M.S., Gomez-Zurita, J., Ribera, I., Barraclough, T.G., Bocakova, M., Bocak, L. & Vogler, A.P. (2007). A comprehensive phylogeny of beetles reveals the evolutionary origins of a superradiation. – Science 318: 1913-1916.

Hurka, K. (1996). Carabidae of the Czech and Slovak Republics. – Kabourek, Zlín.

Hurka, K. (1999). Eizähne als phylogenetisch auswertbares Larval-Merkmal in der Kaferfamilie Carabidae (Coleoptera). – Entomologische Nachrichten und Berichte 43: 169-176.

Hurka, K. & Ruzickova, A. (1999). Classification of the *Amara* (*Amara*) *communis* species aggregate based on the egg and larval stage stage characters (Coleoptera: Carabidae: Amarina). – Acta Societatis Zoologicae Bohemicae 63: 451-461.

Isaac, N.J.B., Mace, G.M. & Mallet, J. (2005). Response to Agapow and Sluys: The reality of taxonomic change. – Trends in Ecology & Evolution 20: 280-281.

Isaac, N.J.B., Mallet, J. & Mace, G.M. (2004). Taxonomic inflation: its influence on macroecology and conservation. – Trends in Ecology & Evolution 19: 464-469.

Jeannel, R. (1941f). Coléoptères Carabiques. – Lechevalier, Paris.

Kane, T.C. (1982). Genetic patterns and population structure in cave animals. – In: Environmental adaptation and evolution (Mossakowski, D. & Roth, G., eds). Gustav Fischer, Stuttgart, p. 131-149.

Kane, T.C., Barr, T.C. & Badaracca, W.J. (1992). Cave beetle genetics – geology and gene flow. – Heredity 68: 277-286.

Kane, T.C., Barr, T.C. & Stratton, G.E. (1990). Genetic patterns and population-structure in Appalachian *Trechus* of the *vandykei* group (Coleoptera, Carabidae). – Brimleyana 1990: 133-150.

Keller, I., Excoffier, L. & Largiader, C.R. (2005). Estimation of effective population size and detection of a recent population decline coinciding with habitat fragmentation in a ground beetle. – Journal of Evolutionary Biology 18: 90-100.

Keller, I. & Largiader, C.R. (2002). Identification of one X-linked and five autosomal microsatellite loci in *Carabus violaceus* (Coleoptera, Carabidae) and their applicability to related taxa. – Molecular Ecology Notes 2: 290-292.

Keller, I. & Largiader, C.R. (2003a). Five microsatellite DNA markers for the ground beetle *Abax parallelepipedus* (Coleoptera, Carabidae). – Molecular Ecology Notes 3: 113-114.

Keller, I. & Largiader, C.R. (2003b). Recent habitat fragmentation caused by major roads leads to reduction of gene flow and loss of genetic variability in ground beetles. – Proceedings of the Royal Society of London Series B-Biological Sciences 270: 417-423.

Keller, I., Nentwig, W. & Largiader, C.R. (2004). Recent habitat fragmentation due to roads can lead to significant genetic differentiation in an abundant flightless ground beetle. – Molecular Ecology 13: 2983-2994.

Kim, C.G., Zhou, H.Z., Imura, Y., Tominaga, O., Su, Z.H. & Osawa, S. (2000). Pattern of morphological diversification in the *Leptocarabus* ground beetles (Coleoptera: Carabidae) as deduced from mitochondrial ND5 gene and nuclear 28S rDNA sequences. – Molecular Biology and Evolution 17: 137-145.

Kirschenhofer, E. (1999). Die paläarktischen, orientalischen und äthiopischen Arten des Subgenus *Nectochlaenius* Antoine, 1959 des Genus *Chlaenius* Bonelli, 1810 (Coleoptera: Carabidae). – Acta Entomologica Slovenica 7: 85-108.

Kirschenhofer, E. (2004). Beitrag zur Kenntnis der Gattung *Chlaenius* Bonelli, 1810 – die paläarktischen und orientalischen Arten der Untergattungen *Chlaenius* (s.str.) Bonelli, 1810, *Amblygenius* La Ferté-Sénectère, 1851 und *Parachlaenius* Jeannel, 1949 (Coleoptera, Carabidae). – Entomologica Basiliensia 26: 1-100.

Kleinfeld, F. & Rapuzzi, I. (2004). Zur Faunistik der *Carabus*- und *Procerus*-Arten im ‚Nahen Osten' (Coleoptera: Carabidae: Carabini). – Lambillionea 54, Supplement: 1-72.

Koch, D. & Thiele, H.U. (1980). Zur ökologisch-physiologischen Differenzierung und Speziation der Laufkäfer-Art *Pterostichus nigrita* (Coleoptera: Carabidae). – Entomologia Generalis 6: 135-150.

Kocher, L. (1963). Catalogue commenté des Coléoptères du Maroc. – Travaux de l'Institut Scientifique Chérifien 27: 1-170.

Krumbiegel, I. (1932). Untersuchungen über physiologische Rassenbildung. Ein Beitrag zum Problem der Artbildung und der geographischen Variation. – Zoologisches Jahrbuch für Systematik und Ökologie 63: 183-280.

Krumbiegel, I. (1936a). Morphologische Untersuchungen über Rassenbildung, ein Beitrag zum Problem der Artbildung und der geographischen Variation. – Zoologisches Jahrbuch für Systematik und Ökologie 68: 105-178.

Krumbiegel, I. (1936b). Sinnesphysiologische Untersuchungen an geographischen Rassen. Zoologisches Jahrbuch für Systematik und Ökologie 68: 179-204.

Krumbiegel, I. (1936c). Untersuchungen über gleichsinnige geographische Variation. Zoologisches Jahrbuch für Systematik und Ökologie 68: 481-516.

Kryzhanovskij, O.L., Belousov, I.A., Kabak, I.I., Kataev, B.M., Makarov, K.V. & Shilenkov, V.G. (1995). A checklist of the ground-beetles of Russia and adjancent lands (Insecta, Coleoptera, Carabidae). – Pensoft, Sofia-Moscow.

Kudoh, K., Abe, A., Kondoh, I.A., Satoh, I.I. & Saitoh, K. (1970). Some cytological aspects of three species of beetles. – Kontyu 38: 232-238.

Larochelle, A. & Larivière, M.-C. (2001). Carabidae (Insecta: Coleoptera): Catalogue. – Fauna of New Zealand 43: 1-285.

Larochelle, A. & Larivière, M.-C. (2003). A Natural History of the Ground-Beetles (Coleoptera: Carabidae) of America north of Mexico. – Pensoft, Sofia-Moscow.

Latreille, P.A. (1810). Considérations générales sur l'ordre naturel des animaux composant les classes des crustacés, des arachnides, et des insectes; avec un tableau méthodique de leurs genres, disposés en familles. – Schoell, Paris.

Ledoux, G. & Roux, P. (2005). *Nebria* (Coleoptera, Nebriidae), Faune mondiale. – Muséum-Centre de Conservation et d'Etude des Collections and Société Linnéenne de Lyon, Lyon.

Lewontin, R.C. & Hubby, J.L. (1966). A molecular approach to the study of genetic heterozygosity in natural populations: II. amount of variation and degree of heterozygosisty in nature. – Genetics 54: 595-609.

Liebherr, J.K. (1986). Cladistic analysis of North American Platynini and revision of the *Agonum extensicolle* species group (Coleoptera: Carabidae). – University of California Publication in Entomology 106: 1-198.

Liebherr, J.K. & Schmidt, J. (2004). Phylogeny and biogeography of the Laurasian genus *Agonum* Bonelli (Coleoptera, Carabidae, Platynini). – Deutsche Entomologische Zeitschrift 51: 151-206.

Liebherr, J.K. & Zimmermann, E.C. (2000). Hawaiian Carabidae (Coleoptera), Part 1: introduction and tribe Platynini. – University of Hawai'i Press, Honolulu.

Lindroth, C.H. (1945). Die fennoskandischen Carabidae. I. Spezieller Teil. – Göteborgs Kungliga Vetenskaps- och Vitterhets-Samhälles Handlingar Sjätte följiden, Ser.B 4 (1): 1-709.

Lindroth, C.H. (1949). Die Fennoskandischen Carabidae III. Allgemeiner Teil. – Göteborgs Kungliga Vetenskaps- och Vitterhets-Samhälles Handlingar Sjätte följiden, Ser.B 4: 1-911.

Lindroth, C.H. (1968). The icelandic form of *Carabus problematicus* Hbst. (Col. Carabidae) – A statistic treatment of subspecies. – Opuscula Entomologica 33: 157-182.

Lindroth, C.H. (1969). The theory of glacial refugia in Scandinavia – Comments on present opinions. – Notulae Entomologicae 39: 178-192.

Lindroth, C.H. (1970). Survival of animals and plants on ice-free refugia during Pleistocene glaciations. – Endeavour 29: 129-134.

Lindroth, C.H. (1971). Disappearence as a protective factor: a supposed case of Bates'ian mimicry among beetles (Coleoptera, Carabidae and Chrysomelidae). – Entomologia Scandinavia 2: 41-48.

Lindroth, C.H. (1974). Coleoptera, Carabidae. Handbooks for the Identification of British Insects IV (2): 1-148.

Lindroth, C.H. (1979). Europe and the circumpolar area: events and personalities. – In: Carabid beetles: their evolution, natural history, and classification (Erwin, T.L., Ball, G.E. & Whitehead, D.R., eds). Dr W Junk, The Hague, Boston, London, p. 9-15.

Lindroth, C.H. (1985f). The Carabidae (Coleoptera) of Fennoscandia and Denmark. – Fauna Entomologica Scandinavica 15: 1-497.

Lipscomb, D., Platnick, N. & Wheeler, Q. (2003). The intellectual content of taxonomy: a comment on DNA taxonomy. – Trends in Ecology & Evolution 18: 65-66.

Löbl, I. & Smetana, A. (Eds) (2003). Catalogue of Palaearctic Coleoptera Vol. 1: Archostemata, Myxophaga, Adephaga. – Apollo Books, Stenstrup.

Lopez, M.A., de la Rosa, J.J. & Baena, M. (2006). Descripción de *Cephalota* (*Taenidia*) *dulcinea* sp. n. de la Península Iberíca (Coleoptera, Cicindelidae). – Boletín Sociedad Entomológica Aragonesa 39: 165-170.

Lorenz, W. (1998a). Nomina carabidarum: a directory of the scientific names of ground beetles (Insecta, Coleoptera „Geadephaga": Trachypachidae and Carabidae incl. Paussinae, Cicindelinae, Rhysodinae). – Lorenz, Tutzing.

Lorenz, W. (1998b). Systematic list of extant ground beetles of the world (Insecta Coleoptera „Geadephaga"; Trachypachidae and Carabidae incl. Paussinae, Cicindelidae, Rhysodinae). – Lorenz, Tutzing.

Lorenz, W. (2005a). Nomina Carabidarum – a directory of the scientific names of ground beetles (Coleoptera „Geadephaga": Trachypachidae and Carabidae, incl. Paussinae, Cicindelinae, Rhysodinae). Second edition. – Lorenz, Tutzing.

Lorenz, W. (2005b). A systematic list of extant ground beetles of the world (Coleoptera „Geadephaga": Trachypachydiae and Carabidae, incl. Paussinae, Cicindelinae, Rhysodinae). Second edition. – Lorenz, Tutzing.

Luff, M.L. (1998). Provisional atlas of the ground beetles (Coleoptera, Carabidae) of Britain. – Biological Records Centre, Huntingdon.

Maddison, D.R., Bake, M.D. & Ober, K.A. (1998). A preliminary phylogenetic analysis of 18 S ribosomal DNA of carabid beetles (Coleoptera). – In: Phylogeny and classification of Caraboidea (Coleoptera: Adephaga) (Ball, G.E., Casale, A. & Vigna-Taglianti, A., eds). Museo Regionale di Scienze Naturali, Torino, p. 229-250.

Maddison, D.R., Baker, M.D. & Ober, K.A. (1999). Phylogeny of carabid beetles as inferred from 18S ribosomal DNA (Coleoptera: Carabidae). – Systematic Entomology 24: 103-138.

Marggi, W. (1992). Faunistik der Sandlaufkäfer und Laufkäfer der Schweiz (Cicindelidae & Carabidae). Text und Verbreitungskarten. – Documenta Faunistica Helvetiae 13/1 and 13/2:1-477 and 471-243.

Marggi, W. & Luka, H. (2001). Laufkäfer der Schweiz. Gesamtliste 2001. – Opuscula Biogeographica Basileensia 1:1-37.

Martinez-Navarro, E.M., Galián, J. & Serrano, J. (2005). Phylogeny and molecular evolution of the tribe Harpalini (Coleoptera, Carabidae) inferred from mitochondrial cytochrome-oxidase I. – Molecular Phylogenetics and Evolution 35: 127-146.

Martinez-Navarro, E.M., Serrano, J. & Galián, J. (2004). Chromosome evolution in ground beetles: localization of the rDNA loci in the tribe Harpalini (Coleoptera, Carabidae). – Journal of Zoological Systematics and Evolutionary Research 42: 38-43.

Matern, A., Drees, C., Meyer, H. & Assmann, T (2008a). Genetic diversity and population structure of the endangered insect species *Carabus variolosus* in its western distribution range: Implications for conservation. – Conservation Genetics (in press). DOI: 10.1007/s10592-008-9606-1

Matern, A., Drees, C., Vogler, A.P. & Assmann, T. (2008b). Linking genetics and ecology: reconstructing the history of relict populations of an endangered semi-aquatic beetle. – In: Survival on changing climate – phylogeography and conservation of relict species (Habel, J.C. & Assmann, T., eds). Springer, Heidelberg, in press.

Mateu, J. (1997). *Tinautius* (n.gen.) *troglophilus* n.sp. nuevo Pterostichini del sur de España (Coleoptera, Carabidae). – Bulletino Museo regionale Science naturale Torino 15: 137-146.

Mateu, J. (2001). *Tinautius exilis* sp.n. (Coleoptera, Carabidae, Pterosctichini) de la Alpujarra almeriense (SE España). – Animal and Biodiversity Conservation 24: 45-49.

Maynard-Smith, J. (1998). Evolutionary genetics. – Oxford University Press, Oxford.

Moravec, P. (2002). Tiger beetles of Madagascar. Volume 2: A monograph of the genus *Physodeutera* (Coleoptera: Cicindelidae). – Kabourek, Zlín

Moret, P. (2005). Los coleópteros Carabidae de páramo en los Andes del Ecuador. Sistemática, ecologia y biogeografía. – Pontificia Universidad Católica del Ecuador, Qito, Ecuador.

Mossakowski, D. (1971). Zur Variabilität isolierter Populationen von *Carabus arcensis* Hbst. (Coleoptera). – Zeitschrift für zoologische Systematik und Evolutionsforschung 9: 81-106.

Mossakowski, D. (2005). Morphological or molecular systematics? A case study of *Carabus*. – DIAS Report 114: 231-241.

Mossakowski, D., Roschen, A. & Vaje, S. (1986). Hybridization in *Chrysocarabus*. – In: Carabid Beetles: Their Adaptations and Dynamics (den Boer, P.J., Luff, M.L., Mossakowski, D. & Weber, F., eds). Fischer, Stuttgart, p. 281-295.

Mossakowski, D. & Weber, F. (1972). Korrelationen zwischen Heterochromratingehalt und morphologischen Merkmalen bei *Carabus auronitens* (Coleoptera). – Zeitschrift für zoologische Systematik und Evolutionsforschung 10: 291-300.

Mossakowski, D. & Weber, F. (1976). Chromosomale und morphometrische Divergenzen bei *Carabus lineatus* und *C. splendens* (Carabidae): I. Ein Vergleich sympatrischer und allopatrischer Populationen. – Zeitschrift für zoologische Systematik und Evolutionsforschung 14: 280-291.

Muergues, G. & Ledoux, G. (1966). Intérêt de l'étute du sac interne dévaginé et en extension. – Annales de la Société entomologique de France N.S. 2: 661-669.

Müller-Kroehling, S. (2006). Ist der Gruben-Laufkäfer *Carabus* (*variolosus*) *nodulosus* ein Taxon des Anhangs II der FFH-Richtlinie in Deutschland? – Waldoekologie online 3: 57-62.

Müller-Motzfeld, G. (Ed.) (2004). Freude, Harde, Lohse, Die Käfer Mitteleuropas, Bd. 2 Adephaga 1: Carabidae (Laufkäfer). – Spektrum, München.

Naviaux, R. (1998). *Ctenostoma* (Coleoptera, Cicindelidae). – Mémoires de la Société Entomologique de France 2:1-186.

Naviaux, R. & Pinratana, B.A. (2004). The tiger beetles of Thailand (Coleoptera, Cicindelidae). – Brothers of St. Gabriel Thailand.

Neculiseanu, Z.Z. & Matalin, A.V. (2000). A catalogue of the ground beetles of the Republic of Moldova (Insecta, Coleoptera: Carabidae). – Pensoft, Sofia-Moscow.

Nettmann, H.K. (1986). Carabid systematics and chromosome numbers. – In: Carabid beetles: Their adaptations and dynamics (den Boer, P.J., Luff, M.L., Mossakowski, D. & Weber, F., eds). Gustav Fischer, Stuttgart, p. 235-254.

Neuwinger, H.-D. (2004). Tödliche Käferlarven: Das Pfeilgift der Kalahari-San. – Biologie in unserer Zeit 34: 88-94.

New, T.R. (1984). Insect Conservation: An Australian perspective. – Junk, Dordrecht.

Nomenclature, I. C. o. Z. (1999). 4th edition. International Trust for Zoological Nomenclature, London.

Nussbaum, Y. (1987). Tiger beetles of Israel and Sinai (Coleoptera: Cicindelidae). – YES Quarterly 4: 7-15.

Obydov, D. (2002). Révision du genre *Callisthenes*. – Magellanes, Andrésy.

Ortuno, V.M. & Toribio, M. (2005). Carabidae de la Península Ibérica y Baleares. Vol. I. Trechinae, Bembidiini. – Argania editio, Barcelona.

Osawa, S., Su, Z.H. & Imura, Y. (2003). Molecular phylogeny and evolution of carabid ground beetles. – Springer, Tokyo.

Pavlicek, T. & Nevo, E. (1996). Genetic divergence in populations of the beetle *Carabus hemprichi* from microclimatically opposing slopes of "Evolution Canyon": A Mediterranean microsite, Mount Carmel, Israel. – Israel Journal of Zoology 42: 403-409.

Pearson, D.L. & Vogler, A.P. (2001). Tiger beetles: The evolution, ecology, and diversity of the cicindelids. – Cornell University Press, Ithaca and London.

Proenca, S.J.R., Collares, M.J. & Serrano, A.R.M. (1999a). Karyological study of *Cicindela trifasciata* (Coleoptera, Cicindelidae) from Cuba: Evidence of B chromosomes. – Genetics and Molecular Biology 22.

Proenca, S.J.R. & Galián, J. (2003). Chromosome evolution in the genus *Cicindela*: physical mapping and activity of rDNA loci in the tiger beetle species *Cicindela littoralis* and *C. flexuosa*. – Journal of Zoological Systematics and Evolutionary Research 41: 227-232.

Proenca, S.J.R., Serrano, A.R.M. & Collares-Pereira, M.J. (1999b). First record on the cytotaxonomy of cicindelids (Insecta, Coleoptera) from an Afrotropical region. – Caryologia 52:37-47.

Prüser, F. & Mossakowski, D. (1998). Low substitution rates in mitochondrial DNA in Mediterranean carabid beetles. – Insect Molecular Biology 7: 121-128.

Prüser, F. & Mossakowski, D. (1998). Conflicts in phylogenetic relationships and dispersal history of the supertribe Carabitae (Coleoptera: Carabidae). – In: Phylogeny and classification of Caraboidea (Coleoptera: Adephaga) (Ball, G.E., Casale, A. & Vigna-Taglianti, A., eds). Museo Regionale di Scienze Naturali, Torino, p. 297-328.

Rasplus, J.Y., Garnier, S., Meusnier, S., Piry, S., Mondor, G, Audiot, P. & Cornute, J.M. (2001). Setting conservation priorities: the case study of *Carabus solieri* (Col. Carabidae). – Genetics, Selection and Evolution 33: 141-175.

Rasplus, J.Y., Meusnier, S., Mondor, G., Piry, S. & Cornuet, J.M. (2000). Microsatellite analysis of genetic population structure in the endangered beetle: *Carabus solieri* (Carabidae). – In: Natural history and applied ecology of Carabid Beetles (Brandmayer, P., Lövei, G., Zetto Brandmayer, T., Casale, A. & Vigna Taglianti, A., eds). Pensoft, Sofia-Moscow, p. 11-24.

Reimann, T., Assmann, T., Nolte, O., Reuter, H., Huber, C. & Weber, F. (2002). Palaeogeography and palaeoecology of *Carabus auronitens* (Coleoptera): characterization and localization of glacial refugia in southern France and reconstruction of postglacial expansion routes by means of allozyme polymorphisms. – Abhandlungen des Naturwissenschaftlichen Vereins in Hamburg (NF) 35: 1-151.

Reitter, E. (1900). Carabidae. Abteilung: Harpalini und Licinini. – Bestimmungs-Tabelle der europäischen Coleopteren 41: 33-155.

Reitter, E. (1919). Bestimmungstabelle der Brachyninae (Col. Carabidae) aus Europa und den angrenzenden Ländern. – Entomologische Blätter 15: 129-146.

Rensch, B. (1943). Studien über Korrelation und klimatische Parallelität der Rassenmerkmalen von *Carabus*-Formen. – Zoologisches Jahrbuch für Systematik und Ökologie 76: 103-170.

Rensch, B. (1948). Histological changes correlated with evolutionary changes of body size. – Evolution 2: 218-230.

Rensch, B. (1950). Die Abhängigkeit der relativen Sexualdifferenz von der Körpergröße. – Bonner Zoologische Beiträge 1: 58-69.

Rensch, B. (1954). Neuere Probleme der Abstammungslehre. – Enke, Stuttgart.

Rensch, B. (1958). Die ideale Artbeschreibung. – Uppsala Universitets Arsskrift 6: 91-103.

Retezár, I. (2008). The *Carabus* of Abkhazia, Caucasus (Coleoptera, Carabidae). – Mesterprint, Budapest.

Ribera, I., Montagud, S., Teruel, S. & Belles, X. (2006). Molecular data supports the inclusion of *Ildobates neboti* Espanol in Zuphiini (Coleoptera: Carabidae: Harpalinae). – Entomologica Fennica 17: 07-213.

Samways, M.J. (2005). Insect Diversity Conservation. – Cambridge University Press, Cambridge (UK).

Sanchez-Gea, J.F., Galián, J. & Serrano, J. (2004). Phylogeny of Iberian *Zabrus* (Coleoptera: Carabidae: Zabrini) based on mitochondrial DNA sequence. – European Journal of Entomology 101: 503-511.

Sanchez-Gea, J.F., Serrano, J. & Galián, J. (2000). Variability in rDNA loci in Iberian species of the genus *Zabrus* (Coleoptera: Carabidae) detected by fluorescence in situ hybridization. – Genome 43: 22-28.

Sander, A.C., Purtauf, T., Holzhauer, S.I.J. & Wolters, V. (2006a). Landscape effects on the genetic structure of the ground beetle *Poecilus versicolor* STURM 1824. – Biodiversity and Conservation 15: 245-259.

Sander, A.C., Purtauf, T., Wolters, V. & Dauber, J. (2006b). Landscape genetics of the widespread ground-beetle *Carabus auratus* in an agricultural region. – Basic and Applied Ecology 7: 555-564.

Sasabe, M., Takami, Y. & Sota, T. (2007). The genetic basis of interspecific differences in genital morphology of closely related carabid beetles. – Heredity 98: 385-391.

Sasakawa, K. & Kubota, K. (2007a). Phylogeny and genital evolution of carabid beetles in the genus *Pterostichus* and its allied genera (Coleoptera: Carabidae) inferred from two nuclear gene sequences. – Annals of the Entomological Society of America 100: 100-109.

Sasakawa, K. & Kubota, K. (2007b). Utility of external morphometrics for cryptic species identification: A case study of two species in *Poecilus* '*samurai*' (Coleoptera: Carabidae). – Applied Entomology and Zoology 42: 353-365.

Saska, P. & Honek, A. (2004). Development of the beetle parasitoids, *Brachinus explodens* and *B. crepitans* (Coleoptera: Carabidae). – Journal of Zoology 262: 29-36.

Schildknecht, H., Winkler, H. & Maschwitz, U. (1968). Vergleichende chemische Untersuchungen der Inhaltsstoiffe der Pygidialdrüsen von Carabiden. – Zeitschrift für Naturforschung 23b: 637-644.

Sciaky, R. & Facchini, S. (2005). Revision of he Chinese *Broscosoma* Rosenhauer, 1846, with description of two new species. – Koleopterologische Rundschau / Coleopterological Revue 75: 1-12.

Scribner, K.T. (2005). Measures of genetic diversity. – In: Principles of conservation biology (Groom, M.J., Meffe, G.K. & Carroll, C.R., eds). Sinauer Associates, Sunderland, Massachusetts U.S.A, p. 378-379.

Sedlmair, D., Gerstmeier, R. & Einspanier, R. (2000). Application of ubiquitin SSCP analysis in taxonomic studies within the subgenus *Orinocarabus* (Coleoptera: Carabidae: *Carabus*). – European Journal of Entomology 97: 387-394.

Serrano, A.R.M. & Aguiar, C.A.S. (2002). The genus *Typhlocharis* Dieck, 1869 (Coleoptera: Carabidae) in Portugal: description of two new species and faunistic notes. – Boletim da Sociedade Portuguesa de Entomologia 198: 181-197.

Serrano, A.R.M. & Aguiar, C.A.S. (2004). Three new species of the genus *Geocharis* Ehlers, 1883 from Portugal (Coleoptera, Carabidae). – Graellsia 60: 71-80.

Serrano, A.R.M. & Aguiar, C.A.S. (2006). New species and new data on the genus *Geocharis* Ehlers 1883 (Coeloptera: Carabidae) from Portugal. – Annales de la Société entomologique de France 42: 79-89.

Serrano, A.R.M., Aguiar, C.A.S.& Carlos, A.S. (2003). The genus *Geocharis* Ehlers, 1883 in Portugal: description of three new species and faunistic notes (Coleoptera, Carabidae). – Nouvelle Revue d'Entomologie 20: 39-50.

Serrano, J. (1980). Diferencias cariotipicas entre *Cicindela maroccana pseudomaroccana* a *C. campestris* (Col. Cicindelidae). – Boletino Asociacion Espana Entomologia 4: 65-68.

Serrano, J. (1980b). *Scarites buparius*, a carabid beetle with an X1X2Y sex-chromosome system. – Experientia 36: 1042-1043.

Serrano, J. (2003). Catalogue of the Carabidae (Coleoptera) of the Iberian Peninsula. – Monografias S.E.A. 9: 1-130.

Serrano, J. & Galián, J. (1998). A review of karyotic evolution and phylogeny of carabid beetles (Coleoptera). – In: Phylogeny and classification of Caraboidea (Coleoptera: Adephaga) (Ball, G.E., Casale, A. & Vigna-Taglianti, A., eds). Museo Regionale di Scienze Naturali, Torino, p. p. 191-228.

Serrano, J. & Yadav, J.S. (1984). Chromosome numbers and sex-determining mechanisms in Adephagan Coleoptera. – Coleopterists Bulletin 35: 335-357.

Shapiro, A.M. & Porter, A.H. (1989). The lock-and-key hypothesis – evolutionary and biosystematic interpretation of insect genitalia. – Annual Review of Entomology 34: 231-245.

Smith, S.G. & Edgar, R.S. (1954). The sex-determining mechanism in some North American Cicindelidae (Coleoptera). – Revue Suisse de Zoologie 61: 657-667.

Sota, T. & Ishikawa, R. (2004). Phylogeny and life-history evolution in *Carabus* (subtribe Carabina: Coleoptera, Carabidae) based on sequences of two nuclear genes. – Biological Journal of the Linnean Society 81: 135-149.

Sota, T., Ishikawa, R., Ujiie, M., Kusumoto, F. & Vogler, A.P. (2001). Extensive trans-species mitochondrial polymorphisms in the carabid beetles *Carabus* subgenus *Ohomopterus* caused by repeated introgressive hybridization. – Molecular Ecology 10: 2833-2847.

Sota, T. & Kubota, K. (1998). Genital lock-and-key as a selective agent against hybridization. – Evolution 52: 1507-1513.

Sota, T., Takami, Y., Monteith, G.B. & Moore, B.P. (2005). Phylogeny and character evolution of endemic Australian carabid beetles of the genus *Pamborus* based on mitochondrial and nuclear gene sequences. – Molecular Phylogenetics and Evolution 36: 391-404.

Sota, T. & Vogler, A.P. (2001). Incongruence of mitochondrial and nuclear gene trees in the Carabid beetles *Ohomopterus*. – Systematic Biology 50: 39-59.

Sota, T. & Vogler, A.P. (2003). Reconstructing species phylogeny of the carabid beetles *Ohomopterus* using multiple nuclear DNA sequences: heterogeneous information content and the performance of simultaneous analyses. – Molecular Phylogenetics and Evolution 26: 139-154.

Stork, N.E. (1988). Insect diversity: facts, fiction and speculation. – Biological Journal of the Linnean Society 35: 321-337.

Streiff, R., Veyrier, R., Audiot, P., Meusnier, S. & Brouat, C. (2005). Introgression in natural populations of bioindicators: a case study of *Carabus splendens* and *Carabus punctatoauratus*. – Molecular Ecology 14: 3775-3786.

Takami, Y. (2003). Experimental analysis of the effect of genital morphology on insemination success in the ground beetle *Carabus insulicola* (Coleoptera Carabidae). – Ethology Ecology & Evolution 15: 51-61.

Takami, Y. (2007). Spermatophore displacement and male fertilization success in the ground beetle *Carabus insulicola*. – Behavioral Ecology 18: 628-634.

Takami, Y. & Suzuki, H. (2005). Morphological, genetic and behavioural analyses of a hybrid zone between the ground beetles *Carabus lewisianus* and *C. albrechti* (Coleoptera, Carabidae): asymmetrical introgression caused by movement of the zone? – Biological Journal of the Linnean Society 86: 79-94.

Tang, Z., Wang, Z., Zheng, C. & Fang, J. (2006). Biodiversity in China's mountains. – Frontiers in Ecology and the Environment 4: 347-352.

Tautz, D., Arctander, P., Minelli, A., Thomas, R.H. & Vogler, A.P. (2003). A plea for DNA taxonomy. – Trends in Ecology & Evolution 18: 70-74.

Terlutter, H. (1990). An allele gradient of an esterase gene locus as a result of recent gene flow: electrophoretic investigations of *Carabus auronitens* F. (Col. Carabidae). – In: The role of ground beetles in ecological and environmental studies (Stork, N.E., Ed.). Intercept, Andover, p. 359-364.

Terlutter, H. (1991). Morphometrische und elektrophoretische Untersuchungen an westfälischen und südfranzösischen *Carabus auronitens*-Populationen (Col. Carabidae): Zum Problem der Eiszeitüberdauerung in Refugialgebieten und der nacheiszeitlichen Ausbreitung. – Abhandlungen aus dem Westfälischen Museum für Naturkunde 53: 3-111.

Thiele, H.-U. (1977). Carabid beetles in their environments. – Springer, Berlin, Heidelberg, New York.

Turanchik, E. & Kane, T.C. (1979). Ecological genetics of the cave beetle *Neaphaenops tellkampfii* (Coleoptera: Carabidae). – Oecologia 44: 63-67.

Turin, H. (2000). De Nederlandse Loopkevers – Verspreiding en oecologie. – Nationaal Natuurhistorisch Museum Naturalis, Leiden.

Turin, H., Penev, L. & Casale, A. (2003). The Genus *Carabus* in Europe. – Pensoft, Sofia-Moscow.

Usami, T., Yokoyama, J., Kubota, K. & Kawata, M. (2006). Genital lock-and-key system and premating isolation by mate preference in carabid beetles (*Carabus* subgenus *Ohomopterus*). – Biological Journal of the Linnean Society 87: 145-154.

Valdecasas, A.G., Williams, D. & Wheeler, Q.D. (2008). 'Integrative taxonomy' then and now: a response to Dayrat (2005). – Biological Journal of the Linnean Society 93: 211-216.

Vogler, A.P., DeSalle, R., Assmann, T., Knisley, B. & Schultz, T.D. (1993). Molecular population genetics of the endangered tiger beetle *Cicindela dorsalis* (Coleoptera: Cicindelidae). – Annals of the Entomological Society of America 86: 142-152.

von Linné, C. (1758). Systema naturae per regna tria naturae, secundum classes, ordines, genera, species, cum characteribus, differentiis, synonymis, locis. Editio decima, reformata. – Tomus I. Laurentii Salvii, Stockholm.

Weber, F. (1966). Beitrag zur Karyotypenanalyse der Laufkäfergattung *Carabus* L. (Coleoptera). – Chromosoma 18: 467-476.

Weber, F. (1967). Zur Polymorphie eines Chromosoms innerhalb von Populationen des Laufkäfers *Carabus auronitens* Fabr. – Experientia 23: 1-4.

Weber, F. (1968). Die interspezifische Variabilität des heterochromatischen Armes eines Chromosoms bei der Gattung *Carabus* L. (Coleoptera). Chromosoma 23: 288-308

Wheeler, Q.D., Raven, P.H. & Wilson, E.O. (2004). Taxonomy: Impediment or expedient? – Science 303: 285-285.

Wiens, J.J. (2007). Species delimitation: New approaches for discovering diversity. – Systematic Biology 56: 875-878.

Winkler, A. (1924-1927). Catalogus Coleopterorum regionis palaearcticae. – Albert Winkler, Wien.

Winston, J.E. (1988). Describing species: practical taxonomic procedure for biologists. – Columbia University Press, New York.

Wrase, D.W. (2005). Nomenclatorial, taxonomic and faunistic notes on some Palaearctic genera and species of ground-beetles (Coleoptera, Carabidae: Apotomini, Chlaenini, Cyclosomini, Harpalini, Lebiini, Licinini, Platynini, Siagonini, Sphodrini). – Linzer Biologische Beitraege 37: 815-874.

Wrase, D.W. & Schmidt, J. (2006). A first representative of the tribe Metriini in the Palaearctic: *Sinometrius turnai* gen. nov., spec. nov. (Insecta: Coleoptera: Carabidae: Metriini). – In: Biodiversität und Naturausstattung im Himalaya II (Hartmann, M. & Weipert, J., eds). Verein der Freunde und Förderer des Naturkundemuseums Erfurt e.V., Erfurt, p. 315-324.

Zaballos, J.P. (2005). Los *Geocharis* Ehlers, 1883 de marruecos y Cadiz (Espana) (Coleoptera, Carabidae, Trechinae, Anillini). – Graellsia 61: 61-81.

Zaballos, J.P. & Ruiz-Tapiador, I. (1997). Nuevos *Typhlocharis* Dieck (Coeloptera, Caraboidea, Trechidae) de Espana. – Graellsia 52: 95-106.

Zarapkin, S.R. (1934). Zur Phänoanalyse von geographischen Rassen und Arten. – Archiv für Naturgeschichte N.F. 3: 161-186.

Zelditch, M.L., Swiderski, D.L., Sheets, H.D. & Fink, W.L. (2004). Geometric morphometrics for biologists – a primer. – Elsevier Academic Press, Amsterdam, Boston, Heidelberg.

Zhang, A.B., Kubota, K., Takami, Y., Kim, J.L., Kim, J.K. & Sota, T. (2005). Species status and phylogeography of two closely related *Coptolabrus* species (Coleoptera: Carabidae) in South Korea inferred from mitochondrial and nuclear gene sequences. – Molecular Ecology 14: 3823-3841.

Zhang, A.B., Kubota, K., Takami, Y., Kim, J.L., Kim, J.K. & Sota, T. (2006). Comparative phylogeography of three *Leptocarabus* ground beetle species in South Korea, based on the mitochondrial COI and nuclear 28S rRNA genes. – Zoological Science 23: 745-754.

Zhang, A.B. & Sota, T. (2007). Nuclear gene sequences resolve species phylogeny and mitochondrial introgression in *Leptocarabus* beetles showing trans-species polymorphisms. – Molecular Phylogenetics and Evolution 45: 534-546.

L. Penev, T. Erwin & T. Assmann (Eds) 2008
Back to the Roots and Back to the Future. Towards a New Synthesis amongst Taxonomic, Ecological and Biogeographical Approaches in Carabidology
Proceedings of the XIII European Carabidologists Meeting, Blagoevgrad, August 20-24, 2007, pp. 77-100.

New Genera of Western Hemisphere Pseudomorphini (Insecta, Coleoptera, Carabidae) with notes on their distributions, ways of life, and hypothesized relationships

Terry L. Erwin & Christy J. Geraci

Hyper-diversity Group, Department of Entomology, MRC-187, National Museum of Natural History, Smithsonian Institution P.O. Box 37012, Washington, DC 20013-7012, USA.
E-mail: erwint@si.edu & geracic@si.edu

SUMMARY

The Western Hemisphere Pseudomorphini was last revised by Notman in 1925 based on only a few known species (22) and paltry few specimens (73). A recent study of collections from throughout the Americas (1360 specimens) has revealed numerous new species contained in four new genera plus the nominate genus, and a change in status of a previously described subgenus. *Manumorpha* **n. gen.** (Type species – *Manumorpha biolat* Erwin & Geraci, new species, Ecuador, Perú), *Samiriamorpha* **n. gen.** (Type species – *Samiriamorpha grace* Erwin & Geraci, new species, Perú), *Yasunimorpha* **n. gen.** (Type species – *Yasunimorpha piranha* Erwin & Geraci, new species, Ecuador), and *Tuxtlamorpha* **n. gen.** (Type species – *Pseudomorpha tuxtla* Liebheer & Will, México) are described and their respective type species designated. *Notopseudomorpha* Baehr 1997, **new status**, is accorded generic rank with *P. laevissima* Chaudoir as type species. A summary of the contained species in each higher-level taxon and their overall distributions are provided. A genus level phylogeny for Western Pseudomorphini is inferred using maximum parsimony based on 33 adult morphology characters.

Keywords: Carabidae, Pseudomorphini, Western Hemisphere, taxonomy, phylogeny, way of life

RESUMEN

Los Pseudomorphini del Hemisferio Occidental fueron revisados por última vez por Notman en 1925 en base a solo unas pocas especies (22) y especimenes (73). Un estudio reciente de colecciones a lo largo de las Américas (1360 especimenes) a revelado numerosas especies nuevas contenidas en cuatro géneros nuevos además del género nominal, y un cambio en el estado de un género previamente descrito. *Manumorpha* n. gen.(Especie tipo- *Manumorpha biolat* Erwin & Geraci, nueva especie, Ecuador, Perú), *Samiriamorpha* n. gen. (Especie tipo - *Samiriamorpha grace* Erwin & Geraci, nueva especie, Perú), *Yasunimorpha* n. gen. (Especie tipo - *Yasunimorpha piranha* Erwin & Geraci, nueva especie, Ecuador), and *Tuxtulamorpha* n. gen. (Especie tipo - *Pseudomorpha tuxtula* Liebheer & Will, México) se describen y se designa respectivamente su especie tipo. Notopseudomorpha Baehr 1997, nuevo estado, se decide como rango genérico con *P. laevissima* Chaudoir como especie tipo. Un resumen de las especies contenidas en cada taxa de nivel superior y su distibución general son provistas. La filogenia a nivel de género para los Pseudomorphini Ocidentales es inferida usando máxima parsimonia basada en 33 caracteres morfológicos de los adultos.

INTRODUCTION

Pseudomorphini Newman 1842 is a Western Hemisphere – Australasian – Afroaustral Tribe of the beetle family Carabidae. Western Hemisphere members of this markedly unusual Tribe, in physical and behavioral attributes, were previously placed in a single genus, *Pseudomorpha* Kirby 1825. Their collective species' distributions encompass both the Nearctic, Neotropical, and northern parts of the Neaustral Realms. In this paper, in preparation for a complete revision of Pseudomorphini of the Western Hemisphere, we describe four new genera and elevate a subgenus status taxon to genus level.

George Horn (1867) wrote of these beetles: "These insects are not easy to obtain, as they are provokingly agile." E.A. Schwarz studied specimens of this Tribe in the early 20th Century and turned his notes over to Howard Notman of the Brooklyn Entomological Society; therein, Schwarz wrote, "... Pseudomorphas are numerous in their habitat, but are difficult to capture. They live in dead leaves and move with great agility, assisted by the numerous setae with which they are provided." We now know that pseudomorphine members are generally found in and around ant or termite nests and their immature stages in the Western Hemisphere are apparently obligatory myrmecophiles (Lenko, 1972, Erwin, 1981), perhaps even isopterophiles, as well (cf. label on a *N. laevissima* specimen). Adults are mostly collected at lights (UV, MV, and White) and while there on the collecting sheet, they scurry about rapidly much like tiny cockroaches, and are often very hard to grab. Notman (1925), in the only revision of the group for the Western Hemisphere, had 73 specimens at hand representing 22 species. Subsequently, Darlington (1935), Van Dyke (1943, 1953), Ogueta (1967), and Liebherr and Will (1997) accounted for six more species represented by 53 additional specimens. Thus, until now, only 126 specimens have been studied and documented in the literature. Currently, we now have borrows at hand that total 1360 specimens, and that fact alone accounts for the extraordinary discoveries noted below under "Accounts of Taxa." Given the state of Carabid taxonomy intensively pursued by our mentors and colleagues over the past century for North America with in-depth knowledge, keys, descriptions, and natural histories of all the Tribes, this discovery of species richness in the Pseudomorphini surely must be regarded as the last "taxonomic goldmine" left in North America. Of course, we can expect many more such taxonomic gold mines in Middle and South America.

At least two Western Hemisphere species, reportedly *P. hubbardi* Notman and *P. augustata* Horn are known to be ovoviviparous (Liebherr & Kavanaugh, 1985). All species included in the genera of the Western Hemisphere, as far as known, are fully winged and male adults of many of them and some females as well have been recorded at lights (UV, MV, and White), thus it is likely that they are very good dispersers. However, most species have fairly restricted known ranges. Whether this has to do with host fidelity to particular ant or termite species, or simply it is a matter of difficulty in collecting them is unknown, however host fidelity could be tested with intensive fieldwork by digging up ant nests to find adults and larvae, rather than merely collecting adults at lights. In his study of *N. laevissima* (Chaudoir), Lenko (1972) reported that 7 out of 32 nests of the ant *Camponotus rufipes* (Fab.) were home to *N. laevissima* larvae.

MATERIALS AND METHODS

Length and width measurements follow the conventions suggested by Ball (1972) and Kavanaugh (1979). Apparent body length (ABL) is measured from apex of labrum to apex of longer elytron. Standardized body length (SBL) is given herein for the Holotype of each type species and is the sum of the lengths of the head (measured from the apex of the clypeus to a point on midline at level of the posterior edge of the compound eyes), pronotum (measured from apical to basal margin along midline), and elytron (measured from apex of scutellum to apex of the longer elytron). In the case of species in the genus *Notopseudomorpha* and *Samiriamorpha*, in which members have hidden mouthparts in dorsal aspect, the measure is taken from the front margin of the head (frons). Total width (TW) is measured across both elytra at their widest point (usually this is a measure of the left elytron doubled because pinned specimens often do not have both elytra contiguous). Pronotum length to width ratios and elytra length to width ratios are given as such following the TW report for each species description below. The habitus images of the adult beetles were made with a Leica M420 microscope and an EntoVision™ system. Precise measures were taken using the Archimed software embedded in the EntoVision™ system. Male genitalia were illustrated using standard pen and ink techniques; an image of one female reproductive system, that of *P. tenebroides* Notman, is provided here and females of an exemplar of each genus will be illustrated and described in the forthcoming monograph of the subtribe mentioned above (Erwin, in prep).

Maximum parsimony analyses of unordered equally weighted multistate morphology characters were performed in PAUP 4.0b10 (Swofford, 1999). A heuristic search was done using TBR branch swapping and a random addition sequence (10 reps). A bootstrap analysis was performed using a full heuristic search algorithm (10,000 replicates) and a random addition sequence (10 reps), retaining groups compatible with the 50 percent majority rule consensus. Characters were mapped onto the most parsimonious topologies recovered by the heuristic search using the "trace all changes" tool in MacClade v.4.08 (Maddison & Maddison, 2000). The genera *Orthogonius* and *Spallomorpha* were chosen *a priori* as outgroups.

ACCOUNTS OF TAXA

Western Hemisphere genera of Pseudomorphini Newman 1842

Manumorpha Erwin & Geraci, **n. gen.**	Ecuador, Perú
Notopseudomorpha Baehr 1997, **new status**	Middle and South America
Pseudomorpha (s. str.) Kirby 1825	USA south to Argentina
Samiriamorpha Erwin & Geraci, **n. gen.**	Perú
Tuxtlamorpha Erwin & Geraci, **n. gen.**	México, Honduras
Yasunimorpha Erwin & Geraci, **n. gen.**	Ecuador

Key to the Western Hemisphere Genera of Pseudomorphini

1 Mouthparts not visible in dorsal aspect. Preocular lobe absent **2**

1' Mouthparts visible in dorsal aspect. Preocular lobe present **3**

2(1) Dorsal surface glabrous, markedly shiny................*Notopseudomorpha* Baehr 1997

2' Dorsal surface finely setiferous, not shiny.. .... *Samiriamorhpa* Erwin& Geraci **new genus**

3(1) Elytra multisetiferous; body form rather broad and subdepressed with elytra not or barely tapered to broadly round apex **4**

3' Elytra with only scutellar and ombilicate setae; body form narrow, somewhat cylindrical with elytra markedly tapered to apex.. .... *Yasunimorpha* Erwin & Geraci **new genus**

4(3) Dorsal surface with dense vestiture, of very long thick erect setae equal in length at lease to basal 4 antennomeres, but no pubescence; body form subconvex, elytra tapered posteriorly...............................*Manumorpha* Erwin & Geraci **new genus**

4' Dorsal surface with sparse or no long vestiture, longer setae equal in length only to at most basal 3 antennomeres, also usually with short pubescence; body form subconvex, elytra slightly tapered posteriorly or not **5**

5(4') Major setae of dorsal surface erect or slightly curved posteriorly .... *Pseudomorpha* Kirby 1825

5' Major setae of elytra posteriorly and markedly decumbent.................................. .... *Tuxtlamorpha* Erwin & Geraci **new genus**

Pseudomorphini Newman 1842

Pseudomorphini Newman, 1842:365.

Proposed English vernacular name. False-form beetles.

Diagnosis: Head ventrally with deeply recess grooves for receiving antennal bases; mandibular scrobe nearly effaced, delimited by row of short setae; mentum and submentum fused; antennal scape partially visible in dorsal aspect. Anterior coxal cavities closed, median coxal cavities conjunct, metepimeron visible. Abdomen with 6 visible sterna, sternum II with medial emargination on posterior edge. Male parameres long, nearly of same length (more or less symmetrical, or not), glabrous or setose, not balteate; phallobase bonnet-shaped.

Classification: According to the Maddison Lab at Tucson (Ober, 2002), Pseudomorphini occupies a position in the higher Carabidae, within the Harpalinae. The male genitalic median lobe has a bonnet-like base as in the lebiomorphs, yet their accompanying parameres are large and nearly symmetrical, as in basal carabid lineages. Although most pseudomorphine lineages are without setae on the parameres, as in the more derived carabids, members of several genera in both the Western and Eastern Hemispheres have one to several short setae, as in some primitive lineages of the family. This is also true for some members of other more derived lineages such as *Orthogonius*, *Graphipterus*, and some Panagaeini. Both DNA sequences and way of life suggest that the orthogonines and pseudomorphines are related; orthogonines are termitophilous. However, all known lineages of Pseudomorphini have been so highly selected for life with ants (and possibly termites) that external structures do not help in finding more normal carabid relatives. Erwin (2007) suggested that the Tribe Xenaroswellianini might be in some way be related to Pseudomorphini; see also Notman (1925) and Baehr (1992, 1997).

Pseudomorpha Kirby 1825

(Figs 1, 7, 13, 15, 16)

Pseudomorpha Kirby, 1825: 98. **Type species**: *Pseudomorpha excrucians* Kirby 1825:101. Original monotypy.

Heteromorpha Kirby, 1825:109. Incorrect subsequent spelling of *Pseudomorpha* Kirby 1825.

Axinophorus Dejean, 1829:174. **Type species**: *Axinophorus lecontei* Dejean & Boisduval 1829, synonym of *Pseudomorpha excrucians* Kirby 1825:101.

Drepanus Dejean, 1831:434. **Type species**: *Axinophorus lecontei* Dejean & Boisduval 1829, synonym of *Pseudomorpha excrucians* Kirby 1825:101.

Heteromorphus Chaudoir, 1852:63. Incorrect subsequent spelling of *Pseudomorpha* Kirby 1825.

Proposed English vernacular name.— Western False-form beetles.

Diagnosis.— Baehr (1997) adequately diagnosed members of this genus as follows: "Body fairly wide to almost cylindrical, elytra posteriorly gently convex; head prognathous; eyes situated laterally, without ventral border, ventral part more or less triangular;

clypeus partly or completely fused to frons; labrum separated from clypeus by a sulcus; supraorbital, clypeal, suborbital, and gular setae present, preorbital seta absent; antennal grooves deep; lateral plate of maxilla not enlarged; antenna elongate, basal antennomere simple; mental tooth elongate, triangular; glossa fused with paraglossae to a wide plate, bisetose but sometimes with additional elongate setae; labial palpi very large, markedly

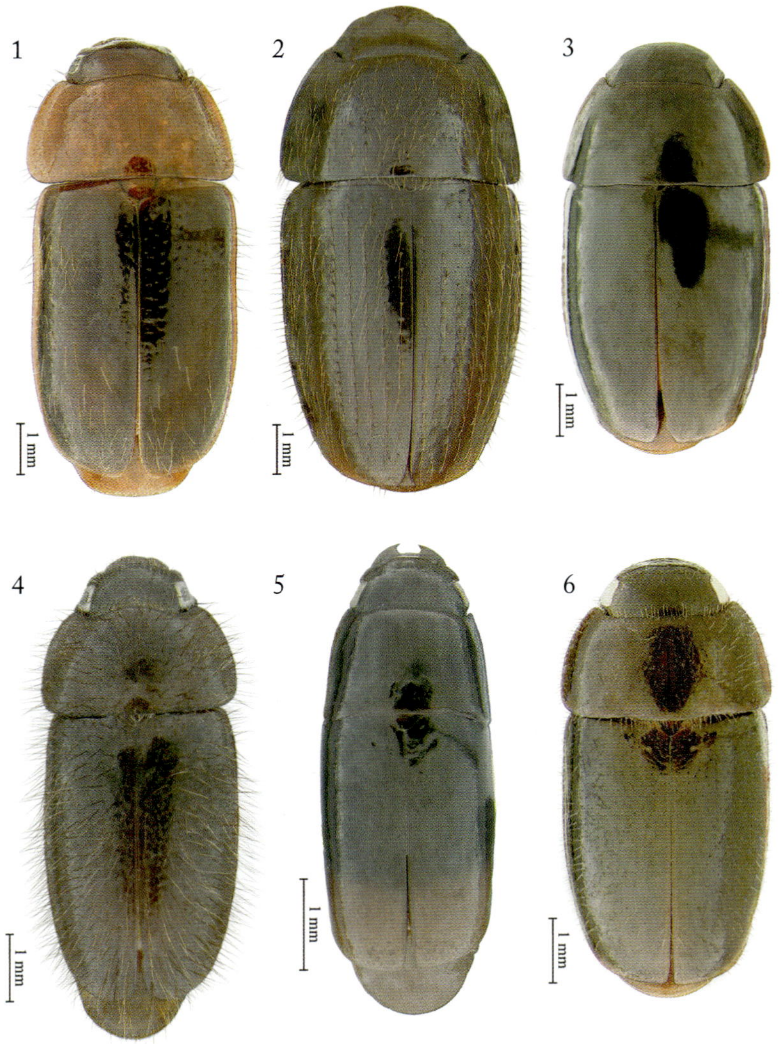

Fig. 1. Habitus, dorsal aspect of *Pseudomorpha* (*Pseudomorpha*) *excrucians* Kirby, Covington, LA.
Fig. 2. Habitus, dorsal aspect of *Tuxtlamorpha tuxtla* (Liebherr & Will), Veracruz, México.
Fig. 3. Habitus, dorsal aspect of *Notopseudomorpha laevissima* (Chaudoir), Brazil.
Fig. 4. Habitus, dorsal aspect of *Manumorpha biolat* Erwin & Geraci, Pakitza, Perú.
Fig. 5. Habitus, dorsal aspect of *Yasunimorpha piranhna* Erwin & Geraci, Yasuni, Ecuador.
Fig. 6. Habitus, dorsal aspect of *Samiriamorpha grace* Erwin & Geraci, Rio Samiria, Perú.

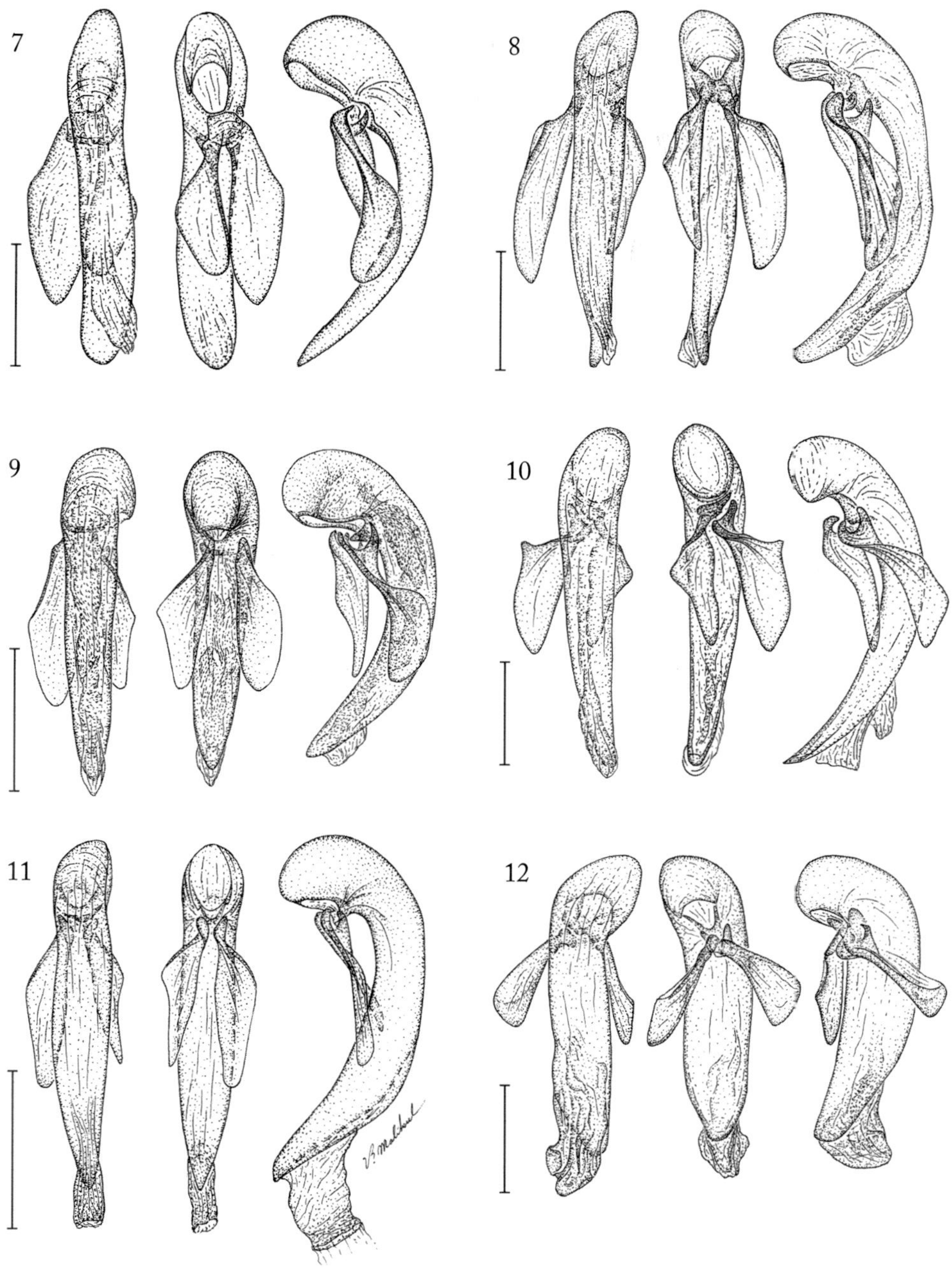

Fig. 7. Aedeagus and parameres of *Pseudomorpha excrucians* Kirby, Georgia, USA.
Fig. 8. Aedeagus and parameres of *Tuxtlamorpha* sp. Guatemala.
Fig. 9. Aedeagus and parameres of *Notopseudomorpha laevissima* (Chaudoir), Brazil.
Fig. 10. Aedeagus and parameres of *Manumorpha biolat* Erwin & Geraci, Pakitza, Perú.
Fig. 11. Aedeagus and parameres of *Yasunimorpha piranhna* Erwin & Geraci, Yasuni, Ecuador.
Fig. 12. Aedeagus and parameres of *Samiriamorpha grace* Erwin & Geraci, Rio Samiria, Perú.

securiform; ventral surface of head large, not concealed by the mouth parts; prosternal process straight, rather short, depressed between coxae; number of umbilical pores of elytra variable; femora moderately or strongly compressed, with deep grooves; tibiae and tarsi not compressed, elongate; 6 protarsus biseriately clothed at 1st and 2nd tarsomere, mesotarsus uniseriately clothed at 1st and 2nd tarsomere or not clothed; S sternum VII not excised; 6 sternum VIII apically divided, highly asymmetric; aedeagus with simply folded internal sac; parameres fairly similar, though left paramere always considerably larger; 2 stylomeres 1 and 2 separated, though shape very variable; no distinct dorsal and ventral ensiform setae present, but nematiform setae present though sometimes very short and not always arising from a pit."

The following is additional information not found in Baehr (1997).

Size small to medium for tribe and family, ABL = 7.0 to 11.1 mm, TW = 2.0 to 5.8 mm.

Male genitalia (Fig. 7): Phallus normal; in ventral aspect (Fig. 7) narrow, basal bulb swollen, crested or not; dorsal surface with short membranous ostium; apex more or less subtruncate, rounded, or acute; in lateral aspect (Fig. 7), with shaft curved ventrad, apical portion narrowed and somewhat acute. Endophallus with or without patches of microtrichia. Parameres (Fig. 7) glabrous or setiferous, left wider than right, both broad and long, equal or subequal in length. *Female genitalia*: As in Fig. 15.

Geographic distribution.— The geographical range of this genus extends from Oregon, Utah, and Colorado, USA in the north, through México and Central America to Catamarca Province, Argentina in the south (Fig. 16). South American species also occur in Brazil and Perú and undoubtedly elsewhere, as well.

Notes.— Ten informal species groups are now recognized (Erwin, in prep) based on sets of shared attributes. In these group, 114 species are currently recognized, 86 of which are new to science. In addition, there are still a few unresolved groups at present and their resolution will add to the list of known species. Whether the ten species groups are truly monophyletic, or not, must await a species level phylogenetic analysis, which is outside the scope of the present paper. In addition, the status of the subgenus *Austropseudomorpha* Baehr 1997 must await further study of the Western Hemisphere fauna.

Pseudomorpha (s. str.) *excrucians* Kirby 1825

(Figs 1, 7)

Pseudomorpha excrucians Kirby, 1825:101.
Axinophorus lecontei Dejean, 1837:176.
Pseudomorpha ruficollis Casey, 1924:148.

Holotype (male). USA – **Georgia**, (D. Francillon)(BMNH:ADP110643).

Derivation of specific epithet.— The word "*excrucians*" is from the Latin, *excrucio*, meaning torture or torment, and likely in reference to the false form of these beetles, i.e. many character states are of the Carabidae, but the general form is not.

Proposed English vernacular name.— Excruciating False-form beetle.

Diagnosis.— With the attributes of the genus as diagnosed above and color piceous, elytra much darker than rufous pronotum and head; and it is the only bicolored species in North America; pronotum (Fig. 1) wider at base than elytra, disk evenly and finely punctate, each puncture with a long seta although the central disc is rubbed in most specimens; elytral interneurs 1-7 well-defined, each with moderately impressed setigerous punctulae; elytral intervals (Fig. 1) randomly punctate with setigerous punctulae wide spread, mostly close to adjacent interneurs.

Description.— (Fig. 1). *Size*: Large. ABL = 8.0 to 9.5 mm; SBL (Holotype) = 7.94 mm; TW = 3.8 to 4.0 mm. Holotype pronotum ratio: 2.50; Holotype elytron ratio: 1.55. *Color*: Head and pronotum rufous, elytra rufopiceous, venter and appendages rufotestaceous. *Luster*: Dorsal surface shiny. *Microsculpture*: Head with very fine isodiametric sculpticells; effaced from pronotal disk; elytra with nearly effaced very fine flat transverse sculpticells. *Head*: Clypeal suture effaced at middle. Frons sparsely and moderately coarsely punctulate in paramedial patches at eye level, setigerous pores with moderately long setae. *Prothorax*: Pronotum (Fig. 1) moderately convex, much wider than long, with fringe of long stout setae along lateral margin, fringe of shorter setae along anterior and medial posterior margins; anterior margin beaded, posterior margin not beaded. Disk with longitudinal shallowly impressed and discontinuous line and with sparse moderately course setigerous punctulae, each with a long erect seta; median disc glabrous. *Pterothorax*: Elytral interneurs moderately coarsely punctate, all interneurs more or less equally impressed, setigerous pores wide-spaced, intervals randomly sparsely punctate. Metepisternum longer than wide, surface sparsely setiferous, setae short. Metasternum sparsely setiferous. Metathoracic wings fully developed. *Abdomen*: Sternum III broadly and shallowly incised medially. All sterna moderately setiferous, IV broadly and densely so, medially; male with dense patch of setae medially on sterna V and VI, their patch width about two-fifths length of posterior trochanter (Fig. 13). *Male genitalia*: (Fig. 7) Phallus slightly arcuate to the left in dorsal aspect, apex broadly rounded, ventral margin nearly straight in apical third. Parameres (Fig. 7): in ventral aspect left shorter than right and slightly narrower, distal margin rounded; distal margin of right acute. Phallobase not crested.

Way of Life.— **Macrohabitat:** Lowlands, 3 – 397 meters altitude in Eastern Deciduous Forest/Pine barrens zones. **Microhabitat:** On sandy substrates likely near ant nests and in the surrounding vicinity. **Dispersal abilities:** Macropterous, capable of flight; swift runner. **Seasonal occurrence:** Adults found active in July (Louisiana), and June – July (South Carolina). **Behavior:** See under genus above.

Geographic distribution.— This species occurs in southeastern USA – AL, GA, LA, MS, SC.

Other specimens examined.— Holotype (male) of *P. ruficollis* Casey – LA: St. Tammany Parrish, Covington (USNM: 48078)(Soltau). Holotype (male) of *P. lecontei* Dejean – North America (MNHN). One male – LA: St. Tammany Parrish, Covington, July (Hubbard & Schwarz)(NMNH:ADP109022). One female – MS: Richton, June (H. Dietrich)(CAS:8111199). One male – SC: Oconee County, Whitewater Falls, July (OL. Cartwright)(NMNH:ADP109020).

Tuxtlamorpha Erwin & Geraci, new genus
(Figs 2, 8, 16)

Type species: *Pseudomorpha tuxtla* Liebherr & Will, 1997:54, here designated.

Proposed English vernacular name.— Combed False-form beetles.

Diagnosis.— With the attributes of the Tribe as described above and dorsal surface with numerous long decumbent and markedly course setae scattered on head and pronotum, and in 9 perfect rows on the elytra (excluding the interrupted ombilicate series). Clypeus and labrum deflected at about 45° angle, frons slightly convex; preocular lobe present produced anteriorly, confluent with eye posteriorly. Antenna short, extended to level of mid-procoxae; antennomeres of equal width distally, each slightly compressed. Gena below eye markedly angulate, sharply beaded. Elytra proportionally smaller in comparison with pronotum and head, and evenly tapered to moderately narrow rounded apex. Male with broad setal patch on sterna VI and VII, set in shallow transverse excavation, this groove posteriorly with a row of long setae that are angulate dorsally at their tips, the setae sigmoid in shape from lateral aspect.

Notes.— Two species are now recognized (Erwin, in prep) based on sets of shared attributes. One of these is new to science.

Geographic distribution.— The geographical range of this genus extends from Vera Cruz, México to Honduras (Fig. 16).

Tuxtlamorpha tuxtla (Liebherr & Will) 1997
(Figs 2, cf. 8)

Pseudomorpha tuxtla Liebherr & Will, 1997:54.

Holotype (female). México – **Vera Cruz,** "Est. Biol. "Los Tuxtlas," 26-VII-1990, 150m el., at light (J.K. Liebherr)(UNAM)" according to Liebherr & Will, 1997:54.

Derivation of specific epithet.— The word "tuxtla" is derived from the name of the biodiversity station in Vera Cruz.

Proposed English vernacular name.— Tuxtla False-form beetle.

Diagnosis.— See under genus above.

Description.— (Fig. 2). *Size*: Large. ABL = 10.6 to 12.0 mm; SBL (Holotype) = 9.04 mm; TW = 6.0 to 6.2 mm. Holotype pronotum ratio: 2.61; Holotype elytron ratio: 1.19. *Color*: Head, pronotum and elytra dark brown, venter and appendages piceous. *Luster*: Dorsal surface moderately shiny. *Microsculpture*: Dorsal surface of pronotum with very fine slightly stretched sculpticells; that of head and elytra very fine isodiametric sculpticells. *Head*: Clypeus glabrous; frons irregularly and sparsely micropunctulate, setigerous pores with short setae; vertex glabrous. Occiput medial to hind margin of eye without small isolated group of coarse setiferous pores. *Prothorax*: Pronotum (Fig. 2) markedly convex, not depressed along midline, wider than long, without fringe of long stout setae along lateral and anterior margins; with such setae in a group of five at hind angle and with sparse marginal setae on posterior margin; entire disk with sparse vestiture of stout decumbent setae; anterior and lateral margins beaded, bead of lateral margin efface at posterior angle, posterior margin somewhat discolored but not beaded; disk without longitudinal shallowly impressed midline. *Pterothorax*: Elytral interneurs finely punctate, not striate, each with long posteriorly decumbent seta; intervals with occasional setigerous pores adjacent to interneurs. Metepisternum longer than wide, surface not setiferous. Metasternum markedly convex medially, surface not setiferous. Metathoracic wing fully developed. *Abdomen*: Sternum III broadly and shallowly arcuate medially. All sterna sparsely setiferous, IV broadly and more densely so medially; male unknown. *Male genitalia*: Unknown (see note below). The male genitalia of an undescribed species from Guatemala is illustrated (Fig. 8)

Way of Life.— Macrohabitat: Lowlands, 150 meters altitude. **Microhabitat:** Unknown. **Dispersal abilities:** Macropterous, capable of flight; swift runner. **Seasonal occurrence:** Adults found in July. **Behavior:** See under genus above. Adults found at lights at night.

Other specimen examined.— Female. México – **Vera Cruz**, Estac. Biol. Los Tuxtlas, 1/9-VII-1988, 150m el., at light (J.A. Chemsak)(UCBC:EMEC61656).

Geographic distribution.— (Fig. 16). This species is known from southeastern México — VC.

Notes.— Both known specimens of *T. tuxtla* are females, however the single specimen of an undescribed species from Honduras is a male (Erwin, in prep). An illustration of that male's aedeagus is presented here, as the likelihood of similarity is great, as exemplified in the other species groups of pseudomorphines. *Male genitalia*: (Fig. 8) Phallus very slightly arcuate to the right in dorsal aspect, apex recurved to the left and narrowly rounded, ventral margin markedly arcuate throughout its length, apex broad, truncate. Parameres (Fig. 8): in ventral aspect left slightly shorter, right paramere somewhat broader than left and somewhat broader distally, distal margins of both acute, rounded at tip.

Notopseudomorpha Baehr 1997, new status

(Figs 3, 9, 14, 16)

Type species. *Pseudomorpha laevissima* Chaudoir 1852:63, Brazil, designated by Baehr (1997:42).

Proposed English vernacular name.— False False-form beetles.

Diagnosis.— With the attributes of the Tribe as described above and dorsal surface devoid of setae except umbilicate series of elytron. Mouthparts mostly hidden in dorsal aspect; clypeus very small, deflected at about 45° angle, frons slightly convex; preocular lobes absent. Antenna short, extended to middle of procoxae; antennomeres of uniform width throughout, each slightly compressed. Gena below eye markedly angulate; beaded. Elytra proportionally small in comparison with pronotum and moderately tapered to narrowly rounded apex. Male with dense and divided setal patches on sterna V and VI, set in two shallow excavations. Posterior-most setae of each patch shallowly elbowed ventrally.

NOTES.— Nine species are now recognized (Erwin, in prep) based on specimens ranging from Argentina north to Costa Rica. Of these nine species, six are new to science.

GEOGRAPHIC DISTRIBUTION.— The geographical range of this genus extends from Costa Rica to Argentina (Fig. 16).

Notopseudomorpha laevissima (Chaudoir) 1852

(Figs 3, 9)

Pseudomorpha laevissima Chaudoir 1852:63.

Holotype (female). BRAZIL, nr. Novo Friburgo (Mniszech)(MNHP). [labeled by G.E. Ball, 2007].

DERIVATION OF SPECIFIC EPITHET.— The word "*laevissima*" is a Latin adjective, meaning smooth, and refers to the entire dorsal surface which is devoid of setae or any type of blemish.

Fig. 13. Abdominal segments of male *Pseudomorpha excrucians* Kirby, Covington, LA; V and VI with modified setae.
Fig. 14. Abdominal segments of male *Notopseudomorpha* sp. Costa Rica; V and VI with modified setae.
Fig. 15. Female gentalia of *Pseudomorpha tenebroides* Notman, Florida Canyon, AZ.

Proposed English vernacular name.— Smooth False-form beetle.

Diagnosis.— See under genus above.

Description.— (Fig. 3). *Size*: Medium. ABL = 9.1 to 9.5 mm; SBL (Holotype) = 7.41 mm, TW = 5.1 to 5.2 mm. Holotype pronotum ratio: 2.55; Holotype elytron ratio: 1.39. *Color*: Head, pronotum and elytra dark rufous brown, venter and appendages dark yellowish-brown. *Luster*: Dorsal surface very shiny. *Microsculpture*: Dorsal surface with very fine isodiametric and slightly stretched sculpticells, these nearly effaced on some individuals. *Head*: Frons and vertex finely micropunctulate. Occiput medial to hind margin of eye without small isolated group of coarse setiferous pores; outer angle of gena at corner of eye 4-setose. *Prothorax*: Pronotum (Fig. 3) markedly convex, not depressed along midline, wider than long, without fringes of setae along margins, some individuals with one or two marginal setae at hind angle; anterior and lateral margins beaded, bead of lateral margin efface at posterior angle, posterior margin somewhat discolored but not beaded; disk with longitudinal shallowly impressed midline. *Pterothorax*: Elytral disc featureless; ombilicate setal series present; margin with fringe of stout setae. Metepisternum longer than wide, surface not setiferous. Metasternum markedly convex medially, sparsely setiferous throughout. Metathoracic wing fully developed. *Abdomen*: Sternum III broadly and shallowly incised medially. All sterna at least sparsely setiferous, IV broadly and more densely so medially; male see above under genus. *Male genitalia*: (Fig. 9) Phallus straight in dorsal aspect, apex acutely rounded, ventral margin markedly tapered throughout its length; basal bonnet relatively quite large. Parameres (Fig. 9): in ventral aspect left and right coequal in length, left slightly smaller and somewhat narrower distally, distal margins of both narrowly rounded.

Way of Life.—**Macrohabitat:** Lowlands to midlands, 150 – 1000 meters altitude, in the Cerrado vegetation zone in Brazil and the Chaco and Yungas zones in Argentina. **Microhabitat:** Adults are found in nests of the ant *Camponotus rufipes* (Fab.) in Brazil. **Dispersal abilities:** Macropterous, capable of flight (Lenko, 1972); swift runner. **Seasonal occurrence:** Adults found in January – February, and December. **Behavior:** See under genus above. Adults found at lights at night. Larvae eat larvae of the ant host in the core of the nest, then retreat to the nest periphery at the time for pupation. Lenko (1972) did not discover the food of adults.

Other specimens examined.— Female, Brazil, Chapada, (CMNH:ADP109125); male, São Paulo, January (Parker)(NMNH:ADP110380); female, Minas Gerais, Viçosa, Corrego da Paraiso (Mata do Prefeitura) (Mata do Paraiso), 703 m, 20.768° S, 42.877° W, February (T.J. Henry)(NMNH:ADP110357).

Additional specimens.— Ogueta (1967) lists the following localities: Brazil: Pirassunga, Edo. São Paulo, Nova Teutonia, Novo Friburgo. Argentina: Misiones (Puerto Bemberg), Tucumán (Las Cuchillas), Catamarca.

Geographic distribution.— (Fig. 16). This species occurs in eastern and southeastern Brazil and northern Argentina.

Notes.— The larval description and some notes on life history were published by Lenko (1972). A second species assigned by Baehr (1992) to this genus (*P. glabra* Ogueta, now *N. glabra* (Ogueta)) is found in the Argentine province of Santiago del Estero.

Manumorpha Erwin & Geraci, n. gen.

(Figs 4, 10, 16)

Type species: *Manumorpha biolat* Erwin & Geraci, **sp. n.,** Perú, present designation.

Proposed English vernacular name.— Hairy False-form beetles.

Diagnosis.— With the attributes of the Tribe as described above and dorsal surface with numerous long and erect markedly course setae densely located on head and pronotum, and in 9 perfect rows on the elytra (excluding the interrupted ombilicate series). Clypeus deflected at about 45° angle, frons slightly bulbous; preocular lobes present, hind angles not flush with eye. Antenna long, extended beyond prosternal process; middle antennomeres broad, decreasing in size proximally and distally, each moderately compressed. Gena below eye markedly angulate; suboptical ridge beaded. Elytra proportionally not small in comparison with pronotum and markedly tapered to narrowly rounded apex. Male with slightly denser setal patch on sternum VI, but this not set in shallow excavation; no denser setal patch on sternum V.

Behavior.— Many species of ants nest in suspended dried palm fronds where members of this genus are found and given the known life history of members of other genera in Pseudomorphini, it is like that *M. biolat* adults frequent these ant nest and the larvae are myrmecophilus there.

Notes.— The genus at present contains three species, all of which are new to science.

Geographic distribution.— The geographical range of this genus extends from Ecuador to Perú (Fig. 16).

Manumorpha biolat Erwin & Geraci, sp. n.

(Figs 4, 10)

Holotype.— Perú: **Madre de Dios Department**, Manu National Reserve, Rio Manu, Pakitza, Trocha Pacal – 25, 11.941° S, 071.303° W, 356m, 14 October 1991 (T.L. Erwin & M.G. Pogue)(NMNH:FOG14644, male). Paratypes listed below under other specimens examined.

Derivation of specific epithet.— The word "*biolat*" is an acronym for the program "Biodiversity in Latin America" run by the Smithsonian Institution in the 1980/90's and under which this species was discovered on sponsored expeditions.

Proposed English vernacular name.— Biolat False-form beetle.

Diagnosis.— With the attributes of the genus as described above and color rufous brown (Fig. 4), elytra darker than pronotum; pronotum (Fig. 4) wider at base than elytra across humeri; elytral intervals 1-9 well-defined, each with markedly coarse setigerous punctulae; elytral (Fig. 4) interneurs without, or rarely with a single random setigerous puncture.

Description.— (Fig. 4). *Size*: Medium. ABL = 7.1 to 8.2 mm; SBL (Holotype) = 6.26 mm; TW = 2.4 to 4.2 mm. Holotype pronotum ratio: 1.54; Holotype elytron ratio: 1.39. *Color*: Head, pronotum, elytra and venter rufous brown, appendages somewhat testaceous. *Luster*: Dorsal surface moderately shiny. *Microsculpture*: Dorsal surface with very fine slightly stretched sculpticells. *Head*: Frons micropunctulate, setigerous pores with short and long setae scattered except on vertex where they form more or less an uneven transverse line. Occiput medial to hind margin of eye without small isolated group of coarse setiferous pores. *Prothorax*: Pronotum (Fig. 4) moderately convex, depressed along midline, wider than long, with fringe of long stout setae along lateral and anterior margins, and over entire disk; anterior and lateral margins beaded, bead of lateral margin efface at posterior angle, posterior margin somewhat discolored but not beaded; disk with longitudinal shallowly impressed midline. *Pterothorax*: Elytral interneurs impunctate, striae barely traceable, intervals with setigerous pores closely spaced, each slightly raised, coarsely impressed. Metepisternum longer than wide, surface sparsely setiferous. Metasternum markedly convex medially, sparsely setiferous throughout. Metathoracic wing fully developed. *Abdomen*: Sternum III broadly and shallowly incised medially. All sterna sparsely setiferous, IV broadly and more densely so medially; male with small denser patch of setae medially on sterna VI, the patch width less than half that of the length of posterior basitarsus. *Male genitalia*: (Fig. 10) Phallus slightly arcuate to the right in dorsal aspect, apex narrowly rounded, ventral margin markedly arcuate throughout its length. Parameres (Fig. 10): in ventral aspect left shorter and slightly smaller than right and somewhat narrower distally, distal margins of both narrowly rounded.

Way of Life.— Macrohabitat: Lowlands, 356 meters altitude, in tropical rain forest. **Microhabitat:** Adults are found in dry attending fronds of palm trees (*Astrocaryum chonta* Mart.). **Dispersal abilities:** Macropterous, capable of flight; swift runner. **Seasonal occurrence:** Adults found in October – November. **Behavior:** See under genus above.

Other specimens examined.— Perú: **Madre de Dios**, Manu National Reserve, Rio Manu, Pakitza, Trocha Pacal – 25, 11.941° S, 071.303° W, 356m, 14 October 1991 (T.L. Erwin & M.G. Pogue)(NMNH:FOG 14644, 14641, 14605, male paratypes; 14642, 14643, female paratypes). The holotype will be deposited at Museo de Historia Natural, Lima, Peru.

Geographic distribution.— (Fig. 16). This species occurs in southeastern Perú.

Yasunimorpha Erwin & Geraci, n. gen.

(Figs 5, 11, 15)

Type species: *Yasunimorpha piranha* Erwin & Geraci, **sp. n.** Ecuador. Present designation.

Proposed English vernacular name.— Narrow False-form beetles.

Diagnosis.— With the attributes of the Tribe as described above, form subcylindrical, and dorsal surface of head and pronotum devoid of setae and pubescence, elytra with only two setae (excluding the interrupted ombilicate series). Clypeus continuous with frons, not deflected at an angle, frons flat; preocular lobes defined but not produced. Antenna very short, not extended to level of procoxal process; antennomeres

3-10 quadrate, flattened, increasing in size distally. Gena below eye not produced, edge barely visible in dorsal aspect. Elytra proportionally small in comparison with length of pronotum and somewhat tapered to narrowly rounded apex. Abdomen of both sexes devoid of pubescence except in male with dense double setal patches on sternum V and VI; these not set in shallow excavation and with very small separation between patches.

Behavior.— Many species of ants build nests in the rain forest canopy and given the known life history of members of other genera in Pseudomorphini, it is like that *Y. piranha* adults frequent these ant nest and the larvae are myrmecophilus there.

Notes.— The genus at present is monotypic.

Geographic distribution.— The presently known location of this genus is eastern Ecuador (Fig. 16).

Yasunimorpha piranha Erwin & Geraci, sp. n.

(Figs 5, 11)

Holotype.— Ecuador: **Orellana Province**, nr. Yasuni National Park, Onkone Gare Station, 0.657° S, 076.452° W, 236m, 16 January 1994 (T.L. Erwin, et al.) (NMNH:ADP110379, male). One paratype is listed below under other specimens examined.

Derivation of specific epithet.— The word "*piranha*" is in reference to the general area in which the holotype and paratype were collected at Onkone Gare Station. Onkone Gare are the Huaorani words for Piraña (piranha) and is the name of the stream near the fogging plot from which specimens were collected.

Proposed English vernacular name.— Yasuni False-form beetle.

Diagnosis.— With the attributes of the genus as described above and color black with rufinistic highlights (Fig. 5), elytra two-toned, basally black, apex rufous; pronotum (Fig. 5) at base subequal to width of elytra across middle; neither elytral interneurs nor intervals evident, disk with only two moderately coarse setigerous punctulae, one near scutellum and the other at basal third midway between suture and lateral side, ombilicate series present.

Description.— (Fig. 5). *Size*: Small. ABL = 5.8 to 6.0 mm; SBL (Holotype) = 4.37 mm; TW = 1.9 to 2.0 mm. Holotype pronotum ratio: 1.67; Holotype elytron ratio: 1.25. *Color*: Head, pronotum and elytra black with rufinistic highlights, elytra two-toned, basally black, apex rufus, venter rufous, appendages testaceous. *Luster*: Dorsal surface moderately alutaceous. *Microsculpture*: Sculpticells effaced from dorsal surface. *Head*: Frons densely micropunctulate, setae, including clypeal and supraorbital setae, and pubescence absent. Occiput medial to hind margin of eye without small group of coarse setiferous pores. Eyes flat, asetiferous. *Prothorax*: Pronotum (Fig. 5) markedly convex, longer than wide, devoid or setae and pubescence, surface densely micropunctate; only lateral margin beaded, that not complete basally, posterior margin discolored but not beaded; disk with longitudinal impressed line nearly effaced. *Pterothorax*: Elytra markedly convex, surface densely micropunctate, interneurs and intervals effaced, two setigerous pores, one near scutellum, the other at middle at basal third of elytron. Metepisternum longer than wide, surface sparsely setiferous. Metasternum short, markedly convex medially, without vestiture. Metathoracic wing fully developed. *Abdomen*: Sternum III broadly and shallowly incised medially. All sterna devoid of vestiture except VII with one pair of wide-spaced setigerous pores near apical margin and male with dense double patch of setae medially on sterna V and VI, their patch width slightly less that length of posterior trochanter and narrowly separated medially. *Male genitalia*: (Fig. 11) Phallus slightly arcuate to the right in dorsal aspect, apex narrowly pointed and thick, ventral margin slightly arcuate throughout its length. Parameres (Fig. 11): in ventral aspect left shorter than right and somewhat narrower, its distal margin pointed, that of the left rounded.

Way of Life.— Macrohabitat: Lowlands, 236 meters altitude, in tropical rain forest. **Microhabitat:** Mixed rain forest canopy crowns of the following tree species: *Astrocaryum chambira* Burret, *Pouteria reticulata* (Engl.) Eyma, *Inga capitata* Desv., *Cecropia ficifolia* Warb. ex Snethl.). **Dispersal abilities:** Macropterous, probably capable of flight. **Seasonal occurrence:** Adults found in January, the dry season. **Behavior:** See under genus above.

Other specimen examined.— Ecuador: **Orellana Province**, nr. Yasuni National Park, Onkone Gare Station, 0.657° S, 076.452° W, 236m, 24 January 1994 (T.L. Erwin, et al.)(NMNH:ADP110292, paratype female). The holotype will be deposited in the National Natural History Museum in Quito, Ecuador.

Geographic distribution.— (Fig. 16). This species is known presently from eastern Ecuador.

Samiriamorpha Erwin & Geraci, n. gen.

(Figs 6, 12, 16)

Type species: *Samiriamorpha grace* Erwin & Geraci, **sp. n.** Perú, present designation.

Proposed English vernacular name.— Flat False-form beetles.

Diagnosis.— With the attributes of the Tribe as described above and dorsal surface with numerous moderately short and markedly fine setae scattered on head and pronotum, and in both interneurs and intervals on the elytra. Form somewhat depressed. Mouthparts not visible from above; clypeus vertical, frons subtly convex; preocular lobes absent. Antenna very short, extended only to level of middle of prosternal process; antennomeres very broad decreasing in size distally from antennomere 9, each moderately compressed. Gena below eye markedly angulate; subocular ridge beaded. Elytra not proportionally small in comparison with pronotum and moderately tapered to narrowly rounded apex. Male with two dense setal patches on sternum V and VI, these divided by a narrow space.

Notes. This genus is at present monotypic.

Geographic distribution.— The presently known location of this genus is north-central Perú (Fig. 16).

Fig. 16. General distribution patterns of overall ranges of the taxa included.

Samiriamorpha grace Erwin & Geraci, sp. n.
(Figs 6, 12)

Holotype.— PERÚ: **Loreto Department**, nr. Pacaya-Samiria National Reserve, Cocha Shinguito, nr. Rio Samiria, 05.179° S, 074.654° W, 119 m, 13 June 1990 (T.L.

Erwin, G.P. Servat, et al.)(NMNH:ADP110418, male). Paratypes are listed below under other specimens examined.

Derivation of specific epithet.— The word "*grace*" is an eponym and used in reference to the first name of one of the collectors of the type series, Grace P. Servat.

Proposed English vernacular name.— Grace's False-form beetle.

Diagnosis.— With the attributes of the genus as described above and color rufopiceous, the head and pronotum more rufous that elytra, venter and appendages rufotestaceous; pronotum (Fig. 6) at base broader than width of elytra across humeri; neither elytral interneurs nor intervals evident although setae appear in lines likely marking the interneurs, and therefore intervals with some scattered setae; ombilicate series present.

Description.— (Fig. 6). *Size*: Medium (all four studied specimens are the same size). ABL = 8.0 mm; SBL (Holotype) = 6.32 mm; TW = 4.3 mm. Holotype pronotum ratio: 2.65; Holotype elytron ratio: 1.13. *Color*: Head, pronotum and elytra rufopiceous, elytra darker in tone, venter and appendages rufotestaceous. *Luster*: Dorsal surface shiny. *Microsculpture*: Dorsal surface with finely impressed and transverse sculpticells; these partially effaced from disc of pronotum. *Head*: Frons sparsely micropunctulate, setae moderately long and very fine; vertex glabrous. Occiput medial to hind margin of eye without small group of coarse setiferous pores. Eyes nearly flat, asetiferous. *Prothorax*: Pronotal disc (Fig. 6) nearly flat, wider than long, surface densely micropunctate, setae moderately long and very fine, all margins with a fringe of moderately coarse setae; anterior and lateral margins beaded, the lateral margin not complete basally; posterior margin discolored but not beaded; disk with longitudinal impressed line shallowly impressed. *Pterothorax*: Elytral disc nearly flat, surface densely micropunctate, interneurs and intervals effaced, traceable only by following rows of setae. Metepisternum longer than wide, surface without vestiture. Metasternum short, markedly convex medially, with patch of vestiture on the convexity. Metathoracic wing fully developed. *Abdomen*: Sternum III and IV fused medially. All sterna with scattered vestiture, no central patch on III; male with dense double patch of setae medially on sterna V and VI, their total patch width subequal to length of posterior trochanter. *Male genitalia*: (Fig. 12) Phallus robust, very slightly arcuate to the right in dorsal aspect, apex broadly rounded and thick, ventral margin slightly almost straight throughout its length. Parameres (Fig. 12): in ventral aspect both nearly of same length, left somewhat narrower, its apex narrowly rounded, that of the left truncate.

Way of Life.— **Macrohabitat:** Lowlands, 119 meters altitude, in rainforest surrounded by black water swamps and rivers. **Microhabitat:** The series of 4 specimens were found by insecticidal fogging a big tree with vines, epiphytes and *Azteca* ant nests. **Dispersal abilities:** Macropterous, probably capable of flight; swift runner. **Seasonal occurrence:** Adults found in June. **Behavior:** See under genus above.

Other specimens examined.— Two females, one male with same data as Holotype, above. The holotype will be deposited at Museo de Historia Natural, Lima, Peru.

Geographic distribution.— (Fig. 16). This species occurs in north-central Perú.

PHYLOGENY

Bootstrap analysis of 33 morphology characters for Pseudomorphini genera produced only moderate support for relationships among genera (Fig. 17a). A heuristic search recovered one most parsimonious tree topology with *Orthogonius* and *Spallomorpha* at the base (tree length = 78, consistency index = 0.667, retention index = 0.518, rescaled consistency index = 0.346). Two synapomorphies support the clade defined at Node A (Fig. 17b): these genera all possess elytra with setose lateral margins and pubescent abdominal sterna. Node B has a moderately strong bootstrap value and is supported by seven characters. Only one of those characters is a an unreversed synapomorphy, however (Fig. 17b, Appendix I). The recovery of *Tuxtlamorpha* and *Samiriamorpha* as sister taxa is supported by a uniquely shared apically blunt phallus, but the placement of *Xenaroswelliana* in relation to *Manumorpha* and *Pseudomorpha* is unclear. Character traces revealed six autapomorphic character states for *Xenaroswelliana* (Fig. 17b) that give the genus a noticeably different gross appearance than other Pseudomorphini (Fig. 17a). This supports Erwin's interpretation of this genus as a separate tribe (Erwin, 2007) that is perhaps related in some way to the Pseudomorphini proper because of the shared and completely unique carabid character, that of a ventral sulcus into which the antennal base is tucked away. In *Xenaroswelliana* members, the sulcus is only partially developed indicating that this taxon should be basal to all the Pseudomorphini, if the character system is evolving toward its sophisticated appearance in the pseudomorphines. Molecular data from multiple gene fragments are needed to confirm the placement of *Xenaroswelliana* in relation to other Western Hemisphere Pseudomorphini genera.

DISCUSSION

The Western Hemisphere Pseudomorphini ranks as one of the poorest known carabid Tribes. Although the North American species have been well collected, they have not been integrated into a synthetic taxonomic treatment and numerous species remain undescribed. Middle and South America are woefully under-collected for members of this Tribe. The lack of tropical specimens collected at lights may mean that adults are not attracted and therefore must be dug from ant nests or hand-collected in the vicinity of nests. Our canopy fogging program has found them, but not commonly, and only so when we actually targeted ant nests. Many ants nest in trees in the tropics, so focused fogging will surely garner specimens. We predict our Neotropical fauna will rival the diversity found in Australia (cf. Baehr, 1992, 1997). A species level revision of the Tribe in the Western Hemisphere is presently underway (TLE) based on 1360 specimens borrowed from many institutions. In this material, more than 120 species and 5 genera are represented.

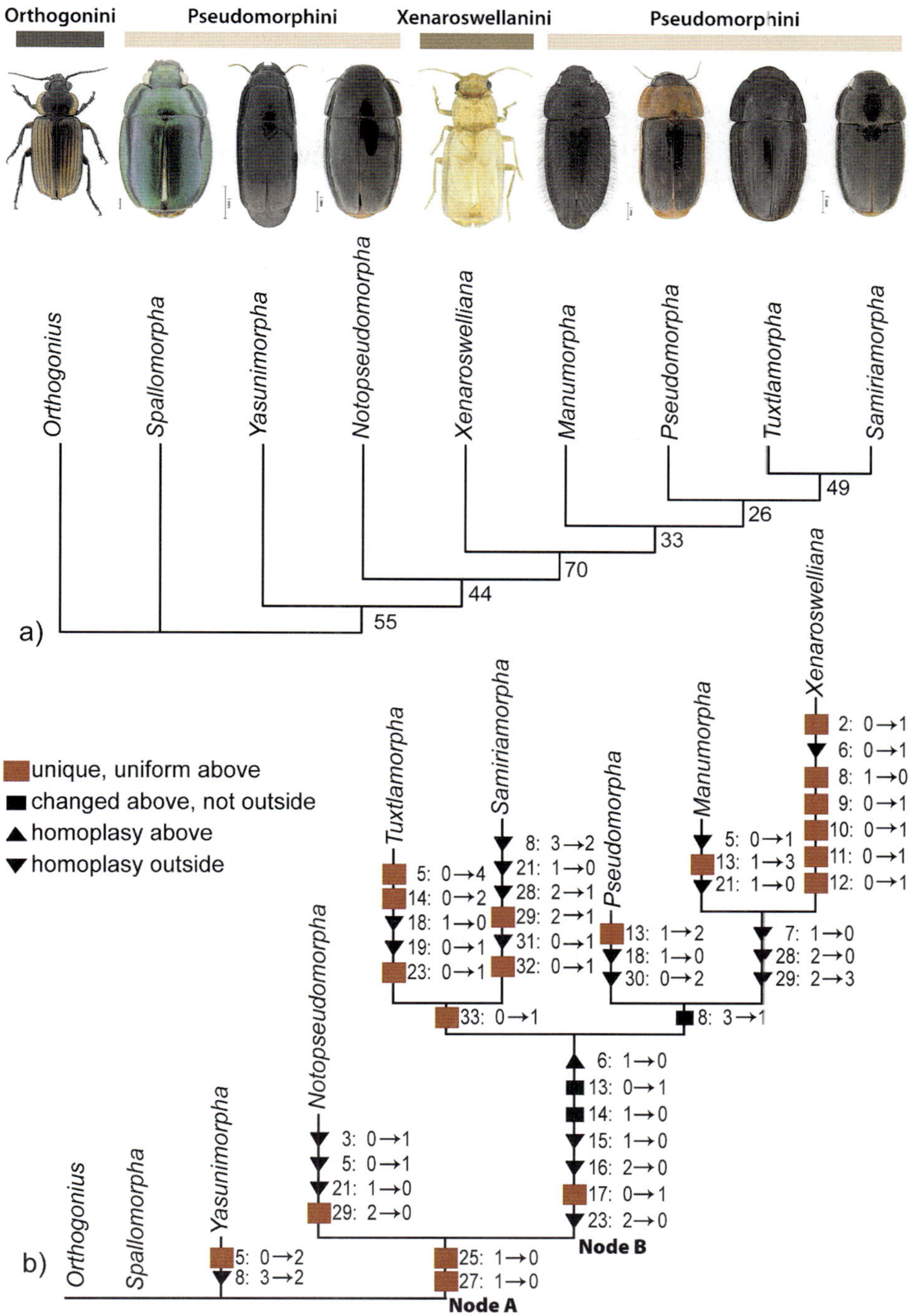

Fig. 17. a. Topology recovered from a bootstrap analysis of 33 structural attributes (bootstrap values presented at appropriate nodes). b. Structural attributes traced on to the most parsimonious tree recovered from a heuristic search: state transitions for corresponding characters are indicated by arrows (see Appendix 1).

ACKNOWLEDGEMENTS

We thank the following curators for the loan of specimens from their institutions that made the present study possible: David H. Kavanaugh, California Academy of Sciences, San Francisco (CAS); Thierry Deuve (with George Ball), Muséum National d'Histoire Naturelle, Paris (MNHP); Conrad Gillett, The Natural History Museum, London (BMNH); additional specimens noted herein are from the NMNH, Washington, DC.

We also thank Warren Steiner who prepared and dissected the male specimens and Nilanjana Saha who imaged the specimens, and Vichai Malikul who prepared the illustrations of the male genitalia, and to Karolyn Darrow who took the images and illustrations and fixed them and arranged them on plates; Darrow, Steiner, and Malikul are staff members in the NMNH Entomology Department, NMNH, and Nilanjana Saha was our winter Intern from Butler University in 2007-2008. We also express our deep appreciation to Valeria Aschero (CRICyT, Mendoza, Argentina) for translating the abstract into Spanish and to Stephen McJonathan of GT Vision for setting up our hardware and software necessary to make the extended focus images. The publication of this paper was funded by the NMNH, Smithsonian Institution.

LITERATURE CITED

Baehr, M. (1992). Revision of the Pseudomorphinae of the Australian region. 1. The previous genera *Sphallomorpha* Westwood and *Silphomorpha* Westwood. Taxonomy, phylogeny, zoogeography. (Insecta, Coleoptera, Carabidae). – Spixiana, Supplement 18: 1-439.

Baehr, M. (1997). Revision of the Pseudomorphinae of the Australian region. 2. The genera *Pseudomorpha* Kirby, *Adelotopus* Hope, *Cainogenion* Notman, *Paussotropus* Waterhouse, and *Cryptocephalomorpha* Ritsema. Taxonomy, phylogeny, zoogeography. (Insecta, Coleoptera, Carabidae). – Spixiana, Supplement 23: 1-508.

Ball, G.E. (1972). Classification of the species of *Harpalus* subgenus *Glanodes* Casey (Carabidae: Coleoptera). – The Coleopterists Bulletin 26: 179-204.

Chaudoir, M. de (1852). Mémoire sur la familla des carabiques, 3e partie. – Bulletin de la Société Impériale des naturalistas de Moscou, 25(1): 3-104.

Dejean, P.F.M.A. (1829). Species général des coléoptères, de la collection de M. le Comte Dejean, 4, vii + 520 pp. – Méquignon-Marvis, Paris.

Dejean, P.F.M.A. (1831). Species général des coléoptères, de la collection de M. le Comte Dejean, 5, viii + 883 pp. – Méquignon-Marvis, Paris.

Erwin, T.L. (1981). A synopsis of the immature stages of Pseudomorphini (Coleoptera: Carabidae) with notes on tribal affinities and behavior in relation to life with ants. — The Coleopterists Bulletin, 35(1): 53-68.

Erwin, T.L. (2007). Xenaroswellanini, *Xenaroswelliana deltaquadrant*, New Tribe, New Genus, and New Species from the Cerrado of Estado de Goiás, Brasil (Insecta: Coleoptera: Carabidae). – Proceedings of the California Academy of Sciences, ser. 4, 58, No.27: 575–581.

Horn, G. (1867). Description of a new *Pseudomorpha* from California, with notes on the

Pseudomorphidae. – Transactions of the American Entomological Society 1: 151-154.

Kavanaugh, D.H. (1979). Studies on the Nebriini (Coleoptera: Carabidae), III. New Nearctic Nebria species and subspecies, nomenclatural notes, and lectotype designations. – Proceedings of the California Academy of Sciences 42: 87-133.

Kirby, W. (1825). A description of some insects which appear to exemplify Mr. William S. MacLeay's doctrine of affinity and analogy. – Transactions of the Linnean Society of London 14: 93-110.

Lenko, K. (1972). *Pseudomorpha laevissma*, un Carabideo mirmecofilo (Coleoptera: Carabidae). – Studia Entomologica 15: 439-444.

Liebherr, J. & Will, K. (1997b). A new *Pseudomorpha* Kirby (Coleoptera: Carabidae: Pseudomorphini) from the Sierra de los Tuxtlas, Veracruz, Mexico. – Folia Entomologica Mexicana 98 (1996): 53-58.

Maddison, D.R. & Maddison, W.P. (2000). MacClade 4: analysis of phylogeny and character evolution, version 4. – Sinauer Associates, Sunderland, MA.

Notman, H. (1925). A review of the beetle family Pseudomorphidae, and a suggestion for a rearrangement of the Adephaga, with descriptions of a new genus and new species. – Proceedings of the United States National Museum 67(14): 1-34.

Newman, E. (1842). List of Insects collected at Port Philipp, South Australia, by Edmund Thomas Higgins, Esq. – Entomologist 23: 361-369.

Ober, K.A. (2002). Phylogenetic relationships of the carabid subfamily Harpalinae (Coleoptera) based on molecular sequence data. – Molecular Phylogenetics and Evolution 24: 228–248.

Ogueta, E. (1967). Las espèces argentinas de la subfamilia Pseudomorhpinae G. Horn, 1881. – Acta Zoológica Lilloana 23: 217-232.

Swofford, D.L. (1999). PAUP* - Phylogenetic analysis using parsimony and other methods, version 4. Sinauer. – Sinauer Associates, Sunderland, MA.

Van Dyke, E.C. (1943). New species and subspecies of North American Carabidae. – Pan-Pacific Entomologist 19: 17-30.

Van Dyke, E.C. (1953c). New Coleoptera from western North America (Carabidae, Throsidae, Curculionidae). – Pan-Pacific Entomologist 29: 98-101.

Appendix I. Character matrix and states for generic representatives of Western Pseudomorphini and outgroup taxa.

	111111111122222222223333 123456789012345678901234567890123
Manumorpha	301010010000300011020100000030000
Notopseudomorpha	301011130000011201000122000200000
Orthogonius	100131030000011001101001111031000
Pseudomorpha	200000110000200010021000010222000
Samiriamorpha	300100120000100011010100010110111
Spallomorpha	000101030000030201101020101032100
Tuxtlamorpha	300140130000120010111110000220001
Xenaroswelliana	212001001111100011011000000030000
Yasunimorpha	300121120000011201001122111120000

(1) Eye: 0 = Hemispherical, 1 = Normal, round, 2 = Angulate, convex, 3 = Angulate, flat; (2) Eye: 0 = Glabrous, 1 = Setiferous; (3) Mouthparts: 0 = Visible in dorsal view, 1 = Not visible in dorsal view, 2 = Semi-hypognathus; (4) Mandibles: 0 = Margin setose, 1 = Margin asetose; (5) Labrum: 0 = Quadrisetose, 1 = Quadrisetose plus minor dorsal setae, 2 = Asetose, 3 = Hexisetose, 4 = Bisetose; (6) Clypeus: 0 = Suture entire, 1 = Suture interrupted; (7) Antenna: 0 = Long, extended beyond prosternal process apex, 1 = Short, not extended beyond prosternal process apex; (8) Antenna: 0 = Antennomeres more or less cylindrical, slender, 1 = Antennomeres compressed, robust, elongate, 2 = Antennomeres compressed, quadrate, 3 = Antennomeres compressed, slender, elongate; (9) Body form, neck: 0 = Broad, hidden, 1 = Constricted, visible; (10) Body form, pronotum: 0 = Rectangulate, 1 = Bow-form; (11) Body form, pronotal margins: 0 = Beaded, 1 = Explanate; (12) Body form, elytra: 0 = Hard, pigmented, 1 = Soft, depigmented; (13) Head vertex: 0 = Glabrous, 1 = With scattered fine setae, 2 = With out-of-line setal row, setae robust, plus scattered setae, 3 = With numerous robust setae; (14) Pronotum: 0 = Lateral margins setose, 1 = Lateral margins asetose, 2 = Lateral margins setose only at hind angle, 3 = Lateral margins setose at anterior and hind angle; (15) Pronotum: 0 = Disc setiferous, 1 = Disc glabrous; (16) Pronotum: 0 = Finely punctuate, 1 = Coarsely punctuate, 2 = Not punctuate; (17) Elytron: 0 = Glabrous, 1 = Multisetiferous; (18) Elytron: 0 = One or more rows of courser setiferous punctures, 1 = No rows of courser setiferous punctures; (19) Elytron: 0 = Shiny, 1 = Alutaceous; (20) Elytron: 0 = Smooth, 1 = Generally finely punctuate, 2 = Generally coarsely punctuate; (21) Elytron: 0 = Short, 1 = Long; (22) Elytron: 0 = Parallel-sided, 1 = Tapered apically; (23) Elytron: 0 = Setae erect, 1 = Setae decumbent, 2 = Setae absent; (24) Elytron: 0 = Intervals flat, 1 = Intervals convex, 2 = Intervals absent; (25) Elytron: 0 = Lateral margins setose, 1 = Lateral margins asetose; (26) Elytron: 0 = Intervals smooth, 1 = Intervals punctuate; (27) Abdomen: 0 = Pubescent, 1 = Glabrous; (28) Male abdomen, sterna 5 and 6: 0 = Absent, 1 = Comb setae short, 2 = Comb setae long; (29) Male abdomen, sterna 5 and 6: 0 = Comb setae divided wide, 1 = Comb setae divided close, 2 = Comb not divided, 3 = No comb; (30) Male parameres: 0 = Glabrous, 1 = Setiferous, 2 = Various; (31) Male parameres: 0 = Long, more than 1/3 phallus length, 1 = Short, less than 1/3 phallus length; (32) Male phallus: 0 = Slender, 1 = Robust; (33) Male phallus: 0 = Apically acute, 1 = Apically blunt

L. Penev, T. Erwin & T. Assmann (Eds) 2008
BACK TO THE ROOTS AND BACK TO THE FUTURE. TOWARDS A NEW SYNTHESIS AMONGST TAXONOMIC, ECOLOGICAL AND BIOGEOGRAPHICAL APPROACHES IN CARABIDOLOGY
Proceedings of the XIII European Carabidologists Meeting, Blagoevgrad, August 20-24, 2007, pp. 101-123.

Larval chaetotaxy in the genus *Rhysodes* Dalman, 1823 and the position of Rhysodidae within Adephaga (Coleoptera)

Kirill V. Makarov

Zoology & Ecology Department, Moscow State Pedagogical University, Kibalchicha Str. 6, Bld. 5, Moscow 129164, Russia. E-mail: kvmac@inbox.ru

SUMMARY

A thorough study of larval morphology in *Rhysodes sulcatus* (Fabricius, 1787) and *Rh. comes* (Lewis, 1888) (Rhysodidae) revealed a unique arrangement of the mouthparts and allowed for homologies of their elements to be elucidated. A special microporous structure was found in the labiomaxillar complex. An analysis of the larval characters resulted in rejection of all the hypotheses treating the Rhysodidae as a taxon subordinate to the Carabidae. The similarity of Rhysodidae larvae to those of the remaining Geadephaga was found to be insignificant. Instead they share some essential larval features with the suborder Archostemata. The hypothesis was put forth that Rhysodidae and Paussinae could have originated from the common ancestor within Archostemata.

Keywords: larvae, morphology, chaetotaxy, mouthparts, Coleoptera, Rhysodidae, Carabidae, relationship

INTRODUCTION

The beetle family Rhysodidae, encompassing about 350 species, is widespread in the tropical and temperate belts (Bell & Bell, 1978). Trophically, rhysodiids are suggested to be intimately associated with Myxomycetes in their ameoboid stage of development (Bell, 1998). The larvae live inside dead wood (Burakowski, 1975; Mamaev & Pototskaya, 1979) and regularly occur together with adults. Their life history and feeding remain poorly-known.

In the 19th century, this family was regarded as being close to Cucujidae and Colydidae (Reitter, 1882), but later its placement within Adephaga was justified (Ganglbauer, 1892; Peyerimhoff, 1903; Böving, 1929). Inside this suborder, Rhysodidae were considered as an independent, rather primitive family (Jeannel, 1941; Crowson, 1955; Ponomarenko, 1995), as the sister-group to (Beutel, 1990, 1992a), or a specialized derivative of, Carabidae (Beutel, 1992b, 1993, 1995), sometimes also as a member of Carabidae in the rank of a subfamily, tribe or even subtribe (Bell & Bell, 1962; Erwin & Sims, 1984; Erwin, 1985; Bell, 1998).

For the first time, larvae of Rhysodidae were briefly described in the early 20th century, without precise species identification (Peyerimhoff, 1903). Later, some larval characters of *Clinidium sculptile* were used by Böving (1929) in discussing the taxonomic position of the family Rhysodidae. However, most of larval morphological evidence was published rather recently (Grandi, 1956, 1972; Burakowski, 1975; Vanin & Costa, 1978; Mamaev & Pototskaya, 1979; Costa et al., 1988). These papers included rather detailed accounts of the external morphology of larvae and, partly, of their anatomy (Beutel, 1992b). However, no special studies on larval chaetotaxy in rhysodids have hitherto been conducted. My research on *Rhysodes sulcatus* (Fabricius, 1787) and *Rh. comes* (Lewis, 1888) larvae allows for a detailed description of larval chaetotaxy in the genus to be made, also suggesting a new view of Rhysodidae relationships.

MATERIAL AND METHODS

A total of 58 larvae of all stages belonging to 2 species of *Rhysodes* were studied.

Rhysodes sulcatus (Fabricius, 1787: 165)

Poland, Białowieża Primeval Forest (National Park), in yellowish-rotting, damp sapwood of *Populus tremula* L., 2.VII.1968, leg. B. Burakowski (MIZ 80475-MIZ 80478) – 3 L3, 1 pupa (*Al*); Northern Caucasus, Krasnodar Territory, Guzeripl, 17.VI.1988, leg. N. Nikitsky – 1 L3 (*Eu*); Northwestern Caucasus, Adygeya, 4 km E of Filimonov's Mt., in *Abies* wood, 25.VI.2007, leg. A. Zaitsev – 3 L1 (2 – *Al*, 1 – *Eu*), 8 L2 (7 – *Al*, 1 – *Eu*).

Rhysodes comes (Lewis, 1888: 79)

Primorye, Southern Sikhote-Alin Mts, Lazovsky Nature Reserve, cordon Korpad', floodplane of Kedrovaya River, in rotten wood of *Betula* sp., 9.VIII.2007, leg. A. Zaitsev & K. Makarov – 1L1, 7L3, 1 pupa (*Al*); same location, in *Ulmus* sp. wood, 10.VIII.2007, leg. A. Zaitsev & K. Makarov – 2L1 (1 – *Al*, 1 – *Eu*), 12L2 (11 – *Al*, 1 – *Eu*), 9L3 (8 – *Al*, 1 – *Eu*); same district, cordon Prosyolochnaya, in rotten wood of *Alnus* sp., 17.VIII.2007, leg. A. Zaitsev & K. Makarov – 6L3, 2 pupa (*Al*); same location, in *Alnus* wood, 20.VIII.2007, leg. A. Zaitsev & K. Makarov – 5L3, (4 – *Al*, 1 – *Eu*), 1L2 (*Al*).

Most specimens are preserved in 70% alcohol (*Al*), and deposited in the collection of the Department of Zoology and Ecology of the Moscow Pedagogical University and in the Museum and Institute of Zoology of the Polish Academy of Sciences. Some specimens were mounted in Euparal microscopic slides (*Eu*) for chaetome investigations.

The external larval morphology of Rhysodidae, including that of both studied species of *Rhysodes*, have been described many times (Grandi, 1972; Burakowski, 1975; Vanin & Costa,

1978; Mamaev & Pototskaya, 1979; Costa et al., 1988). This is why below I mainly give chaetotaxy characteristics followed by the necessary comments. The sensillar nomenclature developed by Bousquet & Goulet (1984) was used, with minor modifications (Makarov, 1996).

RESULTS

Genus *Rhysodes* Dalman, 1823: 93

LARVAL CHAETOTAXY. Head capsule (Figs 1-10) with a reduced set of sensilla. Frontale with neither an antediscal sensillar complex (FR_4, FR_5 and $FR_{c,e}$) nor sensilla of anterior margin of paraclypeus FR_{8-9}, FR_g. Sensilla FR_1 and FR_a, commonly associated

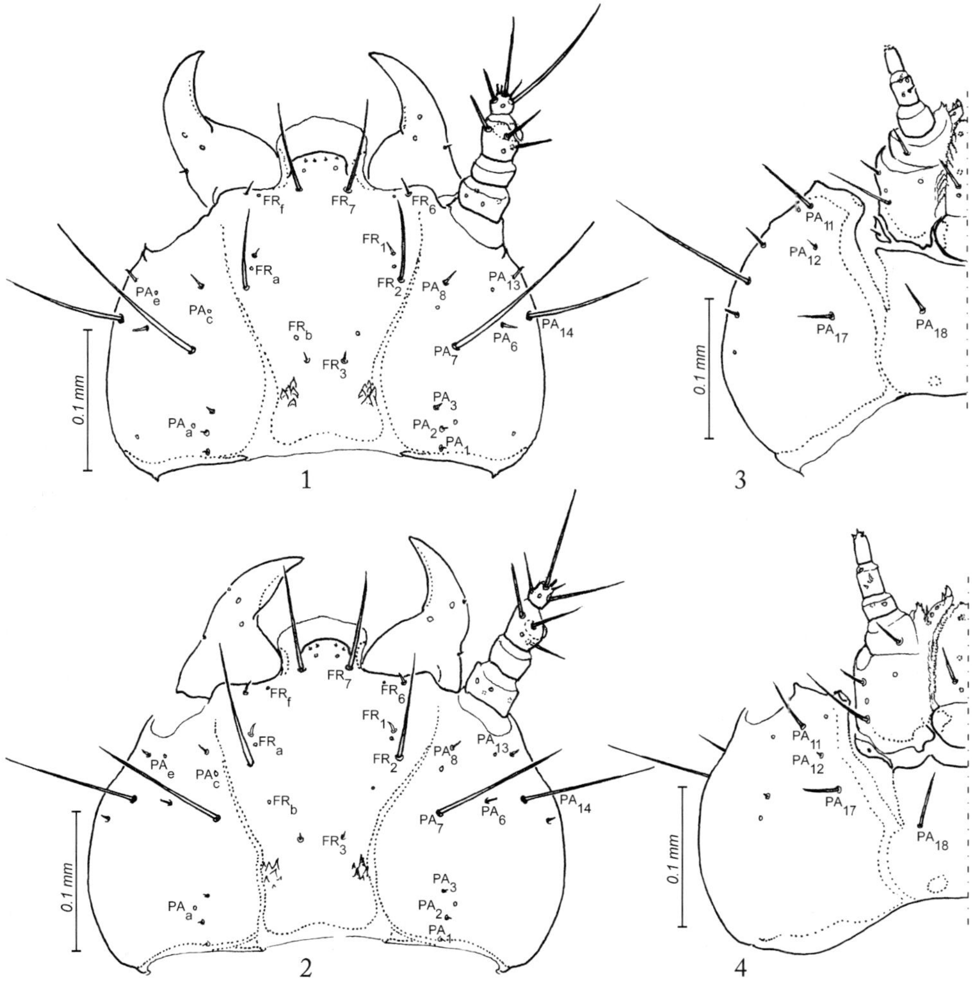

Figs 1-4. *Rhysodes* spp., first instar larvae: 1, 3 – *Rh. sulcatus*; 2, 4 – *Rh. comes*; 1-2 – head, dorsal view, left antenna and labiomaxillar complex not show; 3-4 – right half of head, ventral view.

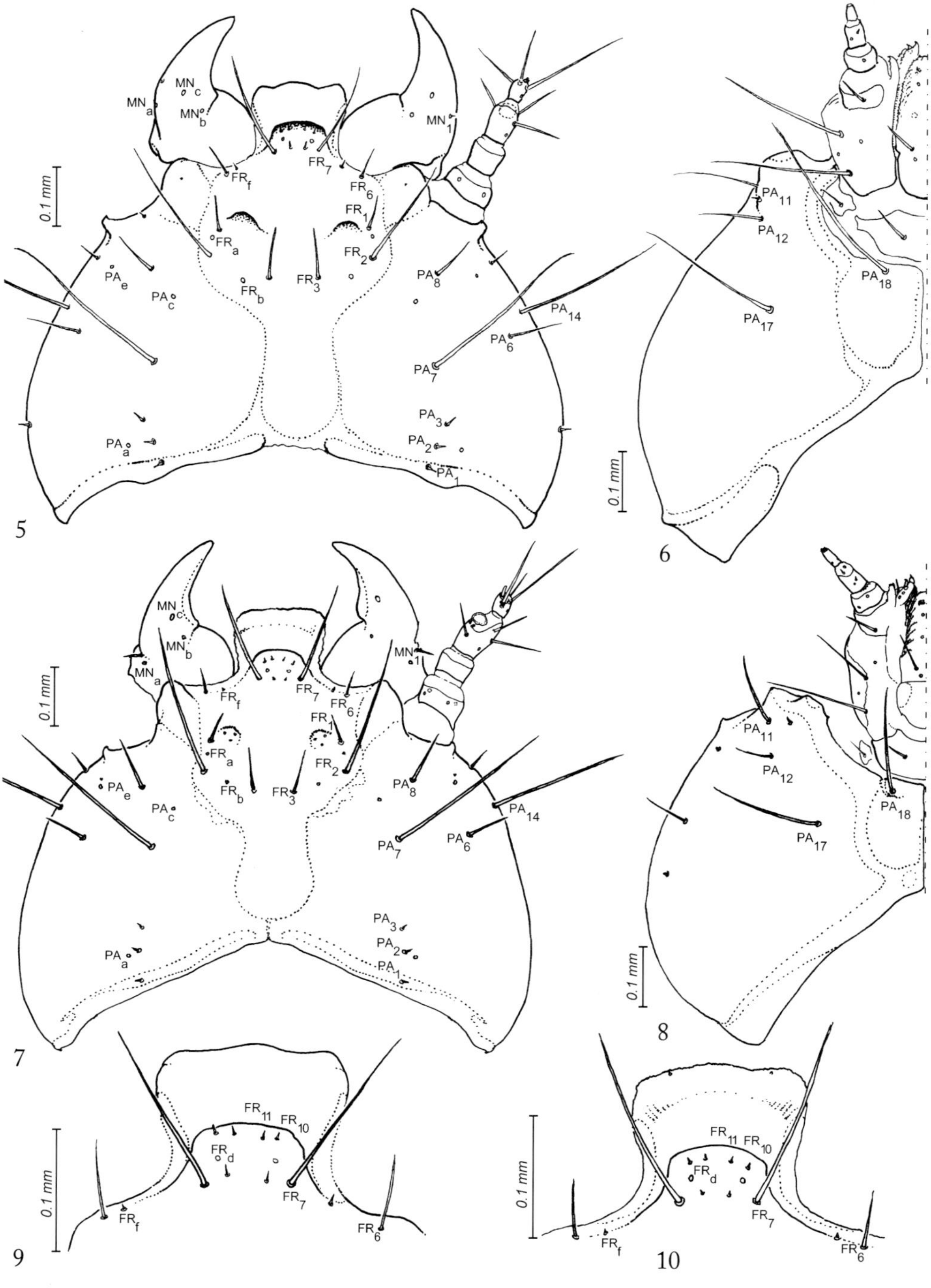

Figs 5-10. *Rhysodes* spp., third instar larvae: 5-6, 9 – *Rh. sulcatus*; 7-8, 10 – *Rh. comes*; 5, 7, – head, dorsal view, left antenna and labiomaxillar complex not show; 6, 8 – right half of head, ventral view; 9-10 – nasale, dorsal view.

with lateral sclerite corners, displaced mediad and positioned anterior to macrosetae FR_2. Location of setae FR_7 also unusual, they being close together so that the distance between their insertions is equal to nasale width. Both FR_3 and FR_b thereby in usual positions, posterior to line FR_2-FR_2. Macrosetae FR_2 and FR_7 2-7 times longer than other setae (Figs 1, 3, 5, 7). Nasale setae FR_{10}-FR_{11} tiny, visible only in cleared specimens (Figs 9-10). Some setae of parietale fully reduced, i.e. dorsal PA_4, PA_5, $PA_ç$ and PA_{10}, and ventral PA_{15} and PA_{16}. Gular area with one pair of setae in anterior part (versus two pairs of setae, PA_{18} and PA_{19}, typical of carabids), their identification as PA_{18} being tentative because of lack of additional markers in this area (Figs 2, 4, 6, 8). Many sensilla absent, including PA_b, PA_d and most of ventrolateral non-trichoid sensilla.

Antenna of typical structure, its chaetotaxy without essential distinctions from basal carabid pattern (Figs 11-13). It is significant that antennomere 3 is with a flattened sensorium in ventral position; apical and subapical sensillar complexes rather poor, latter including only 1-2 basiconical sensilla.

Mandible without penicillus, seta MN_2 absent, seta MN_1 short, in ventrolateral position; sensilla MN_a and MN_b present (Figs 1, 3, 5 ,7).

Maxilla considerably membranous, forming together with labium a functionally entire labiomaxillar complex (Figs 14-15). Its chaetotaxy original: internal stipes margin without gMX so very typical of carabids, serving as a filtration organ. It is replaced by a system of oblique folds covered with rows of cuticular spinules. Coupled with complementary folds of labium, they make a microporous structure. Besides this, galea and lacinia fused with top of stipes and internal margin of palpifer forming a distal extension of microporous structure. Though galea and lacinia cannot be distinguished, sensilla MX_6, $MX_{7,9}$, and MX_d easily recognizable, as well as apical conical sensilla of galeomere 2 (Fig. 16). In spite of such essential transformations, sensillar set of external surface of stipes and maxillar palp almost the same as in carabids, except for reduction of MX_4, MX_5 and total absence of digitiform sensilla in subapical sensorial complex.

Labium of *Rhysodes* larvae representing a merged subcylindrical structure with lateral surfaces tightly adjacent to maxillar stipites and covered with cuticular spinules (Figs 14-15). Labial chaetome reduced strongly enough, but all of its elements corresponding to generalized pattern. Like on maxilla, a number of lateral and apical setae ($LA_{3,4,5,6,7}$) replaced by a microporous formation. Digitiform sensilla totally absent, basal seta of mentum LA_1 misplaced. Sensillar composition of labial palps thereby without modifications, i.e. LA_a, as well as LA_b, LA_c and a ring-shaped complex of conical sensilla usually located on palpomeres, all clearly recognizable (Fig. 17); latter complex forming a distal sensory area.

Thoracal segments with a reduced generalized sensillar set (Figs 18, 26), the number of setae only on tergites of older instar larvae being increased (Figs 20, 28). Pronotum with distinct setal complexes situated medially near anterior and posterior sclerite margins. Some stable combination of different types of sensilla characteristic of basal chaetotaxic pattern recognizable: PR_2-PR_a, PR_3-PR_b near anterior margin, PR_{13}-PR_l, PR_{12}-PR_j, PR_{11}-PR_k near posterior one. In contrast, chaetome of lateral sclerite part, especially in anterior corner area, remarkably modified, with questioned homologies of sensilla. Obviously, generalized chaetotaxic set represented only by macroseta PR_6 and mircosetae PR_5, PR_7. Pronotal disc without medial seta PR_{14}, but with a lateral complex PR_8-PR_f. Pronotal macrosetae (PR_6, PR_{11}, PR_{12}) 7-10 times longer than microsetae.

Prosternite with an ordinary setal set of PS_1 and PS_2, epimeron without seta EM_1, episternum without distal setae ES_3, ES_4, most of episternal setae short, only ES_1 large (Figs 22, 24, 30, 32).

Chaetome of meso- and metathorax modified in a similar way (Figs 19, 21, 27, 29). Microsetal groups of pretergite (ME_3, ME_4, ME_5, ME_6, ME_7), sensilla associated both with anterior tergal keel (ME_1-ME_a, ME_2) and posterior sclerite margin (ME_{14}-

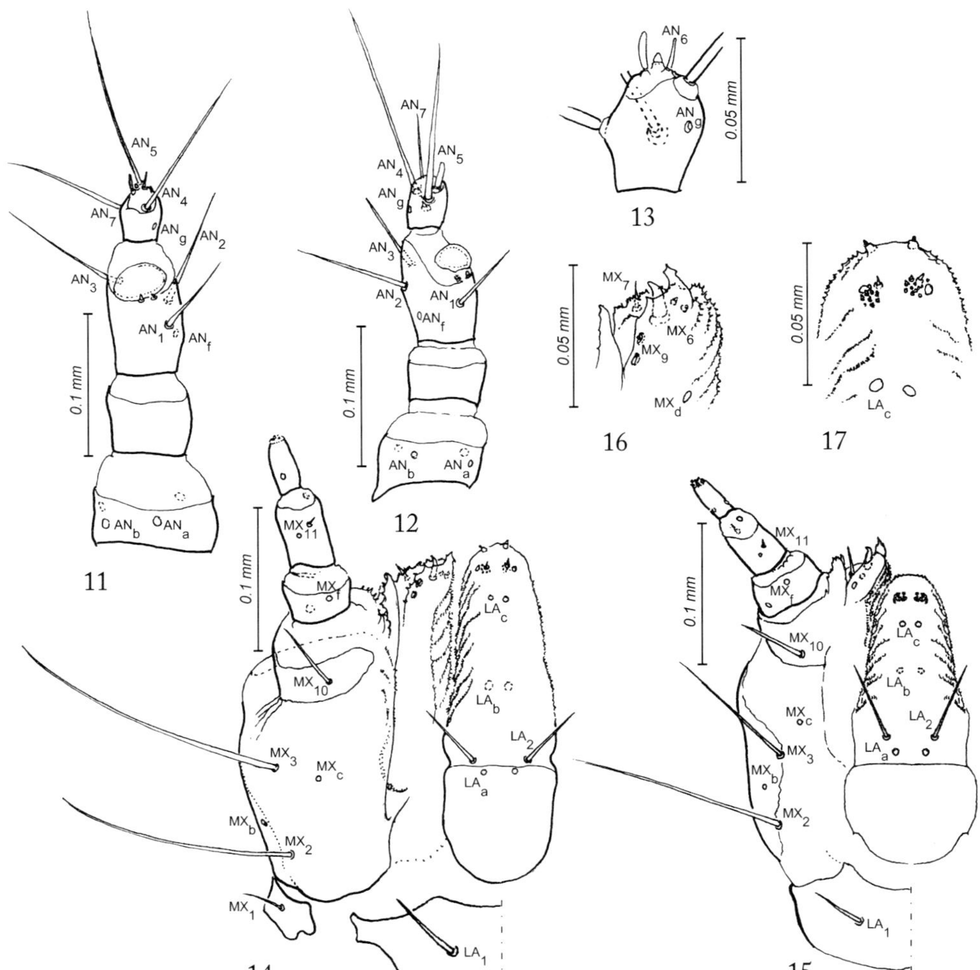

Figs 11-17. *Rhysodes* spp., third instar larvae: 11, 13, 14, 16-17 – *Rh. sulcatus*; 12, 15 – *Rh. comes*; 11-12 – left antenna, dorsal view; 13 – antennomere 4, dorsal view; 14-15 – labiomaxillar complex, ventral view; 16 – apical part of stipes, ventral view; 17 – apical part of labium, ventral view.

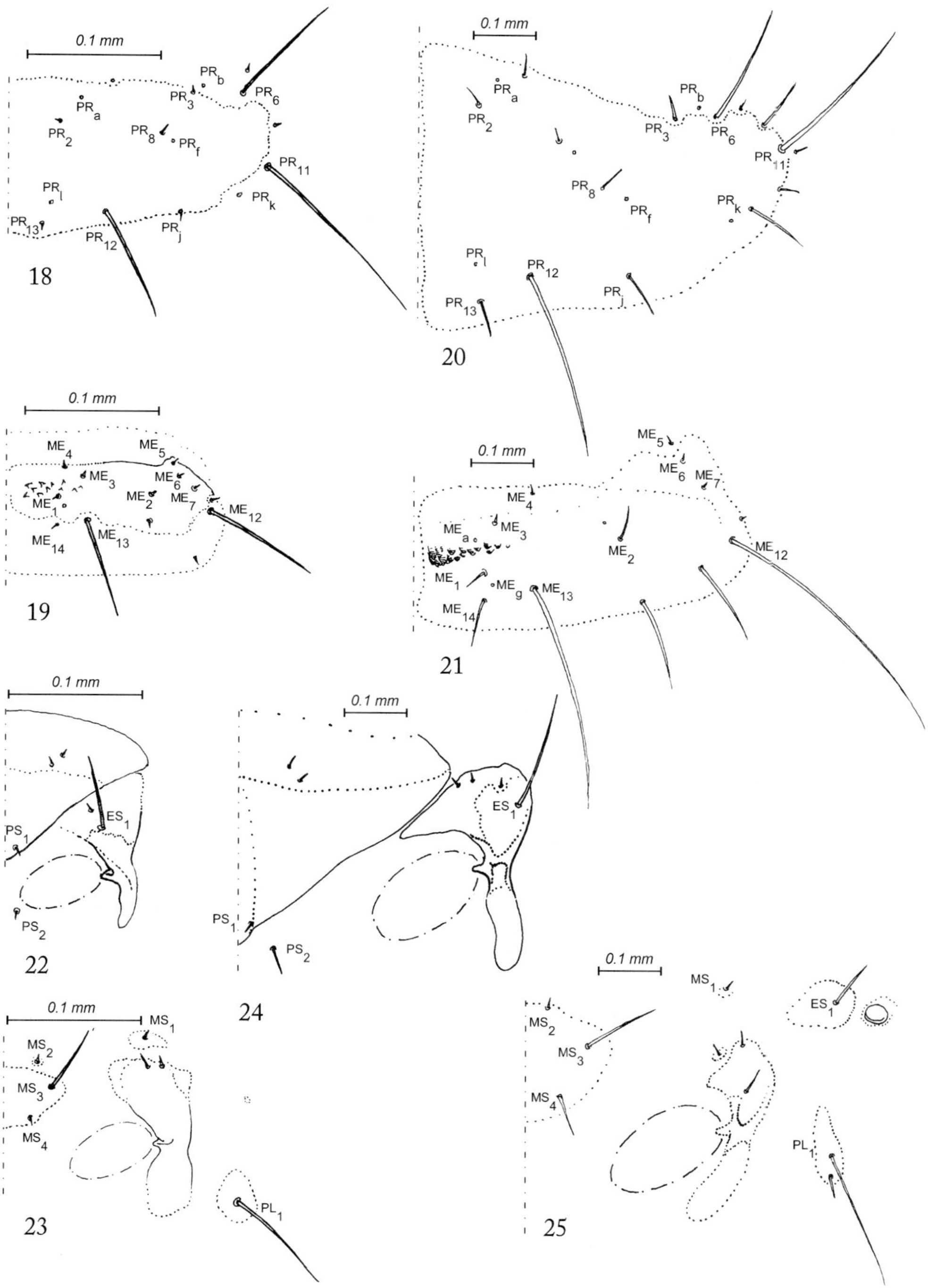

Figs 18-25. *Rhysodes sulcatus*: 18-19, 22-23 – first instar larvae; 20-21, 24-25 – third instar larvae; 18, 20 – left half of pronotum, dorsal view; 19, 21 – left half of mesonotum, dorsal view; 22, 24 – left half of prosternum, ventral view; 23, 25 – left half of mesosternum, ventral view.

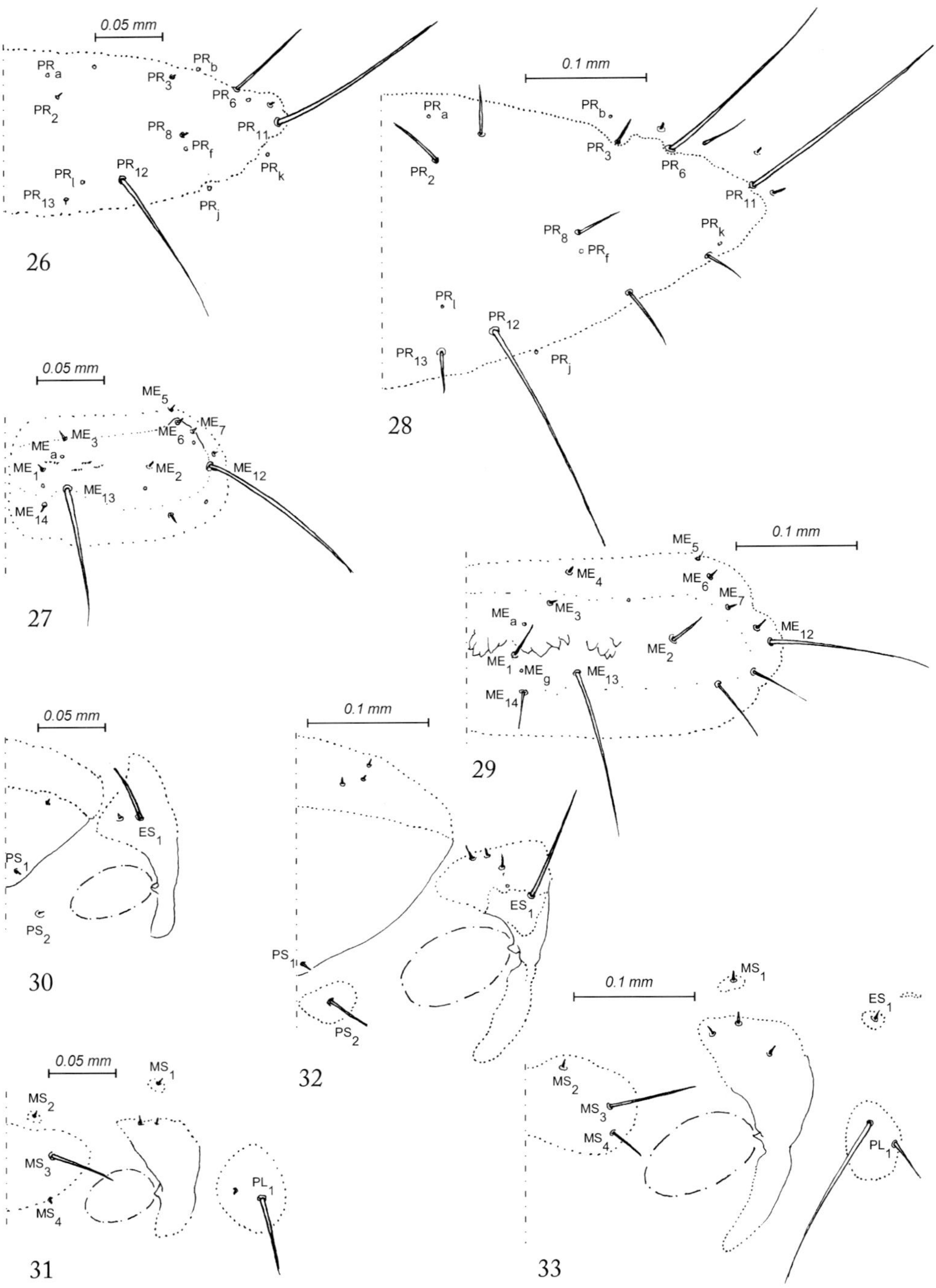

Figs 26-33. *Rhysodes comes*: 26-27, 30-31 – first instar larvae; 28-29, 32-33 – third instar larvae; 26, 28 – left half of pronotum, dorsal view; 27, 29 – left half of mesonotum, dorsal view; 30, 32 – left half of prosternum, ventral view; 31, 33 – left half of mesosternum, ventral view.

ME_9, ME_{13}, ME_{12}) retained from generalized set. Apparently, only ME_8 retained from lateral group. In contrast, chaetome of sternites and pleurites (Figs 23, 25, 31, 33) close to generalized pattern and differing by a few setae reduced, namely EM_1 and MS_4; most of setae tiny, only PL_1 developed as a macroseta.

Leg of structure typical of Adephaga, set of chaetotaxic elements similar to generalized one (Figs 34-37); remarkable differences lying only in trochanter chaetome: trichoid

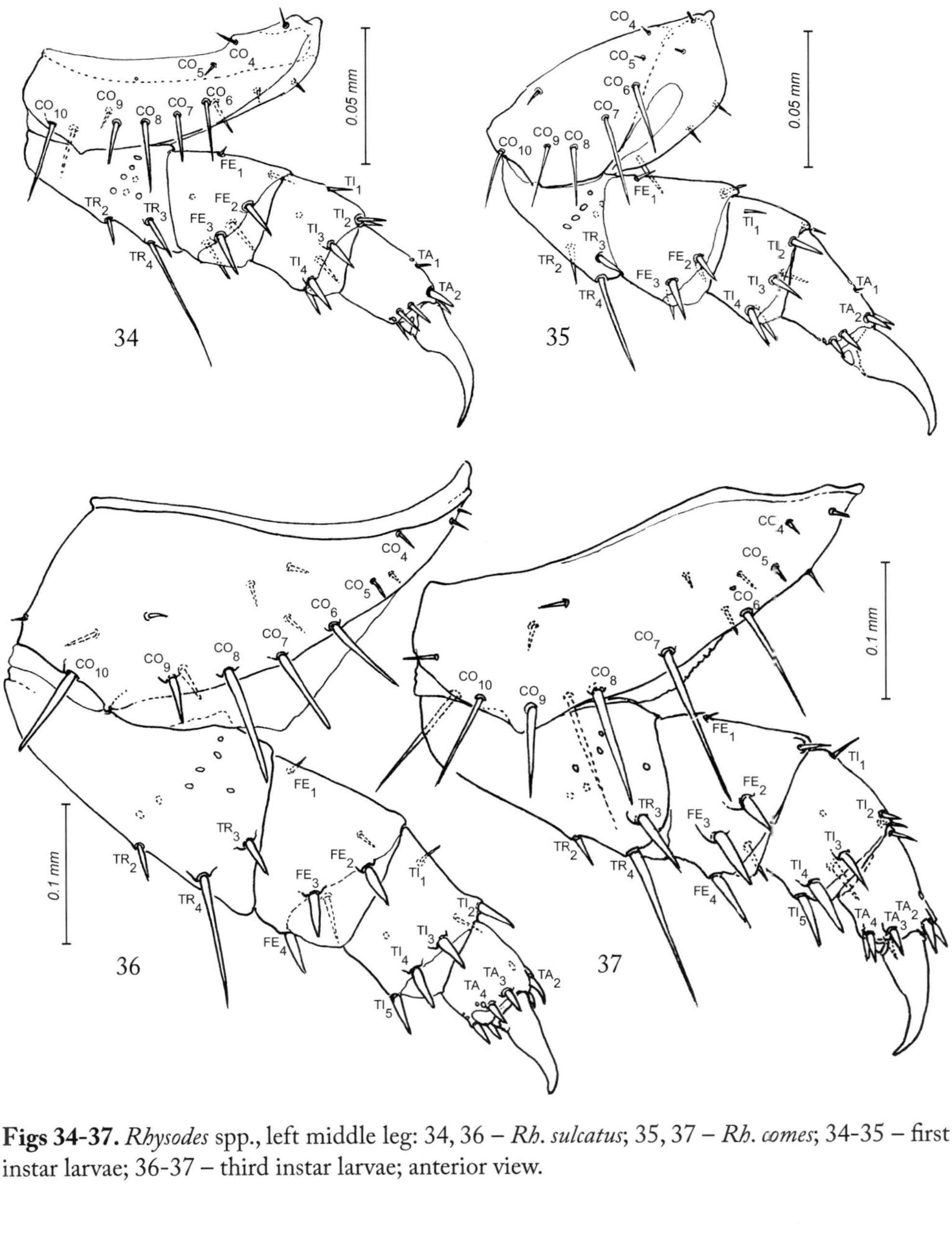

Figs 34-37. *Rhysodes* spp., left middle leg: 34, 36 – *Rh. sulcatus*; 35, 37 – *Rh. comes*; 34-35 – first instar larvae; 36-37 – third instar larvae; anterior view.

sensilla TR_1, TR_8 and spiniform seta TR_6 lacking. Besides this, location of macrosetae on anterior surface of coxa (CO_6, CO_7, CO_8 and CO_9) original, they being clustered and forming a row near external margin; CO_{17} missing. Pretarsus simplified: a single claw present, setae UN_1, UN_2 lacking.

Structure and chaetotaxy of sclerites of different abdominal segments notably distinct. Tergites 1-7 with groups of cuticular tubercles and spines, forming rather spacious fields in first instar larvae (Figs 38-39) and building compact transverse ridges in older instar larvae. Chaetotaxy of these tergites, unlike generalized pattern, possessing some peculiarities: (i) anterior and posterior rows of setae close together while lateral seta TE_6 sometimes placed almost at posterior margin of tergal disc; (ii) setae of anterior row (TE_1, TE_6) generally much shorter than those of posterior row, including macrosetae TE_{10} and TE_9; (iii) like on thoracic segments, sensilla of lateral complex, except TE_9, visibly reduced or absent. Older instar larvae often with 1-2 additional setae in TE_{10}-TE_9 area; the number and position of additional setae varied (Figs 42-43, 46-49). Thus, segment 7 usually with one additional seta, whereas segment 8 sometimes without additional setae.

Tergite 8 with a few cuticular spinules not arranged in transverse rows, tergite 9 without spinules. Setal composition of tergite 8 similar to generalized pattern (Fig. 49), differing mainly in diminished setae TE_6 and TE_7.

Tergite of penultimate segment lacking urogomphi, its chaetome particular and distinct from that of preceding segments (Figs 50-53). Posterior margin of penultimate tergite with 2-4 macrosetae, disc with 1-2 macrosetae. The absence here of urogomphi leads us to the use of the same sensillar nomenclature as that developed for other abdominal tergites. Thus, TE_1, TE_2 and TE_a can be identified near anterior margin of segment, TE_{10} and TE_{11} near its posterior margin. Lateral group, by analogy with other segments, considered as including TE_7 (?TE_8) and TE_9.

Sternites and pleurites 1-9 similar in structure and chaetotaxy. Larvae of both examined species of *Rhysodes* without sternella interior; as a result, setae ST_3 and ST_4 lacking (Figs 40-41, 44-45). In all other respects, the set of sensilla and setae not differing from generalized pattern. Older instar larvae bearing additional setae on hypopleurite, usually one microseta in anterior half and 1-2 macrosetae in posterior half (Figs 44-45).

Segment 10 (pygidium) short (Fig. 54), with a simplified chaetome: besides basal PY_1, very stable in Adephaga, only 3 pairs of setae forming an apical crown (PY_7, PY_3, PY_4) present, as well as dorsolateral placoid sensilla PY_a, PY_e and PY_d.

It is noteworthy that the larval chaetomes of both examined species appear to be strongly variable. In most cases, the topology, infrequently also the composition, of the sensilla on the left and right sides of sclerites are different, this greatly complicating the elucidation of a chaetotaxy pattern.

BETWEEN-INSTAR DIFFERENCES. The first instar larvae differ in the presence of egg busters represented by groups of cuticular spinules (Figs 1, 3), by altered arrangements of spines on the abdominal tergites (not forming distinct ridges, but covering almost the entire sclerite disc) and in chaetotaxic details. Length differences of micro- and macrosetae are more pronounced; setae FR_3 are small, shifted into the

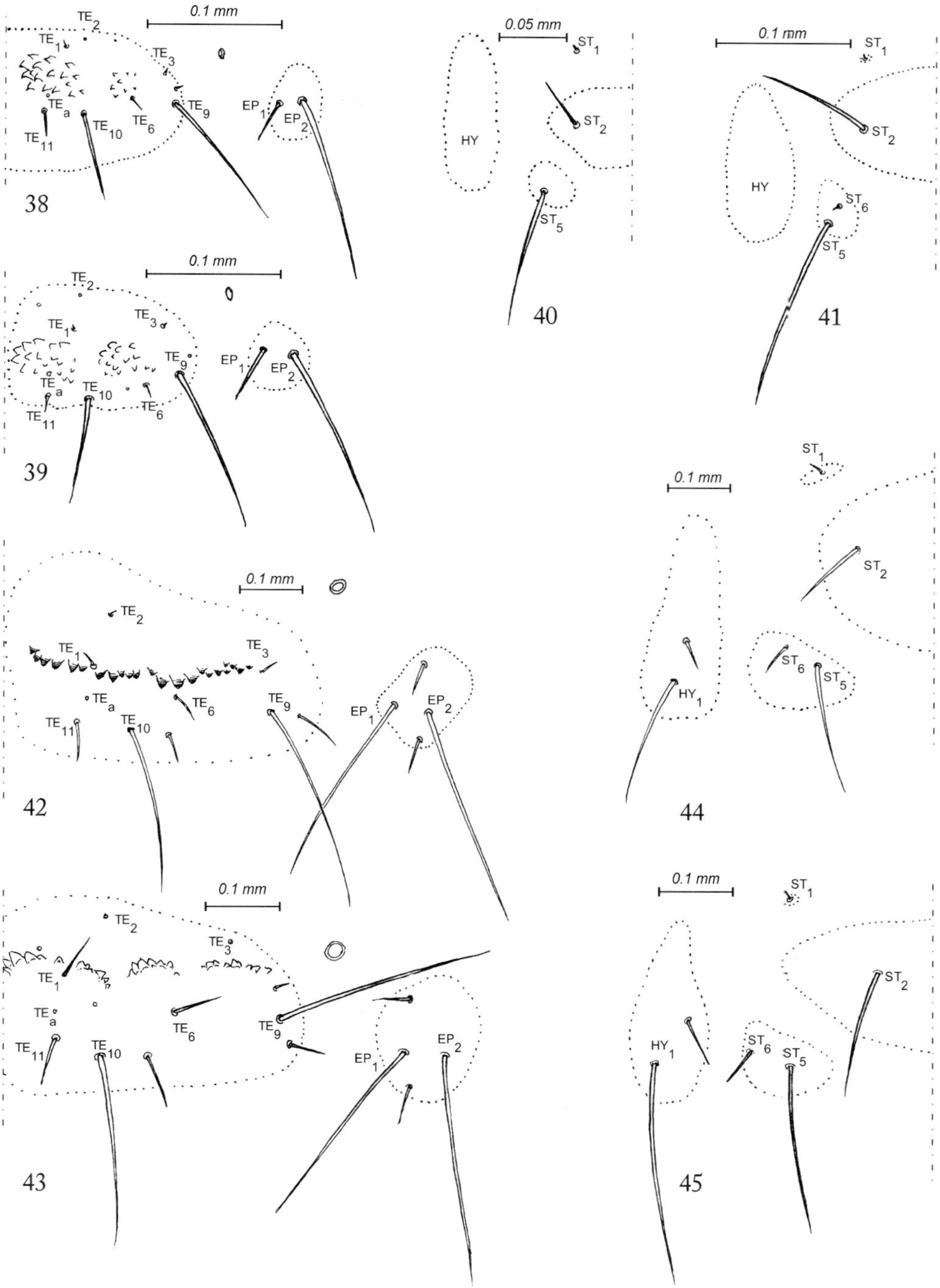

Figs 38-45. *Rhysodes* spp., right half of 4th abdominal segment: 38, 40, 42, 44 – *Rh. sulcatus*; 39, 41, 43, 44 – *Rh. comes*; 38-41 – first instar larvae; 42-45 – third instar larvae; 38-39; 42-43 – tergite and epipleurite, dorsal view; 40-41, 44-45 – sternite and hypopleurite, ventral view.

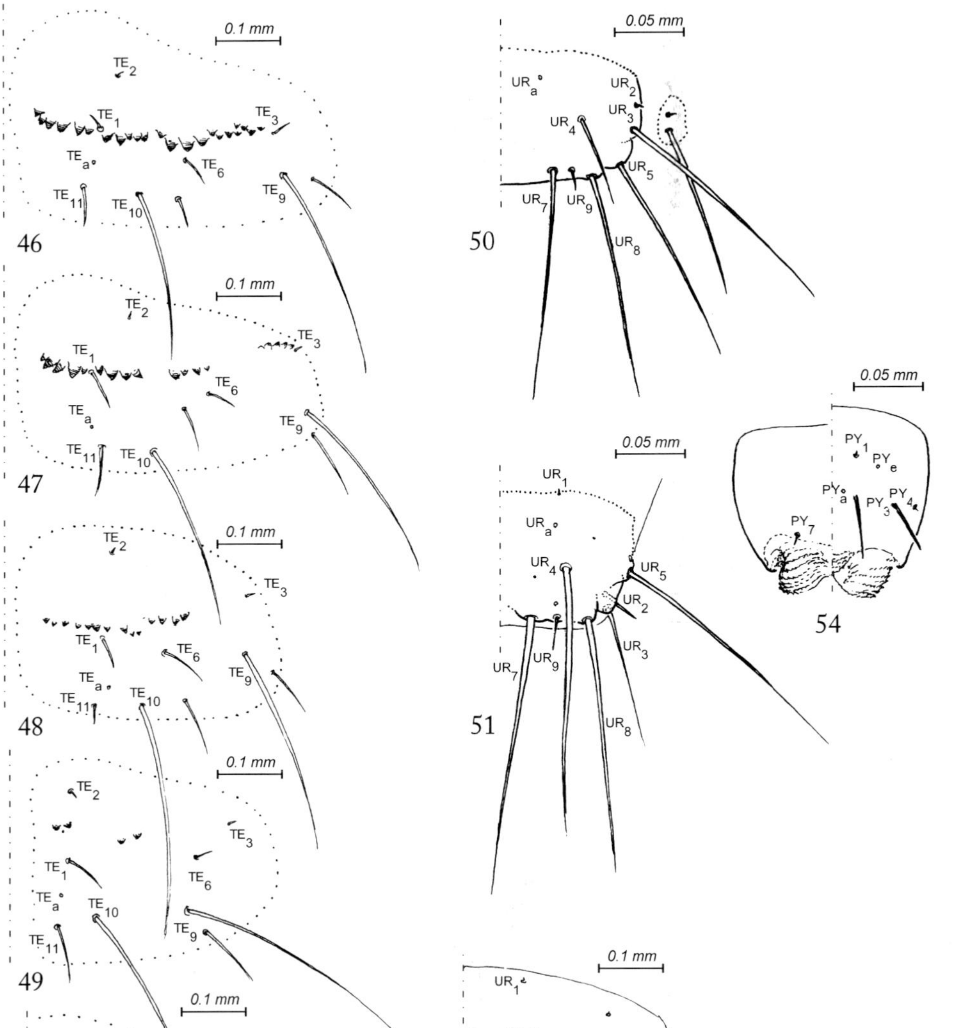

Figs 46-54. *Rhysodes* spp., right half of abdominal segments: 46-49, 50, 53-54 – *Rh. sulcatus*; 51-52 – *Rh. comes*; 46-49, 52-53 – third instar larvae; 50-51, 54 – first instar larvae; 46-49 – tergite of 5th-8th segments respectively, dorsal view; 50-51, 52-53 – tergite of 9th segment, dorsal view; 54 – 10th segment, left – dorsal view, right – ventral view.

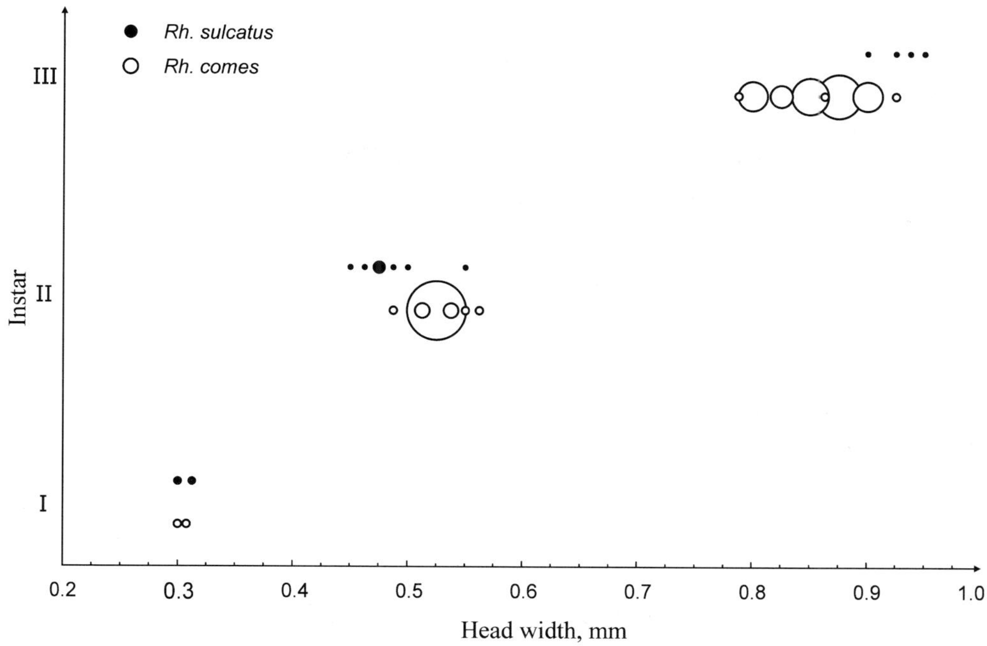

Fig. 55. Head width in different instars of *Rhysodes* spp. larvae. Circle size corresponds to the number of specimens.

basal part of the frontal sclerites (Figs 1, 3); additional setae at the posterior margin of tergites and on the hypopleurites are absent (Figs 38-41); seta TR_4 (Figs 34-35) is thin and long (spiniform in older instar larvae).

Older instar larvae are hardly different, variation concerning the number of additional setae in groups PR_{12}-PR_{11} (Figs 20, 28) and TE_{10}-TE_9 (Figs 42-43). To securely distinguish the older instars, head capsule measurements are suitable (Fig. 55). In both species, head width of first instar larvae is 0.30-0.312 mm (M=0.305, SD=0.006, SE=0.003; n=4), second instar larvae – 0.45-0.56 mm (M=0.514, SD=0.029, SE=0.006, n=22), third instar larvae – 0.79-0.95 mm (M=0.864, SD=0.044, SE=0.008, n=29).

BETWEEN-SPECIES DIFFERENCES. The first instar larvae of *Rh. sulcatus* and *Rh. comes* are very similar, differing only in some chaetotaxic details: in *Rh. sulcatus,* the poststernites are usually with a retained microseta ST_6, the disc of abdominal tergite 9 without macroseta UR_1, as a rule.

Old instar larvae are distinguishable by the size of tubercles on the frontal sclerite (in *Rh. sulcatus*, they are much larger than in *Rh. comes*), in structure of the cuticular ridges on abdominal tergites (usually entire in *Rh. sulcatus*, but subdivided into 2-3 tubercles in *Rh. comes*) and in the position of seta PA_7 (equidistant from PA_3 and PA_8 in *Rh. sulcatus*, close to PA_8 in *Rh. comes*).

DISCUSSION

It is important to notice that larval morphology of other genera of Rhysodidae, such as *Omoglymmius* Ganglbauer, 1892 and *Rhysodiastes* Fairmaire, 1895, known from literature (Grandi, 1972; Vanin & Costa, 1978; Costa et al., 1988), often in details coincides with the data on *Rhysodes* larvae. It makes it possible to extrapolate further deductions on an entire family.

Special features of larval morphology and chaetotaxy in Rhysodidae

Their chaetome is distinctly of carabid type and the considerable part of setae can be surely homologized with basic elements of Carabidae' chaetome (Bousquet & Goulet, 1984).

Generally, the chaetotaxy of *Rhysodes* larvae is characterized by moderate oligochaetosis. Some chaetotaxic peculiarities are obviously associated with exoskeleton transformations. Thus, the partly reduced frontal sensilla and the unusual location of FR_7, FR_1 and FR_a are probably accounted for by the diminished area of the anterior part of the frontale and by the vanished paraclypeus lobes (Figs 1, 3, 5, 7, 9-10). The absence of eyes and the enlarged sigilla of the mandibular adductor (m. craniomandibularis internus) result in disproportions of the parietal sclerites, this being reflected by the reduction of some setae and sensilla (PA_{4-5}, PA_{9-10}, PA_{15}-PA_{16}; PA_b, PA_d and others). Therefore, the basal part of the head capsule became almost bare, with the exception of the complex $PA_{1,2,3,a}$ which is highly stable not only in carabid larvae, but also in larvae of other coleopteran families. In general, no such head chaetome modifications have hitherto been known among ground-beetle larvae.

The microporous structure formed in the place of contact of the stipes and labium probably serves not as a filter, but a capillary sponge facilitating liquid food consumption. Its functional analogue is known in mycetophagous larvae of the genus *Sepedophilus* Gistel, 1856, Staphylinidae (Leschen & Beutel, 2001).

In previous publications concerning the larval morphology of Rhysodidae (Grandi, 1972; Burakowski, 1975; Vanin & Costa, 1978; Mamaev & Pototskaya, 1979; Costa et al., 1988), the structure of the maxilla and labium was repeatedly discussed. It was always suggested thereby that the lacinia, galea, ligula and labial palp in rhysodid larvae were absent, with the exception of small rudiments in *Omoglymmius* (Grandi, 1972). The chaetotaxy, in particular the topology of placoid and conical sensilla (Figs 16-17), allows to conclude that, in this case, I deal not with simple reductions but with fusion of all these structures, resulting in the formation of a sucking labiomaxillar complex; the degree of this fusion varies in different genera. The labium of the *Omoglymmius* larva seems to be the least modified: the distal segments of the labial palp remain separate while proximal ones fused, but their articulation with the mentum is mobile. Earlier, these fused proximal segments were mistakenly interpreted as a prementum (Mamayev & Pototskaya, 1979).

The peculiar structure and chaetotaxy of the leg seem to mainly be accounted for by the beetles' xylobiotic life-style. Among these features, there are the shortening of distal parts, the retention of one claw, the reduction of sensilla (TR_1, UN_1, UN_2), and the unification of macrosetae. Most of the macrosetae become enlarged, spiniform except for the typically trichoid TR_4 in the first instar larva.

The investigation of the chaetotaxy of abdominal tergites brings two interesting conclusions. Firstly, the groups or rows of cuticular spinules on tergites highly characteristic of the family are almost always located between sensilla TE_1-TE_a in the medial area and between sensilla TE_3-TE_6 in the lateral area (Figs 42-48). This proves their homology to the transverse keel separating the pretergite from the tergite in carabid larvae. Secondly, the missing urogomphi allow to ascertain the serial homology of the setae of abdominal tergite 9 to the setae on the tergites of preceding segments and to suggest correspondence between the setae of tergite 9 and those on the urogomphi (Figs 42-49 *vs* 50-53). Thus, sensilla UR_a corresponds to TE_a, seta TE_2 corresponds to UR_1, TE_1 – UR_4, TE_3 – UR_2, TE_9 – UR_3, TE_6 – UR_5, TE_{10} – UR_8, and TE_{11} – UR_7.

The composition and location of the sternites and pleurites in *Rhysodes* larvae have not been discussed yet, apparently due to their very weak sclerotization. The study of their chaetotaxy reveals full reduction of the internal poststernites, including their setae. Larvae of older instars are also without setae in this area, despite the development of some additional setae in the sternopleural region.

Thus, I can homologize most of the chaetome elements of *Rhysodes* larvae typical of Adephaga. A number of features of rhysodid macromorphology (the formation of a labiomaxillar complex accompanied by fusion of the labial palps, the modification of the tergal keel into supporting structures), as well as chaetotaxic traits (the unusual chaetome of the anterior margin of the frontale, the reduction of numerous parietal setae; the incomplete set of trochanteral setae) are unique in this family within the suborder.

Taxonomic position of Rhysodidae

At present, the placement of the Rhysodidae within the group Geadephaga of the suborder Adephaga is proved by numerous morphological data drawn from both the adults (Forbes, 1926; Baehr, 1979; Beutel, 1995) and larvae (Böving, Craighead, 1930; Beutel, 1992b, 1993), being currently regarded as doubtless. However, the understood taxonomic rank varies greatly, ranging from an independent family down to a subtribe within Carabidae (Böving, 1929; Crowson, 1955; Ponomarenko, 1995; Jeannel, 1941; Beutel, 1990, 1992a, b, 1993, 1995; Erwin & Sims, 1984; Erwin, 1985; Bell & Bell, 1962; Bell, 1998). Different aspects of this problem are discussed below, based on new information on larval chaetotaxy.

Subtribe Rhysodina within the tribe Scaritini?

Integration of Rhysodidae and Carabidae was first suggested by Bell & Bell (1962), based mainly on adult external morphological features (structure of meso- and metacoxae, metendosternite, and fore tibia) and related to locomotion in a dense substrate. In this case, the

Rhysodini and the Scaritini were regarded as sister-groups, whose formation was accounted for by specializations to different environments. Later, this viewpoint was supported by the knowledge of the structure of the repugnatorial glands (Forsyth, 1972). Recently, further development of this approach (Bell, 1998) resulted in a still greater decrease in rhysodine rank, namely, the Rhysodina was accepted as a specialized subtribe of Scaritini. This was due to more data accumulated (Adis, 1981; Dostal, 1993) as regards the morphology of some rare and highly specialized tropical Salcediina (now regarded as a subtribe of Clivinini; see Balkenohl, 2001) which show some features determining their habitual similarity to Rhysodidae.

In such a situation, it is important to evaluate the features shared by larvae of Rhysodidae and Scaritini s.l. (including Clivinini and Dyschiriini). The caraboid larvae of Scaritini show not a single characteristic of Rhysodidae in structure of the mouthparts, as well as in the chaetotaxy of the head and body tergites. The single feature in common is the presence of one claw on the pretarsus in some scaritins (Clivinini and Dyschiriini). But this cannot be interpretated as the proof of a relationship since a reduced number of claws is known in many not so closely related carabid groups (Broscini, Trechini, Bembidiini and Pogonini, Orthogoniini, Brachinini). It is noteworthy that, among carabid larvae, there are numerous cases of specialization to moving through thick substrates, including wood (Morionini). All of them are developed on the basis of rather insufficient transformations of the chaetome and exoskeleton. Undoubtedly, both the known and newly revealed differences in the traditional morphology and chaetotaxy of Rhysodidae and Scaritini larvae fail to correspond to the level of distinctions between tribes and, especially, subtribes.

Subfamily Rhysodinae within the family Carabidae?

Since the grounds for the incorporation of the Rhysodidae into the Carabidae mainly lay in similarities shown by adaptive features (e.g. Bell, 1970; Hlavac, 1975), the position of Rhysodidae within this family was repeatedly revised. In particular, the presence of shared characters was revealed, such as disjunct middle coxal cavities both in Rhysodidae and a number of basal carabid groups (Paussini, Cicindelini, Loricerini, Elaphrini, Scaritini and so on). The viewpoint of Erwin & Sims (1984) and Erwin (1985) was particularly meaningful in suggesting similarities between Rhysodidae and some basal representatives of Pterostichini, Morionini and, especially, Psydrini. Erwin considered this group as a disjunct supertribe which, together with Psydritae and Trechitae, form the subfamily Psydrinae of a separate division, Psydriformes. Any further discussion of the rank and position of Rhysodidae in this context is only possible if the following question is answered positively: Do Rhysodidae really belong in the family Carabidae?

In general, the larval structure in Rhysodidae, including chaetotaxy, is indeed similar in many ways to that of Carabidae. This is reflected in the same groundplan of the chaetome (setae and sensilla as homologized above being the proof), antennae (4-segmented with a developed sensorium on antennomere 3) and legs (moveably articulated tibia and tarsus, a developed pretarsus with a claw). However, none of these characters can be treated as synapomorphies of Rhysodidae and Carabidae. Thus, the structure of the antennae and legs is shared by most of the Adephaga, whereas among the securely homologized setae there are many (e.g. PA_1-PA_2-PA_3-PA_a and EP_1-EP_2 complexes, etc.) that are also known in a wide range of beetle families, both Adephaga and Polyphaga.

At the same time, some of the larval features of *Rhysodes* can securely be regarded as high-level autapomorphies, since none of them shows any analogs amongst Carabidae while some are even unique to the Coleoptera as a whole.

This primarily concerns the structure of the labiomaxillar complex (Figs 14-15) characterized by general consolidation. Thus, in the maxilla, the fused galea and lacinia are merged both with the stipes apex and palpifer, whereas all the appendages of the labium are fused to the mentum. The labiomaxillar complex is enlarged, fully covering both the oral opening and the ventral surface of the mandibles, with a particular porose structure formed in the narrow fissures between the labium and the maxillae. This structure can be suggested to provide fluid food, possibly myxomycetes, to be sucked in due to capillary powers. Carabidae do show some cases of feeding on fluid or fungal food (Mormolycini), but no such structures appear (Lieftinck & Wiebes, 1968). Furthermore, liquid food consumption in Carabidae is always accompanied by the formation of pharyngeal pump musculature, whereas in *Omoglymmius* larvae these muscles are relatively poorly developed (Beutel, 1990b). Larval mandibles in Rhysodidae probably do not take part in feeding, because, when closed, they stay isolated from the antebuccal cavity through a long lobe of the epipharynx. So the mouthparts of Rhysodidae, certainly being homologs of those in Carabidae, show nonetheless no functional similarities to any of the mouthpart types occurring in Carabidae. Obviously, some analogs can only be traced to the larval mouthparts of the some mycetophagous Staphylinidae (Leschen & Beutel, 2001).

The next highly specific feachure of Rhysodidae larvae is the total absence of digitiform sensilla in the sensory complex of mouthpart appendages (Figs 14-17). This type sensilla are known in all studied Carabidae larvae, including such disjunct specialists as Paussini, Cicindelini, Mormolycini (Lieftinck & Wiebes, 1968), Peleciini (Liebherr & Ball, 1990), Brachinini, Pseudomorphini (Erwin, 1981), etc. At the same time, they are lacking in most of the Hydradephaga (Noteridae, Dytiscidae, Gyrinidae, Haliplidae), but occur in Trachypachidae. As far as possible to judge, the presence or absence of digitiform sensilla is not related to a xylobiotic way of life. In any case, sensilla of similar types are met with in wood-dwelling larvae of Archostemata (Grebennikov, 2004) and numerous Polyphaga families.

Finally, there are some more special peculiarities to differ the larvae of Rhysodidae and Carabidae. These concern the retention in Rhysodidae of only a single pair of setae in the gular area, the original chaetotaxy of the frontale (see above), the total reduction of internal poststernites of the abdomen.

Therefore, a whole complex of larval traits emphasizes a sufficiently high degree of singularity of Rhysodidae to prevent them from being incorporated within Carabidae. It is noteworthy that most of the students who based their results on adult characters

(Bell & Bell, 1962; Hlavac, 1975; Bils, 1976; Bell, 1978; Baehr, 1979; Beutel, 1990, 1992a, 1995, 1998; Liebherr & Will, 1998) considered the rhysodines as only a taxon subordinate to Carabidae, whereas those who investigated the larvae (Böving, 1929; Böving & Craighead, 1930; Burakowski, 1975; Beutel, 1992b; Arndt, 1998) invariably arrived at the opposite conclusions.

Family Rhysodidae within the suborder Adephaga?

Recognition of the independence of Rhysodidae as a family of their own allows for a discussion of their relationships within the suborder Adephaga.

As noted above, opinions about the taxonomic position of Rhysodidae differ greatly. Thus, Crowson (1960) suggesed that Rhysodidae are one of the most primitive groups within Adephaga and therefore can be considered as the sister-group to the other families of the suborder. At present, the attribution of Rhysodidae to the Geadephaga remains unchallenged (Böving, 1929; Kryzhanovsky, 1983; Baehr, 1979; Beutel, 1990), but the extent of their interrelations with Carabidae is seen differently. In some cases, they are treated as primitive Geadephaga (Böving & Craighead, 1930; Jeannel, 1941; Kryzhanovsky, 1983; Ponomarenko, 1995), in other cases as the sister-group to (Beutel, 1990, 1992a) or a specialized derivative of Carabidae (Beutel, 1992b, 1993, 1995).

The larvae of Rhysodidae show a large set of highly specialized characters missing in carabids and partly unique among the beetles as a whole. The above larval features about equally well distinguish Rhysodidae from the remaining Recent families of Geadephaga, namely, Trachypachidae and Carabidae. The lack in Rhysodidae of digitiform sensilla and the presence of a labiomaxillar complex appear to be especially important distinctions. The labiomaxillar complex in Rhysodidae is associated with paired glands (Beutel, 1990b) which have no analogs amongst the larvae of Adephaga.

Along with these apomorphies, the features uniting the Rhysodidae with Trachypachidae or Carabidae are rather insignificant. The poorly delineated palpifer fused with the galea and, partly, lacinia is known in the Trachypachidae as well as some basal groups of Carabidae, such as Cicindelini, Paussini, Ozaenini, Metriini (Bousquet, 1986; Arndt & Beutel, 1995; Di Giulio & Moore, 2004, Moore & Di Giulio, 2006). In some Paussinae (Ozaenini), the nasale with setae FR_{10}, FR_{11} can be shifted proximally, so that a plate devoid of setae is formed before it (Di Giulio et al., 2003). These carabids possess a strongly enlarged labium, although no labiomaxillar complex is developed. The full absence of urogomphi being characteristic of Rhysodidae is probably not so important, because this feature is found in different, often not closely related carabid groups, namely, Cicindelini, Cychrini, Peleciini, Orthogoniini, Brachinini and some Harpalini. Thus, the Rhysodidae reveals the greatest similarity to the highly specialized, partly myrmecophilous and relatively primitive carabids, viz. Paussini and Ozaenini. The resemblance of these taxa was mentioned as early as in the end of 19th century, when Wasmann (1896) suggested even the family Rhysopaussidae. Later the viewpoint was rightly criticized (Escherich, 1898) and was left without further development.

At the same time, a number of features are shared by the larvae of Rhysodidae and several families of the suborder Archostemata (Grebennikov, 2004) which show a strongly enlarged labium forming, together with the maxillae, a functionally integrated structure; the labial palps are located ventrally and partly merged with the prementum, the shortened gular area, the galea and lacinia partly fused with the prementum. Some more special characters can be mentioned. Thus, the leg chaetotaxy, the structure of the basal part of the frontale and, partly, its chaetotaxy in *Priacma* LeConte, 1874, Cupedidae in general resemble the respective conditions observed in Rhysodidae; the chaetotaxy of the thoracic tergites in *Distocupes* Neboiss, 1984, Cupedidae and *Rhysodes* also show some common traits. The remaining larval features quite evidently separate these taxa, but none of the other Recent groups among Adephaga displays such a variety of features common with Archostemata. It is noteworthy that all known larvae of Archostemata, except for Micromaltidae, are highly specialized saproxylic forms adapted to feeding on hard xylem. This alone prevents from any possible adaptive similarity in mouthparts structure of these groups to Rhysodidae.

The similarity of the larvae of the Rhysodidae and Archostemata, on the one hand, and the cardinal differences between Rhysodidae and Geadephaga, on the other, suggest at least two hypotheses of rhysodine origin: (i) from an ancestor common with Geadephaga (meaning a revived viewpoint of Crowson), and (ii) from some group within Archostemata which was poorly specialized to a xylobiotic life-style.

One must remember that all known fossil larvae of Geadephaga are Mesozoic in age (Ponomarenko, 1985; Makarov, 1995) and they already possessed a typical caraboid structure down to the details of chaetotaxy. Unfortunately, the fossil records of Rhysodidae are represented only by an adult beetle from the Eocene and Miocen amber (Grimaldi & Engel, 2005; Kirejtshuk & Ponomarenko, 2007). Therefore, a relatively late appearance of this highly specialized group can be suggested, apparently in the times when the Geadephaga had contained already several fully developed subtaxa, infequently Recent ones. Thus, the formation of the highly disjunct larvae of Rhysodidae based on any of the then existing typical, specialized prototypes of Geadephaga larvae seems to be improbable, being supported by no morphological evidence.

Our knowledge of the larval stages of Archostemata is restricted to recent organisms only, but their structural details and life-style do not contradict a hypothesized "archostematan" ancestor of Rhysodidae. At the same time, the larvae of modern Archostemata show such a substantial number of apomorphies (endocarina presence, labral structure etc.) that this group could hardly be considered as direct ancestors of Rhysodidae. Despite this, based on morphological larval similarities, the hypothesis of Rhysodidae ancestry shared with Archostemata is certainly preferabale.

In this connection, Wasmann's (1896) assumption concerning the close relationship between the Rhysodidae and the Paussinae acquires a new dimension. One cannot exclude that the latter taxon could be derived from the same archostematan group which gave rise to Rhysodidae. In this case, Carabidae are to be considered as a paraphyletic group. Any further discussion of this still poorly-grounded hypothesis requires new evidence to be brought in, but its development seems to be fruitful in the polemics concerning the placement of Paussini and related groups within the Carabidae. At present this debate comprises highly contradictory and questionable assumptions. In particular, within the framework of this hypothesis both the highly specialized and plesiomorphic features occurring in combination in such a relatively young (Nagel, 1997) group as Paussinae can be accounted for.

The ideas on the young phylogenetic age of Paussinae are based only upon the obvious myrmecophily in certain groups within this subfamily that force to suggest their relatively late appearance after the true ants (Nagel, 1997). However the recent study of morphology and life style of the Ozaenini larvae (Di Giulio & Vigna Taglianti, 2001; Di Giulio et al, 2003) clearly show that the majority of specialized "myrmecophilous" features are related to the specific hunting way, namely from the shelter. Thus, in general Paussinae might appear to be significantly more ancient group (due to the great number of imago plesiomorphies) that makes my "archostemat" hypothesis even more probable.

ACKNOWLEDGMENTS

I am grateful to all colleagues who provided additional material: W. Tomaszewska (Museum and Institute of Zoology of the Polish Academy of Sciences, Warsaw), N.B. Nikitsky (Zoological Museum of the Moscow University) and A.A. Zaitsev (Moscow Pedagogical State University). English translation by Dr. O.L. Makarova. The help of S.I. Golovatch in checking the English is also highly appreciated. The work was financially supported by the Russian Foundation for Basic Research No 06-04-49456-a.

REFERENCES

Adis, J. (1981). The systematic and natural history of *Solenogenys* Westwood (Coleoptera: Carabidae: Scaritini), with a description of a new species from the central Amazon, Brazil. – The Coleopterists Bulletin (Washington, D.C.) 35: 153-166.

Arndt, E. & Beutel, R.G. (1995). Larval morphology of *Systolosoma* Solier and *Trachypachus* Motschulsky (Coleoptera: Trachypachidae) with phylogenetic considerations. – Entomologica Scandinavica 26: 439-446.

Baehr, M. (1979). Vergleichende Untersuchungen am Skelett und an der Coxalmuskulatur des Prothorax der Coleoptera. Ein Beitrag zur Klärung der phylogenetischen Beziehungen der Adephaga (Coleoptera, Insecta). – Zoologica (Stuttgart) 44: 1-76.

Balkenohl, M. (2001). Key and catalogue of the tribe Clivinini from the Oriental Realm with revisions of the genera *Thliboclivina* Kult, and *Trilophidius* Jeannel (Insecta, Coleoptera, Carabidae, Scaritini, Clivinini). Pensoft Series faunistica No.21. – Pensoft, Sofia-Moscow.

Bell, R.T. & Bell, J.R. (1962). The taxonomic position of the Rhysodidae. –The Coleopterists Bulletin (Washington, D.C.) 15: 99-106.

Bell, R.T. & Bell, J.R. (1978). Rhysodini of the World. Part I. A new classification of the tribe, and a synopsis of *Omoglymmius* subgenus *Nitiglymmius*, new subgenus (Coleoptera: Carabidae or Rhysodidae). – Quaestiones Entomologicae14: 43-88.

Bell, R.T. & Bell, J.R. (1979). Rhysodini of the World. Part II. Revisions of the smaller genera (Coleoptera: Carabidae or Rhysodidae). – Quaestiones Entomologicae15: 377-446.

Bell, R.T. & Bell, J.R. (1982). Rhysodini of the World. Part III. Revision of *Omoglymmius* Ganglbauer (Coleoptera: Carabidae or Rhysodidae) and substitutions for preoccupied generic names). – Quaestiones Entomologicae18: 127-259.

Bell, R.T. & Bell, J.R. (1985). Rhysodini of the World. Part IV. Revisions of *Rhyzodiastes* and *Clinidium*, with new species in other genera (Coleoptera: Carabidae or Rhysodidae). – Quaestiones Entomologicae21: 1-172.

Bell, R.T. (1966). *Trachypachys* and the origin of the Hydradephaga (Coleoptera, Carabidae). – The Coleopterists Bulletin (Washington, D.C.) 20: 107-112.

Bell, R.T. (1975). *Omoglymmius* Ganglbauer, a separate genus (Coleoptera: Carabidae or Rhysodidae). – The Coleopterists Bulletin (Washington, D.C.) 29: 351-352.

Bell, R.T. (1998). Where do the Rhysodini (Coleoptera) belong? – In: Phylogeny and Classification of Caraboidea. XX I.C.E. (1996, Firenze, Italy) (Ball, G.E., Casale, A. & Vigna Taglianti, A., eds). Museo Regionale di Scienze Naturali, Torino, p. 261-272.

Beutel, R.G. (1990). Metathoracic features of *Omoglymmius hamatus* and their significance for classification of Rhysodini (Coleoptera: Adephaga). – Entomologia Generalis 15: 185-201.

Beutel, R.G. (1992a). Phylogenetic analysis of thoracic structures of Carabidae (Coleoptera: Adephaga). – Zeitschrift für Zoologische Systematik und Evolutionsforschung 30: 53-74.

Beutel, R.G. (1992b). Larval head structures of *Omoglymmius hamatus* and their implications for the relationships of Rhysodidae (Coleoptera: Adephaga). – Entomologica Scandinavica. 23: 169-184.

Beutel, R.G. (1993). Phylogenetic analysis of Adephaga (Coleoptera) based on character of the larval head. – Systematic Entomology 18: 127-147.

Beutel, R.G. (1995). The Adephaga (Coleoptera): phylogeny and evolutionary history. – In: Biology, Phylogeny, and Classification of Coleoptera. Papers celebrating the 80th Birthday of Roy A. Crowson (Pakaluk, J. & Ślipiński, A., eds). Muzeum i Instytut Zoologii PAN, Warsaw, p. 173-217.

Bousquet, Y. & Goulet, H. (1984). Notation of primary setae and pores on larvae of Carabidae (Coleoptera, Adephaga). – Canadian Journal of Zoology 62: 573-588.

Bousquet, Y. (1986). Description of first-instar larva of *Metrius contractus* Eschscholtz (Coleoptera: Carabidae) with remarks about phylogenetic relationships and ranking of the genus *Metrius* Eschscholtz. – Canadian Entomologist 118: 373-388.

Bousquet, Y. (2001). Larval features of Morionini (Coleoptera, Carabidae) discussed: is the tribe more closely related to Scaritini or Pterostichini? – Russian Entomological Journal 10: 253-260.

Böving, A.G. (1929). Classification of beetles acciording to larval characters. – Bulletin of the Brooklyn Entomological Society (N.S.) 24: 55-80.

Böving, A.G. & Craighead, F.C. (1931). An illustrated synopsis of the principal larval forms of the order Coleoptera. – Entomologica Americana (n.s.) 11: 1- 272, 125 pls.

Burakowski, B. (1975). Description of larva and pupa of *Rhysodes sulcatus* (F.) (Coleoptera, Rhysodidae) and notes on the bionomics of this species. – Annales Zoologici. 32: 271-287.

Costa, C., Vanin, S.A. & Casari-Chen, S.A. (1988). Larvas de Coleoptera do Brasil. –Museo de Zoologia, Sao Paolo.
Crowson, R.A. (1955). The natural classification of the families of Coleoptera. – Nathaniel Lloyd, London.
Crowson, R.A. (1960). The phylogeny of Coleoptera. – Annual Review of Entomology5: 111-134.
Di Giulio, A. (1998). Implicazioni filogenetiche dello studio delle larve dei Carabidi: Il caso dei Paussidi (Coleoptera: Carabidae). – Atti dell' Accademia Nazionale Italiana di Entomologia 46: 278-308.
Di Giulio, A., Fattorini, S., Kaupp, A., Vigna Taglianti, A. & Nagel, P. (2003). Review of competing hypotheses of phylogenetic relationschips of Paussinae (Coleoptera: Carabidae) based on larval characters. – Systematic Entomology 28: 509-537.
Di Giulio, A. & Moore, W. (2004). The first-instar larva of the genus *Arthropterus* (Coleoptera: Carabidae: Paussinae): implications for evolution of myrmecophily and phylogenetic relationships within the subfamily. – Invertebrate Systematics 18: 101-115.
Di Giulio, A. & Vigna Tagliantii, A. (2001). Biological observations on *Pachyteles* larvae (Coleoptera Carabidae Paussinae). – Tropical Zoology 14: 157-173.
Dostal, A. (1993). Description of *Androzelma gigas* (Coleoptera: Carabidae: Salcediina). – Zeitschrift der Arbeitsgemeinschaft Österreichischer Entomologen 45: 117-121.
Erwin, T.L. (1981). A synopsis of the immature stages of Pseudomorphini (Coleoptera: Carabidae) with notes on tribal affinities and behavior in relation to life with ants. – The Coleopterists Bulletin (Washington, D.C.) 35: 53-68.
Erwin, T.L. (1985). The taxon pulse: a general pattern of lineage radiation and extinction among carabid beetles. – In: Taxonomy, phylogeny and zoogeography of beetles and ants. A volume dedicated to the memory of Philip Jackson Darlington, Jr. (1904-1983). Series entomologica, 33 (Ball, G.E., ed.). The Hague, Dr. W. Junk Publishers, p. 437-472.
Erwin, T.L. & Sims, L.L. (1984). Carabid Beetles of the West Indies (Insecta, Coleoptera): a synopsis of the genera, and checklists of the tribes of Caraboidea, and of the West Indian species. – Quaestiones Entomologicae20: 351-466.
Escherich, K. (1898). Beitrag zur Morphologie und Systematik der Coleopteren-Familie der Rhysodiden. – Wiener Entomologische Zeitung 17 (2): 41-50.
Forbes, W.T.M. (1926). The wing folding patterns of the Coleoptera. – Journal of the New York Entomological Society 34: 42-139.
Forsyth, D.J. (1972). The structure of the pygidial defence glands of Carabidae (Coleoptera). – Transaction of the Zoologica Society of London 32: 249-309.
Ganglbauer, L. (1892). Die Käfer von Mitteleuropa. Die Käfer der österreichisch-ungarischen Monarchie, Deutschlands, der Schweiz, sowie des französischen und italischen Alpengebietes. Bd.1. Carl Gerols's Sohn, Vienna.
Grandi, G. (1956). *Rhysodes germari* Ganglbauer, documenti morfologici et eco-ecologici – Bollettino dell' Istituto di Entomologia dell' Università degli studi di Bologna 21: 179-196.
Grandi, G. (1972). Comparative morphology and ethology of insects with a specialized diet, *Rhysodes germari* Ganglb. – Bollettino dell' Istituto di Entomologia dell' Università degli studi di Bologna 30: 21-47.
Grimaldi, D. & Engel, M.S. (2005). Evolution of the Insects. University Press, Cambridge.

Hlavac, T.F. (1975). The prothorax of Coleoptera (except Bostrichiformia-Cucujiformia). – Bulletin of the Museum of Comparative Zoology Harvard 147: 137-183.

Grebennikov, V.V. (2004). Review of larval morphology of beetles of the suborder Archostemata (Insecta: Coleoptera) including first-instar chaetotaxy. – European Journal of Entomology 101: 273-292.

Jeannel, R. (1941). Coléoptères Carabiques. I. Faune de France, 39. Lechevalier, Paris.

Kirejchuk, A.G. & Ponomarenko, A.G. (2007). Taxonomic list of fossil beetles of suborders Cupedina, Carabina and Scarabaeina (Part 1). – http://www.zin.ru/Animalia/Coleoptera/eng/paleosy0.htm.

Kryzhanovsky, O.L. (1983). Fauna SSSR. Zhestkokrylye. T.1. Vyp.2. Zhuki podotryada Adephaga: semeistva Rhysodidae, Trachypachidae; semeistvo Carabidae (vvodnaya chast, obzor fauny SSSR). – Leningrad, Nauka. [In Russian].

Leschen, R.A.B. & Beutel R.G. (2001). Pseudotracheal tubes, larval head, and mycophagy in *Sepedophilus* (Coleoptera: Staphylinidae: Tachyporinae). – Journal of Zoological Systematics and Evolutional Research 39: 25-35.

Liebherr, J.K. & Ball, G.E. (1990). The first instar larva of *Erypus oaxacanus* Straneo & Ball (Coleoptera: Carabidae: Peleciini): indicator of affinity or convergence? – Systematic Entomology 15: 69-79.

Lieftinck, M.A. & Wiebes, J.T. (1968). Notes on the genus *Mormolyce* Hegenbach (Coleoptera, Carabidae). – Bijdragen tot de Dierkunde 38: 59-68.

Lindroth, C.H. (1960). The larvae of *Trachypachus* Motsch., *Gehringia* Darl., and *Opisthius* Kby. (Col, Carabidae). – Opuscula Entomologica 25: 30-42.

Makarov, K.V. (1995) New data on the larvae of the Jurassic Carabomorpha (Coleoptera, Adephaga). – Paleontologicheskij Zhurnal 1: 122-125. [In Russian]

Makarov, K.V. (1996). Patterns of chaetome modifications in ground-beetle larvae (Coleoptera: Carabidae). – Acta Societatis Zoologicae Bohemicae 60: 391-418.

Mamaev, B.M. & Pototskaya, V.A. (1979). Larvae of Palaearctic species of the genus *Rhysodes* Dalm. (Coleoptera, Rhysodidae). – In: Stem boring insects and their entomophages (Pravdin, F.N. ed.). Nauka, Moscow, p. 199-204. [In Russian].

Moore, W. & Di Giulio, A. (2006). Description and behaviour of *Goniotropis kuntzeni* larvae (Coleoptera: Carabidae: Paussinae: Ozaenini) and key to genera of Paussinae larvae. – Zootaxa 1111: 1-19.

Nagel, P. (1997). New fossil pausside from Dominican amber with notes of the phylogenetic systematics of the paussine complex (Coleoptera: Carabidae). – Systematic Entomology 22: 345-362.

Peyerimhoff, P. de (1903). Position systématique des Rhysodidae. – Revue d'Entomologie, Caen. 22: 80-84.

Ponomarenko, A.G. (1969). The historical development of archostematan beetles. – Trudy Paleontologicheskogo Instituta Akademii Nauk SSSR 125: 1-238. [In Russian]

Ponomarenko, A.G. (1985). Coleoptera from the Jurassic of Siberia and western Mongolia. – In: Yurskie nasekomye Sibiri i Mongolii (Rasnitsyn, A.P., ed). Trudy Paleontologicheskogo Instituta Akademii Nauk SSSR 211: 47-87. [In Russian]

Ponomarenko, A.G. (1995). The geological history of beetles. – In: Biology, Phylogeny, and Classification of Coleoptera. Papers celebrating the 80th Birthday of Roy A. Crowson (Pakaluk, J. & Ślipiński A., eds). Muzeum i Instytut Zoologii PAN, Warsaw, p. 155-171.

Reitter, E. (1882). Bestimmungs-Tabellen der europäischen Coleopteren. VI. Rhysodidae. – Verhandlungen der Naturforschenden Vereins Brünn 20: 140-141.
Vanin, S.A. & Costa, C. (1978). Larvae of Neotropical Coleoptera. II: Rhysodidae. – Papéis Avulsos de Zoologia 31: 195-201.
Wasmann, E. (1896). Neue Termitophilen und Termiten aus Indien. IV. (Nachtrag). – Annali del Musoe Civico di Stortia Naturale di Genova. Ser. 2. 37: 149-152.

L. Penev, T. Erwin & T. Assmann (Eds) 2008
Back to the Roots and Back to the Future. Towards a New Synthesis amongst Taxonomic, Ecological and Biogeographical Approaches in Carabidology
Proceedings of the XIII European Carabidologists Meeting, Blagoevgrad, August 20-24, 2007, pp. 125-141.

Studies on genus *Speluncarius*, with description of a new subgenus and notes on the systematic position of *S.* (*Hypogium*) *albanicus* (Coleoptera, Carabidae, Pterostichini)

Borislav V. Guéorguiev[1] & Roman Lohaj[2]

[1] Natural Museum of Natural History, 1 Blvd. Tzar Osvoboditel, 1000 Sofia, Bulgaria.
E-mail: bobivg@yahoo.com
[2] Institute of Forensic Sciences, Kuzmányho 8, SK-041 02 Košice, Slovakia.
E-mail: lohaj@minv.sk, rlohaj@gmail.com

SUMMARY

The examination of the holotypes of *Speluncarius ponticus* and *Platysma albanicus* demonstrates that the two taxa are members of separate phyletic lineages of the genus *Speluncarius*. On this ground the new subgenus *Pontotapinus* **subgen. nov.** is proposed, based on type species *Speluncarius ponticus* Casale & Giachino, 1991. On the other hand, the study suggests that *Hypogium* represents a well-isolated lineage probably with a basal position within the genus, and not belong to the "molopite complex". The female genitalia of the type species of *Speluncarius* s. str., *Hypogium* and *Pontotapinus* **subgen. nov.** as well as parameres of the genus type are illustrated and described for the first time. Remarks on the systematics of the genus are made based on the present knowledge of the adult morphology. As a result of these monophyly or polyphyly of *Speluncarius* cannot be demonstrated at present while the monophyly of the complex "*Tapinopterus* – *Speluncarius*" is well-supported by three clear synapomorphies. Key to subgenera of *Speluncarius* is also provided.

Keywords: Carabidae, Pterostichini, *Speluncarius*, taxonomy, Eastern Mediterranean, Albania, Turkey

INTRODUCTION

Most species of microphthalmic and eyeless pterostichine beetles from the Western Palaearctic region belong to the genus *Speluncarius* Reitter, 1886. Species from this genus are distributed from north-western Italy through Croatia, Bosnia and Herzegovina, Montenegro, Albania and Greece to the north-eastern Turkey. At present there are 25 valid species in the genus grouped in four subgenera (Bousquet, 2003). Eighteen described species inhabit Greece and Turkey and more than 60 new species from Greece await description (Giachino pers. com., 2007) which coincides with the center of the diversity of the genus *Tapinopterus* Schaum, 1858. Some authors expressed the view that *Speluncarius* may be polyphyletic with respect to *Tapinopterus* and that these two genera are closely related (Jeanne, 1982; Sciaky, 1982; Casale & Giachino, 1991; Casale et al., 1998; Guéorguiev & Guéorguiev, 1999). The presence of distinct transverse apophysis on the left paramere in the type species of the subgenera *Speluncarius* s. str., *Elasmopterus* Kraatz, 1886, *Hypogearius* Jeannel, 1953 as well as in *S. ponticus* Casale & Giachino, 1991 (Jeannel, 1953: 14, Fig. 11; Casale & Giachino, 1991: 218, Fig. 7; Sciaky & Persohn, 1994: 44, Fig. 3; current paper, Fig. 1) is evidence that the genus belongs to the "pterostichite complex" (cfr. Bousquet, 1999: 37). Only the systematic position of *Hypogium* Tschitschérine, 1900, which has been frequently discussed (Jeannel, 1950; Jeannel, 1953; Vigna Taglianti, 1973; Sciaky, 1982; Casale & Giachino, 1991), is not definitely settled since the male of *S. albanicus* is hitherto unknown. However, the study of several previously unconsidered morphological features contribute to the correct systematic position of this remarkable species.

Generally, the work aims to examine and describe new characters in the adults from some lineages of *Speluncarius*, as well as to specify the relationship between this genus and *Tapinopterus*. Comparison and a more detailed examination of *S. albanicus* and *S. ponticus* are provided as we assume that both belong to different phyletic lineages of the genus. Moreover, the former is a species which up to now has been known only from single specimen collected more than 100 years ago.

MATERIAL AND METHODS

The following material has been studied:

Speluncarius (*Speluncarius*) *anophthalmus* (Reitter, 1886)

3 specimens, including 1 syntype (MNHUB); 3 specimens (HNHM); 11 specimens (NMW); 1 male, "Herzegovina, Reitter" / "Zool. Inst. St. Petersburg" (ZISP); 1 female, "Herzegovina, Reitter" / "174" / "Zool. Inst. St. Petersburg" (ZISP); 1 female, "Pl. 4,3. 6.26 1200 m. Trubar Gau" / "*Speluncarius trubarensis*" (BMNH); 79 specimens, Herzegovina, Dikliči, Djediči, Žarbina jama, Jama pod Velikom Cibinom, Jakština pečina (NMP); 15 specimens, Herzegovina, Popovo polje (NMP); 3 specimens Montenegro, Golodražnica, route Risan – Kotor, "*Speluncarius anophthalmus golodraznicensis*"

(NMP); 2 specimens, Dalmatia, Kotor, "*Speluncarius anophthalmus kotorensis*" (NMP); 6 specimens, Bjelašnica, "*Speluncarius anophthalmus radošensis*" (NMP); 1 female, SW Herzegovina, Popovo polje, Zavala env., Orlica pečina Cave, 11.6.2000, R. Mlejnek leg. (cRL); 8 specimens, Herzegovina, Trebinje env., Matuličova pečina Cave, 16.9.2003, G. Dunay, R. Lohaj, J. Lakota and D. Čeplík leg. (cDC; cGD; cJL; cRL); 52 specimens, Herzegovina, Orjen Bjela Gora, Milanov Ocijek, ca 1000 m a.s.l., 1-4.6.2004, R. Lohaj, J. Lakota and D. Čeplík leg. (cDC; cJL; cRL); 1 female, "Hercegovina, Orjen Bjela Gora, Milanov Ocijek, 1 – 4.6.2004 R. Lohaj lgt." (NMNHS); 1 female Hercegovina, Trebinje env., Trebinska šuma (forest), Taleža jama, 16.4.2006, D. Čeplík leg. (cDC).

Speluncarius (*Speluncarius*) *minutulus* J. Müller, 1937

1 female (designated as paralectotype by P.M. Giachino), "Shkëlzën" / "Albania leg. Bischoff 1936" (MNHUB).

Speluncarius (*Speluncarius*) *pesarinii* Bucciarelli, 1979

1 male, 1 female, "M. Grappa (VI) campi solagna 28.VII.1987 R. Monguzzi" / "*Speluncarius pesarinii* Bucc. det. R. Monguzzi" (NMNHS; cRL); 1 male, 2 females (NMW).

Speluncarius (*Speluncarius*) *setipennis* (Apfelbeck, 1899)

2 females (including 1 syntype), 1 male (NMW); 4 specimens, Njeguš Mt., Orjen Mt. (NMP).

Speluncarius (*Speluncarius*) *stefani* (Jureček, 1910)

1 female (NMW); 2 females, Italy, "M. Lessini (VR) Malga Bagorno (Rovere) 3.10.90" / "*Speluncarius stefani* Jur. det. R. Monguzzi" (NMNHS); 1 male, Italy, Verona District, Monti Lessini, Tracchi, 1450 m, E. Ollivier leg. (cRL).

Speluncarius (*Hypogearius*) *boluensis* Schweiger, 1966

1 male (NMW).

Speluncarius (*Elasmopterus*) *leonhardi leonhardi* (Breit, 1914)

2 females, including 1 syntype (DEI); 1 male (NMW).

Speluncarius (*Pontotapinus* **subgen. nov.**) *ponticus* Casale & Giachino, 1991

male holotype, "Turchia vil. Kastamonu M. Yarligöz m 1200 X.1988 R. Roma leg." / "Holotypus *Speluncarius* (*Hypogeobium*) *ponticus* n.sp. A. Casale P.M. Giachino det. 1989" (cGI).

Speluncarius (*Pontotapinus* **subgen. nov.**) sp. cf. *ponticus* Casale & Giachino, 1991

1 female, "Turkey, 10.5.2001 Vil. Zonguldag, Ereğli, Koca Ali, Amaçlar cave entrance, R. Lohaj leg." (cRL).

Speluncarius (*Hypogium*) *albanicus* (Tschitschérine, 1900)

female holotype, "Albania" / "G.C.Champion Coll. B.M.1927-409" / "*Platysma* (*Hypogium*) *albanicum* m. typ. Tschitscherine det" / "syntype" [small round label with blue margin] (BMNH).

The studied material is housed in the following institutional and private collections:

BMNH – Natural History Museum, London, United Kingdom (R. Booth; C. Gillett).

DEI – Deutsches Entomologisches Institut, Deutschen Akademie der Landwirtswissenschaften zu Berlin, Müncheberg, Germany (L. Zerche, M. Behne).

HNHM – Hungarian Natural History Museum (Magyar Természettudományi Múzeum), Budapest, Hungary.

MNHUB – Museum für Naturkunde der Humboldt Universität zu Berlin, Bereich Zoologisches Museum, Berlin, Germany (M. Uhlig, B. Jaeger).
NMNHS – coll. National Museum of Natural History, Sofia (A. Popov).
NMP – National Museum Prague (Národní Museum Praha), Czech republic (J. Hájek).
NMW – Naturhistorisches Museum Wien, Vienna, Austria (H. Schönmann).
ZISP – Zoological Institute, Russian Academy of Sciences, St. Petersburg (B. Kataev).
cDC – coll. David Čeplík, Košice, Slovakia.
cGD – coll. Gejza Dunay, Kráľovce, Slovakia.
cGI – coll. Pier Mauro Giachino, Torino, Italy.
cJL – coll. Ján Lakota, Ružomberok, Slovakia.
cRL – coll. Roman Lohaj, Košice, Slovakia.

The systematics of Pterostichini follows Bousquet (1999), and the configuration of the female sterna and terga is in accordance with Deuve (1993) and Liebherr & Will (1998).

SYSTEMATICS

"*Tapinopterus - Speluncarius*" complex

Remarks. The adults of *Speluncarius* s.l. share three obvious synapomorphies with those of *Tapinopterus* s.l. – absence of basolateral setae of pronotum, subquadrate metaepisternae and presence of medial seta on hind coxae. This combination of apomorphies (Bousquet, 1999: 33, Table 3) is unique among the Palaearctic Pterostichini and supports close relationships and common origin of the species from this complex. Other characters have often been used (separately or together) to distinguish species of these two genera – 1/ presence (*Tapinopterus*) or absence (*Speluncarius*) of ventral setae on onychium, and 2/ presence (*Tapinopterus*) or reduction / absence (*Speluncarius*) of eyes. Both states in *Speluncarius* are hypothesized as probably reflecting the modifications due to the adaptation to hypogean way of life (Guéorguiev & Guéorguiev, 1999: 44). The polarities of both characters have been marked as apomorphic in supraspecific taxa of Nearctic Pterostichini (Bousquet, 1999: 33, Table 3) and in our opinion such assessment is also valid for the Holarctic representatives of the tribe. However, the level of eye reduction with a genus varies, while the absence of ventral seta on onychium occurs in five species from both subgenera of *Tapinopterus* – *Pterotapinus* Heyden, 1883 and *Molopsis* Schatzmayr, 1943 (excluding *T. machardi* Jeanne, 2005).

This complex includes approximately 80 described species, as well as some still undescribed species (P.M. Giachino, unpublished data; present authors, unpublished data).

Speluncarius Reitter, 1886 s.l.

Remarks. Sciaky (1982: 16) generalized at least five characters to distinguish adults of *Speluncarius* from adults of *Tapinopterus*, but in fact any of them is inadequate to separate species of both genera: 1/ depigmented tegument of body (however, the colour of tegument in specimens of *S. anophthalmus* and *S. stefani* is dark brownish like in the most species of *Tapinopterus*, while specimens of several species from the latter genus have depigmented cuticle); 2/ eyes atrophied or absent (however, *Pontotapinus* **subgen. nov.** has minute and presumably still functioning eyes); 3/ absence of basolateral seta of pronotum (however, this character is shared by all species of *Tapinopterus* s.l., see above); 4/ apophysis of prosternum sulcate (however, this character also occurs *Tapinopterus*, e.g. species of the "*balcanicus*" group from Serbia and Macedonia and species of the "*laticornis*" group from the southern Turkey); 5/ absence of ventral seta on onychium (for this character see the comments above).

Besides, all species of the genus, except one belonging to the *Pontotapinus* **subgen. nov.**, share another feature – more or less projecting fore angles of the pronotum. The same characteristic occurs also in *Tapinopterus marani* V. B. Guéorguiev & B. V. Guéorguiev, 1998 (Guéorguiev & Guéorguiev, 1999: 42) and *T.* (*Elasmopterus*) *filigranus* Miller, 1862 (Giachino pers. com., 2007). Eight species of *Speluncarius* have distal part of antennomere 3 pubescent, while another seventeen have antennomere 3 glabrous. The same character occurs intraspecifically in the subgenus *Crisimus* Habelmann, 1885 of *Tapinopterus*. Six species of *Speluncarius* have two anterolateral setae on pronotum, but this is also found in species of the "*extensus*" group of *Tapinopterus*.

The question is whether *Speluncarius* is a genus separate from *Tapinopterus* or not? We studied critically all the characters, which have been used to divide *Speluncarius* and *Tapinopterus*. In reality, none seem constant or stable. The polarities of the selected characters occur in both genera, e.g. they are homoplasious and seemingly reflect morphological changes such as resulting from an adaptation to similar environments. Most probably *Speluncarius* represents a grade which arose from one, two or even more different lineages within *Tapinopterus*. We feel that the question above could be best addressed by phylogenetic study of species from the "*Tapinopterus* – *Speluncarius*" complex using a matrix of DNA sequence data and larval and adult morphology. The small amount of data available seems insufficient to answer the question. A reasonable and conservative approach is to follow the view of Casale & Vigna Taglianti (1999), combining all the subgenera of both genera into a single genus.

Speluncarius (*Speluncarius*) *anophthalmus* (Reitter, 1886)

(Figs 1, 2, 5, 8, 11, 12)

Legs. Hind trochanter round at apex, shorter than half length of hind femur. Hind coxa with posterolateral and medial setae. Onychium with pair of dorsolateral setae, dorsal setae absent.

Male genitalia. Median lobe of aedeagus as figured by Sciaky (1982: 22, Figs 2-3); left paramere with distinct transverse apophysis, right one long and thin (Fig. 1).

Female genitalia. Tergum VIII short, wide, with long "anterolateral apophyses" (Fig. 2). Both lobes of sternum VIII with fairly reduced internal membrane areas, "anterolateral

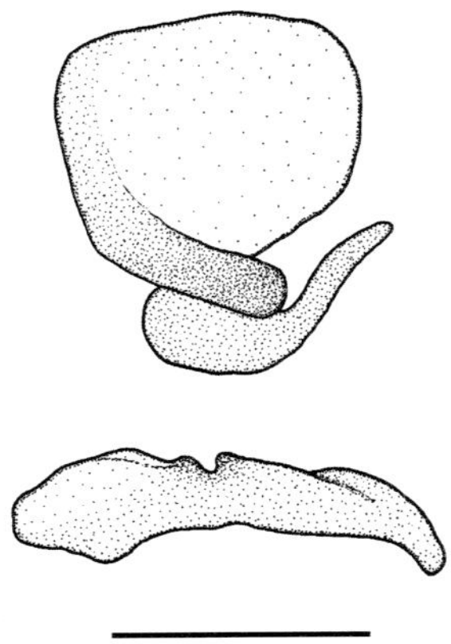

1

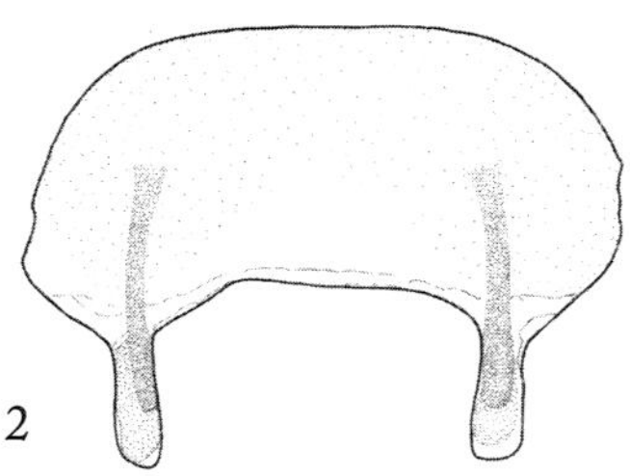

2

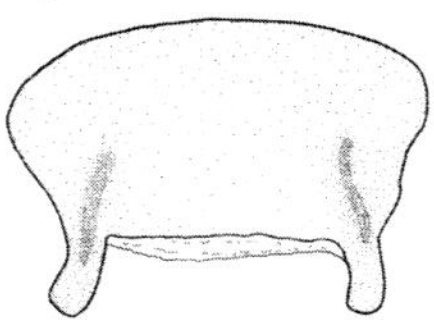

3

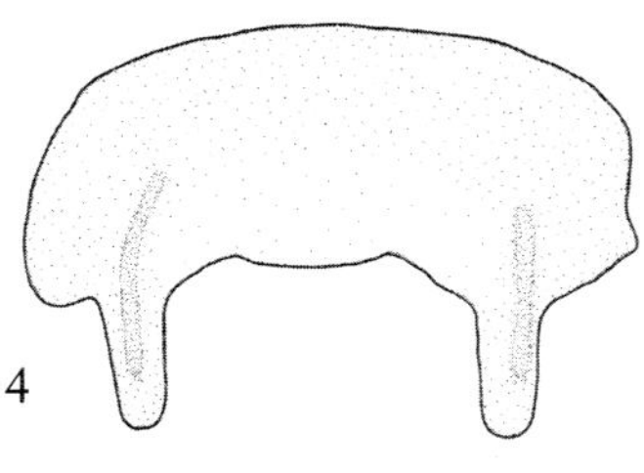

4

5

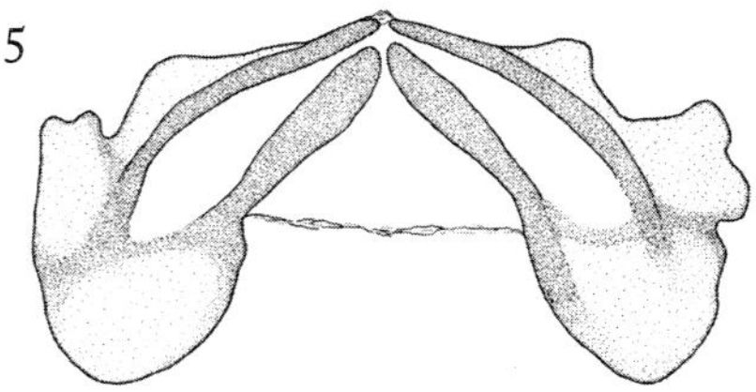

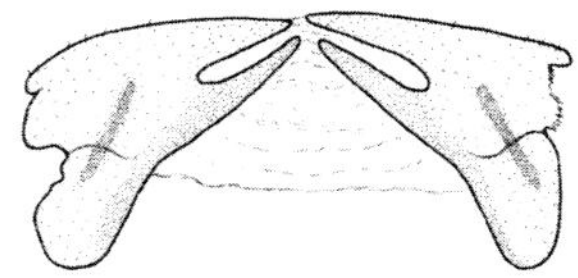

6

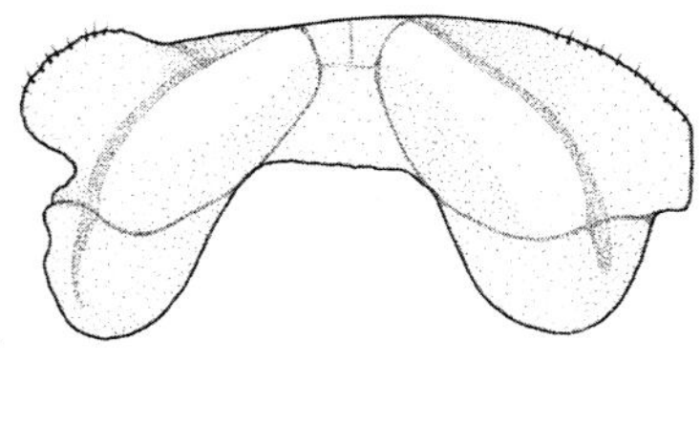

7

8

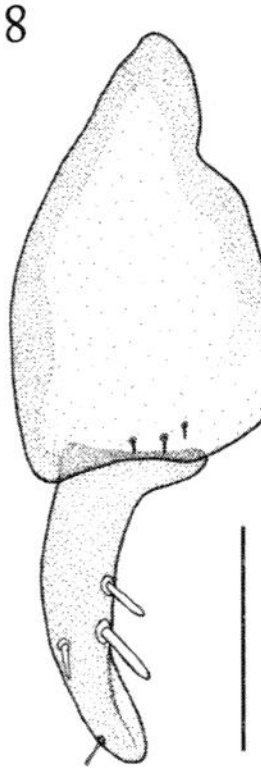

9

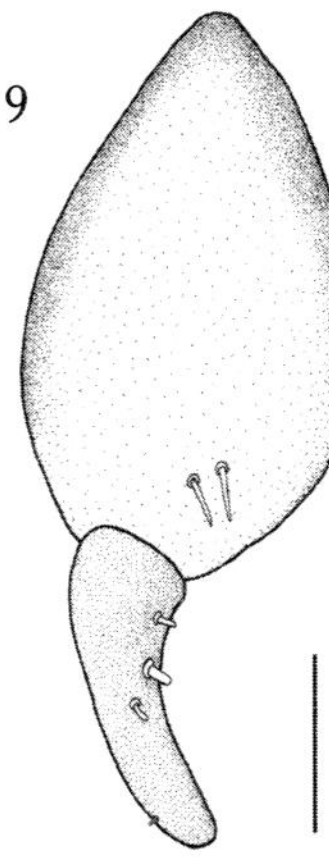

10

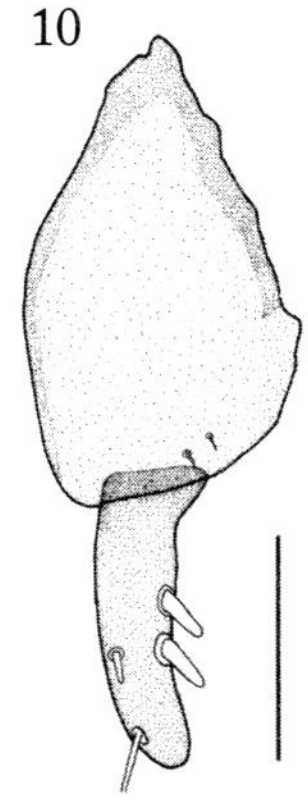

Fig. 1. Drawings of left and right parameres of *Speluncarius anophthalmus* (male from "Herzegovina, Reitter"). Scale line = 0.5 mm.

Figs 2-4. Drawings of tergum VIII. Fig. 2: *Speluncarius anophthalmus* (female from "Herzegovina, Reitter"); Fig. 3: *S.* sp. cf. *ponticus* (female from Amaçlar Cave); Fig. 4: *S. albanicus* (holotype). Scale line = 0.5 mm.

Figs 5-7. Drawings of sternum VIII. Fig. 5: *Speluncarius anophthalmus* (female from "Herzegovina, Reitter"); Fig. 6: *S.* sp. cf. *ponticus* (female from Amaçlar Cave); Fig. 7: *S. albanicus* (holotype). Scale line = 0.5 mm.

Figs 8-10. Drawings of stylus of ovipositor. Fig. 8: *Speluncarius anophthalmus* (female from Herzegovina); Fig. 9: *S.* sp. cf. *ponticus* (female from Amaçlar Cave); Fig. 10: *S. albanicus* (holotype). Scale line = 0.2 mm (Figs 7, 9); = 0.1 mm (Fig. 8).

apophyses" widely round (Fig. 5). Apical stylomere of left stylus of usual (for Pterostichini) size relative to basal stylomere (Fig. 8); dorsal ensiform seta dorsad, removed from distal dorsolateral ensiform seta at distance almost equal to distance between two dorsolateral setae; both dorsolateral ensiform setae long and thin, situated in medially in stylomere; nematiform setae long. Spermatheca with seminal canal and receptaculum differentiated (Figs 11-12); seminal canal long, slightly thinner than receptaculum; receptaculum elongate, sharply curved apically; spermathecal canal inserted proximally on receptaculum just before junction of seminal canal and receptaculum.

Speluncarius (*Pontotapinus* subgen. nov.)

Diagnosis. The new subgenus differs from all the other subgenera of *Speluncarius* in having minute eyes with clear and probably functioning ommatidia (vs. reduced spot-like eyes without ommatidia or eyes totally absent), sub-cordiform pronotum (vs. trapezium-shaped pronotum), and anterior angles of pronotum scarcely projecting (vs. anterior angles of pronotum clearly projecting). *Pontotapinus* **subgen. nov.** is distinguished from *Speluncarius* s. str. in having the distal half of third antennomeres pubescent (vs. distal half of third antennomeres glabrous) and the hind trochanter half the length of hind femur (vs. hind trochanter notably shorter than half the length of the hind femur). The new taxon is distinct from *Elasmopterus* Kraatz, 1886 in having the lateral border of pronotum with only single anterolateral seta (vs. lateral border of pronotum having two anterolateral setae). In addition, *Pontotapinus* **subgen. nov.** differs from *Hypogearius* Jeannel, 1953 in the distal half of third antennomeres pubescent (vs. distal half of third antennomeres glabrous). Finally, the new subgenus differs from the subgenus *Hypogium* Tschitschérine in three more characters: 1/ outer posterolateral seta on hind coxa present (vs. outer posterolateral seta on hind coxa absent); 2/ hind trochanter round at apex (vs. hind trochanter pointed at apex); 3/ onychium without dorsal setae and with single pair of dorsolateral setae (vs. onychium with dorsal setae and with two pairs of dorsolateral setae). A generalization and comparison of selected characters in the subgenera *Speluncarius* s. str., *Hypogium* and *Pontotapinus* **subgen. nov.** is given in Table 1.

Type species. *Speluncarius ponticus* Casale & Giachino, 1991

Etymology. The name of the new subgenus is combination from the Greek based on the name "Pontos", a sea deity worshipped by the ancient Greek (equivalent of the English word "sea") and the Greek adjective "tapeinos", tantamount to the English words "humble" and "meek".

Ecological remarks. The presumably functioning eyes, sub-cordiform pronotum without protruding fore angles, and sutural fusion of the elytra are presumed to be the result of partial adaptation to subterranean environment. Assuming that the edaphobite manner of living is ancestral and that the habitus of *S. ponticus* includes plesiomorphic features, we take this to indicate a somewhat hypogean way of life.

Table 1. Differing states of characters in *Speluncarius* s. str., *Hypogium* and *Pontotapinus* **subgen. nov.**

Character states and ratios	***Speluncarius* s. str.**	***Hypogium***	***Pontotapinus***
Setation on second half of antennomere 3	Absent	Present	Present
Fore angles of pronotum	Projecting, more or less pointed	Projecting, more or less pointed	Neither projecting not pointed
Scutellar stria	Traceable	Absent	Absent
Posterolateral seta on hind coxa	Present	Absent	Present
Hind trochanter	Round at apex, shorter than half length of hind femur	Pointed at apex, as long as half length of hind femur	Round at apex, as long as half length of hind femur
Dorsolateral setae on onychia	One pair	Two pairs	One pair
Dorsal setae on onychia	Absent	Two pairs	Absent
Tergum VIII	Short and wide with long "anterolateral apophyses"	Short and wide with long "anterolateral apophyses"	Long and narrow with short "anterolateral apophyses"
Sternum VIII – reduction of internal membrane areas	Fairly reduced	Not reduced	Slightly reduced
Sternum VIII - "anterolateral apophyses"	Widely round	Widely round	Closely round
Dorsal ensiform seta of apical stylomere of stylus	Dorsal position, removed from distal dorsolateral seta	Dorsal position, removed from distal dorsolateral seta	Subdorsal position, close to distal dorsolateral seta
Dorsolateral ensiform setae of apical stylomere of stylus	Long and thin, both situated in medial part of stylomere	Long and thick, both situated in medial part of stylomere	Short, proximal seta situated in proximal part of stylomere
Nematiform setae of apical stylomere of stylus	Long (well developed)	Long (well developed)	Short (reduced)
Seminal canal of spermatheca	Long	Short	?
Receptaculum of spermatheca	Sharply curved apically	Gradually curved apically	?

Speluncarius (*Pontotapinus*) *ponticus* Casale & Giachino, 1991

(Fig. 16)

References. *Speluncarius* (*Hypogeobium*) *ponticus*: Casale & Giachino, 1991: 215; Lorenz, 1998: 268. *Speluncarius* (*Hypogium*) *ponticus*: Bousquet, 2003: 516.

Remarks. This species was described from single male from the Yaraligözdag Mt. and was provisionally placed in the subgenus *Hypogium* Tschitschérine, 1900 (Casale & Giachino, 1991: 215-216). Lorenz (1998) and Bousquet (2003) also followed this placement. The study of the holotype of *S. ponticus* and the female from the region of Koca Ali Village showed that this species has significant differences in external and internal characters as compared to *S. albanicus*. For this reason these taxa are placed in separate supraspecific groups (see "Diagnosis" above and data in Table 2). The holotype is shown in Fig. 16.

Male genitalia. Median lobe of aedeagus, parameres, and urite figured and described by Casale & Giachino (1991: 218, Figs 4-8).

Legs. Hind trochanter round at apex, as long as half length of hind femur. Hind coxa with posterolateral and medial setae. Onychium with one pair of dorsolateral setae; dorsal setae absent.

Speluncarius (*Pontotapinus*) sp. cf. *ponticus* Casale & Giachino, 1991

(Figs 3, 6, 9, 17)

Description of the locality. Amaçlar Cave is situated west of the Koca Ali Village, ca. 250 m elevation, 14 km southwest of the Ereğli Town, Zonguldag District, Northwestern Anatolia. The cave (known as Amaçlar Cave No. 1) was visited by Coiffait on 7th May 1954 (Coiffait, 1959: 444). In addition to the other fauna of the cave, one female Trechini was found by this author at the entrance of the cave which was subsequently described as *Anillidius coiffaiti* (Jeannel, 1955).

The entrance of the Amaçlar Cave is formed by a 6-7 metres long slope covered by humus, stones and vegetation, which at present is devastated by the local people and is used as dump for garbage. The unique specimen was found by the second author under a rusty piece of barrel fixed ca. 30 cm into the ground on 10th May 2001. In the same place he collected also the carabids *Nebria* (*Nebria*) *brevicollis* (Fabricius, 1792), *Porotachys bisulcatus* (Nicolai, 1822), *Bembidion* (*Sinechostictus*) *lederi anatolicum* (Korge, 1964), *Bembidion* (*Peryphanes*) *dalmatinum* (Dejean, 1831), *Bradycellus* (*Bradycellus*) *verbasci* (Duftschmid, 1812) and the scydmenid *Palaeostigus ruficornis schimitscheki* (Machulka, 1944) (P. Hlaváč det.). During the same visit, two specimens of *Trechus zonguldakensis* Donabauer (2004) were found. The cave was visited for the second time by one of us (RL) on the beginning of June 2003, but no other material of *Speluncarius* has been found. At this time, the slope before the cave was even more disturbed than earlier, with water conduit put ca 0,5 m deep into the soil to lead the water from the cave to the local people.

In the vicinity of Ereğli, *Speluncarius* sp. cf. *ponticus* live sympatrically with another hypogean congener – *S.* (*Hypogearius*) *heracleotes* Jeannel, 1950 (see also Jeannel, 1955: 9). The large massif of the Bolu Daglari is unique considering the occurrence of species from three phyletic lineages of *Speluncarius* and three phyletic lineages of *Tapinopterus* - *Speluncarius* (*Pontotapinus* **subgen. nov.**), *S.* (*Speluncarius*) *minimus* Cerruti, 1977, *S.*

(*Hypogearius*) spp., *Tapinopterus* (*Hoplauchenium*) *minax* (Tschitschérine, 1900), *T.* (*Molopsis*) spp. and *T.* (*Percosteropus* Ganglbauer, 1896) spp. (Vigna Taglianti, 1973; 1980; present data; unpublished data) there.

Remarks. The taxonomic status of the population near Koca Ali Village is based only on female characters. Based on the clear overall similarity and shared details of the external morphology with the holotype of *S. ponticus* the female from Amaçlar Cave these are treated as conspecific. However, the microphthalmia, fused suture of the elytra, and distance between localities (ca. 190 km distance by airline), and several characters shown in Table 2 indicate the possibility of genetically isolated populations or even separate species.

The female examined has minute eyes, distal antennomere entirely exceeding base of pronotum and fast coalesced elytra. The specimen is shown in Fig. 17.

Table 2. Differences between the holotype of *Speluncarius ponticus* and female identified as *S.* sp. cf. *ponticus*.

Character states, measurements and ratios	*Speluncarius ponticus* (holotype)	*S.* sp. cf. *ponticus*
Length of body	7.8. mm	6.5 mm
Maximal width of body	2.3 mm	2 mm
Side of pronotum in basal fifth (before angles)	Slightly sinuate	Right
Hind angles of pronotum	Almost rectangular	Obtuse
HL / HW	1.36	1.28
HL / PL	1.05	0.94
HW / PW	0.72	0.69
PW / PL	1.06	1.07
PW / PA	1.25	1.26
PW / PB	1.5	1.49
PA / PB	1.2	1.18
EW / PW	1.16	1.21
EL / EW	1.7	1.73

Symbols: HL - length of head; HW – maximal width of head; PL - length of pronotum; PW – maximal width of pronotum; PA – width of apical border of pronotum; PB – width of basal border of pronotum; EL - length of elytra; EW – maximal width of elytra.

Female genitalia. Tergum VIII long, narrow, with short "anterolateral apophyses" (Fig. 3). Both lobes of sternum VIII with slightly reduced internal membrane areas, "anterolateral apophyses" closely round (Fig. 6). Apical stylomere of left stylus smaller relative to basal stylomere than usual (for Pterostichini) (Fig. 9); dorsal ensiform seta close to distal ensiform dorsolateral seta (as distance between them smaller than distance between both dorsolateral seta); both dorsolateral ensiform setae minute, distal seta situated medially in stylomere, proximal one situated proximally on stylomere; nematiform setae very short, hardly visible. Spermatheca not studied.

Speluncarius (*Hypogium* Tschitschérine, 1900) [= *Hypogeobium* Tschitschérine, 1903]

Diagnosis. The adults of this subgenus differ from the adults of all the other subgenera of the genus by the presence of the following three characters: 1) outer posterolateral seta on hind coxa absent (Fig. 14); 2) hind trochanter pointed at apex (Fig. 14); 3) onychium with two pairs of dorsal (first one situated in distal half, second one situated in proximal half of article) and two pairs of dorsolateral setae (Fig. 15).

Type species. *Platysma albanicum* Tschitschérine, 1900

Remarks. Varying opinions have been published as to the systematic position of *S. albanicus* (Jeannel, 1950; Jeannel, 1953; Vigna Taglianti, 1973; Sciaky, 1982; Casale & Giachino, 1991). Some have agreed that it belongs to *Speluncarius* rather than to the genera related to *Molops* Bonelli, 1810 (Vigna Taglianti, 1973; Sciaky, 1982; Casale & Giachino, 1991). Müller (1937: 134) proposed a synonymy between *Elasmopterus* and *Hypogeobium* (replacement name for *Hypogium*), and Jeanne (1982) did not include the latter within *Speluncarius*. No doubt, the main reason for these conflicting placements was the lack study material, especially the important male specimens.

We believe that the two morphological features previously deemed important for the distinction of this species from the other taxa of *Speluncarius*, apical half of antennomere 3 paddle-like and presence of small tubercle at the posterior supraorbital pore, are overestimated (Apfelbeck, 1904: 236; Mařan, 1932: 36; Straneo, 1935: 85; Schatzmayr, 1942: 51; Schatzmayr, 1943: 130) and are not useful at the specific or supraspecific level. This is based on the lack distinctive states for these features in the specimen at hand. We think also that the absence of the outer posterolateral seta on the hind coxa is certainly autapomorphy in *Hypogium* relative to other phyletic lineages of "*Tapinopterus – Speluncarius*". The attenuation of the apex of hind trochanter is found in other lineages of this complex, like *Hoplauchenium* Tschitscherine, 1900 and *Hoplodactylella* Strand, 1936, and probably represents an adaptive character to subterranean life. In some pterostichines from the New World this feature is sexually dimorphic (Will, 2004). The additional pubescence on the dorsal part of onychium in *S. albanicus* is unique in the complex "*Tapinopterus-Speluncarius*" but it is not clear if this is a plesiomorphic or apomorphic state.

Speluncarius (*Hypogium*) *albanicus* (Tschitschérine, 1900)

(Figs 4, 7, 10, 13, 14, 15, 18)

References. *Platysma* (*Hypogium*) *albanicum*: Tschitschérine, 1900: 49-50 ("Albanie", loc. typ.). *Pterostichus* (*Hypogium*) *albanicus*: Apfelbeck, 1904: 236-237; Heyden et al., 1906: colum 82. *Platysma* (*Hypogeobium*) *albanicum*: Jakobson, 1907: 353. *Speluncarius* (*Hypogeobium*) *albanicum*!: Reitter, 1914: 262. *Platysma albanicus*: Lutshnik, 1922: 78. *Pterostichus* (*Hypogeobium*) *albanicus*: Winkler, 1924: 175; Schatzmayr, 1930: 329. *Tapinopterus* (*Hypogeobium*) *albanicus*: Mařan, 1932: 36; Müller, 1937: 134; Schatzmayr, 1942:

51; Schatzmayr, 1943: 131. *Hypogeobium albanicum*: Jeannel, 1950: 160-161; Jeannel, 1953: 11; Mateo, 1955: 299; Negre, 1977: 140; Vigna Taglianti, 1973: 356. *Tapinopterus albanicus*: Turin, 1981: 127. *Speluncarius* (*Hypogeobium*) *albanicus*: Sciaky, 1982: 28; Casale & Giachino, 1991: 215-216; Lorenz, 1998: 268. *Speluncarius* (*Hypogium*) *albanicus*: Bousquet, 2003: 516; Vigna Taglianti, 2004.

Remarks. The type locality is not specified, but this species is remarkable for its chetotaxy. The holotype in shown in Fig. 18.

Legs. Hind trochanter pointed at apex, as long as half length of hind femur (Fig. 14). Hind coxa with medial seta, posterolateral seta absent (Fig. 14). Onychium with

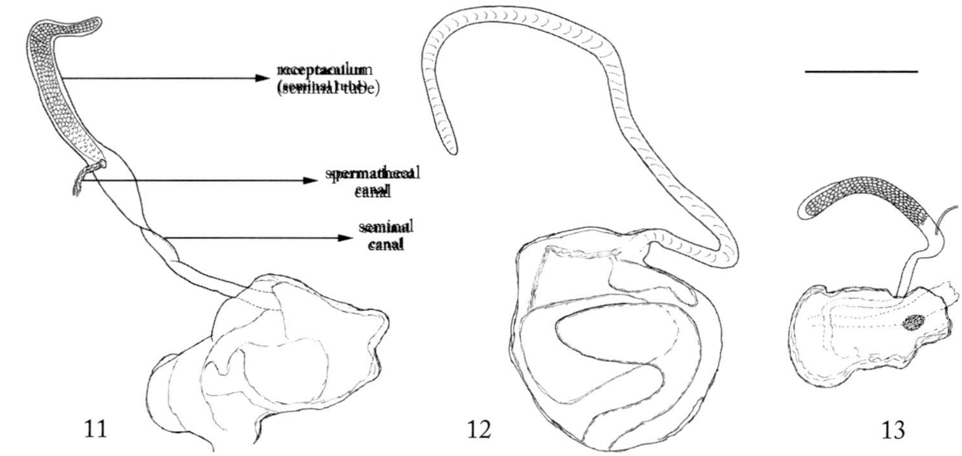

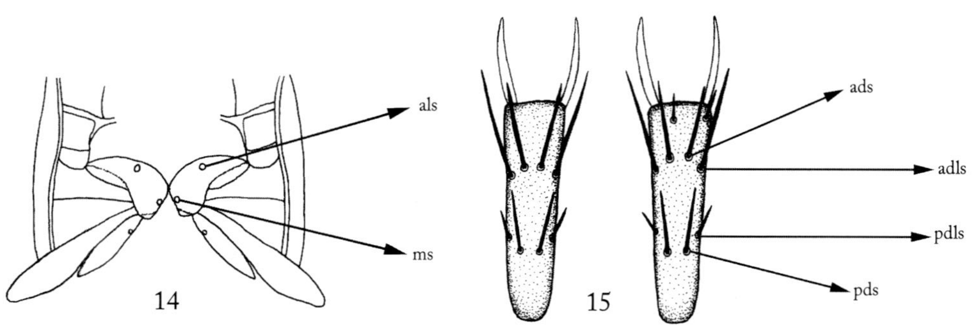

Figs 11-13. Drawings of spermatheca. Fig. 11: *Speluncarius anophthalmus* (female from Herzegovina, Orjen Mt., Bjela Gora, Milanov Ocijek, ventral view); Fig. 12: *Speluncarius anophthalmus* (female from "Herzegovina, Reitter", dorsal view, spermathecal canal not shown); Fig. 13: *S. albanicus* (holotype, ventral view). Scale line = 0.5 mm.

Fig. 14. Sketch of metasternum, part of hind legs and setation of hind coxae, *Speluncarius albanicus* (holotype, ventral view), als – anterolateral seta, ms – medial seta.

Fig. 15. Sketch of left and right hind onychia, *Speluncarius albanicus* (holotype, dorsal view), ads – anterior dorsal seta, adls – anterior dorsolateral seta, pds – posterior dorsal seta, pdls – posterior dorsolateral seta.

16 17 18

Figs 16-18. Photos of habitus. Fig. 15: *Speluncarius ponticus* (holotype); Fig. 16: *S.* sp. cf. *ponticus* (female from Amaçlar Cave); Fig. 17: *S. albanicus* (holotype).

two pairs of lateral and two pairs of dorsal setae as well as with a few additional dorsal setae (Fig. 15).

Female genitalia. Tergum VIII short, wide, with relatively long "anterolateral apophyses" (Fig. 4). Sternum VIII asymmetrical, both lobes with complete internal membrane areas, "anterolateral apophyses" widely round (Fig. 7). Apical stylomere of left stylus of usual (for Pterostichini) size with relation to basal stylomere (Fig. 10); dorsal ensiform seta dorsad, removed from distal dorsolateral ensiform seta at distance longer than distance between two dorsolateral setae; both dorsolateral ensiform setae long and thick, situated in paramedial part of stylomere; nematiform setae long. Spermatheca with seminal canal and receptaculum differentiated (Fig. 13); seminal canal short, thinner than receptaculum; receptaculum elongate, gradually curved apically; spermathecal canal inserted on proximal part of receptaculum just before junction of seminal canal and receptaculum.

Key for identification of the subgenera of *Speluncarius* Reitter, 1886

1(6) Distal part of antennomere 3 pubescent

2(3) Pronotum with two anterolateral setiferous punctures. Greece .... ***Elasmopterus***

3(2) Pronotum with single anterolateral setiferous puncture

4(5) Fore angles of pronotum slightly projecting. Posterolateral seta on hind coxa present. Hind trochanter round at apex. Dorsal setae of onychium absent. Northwest Anatolia .... ***Pontotapinus* subgen. nov.**

5(4) Fore angles of pronotum fairly projecting. Posterolateral seta on hind coxa absent. Hind trochanter pointed at apex. Dorsal setae of onychium present. Albania....... ....***Hypogium***

6(1) Distal part of antennomere 3 glabrous

7(8) Lateral border of pronotum sinuate or rectilinear in apical part, not or hardly sinuate in basal part. Aedeagus strongly flattened laterally. Distal half of elytra with single setiferous puncture at interval 3. Northwest Anatolia ....***Hypogearius***

8(7) Lateral border of pronotum more or less convex in apical part, sinuate in basal part. Aedeagus not flattened laterally. Distal half of elytra with one or more setiferous punctures at interval 3. Italy, Balkan Peninsula (Herzegovina, Montenegro, Albania, Greece, Crete), North Anatolia .... ***Speluncarius* s. str.**

CONCLUSIONS

We arrive at several conclusions based on the study of selected characters (in particular the setation of the legs and the female genitalia) and on the present knowledge of the morphology of adults. The examination of the holotypes of *Speluncarius ponticus* and *Platysma albanicus* demonstrates that the two taxa are members of separate phyletic lineages of the genus *Speluncarius*. On this basis the new subgenus *Pontotapinus* **subgen. nov.** is proposed, based on type species *Speluncarius ponticus* Casale & Giachino, 1991. The study of the female collected near Amaçlar Cave shows that probably it is conspecific with *Speluncarius ponticus* Casale & Giachino, 1991 despite the presence of several differences. On the other hand, the study suggests that *Hypogium* Tschitschérine, 1900 represents well-isolated lineage with probably basal position within the genus, and not belong to the "molopite complex". For the time being, monophyly or polyphyly of *Speluncarius* cannot be demonstrated, while the monophyly of the "*Tapinopterus – Speluncarius*" complex is well-supported by three clear synapomorphies. Excluding the nominotypical subgenus, the monophyly of the subgenera of *Speluncarius* seems obvious. *Speluncarius* s. str. may include at least three separate phyletic lineages – the "*anophthalmus*" group, the *pasquinii*" group, and the "*breuningi*" group (cfr. Jeanne, 1982).

ACKNOWLEDGEMENTS

This study has been realized due to the first author's visits granted by the European Union-funded Integrated Infrastructure Initiative "Synthesys" (Applications DE-TAF-725, AT-TAF-758 and HU-TAF-817). For the work with the collections and the materials on loan, we thank to Petar Beron (NMNHS), Robert Booth (BMNH), Robert Davidson (CMNH), Conrad Gillett (BMNH), Pier Mauro Giachino (Torino, Italy), Bernd Jaeger (ZMHU), Jiří Hájek (NMP), Alexi Popov (NMNHS), Heinrich Schönmann (NMW),

Gyozo Szel (HNHM), Manfred Uglih (ZMHU) and Lothar Zerche (DEI). Authors would like to thank also to Gejza Dunay (Kráľovce, Slovakia) and Katarína Martiňáková (Košice, Slovakia) for the habitus photos. Kipling Will (University of California, Berkeley, USA) and Peter Hlaváč (Košice, Slovakia) read and improved the English language of the manuscript. We are indebted again to Kipling Will, who took the time to revise and make substantial critical comments on the manuscript.

REFERENCES

Apfelbeck, V. (1904). Die Käferfauna der Balkanhalbinsel, mit Berücksichtigung Klein-Asiens und der Insel Kreta. Erster Band: Familienreihe Caraboidea. – R. Friedländer und Sohn, Berlin.

Bousquet, Y. (1999). Supraspecific classification of the Nearctic Pterostichini (Coleoptera: Carabidae). – Fabreries, Supplément 9: 1-292.

Bousquet, Y. (2003). Pterostichini. – In: Catalogue of Palearctic Coleoptera, Vol. 1, Archostemata – Myxophaga – Adephaga (Löbl, I. & Smetana, A., eds). Apollo Books, Stenstrup, p. 469-521.

Casale, A. & Giachino, P.M. (1991). Due nuovi carabidi della fauna sotterranea di Turchia (Coleoptera Carabidae). – Bolletino della Società Entomologica Italiana 122: 211-220.

Casale, A. & Vigna Taglianti, A. (1999). Caraboid beetles (excl. Cicindelidae) of Anatolia, and their biogeographical significance (Coleoptera, Caraboidea). – Biogeographia 20: 277-406.

Casale, A., Vigna Taglianti, A. & Juberthie, C. (1998). Coleoptera Carabidae. – In: Encyclopædia Biospeologica II (Juberthie, C. & Decu, V., eds). Société de Biospéologie, Moulis – Bucarest, p. 1047-1081.

Coiffait, H. (1959). Énumération des Grottes visitées, 1950-1957. (Neuviéme série). – Arch. Zool. exp. génér. 97: 102-465.

Deuve, T. (1993). L'abdomen et les genitalia des femelles de Coléoptères Adephaga. – Mémoires du muséum national d'histoire naturelle, Zoologie 155: 1-184.

Donabauer, M. (2004). Sechs neue Arten der gattung *Trechus* Clairville, 1806 aus der Nord-Türkei (Coleoptera: Carabidae). – Zeitschrift der Arbeitsgemeinschaft Österreichischer Entomologen 56: 43-60.

Guéorguiev, V. & Guéorguiev, B. (1999). Notes on the genus *Tapinopterus* Schaum, 1858 (Coleoptera: Carabidae: Pterostichini). I. *Tapinopterus* (s. str.) *marani* sp. n. – a new polymorphic ground beetle from the islands of Crete and Naxos. – Elytron 12: 39-48.

Heyden, L., Reitter, E. & Weise, J. (1906). Catalogus Coleopterorum Europae, Caucasi et Armeniae rossicae. Editio secunda. – R. Friedländer & Sohn, Berlin; Edmund Reitter, Paskau; Revue d'Entomologie, Caen.

Hůrka, K. (1996). Carabidae of the Czech and Slovak Republics. – Kabourek, Zlín.

Jakobson, G.G. (1907). Zhuki Rossii i Zapadnoi Evropy. – A.F. Devrien, Sankt-Petersburg.

Jeanne, C. (1982). Le genere *Speluncarius* Reitt. Description d'une espèce nouvelle et tableau des espèces connues. – Nouvelle Revue d'Entomologie 12: 37-44.

Jeanne, C. (2005). Un *Molopsis* Schatzmayr nouveau d'Anatolie méridionale (Coleoptera, Carabidae). – Nouvelle Revue d'Entomologie (N.S.) 21 (4): 383-385.

Jeannel, R. (1950). Sur deux ptérostichides cavernicoles de Majorque. – Revue française d'Entomologie 17: 157-165.

Jeannel, R. (1953). Un pterostichide cavernicole de Turquie, et remarques sur la systématique des *Tapinopterus* Schaum et genres voisins. – Notes Biospéologiques 8: 9-15.

Jeannel, R. (1955). Mission de H. Coiffait en Anatolie. – Note Biospéologiques 10: 3-10.

Liebherr, J.K. & Will, K.W. (1998). Inferring phylogenetic relationships within Carabidae (Insecta, Coleoptera) from characters of the female reproductive tract. – In: Phylogeny and classification of Caraboidea (Coleoptera: Adephaga) (Ball, G.E., Casale, A. & Vigna Taglianti, A., eds). Museo Regionale di Scienze Naturali, Torino, p. 107-170.

Lorenz, W. (1998). Nomina Carabidarum – a directory of the scientific names of ground beetles (Insecta, Coleoptera "Geadephaga": Trachypachidae and Carabidae incl. Paussinae, Cicindelinae, Rhysodinae). – W. Lorenz, Tutzing.

Lutshnik, V. (1922). O Platysmatina. – Trudy Stavropolskogo Selskokhozaystvennogo Instituta 1: 67-79.

Mařan, J. (1932). Čtyři noví Carabidi z pohoří Ilgaz-dagh v Malé Asii. Vier neue Carabiden von Ilgaz-dagh in Kleinasien. – Časopis Československé Společnosti Entomologické 28 (1-2): 30-37.

Mateo, J. (1955). Los Molopini Bon. de la Peninsula Iberica. – Eos, Revista Española de Entomologia 31: 20-301.

Müller, J.[G.] (1937). Note su alcuni carabidi della Balcania e delle regione Mediterranea. – Atti del Museo Civico di Storia Naturale di Trieste 13: 119-134.

Negre, J. (1977). Sur les Molopini Hypogés de Catalogne. – In: Communicacions del 6é. Symposium d'Espeleologia. Bioespeleologia, Terrassa, p. 139-141.

Reitter, E. (1914). Beitrag zur Kenntnis der blinden *Tapinopterus*-Arten (Col. Pterostichini). – Wiener Entomologische Zeitung 33: 261-263.

Schatzmayr, A. (1930). I *Pterostichus* italiani. – Bollettino della Società Entomologica Italiana 8: 145-339.

Schatzmayr, A. (1942). Bestimmungs-Tabellen europäischer Kafer. II. Fam. Carabidae. Subfam. Pterostichinae. 65. Gattungen: *Pterostichus* Bon. u. *Tapinopterus* Schaum. – Koleopterologische Rundschau 27: 1-80.

Schatzmayr, A. (1943). Bestimmungs-Tabellen europäischer Käfer. II. Fam. Carabidae. Subfam. Pterostichinae. 65. Gattungen: *Pterostichus* Bon. u. *Tapinopterus* Schaum. – Koleopterologische Rundschau 27: 81-144.

Sciaky, R. (1982). Le attuali conoscenze sul genere *Speluncarius* Reitt. (VII contributo alla conoscenza dei Coleoptera Carabidae). – Giornale italiano di Entomologia 1: 15-33.

Sciaky, R. & Persohn, M. (1994). Description of the male of *Speluncarius oertzeni* Kraatz, 1886 (Coleoptera Carabidae Pterostichinae). – Acta coleopterologica 10 (2): 41-44.

Straneo, S.L. (1935). Note sui Pterostichini palearctici. 1a. Alcune osservazioni sui sottogeneri *Tapinopterus*, *Crisimus*, *Nesosteropus*, *Pterotapinus*. – Bollettino della Società Entomologica Italiana 67: 82-91.

Tschitschérine, T. (1900). Description de deux nouvelles espèces du genre *Platysma* (Bon.). –L'Abeille, Journal d'Entomologie 30: 47-52.

Turin, H. (1981). Provosional checklist of the European ground-beetles (Coleoptera, Cicindelidae & Carabidae). – Monografieën van de Nederlandse Entomologische Vereniging 9: 1-249.

Vigna Taglianti, A. (1973). Considerazioni sui Carabidi cavernicoli ed endogei dell'Asia Minore (Coleoptera, Carabidae). – Int. J. Speleol. 5: 349-360.

Vigna Taglianti, A. (1980). Nouvelles données sur la systematique et la repartition géographique des Coléoptères Carabiques cavernicoles et endogés du Proche-Orient (Coleoptera, Carabidae). – Mém. Biospéol. 7: 163-172.

Vigna Taglianti, A. (2004). Family Carabidae. – In: Coleoptera. Fauna Europaea (Audisio, P., ed.). http://www.faunaeur.org/ (version 4.XI.2004). [Vigna Taglianti is not present at the given address. Please specify the exact URL!]

Will, K.W. (2004). A remarkable new species of *Trirammatus* Chaudoir (Coleoptera: Carabidae: Pterostichini) from the Valdivian Forest of Chile. – Zootaxa 758: 1-9.

Will, K.W. (2007). Four New Species of the Subgenus *Leptoferonia* Casey (Coleoptera, Carabidae, *Pterostichus* Bonelli) from California. – Proceeding of the California Academy of Sciences (series 4) 58: 49-57.

Winkler, A. (1924). Carabidae. Catalogus Coleopterorum regionis palaearcticae. Pars 2. – Winkler & Wagner, Wien.

L. Penev, T. Erwin & T. Assmann (Eds) 2008
Back to the Roots and Back to the Future. Towards a New Synthesis amongst Taxonomic, Ecological and Biogeographical Approaches in Carabidology
Proceedings of the XIII European Carabidologists Meeting, Blagoevgrad, August 20-24, 2007, pp. 143-158.

Comparative antennal morphometry and sensilla distribution pattern in three species of Siagoninae (Coleoptera, Carabidae)

Anita Giglio[1*], Pietro Brandmayr[1], Enrico A. Ferrero[2], Enrico Perrotta[1], Mariastella Romeo[1], Tullia Zetto Brandmayr[1] & Federica F. Talarico[1]

[1] Department of Ecology, University of Calabria I-87036 Arcavacata di Rende (CS), Italy. *E-mail: anita.giglio@unical.it
[2] Department of Biology, University of Trieste, via Giorgieri 7, I-34127 Trieste, Italy

SUMMARY

Antennal sensilla typology, number and distribution pattern were studied in three myrmecophagous species: *S. jenissoni* Dejean 1826, *S. dejeani* Rambur 1837 and *S. europaea* Dejean 1826 using scanning electron microscopy. Morphometric analyses show that the antennae of male and female beetles are similar in their general structure for each of the three species and vary only in size. Five different sensillar types were distinguished: sensilla chaetica, sensilla basiconica, sensillum coelocapitolum, sensillum styloconicum and sensillum "sicula-sickle"-shaped. Sensilla chaetica form four subtypes while sensilla basiconica - three subtypes. So, ten different sensilla were recognised on the antennae of each species of *Siagona*. The possible function of the sensilla was discussed and two types of sensilla (sensilla chaetica types 3 and 4) were considered as mechanoreceptors; sensilla basiconica type 3 - as proprioceptor; sensilla chaetica types 1 and 2, sensilla basiconica types 1 and 2, sensilla styloconica and possibly sensilla sicula-sickle-shaped - as chemoreceptors; sensilla coelocapitula - as thermoreceptors. No differences occur about sensillar typologies while their distribution is susceptible of increasing or diminution in number in all three species.

Keywords: carabid beetles, antennae, sensory organs, myrmecophagous

INTRODUCTION

The ground beetles, with about 40.000 species described, are among the dominant groups of terrestrial predators. The most important studies of carabid beetles refer to their economic importance as natural enemies in agricultural fields and as environmental indicators (Kielty, 1996; Lövei & Sunderland, 1996). On the grounds of these studies, the required knowledge on how the carabids perceive the specific biotic and abiotic external stimuli of the life side implies a detailed investigation of sensory structures involved in environmental stimuli perception. In insects, the discrimination of complex environmental chemical and mechanical cues is mediated by many types of cuticular sensory receptors. In the body plan, sensory structures vary in number and shape and their density are closely related with the corresponding behaviour (predation, reproduction, habitat choice and intraspecific communication) of the species. The antennae are the most important multimodal sensory organs for insects, bearing the sensilla of olfaction, taste, mechano-, hygro- and thermoreception (Keil, 1999). Therefore, they are used as "feelers" (Schneider, 1964; Zacharuk, 1985). Antennal sensilla morphology, distribution pattern and electro-physiology were studied in several ground beetles to prove that the antennae are used in prey detection and habitat selection (Kim & Yamasaki, 1996; Merivee et al., 2000, 2001, 2002, 2003, 2005; Must et al., 2006 a, b; Ploomi, 2003). Behavioural evidences suggest that all carabid predators have a prevalent use of vision or olfaction or contact senses for prey detection (Wheater, 1989; Kielty et al., 1996).

In order to give additional information about antennal sensory structures involved in prey detection we observed the antennae of three myrmecophagous species *S. jenissoni* Dejean 1826, *S. dejeani* Rambur 1837 and *S. europaea* Dejean 1826.

The genus *Siagona*, belonging to the Siagonini tribe, with ca 50 species described, has a wide geographic distribution including India, Arabia, Africa and the Mediterranean region (Andrewes, 1929; Bauer et al., 2005). However, only three species - discussed here - are present in southern Europe. Typically, all Siagonini have a strikingly flat body with a stalk-like constriction between the pro- and mesothorax that are obvious adaptations to life in narrow soil crevices. The short and strong mandibles with large retinacula are well-suited for grasping and chewing arthropod prey with hard cuticles (e.g. ants in *S. europaea*; Bauer et al., 2005). A number of systematic and geographic distribution studies have been conducted on this group but, to date, less is known about the ecological demands and life style of the species of this genus. About *S. jenissoni* and *S. dejeani* there is limited and fragmentary informations. They live in central Spain and Portugal and on the opposite coast of Morocco (Andrewes, 1929; Antoine, 1955). *S. europaea* lives on the open lands of the Mediterranean biome (sclerophylls in Italy, Spain and North Morocco) (Brandmayr & Pizzolotto, 1990). In southern Italy, *S. europaea* occurs in open land on clay soil up to altitude of about 200-600 m a.s.l. In early spring, when soil moisture is high, the beetles are found under stones. From mid-April onwards, when the soil dries out and becomes deeply fissured, they retreat in to deeper crevices, especially during the hot dry hours of the day (Bauer et al., 2005). Recent studies (Zetto Brandmayr et al., 1994, 1998) proved that its diet is exclusively myrmecophagous.

In this study, we investigated by SEM the antennal sensilla typology and the number and distribution pattern in males and females of *S. jenissoni*, *S. dejeani* and *S. europaea*.

MATERIALS AND METHODS

Insects

S. dejeani (Fig. 1A) and *S. jenissoni* (Fig. 1B) specimens were found in large aggregations under stones in Southern Spain meadows in spring 2005 (100-400 m asl) and collected by hand.

S. europaea specimens (Fig. 1C) were collected by bait-traps in southern Italy (Calabria) from open fields and pastures (250 m asl) (Squillace, Catanzaro) in spring 2003 and 2004.

Morphometric analyses

For measures, all specimens were stored in ethanol (70%). The antennae of five males and five females for each of the three species were examined and counting of antennal segments started from beetle's head. Photographs were taken with a stereoscope (Zeiss Stemi SV 11Apo) and acquired by Matrox PC-VCR software (Windows 2000). Measurements were taken using Sigma Scan Pro 5 Software (SPSS® Inc.).

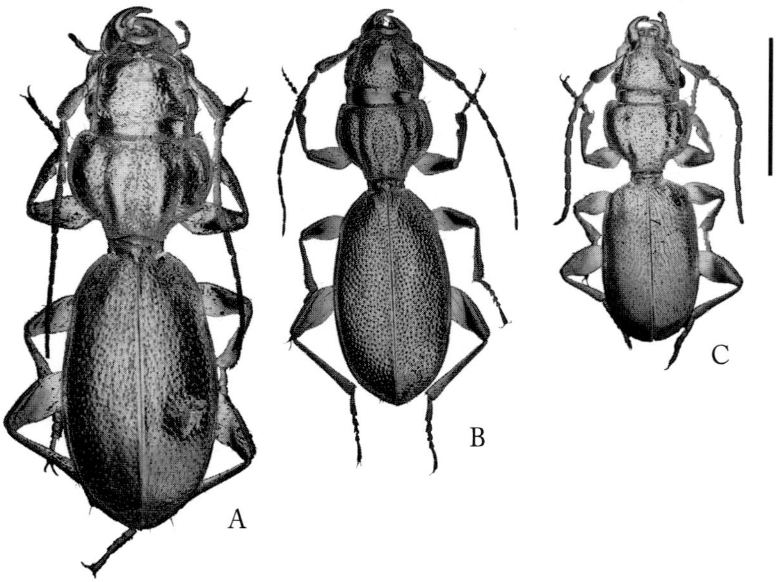

Fig. 1. Male specimens of A) *Siagona dejeani* B) *Siagona jenissoni* C) *Siagona europaea*. Scale bar 4.5 mm

Scanning electron microscopy (S.E.M.)

Carabid beetle specimens were anaesthetized with chloroform and beheaded. Four antennae from individuals of both sexes and for each of the three species were cut and fixed in 3% glutaraldehyde in 0.1 M cacodylate buffer pH 7.2-7.3, dehydrated in a graded ethanol series and critical point-dried. After dehydration, they were attached on stubs, gold coated and examined in a Cambridge Stereoscan 100 scanning electron microscope. The types, number and localization of sensilla on antennal segments were made from scanning electron microscopy montage micrographs. The percentage distribution of the sensilla types was examined on the ventral and dorsal surface. Measurements of sensorial structures were taken with Image-Pro Plus version 4.5 software (Media Cybernetics) on digitised pictures. Sensilla types are described and classified according to Schneider (1964) and Zacharuk (1985).

RESULTS

The antennae of ground beetles *S. europaea* (Figs 2, 3), *S. dejeani* (Figs 4A-C) and *S. jenissoni* (Fig. 4D) are filiform and consist of the scape, pedicel and nine flagellomeres. Table 1 shows antennal segments length and diameter for the three species. The scape, the pedicel and first-second flagellomeres are slightly flattened dorso-ventrally, while from third to ninth, the flagellomeres are round in their cross section (Figs 2A, B, C, 4A, D). The flagellomeres become shorter and wider towards the distal end of the antenna and their cuticular surface appears sculptured. The ball-joints between the scape and the head and the scape and pedicel enable the antenna to move in every direction. The antennae of male and female beetles are similar in their general structure for each of the three species and vary only in size. On the antennae of both male and female of each of the three species were identified according to their size and shape four subtypes of sensilla chaetica (s.ch. 1-4), three subtypes of sensilla basiconica (s.b. 1-3), one type of sensillum coelocapitulum (s.co), one type of sensillum styloconicum (s.st.) and one type of sensillum sicula (sickle-shaped; s.s.).

The density of sensilla per antennomere increases distal wards. Sexes do not differ in typology and number of sensilla while there is a consistent difference in sensilla topography between the three species. Number, percentage and distribution of sensilla along the antennae are summarized in Tables 2, 3 and 4.

Sensilla chaetica subtype 1 (s.ch.1) (Figs 2A, D; 3A, B; 4A, B, D) are long bristles with longitudinal grooves that spiral slightly around their surface. They are a sharp tip and an apical pore (Fig. 3B). The bristle base inserts in a wide socket and projects from the antennal surface at 40-50°. Their length ranges from 30μm to 70μm. It is the most abundant type of all sensilla on the antennae for each species and is represented on all segments. Their number grows towards the tip of the antenna in all species whereas the percentage relative to all sensilla decreases in amount.

Sensilla chaetica subtype 2 (s.ch.2) (Figs 2A, B, C; 4A, C, D) are straight bristles with a blunt tip, longitudinal grooves and a wide basal socket. Their length ranges from 90 μm to 150 μm. The number of s.ch.2 on the pedicel and from first to eighth flagellomeres is extremely consistent in all species. They are six, located in a line around the distal margin of the antennomeres. On the ninth flagellomere of all species, two sensilla stand perpendicularly to the tip and close to an apical sensorial dome (Fig. 4C).

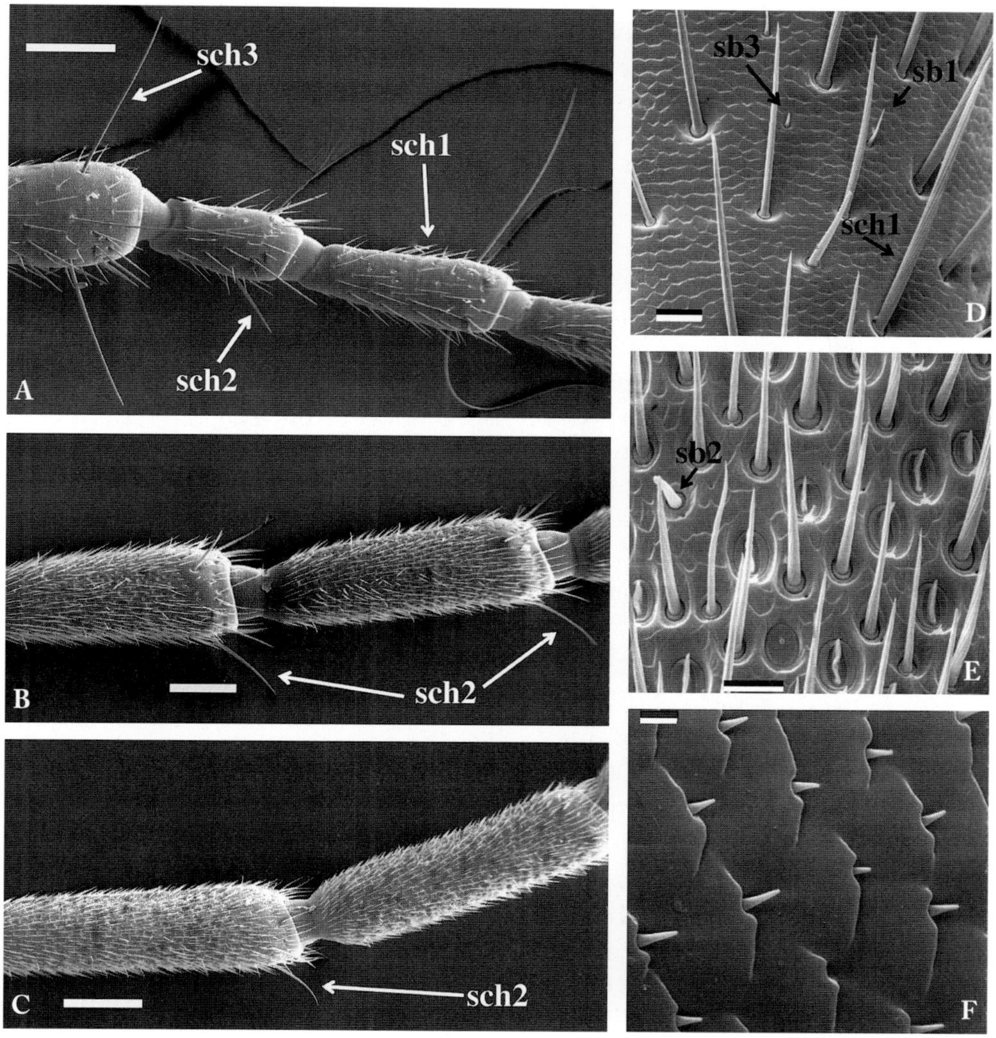

Fig. 2. *Siagona europaea* antenna. A) scape, pedicel and first flagellomere. B) second-third flagellomeres C) eighth flagellomeres. D) detail of scape surface. E) detail of eighth flagellomere surface. F) sensilla basiconica subtype 3 on intersegmental joints between the antennomeres. sb1, 2 and 3: sensilla basiconica subtype 1, 2 and 3; sch1, 2 and 3: sensilla chaetica subtypes 1, 2 and 3. Scale bars 200 μm (A), 100 μm (B and C), 10 μm (D and E) 2 μm (F).

Sensilla chaetica subtype 3 (s.ch. 3) (Figs 2A, 4A) are very long bristles (400-500 µm) with a pointed tip. These bristles have longitudinal grooves and a wide articular socket. Four of them are located in the distal part of the scape and three on the lateral side of the pedicel and first flagellomere.

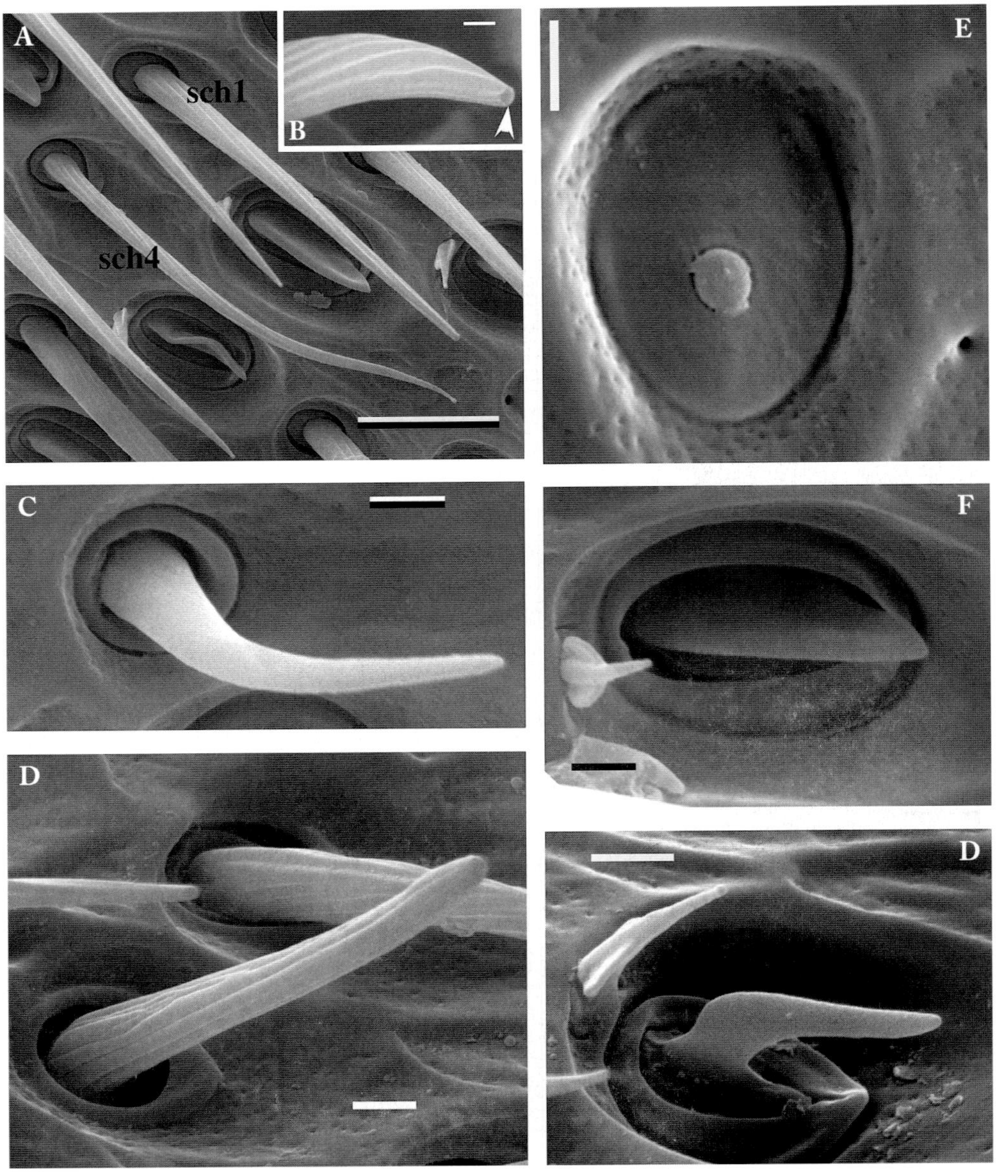

Fig. 3. *Siagona europaea* antenna. A) detail of sensilla chaetica subtypes 1 (sch1) and 4 (sch4). B) apical pore (white head arrow) of sensillum chaeticum subtype 1. C) sensillum basiconicum subtype 1. D) sensillum basiconicum subtype 2. E) sensillum styloconicum. F) sensillum coelocapitula. G) sicula-sickle-shaped sensillum. Scale bars 10 µm (A), 1 µm (B), 2 µm (C, D, E, F and G).

Sensilla chaetica subtype 4 (s.ch. 4) (Figs 3A; 4B, C) are curved bristles, with a grooved wall, a basal socket and a tapering fine point. The shaft length varies from 35μm to 45μm and width and it's base is 2.0-2.5 μm and they projects from the antennal surface at 10-15°. They are located on both ventral and dorsal aspect of all antennomeres and their distribution and density is variable in all three species.

Sensilla basiconica subtype 1 (s.b.1) (Figs 2D, 3C; 4B) are small, 17-19μm long pegs with a smooth wall and a sharp tip curved towards the antennal shaft. The base is

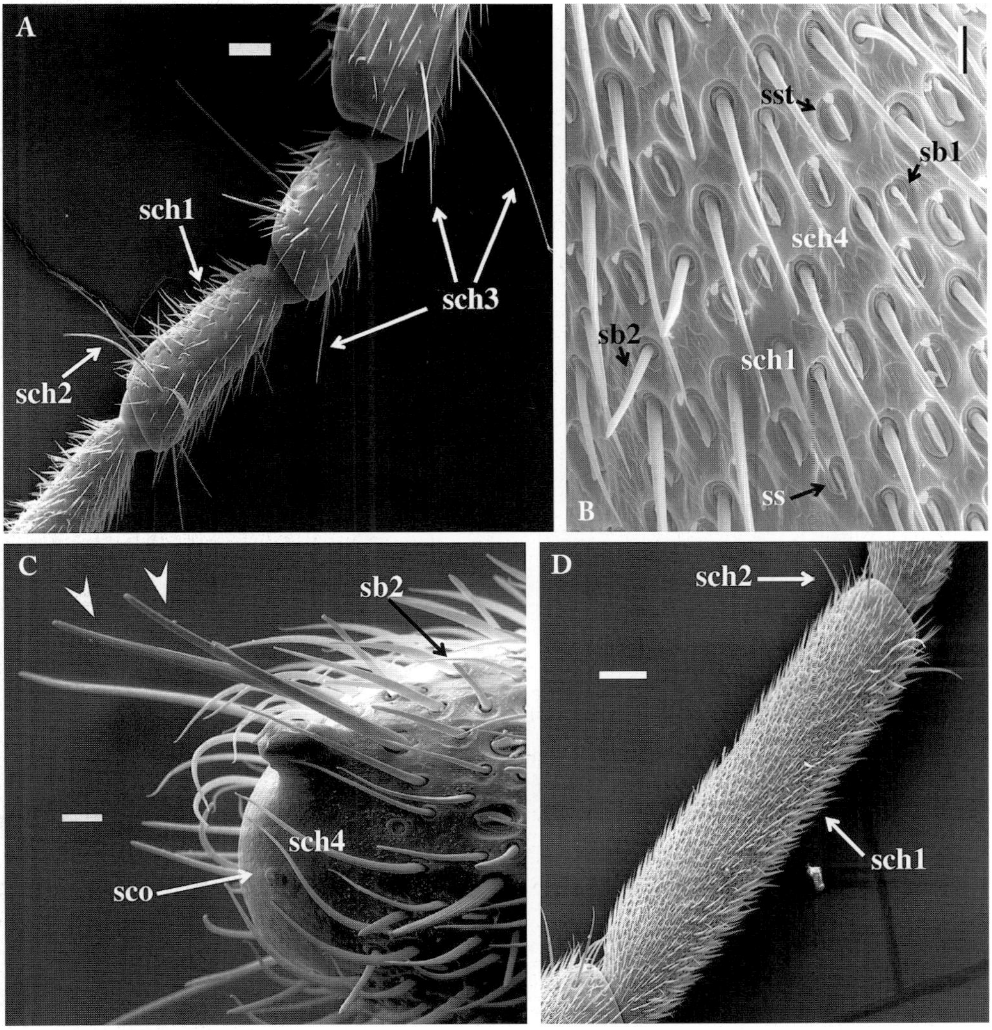

Fig. 4. *Siagona dejeani* A) scape, pedicel and first-second flagellomeres B) detail of sixth flagellomere surface. C) apical dome and sensilla chaetica subtype 2 (white head arrow) on ninth flagellomeres. *Siagona jenissoni* D) eighth flagellomere. sb1 and 2: sensilla basiconica subtypes 1 and 2; sco: sensillum coelocapitula; sch1, 2, 3 and 4: sensilla chaetica subtypes 1, 2, 3 and 4; ss: sensillum siculum; sst: sensillum styloconicum. Scale bars 200 μm (A), 10 μm (B and C), 100 μm (D).

Table 1. Length and diameter measurements (mean ± SD) of antennal segments of three *Siagona* species. Sc: scape; Pe: pedicel.

Species	Sex	Total length (mm)	Antennal segments											
					Flagellomeres									
			Sc	Pe	1	2	3	4	5	6	7	8	9	
S. europaea	Male	6.91	1	0.43±0.03	0.61±0.02	0.56±0.03	0.58±0.02	0.59±0.03	0.61±0.05	0.59±0.05	0.59±0.07	0.56±0.04	0.67±0.03	Length (mm)
	(N=5)	±0.17	0.4±0.04	0.2±0.02	0.19±0.01	0.19±0.01	0.2±0.02	0.19±0.02	0.2±0.01	0.16±0.02	0.16±0.01	0.15±0.01	0.14±0.01	Diameter (mm)
	Female	7.03	1	0.43±0.08	0.56±0.05	0.55±0.05	0.57±0.01	0.56±0.04	0.62±0.05	0.59±0.05	0.57±0.08	0.55±0.1	0.61±0.14	Length (mm)
	(N=5)	±0.66	0.4±0.04	0.23±0.05	0.21±0.03	0.19±0.06	0.2±0.03	0.21±0.02	0.19±0.02	0.16±0.01	0.15±0.01	0.14±0.01	0.13±0.02	Diameter (mm)
S. dejeani	Male	12.41	2.37±0.30	0.88±0.04	1.12±0.03	1.02±0.07	1.10±0.05	1.15±0.05	1.13±0.04	1.07±0.06	1.05±0.05	1.00±0.05	0.98±0.03	Length (mm)
	(N=5)	±0.71	0.67±0.06	0.44±0.04	0.40±0.05	0.35±0.04	0.32±0.03	0.30±0.03	0.25±0.02	0.25±0.02	0.21±0.02	0.19±0.02	0.18±0.02	Diameter (mm)
	Female	12.43	2.39±0.26	0.78±0.11	1.04±0.05	0.95±0.08	1.02±0.04	1.11±0.06	1.06±0.06	1.04±0.06	0.98±0.02	0.98±0.03	0.95±0.06	Length (mm)
	(N=5)	±0.58	0.74±0.05	0.50±0.03	0.45±0.04	0.40±0.05	0.33±0.01	0.31±0.03	0.30±0.01	0.25±0.05	0.21±0.01	0.22±0.02	0.20±0.03	Diameter (mm)
S. jenissoni	Male	9.15	1.44±0.08	0.56±0.03	0.72±0.06	0.68±0.05	0.76±0.03	0.76±0.04	0.77±0.03	0.77±0.05	0.72±0.04	0.7±0.04	0.74±0.03	Length (mm)
	(N=5)	±0.23	0.45±0.03	0.28±0.02	0.23±0.02	0.24±0.02	0.22±0.01	0.21±0.01	0.21±0.02	0.17±0.01	0.16±0.01	0.14±0.01	0.14±0.01	Diameter (mm)
	Female	8.90	1.45±0.09	0.56±0.02	0.76±0.02	0.67±0.02	0.75±0.04	0.75±0.05	0.78±0.04	0.78±0.04	0.73±0.02	0.75±0.03	0.72±0.02	Length (mm)
	(N=5)	±0.41	0.43±0.05	0.27±0.02	0.24±0.01	0.24±0.02	0.23±0.02	0.22±0.02	0.21±0.01	0.18±0.02	0.17±0.01	0.15±0.01	0.15±0.01	Diameter (mm)

Table 2. Percentage numbers and distribution of sensilla on the male (A) and female (B) antennae of *S. europaea* Sc: scape; Pe: pedicel

A)		Antennal segments (male)										
				Flagellomeres								
	Type (%)	Sc	Pe	1	2	3	4	5	6	7	8	9
Ventral side	s.ch.1	82	81	86	81	84	68	61	57	55	53	54
	s.ch.4	5	2	-	19	8	11	8	6	7	6,5	6
	s.st.	-	-	-	-	3	11	15	17	14	14	6,5
	s.b.1	-	-	-	-	2	0,5	1	0,5	0,5	1	4
	s.b.2	-	-	9	-	1	1,5	2	2,5	2	2	1,5
	s.b.3	13	17	5	-	-	-	-	-	-	-	-
	s.co.	-	-	-	-	-	-	-	-	-	-	-
	s.s.	-	-	-	-	2	8	13	17	21,5	23,5	28
Dorsal side	s.ch.1	81	85	83	85	77	62,5	55,5	54	48	50	52
	s.ch.4	-	-	1,5	15	11	9	8,5	8,5	8	7	6
	s.st.	-	-	-	-	5	11	17,5	19,5	20	12	11
	s.b.1	-	-	-	-	-	-	-	-	0,5	1	0,5
	s.b.2	-	-	15,5	-	3	3	3	3	2	1	3
	s.b.3	19	15	-	-	-	-	-	-	-	-	-
	s.co.	-	-	-	-	1	0,5	-	-	0,5	-	-
	s.s.	-	-	-	-	3	14	15,5	15	21	29	27,5
B)												
Ventral side	s.ch.1	87	81	83	88	81	63	65	55	58	54	55,5
	s.ch.4	3	-	3	12	7,5	6	11	6	5	7	4
	s.st.	-	-	-	-	6,5	22	17	25,5	23,5	17	15
	s.b.1	-	-	-	-	-	1	-	0,5	0,5	1,5	1
	s.b.2	-	12	14	-	0,5	1,5	1	1	1	2	1,5
	s.b.3	10	7	-	-	0,5	-	-	-	-	-	0,5
	s.co.	-	-	-	-	-	0,5	-	1	-	0,5	0,5
	s.s.	-	-	-	-	4	6	6	11	12	18	22
Dorsal side	s.ch.1	69	86	76,5	85	77	65	61	59,5	60	57	48
	s.ch.4	4	6	3	15	9	7	7	4,5	5	7	6
	s.st.	-	-	2	-	7	15	20	20	18	11	11
	s.b.1	-	-	-	-	-	-	-	-	-	0,5	-
	s.b.2	-	-	14,5	-	1	2	2	2	2	1	4
	s.b.3	27	8	4	-	-	-	-	-	-	0,5	0,5
	s.co.	-	-	-	-	1	0,5	1	-	-	-	0,5
	s.s.	-	-	-	-	5	10,5	9	14	15	23	30

Table 3. Percentage numbers and distribution of sensilla on the male (A) and female (B) antennae of *S. dejeani* Sc: scape; Pe: pedicel

A)		Antennal segments (male)										
				Flagellomeres								
	Type (%)	Sc	Pe	1	2	3	4	5	6	7	8	9
	s.ch.1	93	80	88	87	86,5	70,5	65	69	68	58	62
	s.ch.4	7	7	7	11	11	7	4	3	3	1	1,5
	s.st.	-	-	-	-	-	20	27	25	26	29	27
Ventral side	s.b.1	-	-	1	-	0,5	1	1,5	1	2	11	8
	s.b.2	-	-	-	-	1	1	2	1,5	0,5	0,5	1
	s.b.3	-	13	-	1	-	-	-	-	-	-	-
	s.co.	-	-	4	1	-	-	-	-	-	-	-
	s.s.	-	-	-	-	1	0,5	0,5	0,5	0,5	0,5	0,5
	s.ch.1	94	80	86	88	75	67	57,5	63	60	58	60
	s.ch.4	6	10	14	12	5	4	5	2,5	3	2	2
	s.st.	-	-	-	-	18	26	34	32	34	31	30,5
Dorsal side	s.b.1	-	-	-	-	0,5	0,5	0,5	0,5	1	7	6
	s.b.2	-	-	-	-	1	1	2	1	1	1	1
	s.b.3	-	10	-	-	-	-	-	-	-	-	-
	s.co.	-	-	-	-	-	1	0,5	0,5	0,5	0,5	-
	s.s.	-	-	-	-	0,5	0,5	0,5	0,5	0,5	0,5	0,5
B)												
	s.ch.1	96	82	86	86	75	58	58	56	48	51	54
	s.ch.4	4	7	8	10	7	4	3	3	3	2	2.5
	s.st.	-	-	-	-	14.5	35	36	36	42	39	37
Ventral side	s.b.1	-	-	2	3	-	1	1	1	4	5	4.5
	s.b.2	-	4	-	-	2.5	1,5	1,5	3	2	2,5	1
	s.b.3	-	7	4	-	-	-	-	-	-	-	-
	s.co.	-	-	-	1	0.5	-	-	-	-	-	0.5
	s.s.	-	-	-	-	0.5	0,5	0,5	1	1	0,5	0,5
	s.ch.1	90	90	77	92	76	62	53	50	46	42	44
	s.ch.4	10	10	14	5	5	4	4	3	3	2	3
	s.st.	-	-	-	-	15	31.5	38	44.5	43	40	40
Dorsal side	s.b.1	-	-	-	-	-	0.5	2	-	6	13	11
	s.b.2	-	-	5	3	2,5	1	2	1	1	1	1
	s.b.3	-	-	4	-	-	-	-	-	-	-	-
	s.co.	-	-	-	-	0,5	0,5	0,5	1	0,5	0,5	0,5
	s.s.	-	-	-	-	1	0.5	0,5	0,5	0,5	1,5	0,5

Table 4. Percentage numbers and distribution of sensilla on the male (A) and female (B) antennae of *S. jenissoni*. Sc: scape; Pe: pedicel

A)		Antennal segments (male)										
				Flagellomeres								
	Type (%)	Sc	Pe	1	2	3	4	5	6	7	8	9
	s.ch.1	97	84	65	84	84	85	68	60	48	42	56
	s.ch.4	3	8	16	13	8	7	3	5	3	6	3
	s.st.	-	-	-	-	4	4	25	30	43	47	36
Ventral side	s.b.1	-			3	1	1	2	1	3	2	2
	s.b.2	-	-	-	-	1	1	1	1	1	0,5	1
	s.b.3	-	8	10	-	-	-	-	-	-	-	-
	s.co.	-		9	-	1	1	0,5	2,5	1	2	1
	s.s.	-	-	-	-	1	1	0,5	0,5	1	0,5	1
	s.ch.1	94	82	77	77	83	68	43	56	44	36	48
	s.ch.4	6	12	10	9	3	5	10	4	3,5	3	3
	s.st.	-	-	-	-	4	21	39	32	47	52	43
Dorsal side	s.b.1	-	-	4	7	1	1	1	2	2	3	2
	s.b.2	-	-	-	3	4	1,5	5	3,5	2,5	3	3
	s.b.3	-	6	5	-	-	-	-	-	-	-	-
	s.co.	-	-	4	4	4,5	3	1,5	2	0,5	2,5	0,5
	s.s.	-	-	-	-	0,5	0,5	0,5	0,5	0,5	0,5	0,5
B)												
	s.ch.1	93	74	80	87	87	68	60	52	47	50	42,5
	s.ch.4	7	14	13	10	6	5	11	8	10	11	4,5
	s.st.	-	-	-	-	-	19	21	29	28	23	35
Ventral side	s.b.1	-	-	-	-	1	3	3	5	10	12	13
	s.b.2	-	-	-	-	3	3	2	3,5	3	3	3,5
	s.b.3	-	12	7	-	-	-	-	-	-	-	-
	s.co.	-	-	-	3	2	1,5	2,5	2	1	0,5	1
	s.s.	-	-	-	-	1	0,5	0,5	0,5	1	0,5	0,5
	s.ch.1	82	77	85	87	72	63	49	50	40	44	44
	s.ch.4	18	13	12	10	6	9	8	7	9	9	4
	s.st.	-	-	-	-	14	21	34	33	37	33	38
Dorsal side	s.b.1	-	-	-	-	2	1	2	3	3	5	6
	s.b.2	-	-	-	-	3,5	3	3,5	5	5	7	7
	s.b.3	-	10	3	-	-	-	-	-	-	-	-
	s.co.	-	-	-	3	2	2	3	1,5	5,5	1,5	0,5
	s.s.	-	-	-	-	0,5	1	0,5	0,5	0,5	0,5	0,5

2.6-2.8 µm wide and sits in a socket. They are found on *S. europaea* from third to ninth flagellomeres in males and from fourth to ninth in females; on *S. dejeani* from first to ninth flagellomeres in both males and females and on *S. jenissoni* from first to ninth flagellomeres in males and from third to ninth in females. Their number on the ventral side of the flagellum is higher than on the dorsal side and in *S. jenissoni* this peg occurs along the middle of flagellomere ventral area.

Sensilla basiconica subtype 2 (s.b.2) (Figs 2E; 3D; 4B, C) are 20-30 µm long with longitudinal grooves on the wall and blunt tip. Their diameter at the base is 3.5-4.5 µm and the peg projects from the antennal surface at 50-70°. They are found on *S. europaea* from first to ninth flagellomeres in males and pedicel and flagellum in females, on *S. dejeani* from third to ninth flagellomeres in males and in females flagellum and on *S. jenissoni* from second to ninth flagellomeres in males and from third to ninth in females.

Sensilla basiconica subtype 3 (s.b.3) (Figs 2D, F) are very thin cones of 6-10µm in length, with a sharp tip and a basal socket. They occur in a small number: a) on scape, pedicel and first, third, eighth, ninth flagellomeres in *S. europaea* female and on scape, pedicel and first flagellomere in male (Fig. 2D); b) on pedicel and first flagellomere in both males and females of *S. jenissoni*; c) on pedicel and first flagellomere in *S. dejeani* female and on pedicel and second flagellomere in male; d) at the base of all antennomeres close to intersegmental joints (Fig. 2F).

Sensilla coelocapitula (s.co) (Figs 2E, 3E; 4C) are small caps (1.5-1.8 µm in diameter) sitting in center of flat-organs (11x6µm in diameter) surrounded by a cuticular ring. They are not great number (from 0.5% to 3.5%), however they are a larger amount in *S. jenissoni* than in *S. dejeani* and *S. europaea*. They occur on from second to ninth flagellomere in *S. europaea* and on from third to ninth flagellomere in *S. dejeani* and *S. jenissoni*.

Sensilla styloconica (s.st.) (Figs 2E, 3A, F; 4B) are pegs with a knife edge and an elliptical base (12-15µm x 6-7.5µm) inserted in the cuticle by in sunk socket. The longest dimension is parallel to the long axis of the antenna. A spine projects at the proximal base edge of this sensillum. They are located both on ventral and dorsal side from third to ninth flagellomere in all three species and their number grows towards the tip of the antenna.

Sensillum sicula-sickle-shaped (s.s.) (Figs 3G, 4B) is sickle shaped pegs 6-10µm long with a smooth wall pointing distally with a smooth wall. The elliptical base is 2-4µm x 3.5-6 µm in diameter. A spine projects from the edge of to the depression where this sensillum sits. They are found from third to ninth flagellomere both in males and females of all three species. This sensillum is in largest amount in *S. europaea* (from 3% to 30% among flagellomeres).

DISCUSSION

The present study shows that - based on morphological features - there are ten distinct sensilla on the antennae of *S. europaea, S. dejeani* and *S. jenissoni*. These sensilla belongs to five types: chaetica, basiconica, coelocapitula, styloconica and sicula-sickle-shaped.

Four subtypes of sensilla chaetica are found on the antennae of each species. In the Coleoptera, sensilla chaetica have been described for the antennae of a large number of species. In most cases, they are classified as trichoid or chetoid sensilla. S.ch.1 are one of the most abundant type of sensilla and they cover the entire surface of the beetle's antennae. The presence of a flexible socket and an apical pore indicate that it's a contact chemoreceptor. Sensilla chaetica type 2, 3 and 4 have a wide articulated socket and a pointed tip and differ mainly in length and distribution pattern. Based on electrophysiological analysis s.ch.2 is a contact chemoreceptor that in *Pterostichus aethiops* and *P. oblongopunctatus* responds to the solutions of nine salts tested and pH variation (Merivee et al., 2004, 2005). Ultrastructural evidence in ground beetle *Nebria brevicollis* indicates that the most probable function of s.ch.3 and 4 is mechanoreception (Daly & Ryan, 1979). In *Loricera pilicornis* they are used to form a trap for capturing collembolans on which it feeds (Hintzpeter & Bauer, 1986).

Three subtypes of pegs are found. Sensilla s.b.1 is an articulated peg located on the ventral side of antennae that resembles s.b.1 and 2 of *Platynus dorsalis* and *Bembidion properans* (Merivee et al., 2000, 2001, 2002) and sensillum trichodeum B of parasitic and non-parasitic bees (Wcislo, 1995). Its function is not known. Sensilla like s.b.2 occur on parasitic Hymenoptera antennae and they are indicated as uniporous gustatory sensilla involved in host recognition by Isidoro et al. (1996). Ultrastructural evidence in the ground beetle *N. brevicollis* indicates that s.b.3, located on antennomere surface, function as olfactory receptors (Daly & Ryan, 1979). Nevertheless, their placement on the antennomere joint of *Siagona* species suggests a role in proprioception of antennal position and movement as in other carabid species investigated by Merivee et al. (2000, 2001, 2002).

Coelocapitulum sensillum is a mushroom-shaped protrusion situated in a shallow depression that appears to be similar in bees and ants where it is involved in hygro- and thermoreception (Yokohari, 1983; Ehmer & Grouenberg, 1997). The morphology, number and distribution along the antennae in all the three species described are similar to sensilla named 'campaniformia' found in other carabid and click beetles described (Merivee et al., 1998, 1999). Electrophysiological tests in *P. aethiops, P. oblongopuntactus* and *Poecilus cupreus* show that this sensillum on the antenna respond to temperature changes as typical cold receptors (Merivee et al., 2003; Must, 2006 a, b).

Sensilla styloconicum and sicula that occur from third to ninth flagellomere have not been found on the antenna of carabid beetles and generally for other Coleoptera described in literature. Sensilla styloconica seem to be similar to multiporous sensilla that occur on parasitic Hymenoptera antenna interpreted as olfactory organs on the basis of ultrastructural investigations (Isidoro et al., 1996, 2001). Sensilla like s. sicula are described in antenna of worker ants but their function is not known (Callahan, 1975).

To conclude, based on purely morphological evidence and on literature data (Nagel, 1979; Daly & Ryan, 1979; Kim & Yamasaki, 1996; Merivee et al., 1997, 1999, 2000, 2001, 2002, 2003), the sensillar pattern of the three species is capable of responding to a variety of stimuli, olfactory, gustatory and tactile as well as of being involved in thermoreception. The nomenclature varies sometimes because of lack of fixed criteria for their identification on the basis of the cuticular shape and the position. Behavioural data have show that *S. europaea* beetles are predators with nocturnal activity (Zetto Brandmayr et al., 1998; Bauer et al., 2005). In absence of light therefore, it is needful a large and diverse sensory complement to enable predation. Our results suggest that the olfactory and gustatory sensorial pattern on the antenna can be related to the predator behaviour of those myrmecophagous species. Indeed, chemically mediated habitat selection and prey detection, involving mainly antennal receptors, has been shown to occur in several species of ground beetles by many authors (Evans, 1983; Kielty et al., 1996; Merivee et al., 2004; Milius et al., 2006), while thermoreception may be correlated with specific ecological preferences of those eurythermic species. We suppose that thermoreception plays an important role in the behaviour of these three species living in open land on clay soil. In fact, after the rain season, when the soil dries out and cracks, microclimatic conditions may be quite disadvantageous in the absence of informative cues about environmental temperature and humidity. No sexual differences in the types of sensilla were found on the antenna of these three species. This rules out sex specific specialized sensilla for pheromonal reception and reproduction. However, their distribution varies in number both on ventral and dorsal side of antennomeres in all three species. Asymmetries in the distribution pattern of sensilla on insect antennae may be due to the peculiarities of their behaviour (waiting, walking, antennal movements).

Further investigations are needed to clarify the function and ultrastructure of the sensilla found.

ACKNOWLEDGMENTS

This research was funded by a MURST (ex 60%) grant "Ricerche eco-etologiche e morfofunzionali in Coleotteri carabidi" to T. Zetto.

REFERENCES

Andrewes, H.E. (1929). Coleoptera, Carabidae. I. Carabinae. – In: The fauna of British India, including Burma and Ceylon (J. Stephenson ed.). Taylor and Francis, London.

Antoine, M. (1955). Coléoptères Carabiques du Maroc (primière partie). – La Rose, Paris. 177 pp.

Bauer, T., Talarico, F.F., Mazzei, A., Giglio, A., Zetto Brandmayr, T., Brandmayr, P. & Betz, O. (2005). Hunting ants in Mediterranean clay soils: life history of *Siagona europaea* (Coleoptera, Carabidae). – Italian Journal of Zoology 73: 33-42.

Brandmayr, P. & Pizzolotto, R. (1990). Ground beetle coenoses in the landscape of the Nebrodi Mountains, Sicily (Coleoptera, Carabidae). – Il Naturalista Siciliano XIV (suppl.): 51-64.

Callahan, P.S. (1975). Insect antennae with special reference to the mechanism of scent detection and the evolution of the sensilla. – International Journal of Insect Morphology & Embryology 4 (5): 381-450.

Daly, P.J. & Ryan M.F. (1979). Ultrastructure of antennal sensilla of *Nebria brevicollis* Fab. (Coleoptera, Carabidae). – International Journal of Insect Morphology 8: 169-181.

Ehmer, B. & Grouenberg, W. (1997). Proprioceptors and fast antennal reflexes in the ant *Odontomachus* (Formicidae, Ponerinae). – Cell Tissue Res. 290:153-165.

Evans, W.G. (1983). Habitat selection in the Carabidae. – The Coleopterists Bulletin 37 (2): 164-167.

Hintzpeter, U. & Bauer, T. (1986). The antennal setal trap of the ground beetle *Loricera pilicornis*: a specialization for feeding on Collembola. – J. Zool. Lond. 208: 615-630.

Isidoro, N., Bin, F., Colazza, S.& Vinson, S.B. (1996). Morphology of antennal gustatory sensilla and glands in some parasitoid Hymenoptera with hypothesis on their role in sex and host recognition. – J. Hym. Res. 5: 206-239.

Isidoro, N., Romani, R. & Bin, F. (2001). Antennal multiporous sensilla: their gustatory features for host recognition in female parasitic wasps (Insecta, Hymenoptera: Platygastroidea). – Microscopy Research and Technique 55: 350-358.

Keil, T. A. (1999). Morphology and development of the peripheral olfactory organs. – In: Insect olfaction (Hansson, B.S. ed.). Springer-Verlag, Berlin-Heidelberg, p 5-48.

Kielty, J.P., Allen-Williams, L.J., Underwood, N. & Eastwood, E.A. (1996). Behavioral responses of three species of Ground beetles (Coleoptera: Carabidae) to olfactory cues associated with prey and habitat. – Journal of Insect Behavior 9 (2): 237-250.

Kim, J.L. & Yamasaki, T. (1996). Sensilla of *Carabus (Isiocarabus) fiduciarius saishutoicus* Csiki (Coleoptera: Carabidae). – Int. J. Insect Morphol. & Embryol. 25 (1/2): 153-172.

Lövei, G.L. & Sunderland, K.D. (1996). Ecology and Behaviour of ground beetles (Coleoptera: Carabidae). – Annual Review of Entomology 41: 231-256.

Merivee, E., Ploomi A., Milius, M., Luik, A. & Heidemaa, M. (2005). Electrophysiological identification of antennal pH receptors in the ground beetles *Pterosticus oblongopunctatus*. – Physiological Entomology 30: 122-133.

Merivee, E., Ploomi, A., Rahi M., Bresciani, J., Ravn, H.P., Luik A. & Sammelselg, V. (2002). Antennal sensilla of the ground beetle *Bembidion properans* Steph. (Coleoptera, Carabidae). – Micron 33: 429-440.

Merivee, E., Ploomi, A., Rahi M., Luik A. & Sammelselg V. (2000). Antennal sensilla of the ground beetle *Bembidion lampros* Hbst (Coleoptera, Carabidae). – Acta Zoologica (Stockholm) 81: 339-350.

Merivee, E., Ploomi, A., Luik A., Rahi M. & Sammelselg V. (2001). Antennal sensilla of the ground beetle *Platynus dorsalis* (Pontoppidan, 1763) (Coleoptera, Carabidae). – Microscopy Research and Technique 55: 339-349.

Merivee, E., Rahi, M., Bresciani, J., Ravn, H.P. & Luik A. (1998). Antennal sensilla of the click beetle, *Limonius aeruginosus* (Olivier) (Coleoptera: Elateridae). – Int. J. Insect Morphol. & Embryol. 27 (4): 311-318.

Merivee, E., Rahi M. & Luik A. (1997). Distribution of olfactory and some other antennal sensilla in the male click beetle *Agriotes obscurus* L. (Coleoptera: Elateridae). – Int. J. Insect Morphol. & Embryol. 26 (2): 75-83.

Merivee, E., Rahi, M. & Luik A. (1999). Antennal sensilla of the click beetle, *Melanotus villosus* (Geoffroy) (Coleoptera: Elateridae). – Int. J. Insect Morphol. & Embryol. 28: 41-51.

Merivee, E., Renou, M., Mänd, M., Luik, A., Heidemaa, M. & Ploomi A. (2004). Electrophysiological responses to salts from antennal chaetoid taste sensilla of the ground beetle *Pterosticus aethiops*. – Journal of Insect Physiology 50: 1001-1013.

Merivee, E., Vanatoa, A., Luik, A., Rahi, M., Sammelselg, V. & Ploomi, A. (2003). Electrophysiological identification of cold receptors on the antennae of the ground beetle *Pterostichus aethiops*. – Physiological Entomology 28: 88-96.

Milius, M., Merivee, E., Williams I., Luik, A., Mänd, M. & Must, A. (2006). A new method for electrophysiological identification of antennal pH receptors cells in ground beetles: the example of *Pterostichus aethiops* (Panzer, 1796) (Coleoptera; Carabidae). – Journal of Insect Physiology 52: 960-967.

Must, A., Merivee, E., Luik, A., Mänd, M. & Heidemaa M. (2006a). Responses of antennal campaniform sensilla to rapid temperatures changes in ground beetles of the thibe platynini with different habitat preferences and daily activity rhythms. – J. Insect Ph. 52: 506-513.

Must, A., Merivee, E., Mänd, M., Luik, A. & Heidemaa M. (2006b). Electrophysiological responses of the antennal campaniform sensilla to rapid changes of temperatures in the ground beetles *Pterostichus oblongopunctatus* and *Poecilus cupreus* (Tribe Pterostichini) with different ecological preferences. – Ph. Ent. 31 (3): 278-285.

Nagel, P. (1979). Aspects of the evolution of myrmecophilous adaptation in Paussinae (Coleoptera, Carabidae). – In: On the evolution of behaviour in Carabid Beetles (Den Boer et al. eds).Gustav Fischer, Stuttgart-New York. p. 15-34.

Ploomi, A., Merivee, E., Rahi, Bresciani, J., Ravn, H.P., M., Luik, A., M. & Sammelselg, V. (2003) Antennal sensilla in ground beetles (Coleoptera, Carabidae). – Agronomy Res. 1 (2): 221-228.

Schneider, D. (1964). Insect antennae. – Annual Review of Entomology 9: 103-122.

Wcislo, T.W. (1995). Sensilla numbers and antennal morphology of parasitic and non-parasitic bees (Hymenoptera: Apoidea). – Int. J. Insect Morphol. & Embryol. 24 (1): 63-81.

Wheater C.P. (1989). Prey detection by some predatory Coleoptera (Carabidae and Staphylinidae). – Journal Zoological Society of London 218: 171-185.

Yokohari F., (1983). The coelocapitular sensillum, an antennal hygro-and thermoreceptive sensillum of the honey bee, *Apis mellifera* L. Cell Tissue Res. 233: 355-365.

Zacharuk, R.Y. (1985). Antennae and sensilla. – In: Comparative Insect Physiology, Biochemistry and Pharmacology vol. 6 (Kerkut, G.A. & Gilbert, L.I, eds). Pergamon Press, Oxford, p. 1-69.

Zetto Brandmayr, T., Giglio, A. & De Rose, E. (1998). Feeding behaviour and food preference of *Siagona europaea* Dejean, a myrmecophagous carabid beetle (Coleoptera, Carabidae). – Ins. Soc. Life 2: 203-207.

Zetto Brandmayr, T. & Pizzolotto, R. (1994). *Siagona europaea* Dejean: first result from field collecting, life cycle and the evidence of a possible myrmecophagous diet (Coleoptera, Carabidae, Siagonini). – The Entomologist 113 (2): 120-125.

L. Penev, T. Erwin & T. Assmann (Eds) 2008
BACK TO THE ROOTS AND BACK TO THE FUTURE. TOWARDS A NEW SYNTHESIS AMONGST TAXONOMIC, ECOLOGICAL AND BIOGEOGRAPHICAL APPROACHES IN CARABIDOLOGY
Proceedings of the XIII European Carabidologists Meeting, Blagoevgrad, August 20-24, 2007, pp. 159-171.

Thoracic endoskeleton of carabid larvae (Coleoptera, Carabidae)

Artjem A. Zaitsev

Zoology & Ecology Department, Moscow Pedagogical State University, Kibalchicha Str. 6, Bld. 5, Moscow 129164, Russia. E-mail: ztema@mail.ru

SUMMARY

The thoracic endoskeleton of larvae of 17 genera belonging to 11 tribes of Carabidae, was examined. Larvae of Trachypachidae, Gyrinidae, Haliplidae, Noteridae, Dytiscidae and Rhysodidae were studied for comparison. Adaptive and taxonomic significance of endoskeletal characters are discussed.

Keywords: Adephaga, Carabidae, endoskeleton, larva, thorax

INTRODUCTION

The adaptive features of external structures of Coleoptera larvae engaged in locomotion within different habitats (modifications of urogomphi, X abdominal segment, etc.) are well studied. Apart from morpho-functional analysis, these characters are often used in systematic studies at various taxonomical levels. At the same time, the data on the muscular system and structure of the endoskeleton in particular, are fragmentary and rarely employed in the taxonomy of the order. For a clarification of the taxonomic value of endoskeletal characters it is important to recognize the presence of parallel development in unrelated taxa. The best way to find it out is the study of larvae belonging to different life forms within one family. Among Coleoptera, carabids are notable for a large diversity of larval life forms (Sharova, 1981; Brandmayr et al., 1998); that is why they were chosen as the model taxon for this study.

The structure of the thoracic endoskeleton of larval Coleoptera, and particularly, Carabidae, is poorly examined. The most detailed description for a Carabidae larva was provided by J. Barlet (1992), but only for *Cicindela hybrida* L., 1758. Other data are rather fragmentary and lack the comparative analysis within the family and between related taxa.

MATERIAL AND METHODS

Larvae of 17 genera belonging to 11 tribes of Carabidae were studied: *Cicindela campestris* L., 1758; *Megacephala euphratica* Dej., 1822; *Lophyridia littoralis* Dej., 1831 (Cicindelini); *Loricera pilicornis* Fabr., 1775 (Loricerini); *Pelophila borealis* Payk., 1790 (Pelophilini); *Calosoma auropunctatum* Herbst, 1874; *Carabus nemoralis* Muller,1764; *C. prometheus* Reitt., 1887 (Carabini), *Cychrus caraboides* L., 1758 (Cychrini), *Broscus* sp. (Broscini), *Scarites terricola* Bon., 1813 (Scaritini), *Agonum muelleri* Herbst, 1784; *Platynus assimile* Payk., 1790 (Platynini), *Zabrus tenebrionoides* Goeze, 1777 (Zabrini), *Anisodactylus binotatus* Fabr., 1787; *Harpalus rufipes* DeGeer, 1774; *Ophonus azureus* Fabr., 1775 (Harpalini), *Orthogonius* sp. (Orthogonini). For comparison larvae of other Adephaga were examined: *Trachypachus holmbergi* Mnh., 1853 (Trachypachidae); *Dineutus* sp., *Gyrinus marinus* Gyll., 1808 (Gyrinidae); *Acilius sulcatus* L., 1758; *Dytiscus marginalis* L., 1758 (Dytiscidae); *Noterus crassicornis* Muller, 1776 (Noteridae); *Haliplus* sp., *Peltodytes caesus* Duft., 1805 (Haliplidae); *Rhysodes comes* Lew., 1888 (Rhysodidae). All material is deposited in the MPSU collection.

For study of endoskeletal structure, all specimens were macerated in hot KOH solution, then examined under a Leica MZ6 stereo microscope.

All drawings were executed in pencil using a drawing tube and then processed in Corel Draw 12.

Nomenclature of sclerites and apodemes follows Barlet (1992).

Following abbreviations used in publication:

aem1-3 – epimeral apodeme of I-III thoracic segments; *aes1-3* – episternal apodeme I-III, *ais1-3* – intersegmental apodeme I-III; *apl1* – pleural apodeme I, *cd1-3* – coxal condyle I-III; *cxap2-3* – coxal apodeme II-III, *em1-3* – epimeron I-III; *es1-3* – episternite of I-III; *f1-3* – furca I-III; *pp1-3* – pleural process of esI-III; *sp2-3* – spina II-III, *spap* – apodeme of spiracle sclerite, TH1-3 – thoracic segments I-III.

RESULTS

Generalized plan of thoracic endoskeleton of Carabidae larvae

In general, the thoracic endoskeleton of Carabidae larvae consists of pleural, sternal and intersegmental apodemes (Fig. 1). The pleural ones are episternal apodemes (*aes1-3*), with

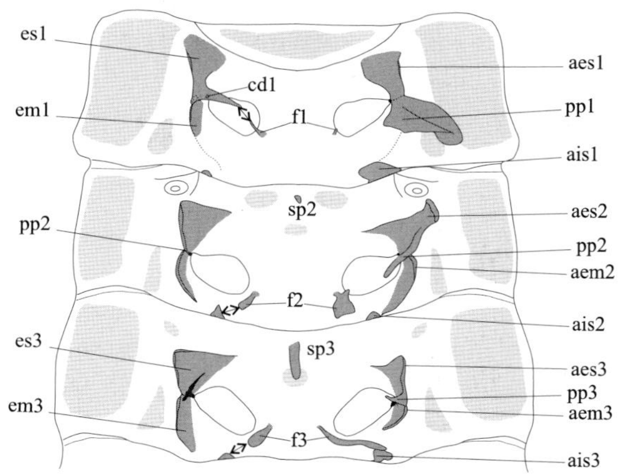

Fig. 1. Generalized plan and some variations of thoracic endoskeleton of Carabidae larvae (see abbreviations in text)

a long large process (*pp1*) on the prothorax, and epimeral apodemes (*aem1-3*). *Pp2-3* poorly developed in the majority of Carabidae larvae, with the exception of Cychrini, Cicindelini and Orthogonini. Sternal apodemes of endoskeleton include long paired furcas on TH1-3 (*f1-3*), and short unpaired medial apophyses (spina, by Barlet, 1992) on TH2-3 (*sp2-3*). *F1* connected with *pp1*; *f2-3* – with intersegmental apodemes (*ais2-3*) to provide the additional area for muscle attachment. *Ais1* is usually smaller than *ais2-3* (except of Orthogonini), and also serves as a muscle insertion point.

Thoracic endoskeleton of examined Carabidae larvae (excl. Cicindelini, Orthogonini)

In the process of doing the research upon which this paper is based, it was found that the structure of thoracic endoskeleton is very similar within the tribe (Figs 2-15). Common features of the thoracic endoskeleton of larval Carabidae (excl. Cicindelini and Orthogonini) are the following.

Prothoracic endoskeleton: both *aes1* and *aem1* poorly developed. *Pp1* of various shape, always enlarged and broad at base. *F1* smaller than *f2-3* and connected with *pp1*. *Ais1* triangular or rounded, usually smaller than *ais2-3*.

Meso- and metathoracic endoskeleton almost identical. It consist of well-defined *aes2-3* and *aem2-3*; *pp2-3* usually poorly developed, and delimit *es* from *em*. *F2-3* larger than *f1*, connected with *ais2-3*. *Spina2-3* small and commonly rounded apically or triangular.

At the same time, examined tribes can be distinguished for certain characters of single apodemes, especially *pp1*, remaining rather similar in general. Peculiarities of each tribe are the following:

Loricerini (Fig. 2): *pp1* bifurcated, *f1* much smaller, comparative to other tribes. *Pp2-3* not distinct, *ais1-3* triangular, *sp2-3* rounded apically.

Pelophilini (Fig. 3): *pp1* massive, triangular. *Ais1-3* rather broad, *ais1* bigger than *ais2-3*. *Sp2-3* triangular.

Carabini (Figs 4-6): the structure of thoracic endoskeleton is similar to Pelophilini, except more narrow, triangular *ais1-3* and smaller rounded apically *sp2-3*.

Cychrini (Fig. 7): *pp1* is identical to Carabini and Pelophilini, *f1-3* and *ais1-3* much smaller. *Pp2-3* distinct, triangular.

Broscini (Fig. 8): *pp1* flat, broad, declinate anteriad and rounded apically. *Ais 1-3* elongated, triangular. *F1* relatively small and thin, *f2-3* heavily broad apically. *Sp2-3* rounded.

Scaritini (Fig. 9): *pp1* flat, triangular, slightly declinate laterally; *f3* of uniform wight, blunt at apex. *Ais2-3* slightly bifurcated, sp2-3 rounded.

Platynini (Figs 10-11): *pp1* flat, rather broad, narrowed anteriad, *ais1-3* triangular, *sp2-3* rounded apically.

Zabrini (Fig. 12): the structure of thoracic endoskeleton identical to Platynini.

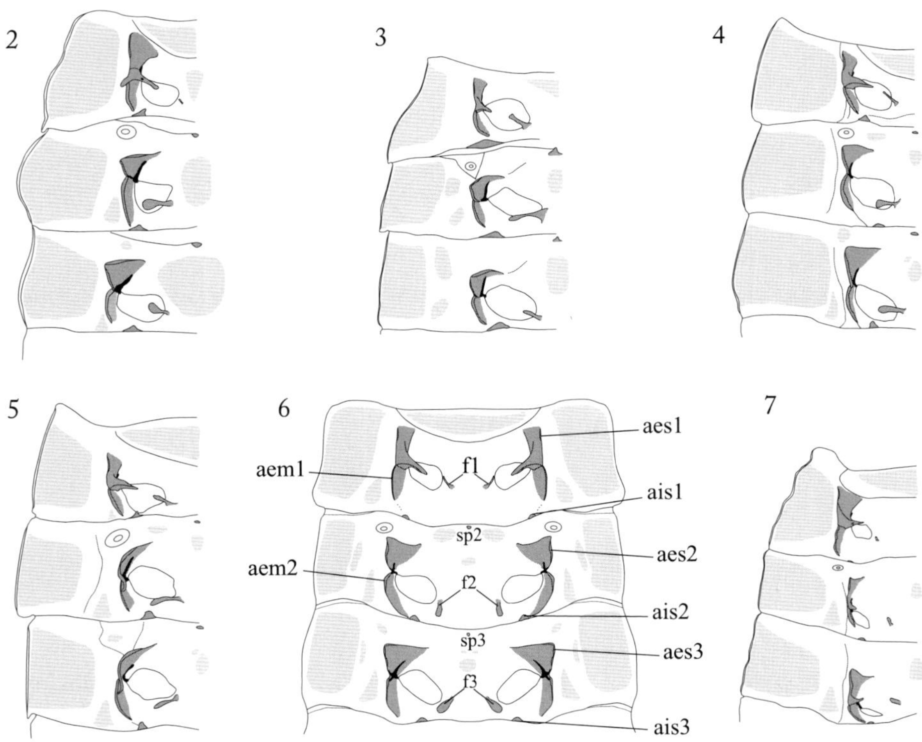

Figs 2-7. Thoracic endoskeleton of Carabidae larvae (see abbr. in text): 2 – *Loricera pilicornis*; 3 – *Pelophila borealis*; 4 – *Carabus nemoralis*; 5 – *C. prometheus*; 6 – *Calosoma auropunctata*; 7 – *Cychrus caraboides*.

Harpalini (Figs 13-15): identical to Platynini and Zabrini, but *pp1 broader* and *ais 2-3* slightly more elongated.

Thus, the structure of the thoracic endoskeleton in examined tribes varies mainly in modification of *pp1* (Fig. 27, see discussion below).

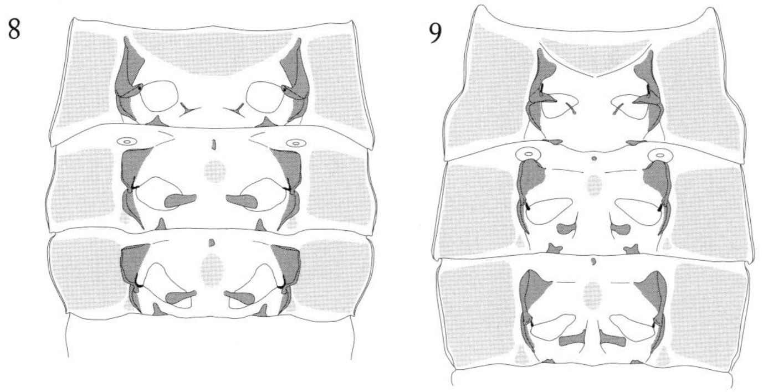

Figs 8-9. Thoracic endoskeleton of Carabidae larvae (see abbr. in text): 8 – *Broscus* sp.; 9 – *Scarites terricola.*

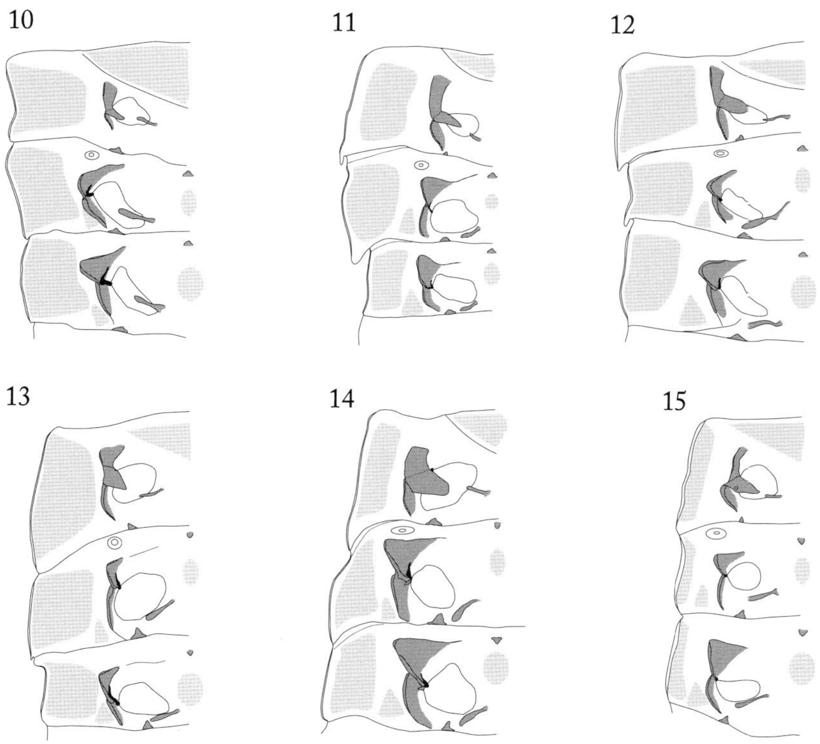

Figs 10-15. Thoracic endoskeleton of Carabidae larvae (see abbr. in text): 12 – *Platynus assimile,* 13 – *Agonum muelleri*; 14 – *Zabrus tenebrionoides*; 13 – *Ophonus azureus*; 14 – *Harpalus rufipes*; 15 – *Anisodactylus binotatus.*

Highly specialized larvae of Cicindelini and Orthogonini are characterized by considerable changes in shape and relative size of the majority of thoracic apodemes. Moreover, there were found unique endoskeletal elements in Cicindelini larvae (see below).

Thoracic endoskeleton of Cicindelini (Figs 16-18). ***Prothoracic endoskeleton*** notable for very large, bilobed triangular *pp1*. *F1* broad anteriad (*Cicindela campestris*, *Lophyridia littoralis*), or, on the contrary, narrowed (*Megacephala*). *Ais1* broad. ***Mesothoracic endoskeleton*** with well developed, elongated *pp2*. *Sp2* increase in size in sequence *Cicindela-Lophyridia-Megacephala*. *F2* of unusual form for Carabidae, widened anteriad in *Cicindela*, and greatly rounded in other examined genera. *Ais2* rather small in *Cicindela*; distinctly widened in *Lophyridia* and *Megacephala*. Characteristic feature is the presence in all examined Cicindelini larvae of additional paired apodemes of spiracle sclerite (*spap*). ***Metathoracic endoskeleton*** rather different from mesothorax: *pp3* less developed, *f3* with distinct elongated branch, and *sp3* distinctly enlarged.

Thoracic endoskeleton of Orthogonini (Fig. 19). ***Prothoracic endoskeleton***: characteristic features are very broad, large, rectangular *pp1* and *ais1*, *f1* thin and elongated. ***Meso- and metathoracic endoskeleton:*** extremely elongated *aes2-3*, *pp2-3* and *f2-3*, large *ais2-3*; *sp2-3* small and rounded.

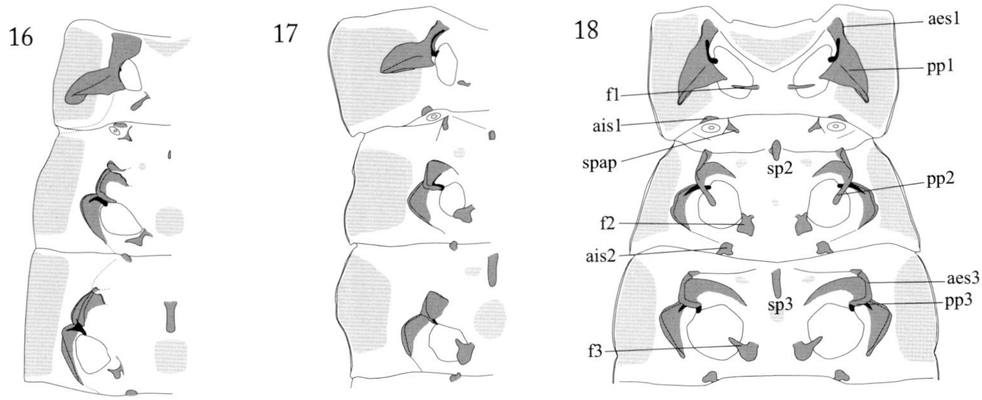

Figs 16-18. Thoracic endoskeleton of Carabidae larvae (see abbr. in text): 16-*Cicindela campestris*; 17 – *Lophyridia littoralis*; 18 – *Megacephala euphratica*.

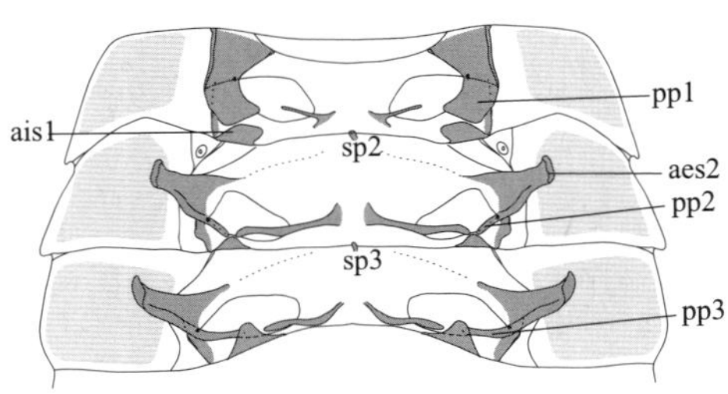

Fig. 19. Thoracic endoskeleton of *Orthogonius* sp. larva

The thoracic endoskeleton of highly specialized Cicindelini and Orthogonini larvae differs from other Carabidae because of enlargement and modification of the majority of apodemes. Moreover, unique apodemes were found in Cicindelini.

Thoracic endoskeleton of other Adephaga larvae

Trachypachidae: the endoskeleton of *Trachypachus holmbergi* larva was studied on damaged material, so only pleural apodemes were observed (Fig. 20). However, the structure of *pp1* seen, and can be considered as groundplan for Carabidae (Fig. 27). Further research is needed to reveal its peculiarities.

Gyrinidae (Fig. 21)**:** thoracic endoskeleton of *Dineutus* and *Gyrinus* identical. ***Prothoracic endoskeleton*:** *aes1* and *aem1* poorly developed, *pp1* relatively narrow at base, dilated anteriad and connected with *f1*, which is rather short and situated very close to coxal cavity. *Ais1* absent.

***Meso- and metathoracic endoskeleton*:** *aes2-3* and *aem 2-3* well developed, *aes2-3* with long and broad *pp2-3*. *F2-3* similar to *f1*, and have the same point of origin. *Sp2-3* elongated, *ais2-3* absent. Unique paired coxal apodemes (*cxap2-3*) are present.

Dytiscidae: *Acilius sulcatus* and *Dytiscus marginalis* larvae have an identical thoracic endoskeleton (Fig. 23). ***Prothoracic endoskeleton*** includes long and well developed *aes1*, *pp1* and elongated *f1*. Also additional paired pleural apodeme is found (*apl1*), not known for other examined Adephaga larvae. ***Meso- and metathoracic endoskeleton*:** *aes2-3* well developed, with long, apically broadened *pp2-3*. *F2-3* elongated, *sp2-3* rather small and rounded, *ais1-3* absent.

Noteridae: ***Prothoracic endoskeleton*** (Fig. 22) is poorly developed, *f1*, *ais1* absent. *Pp1* enlarged and almost reach the prothoracic segment border. ***Meso- and metathoracic***

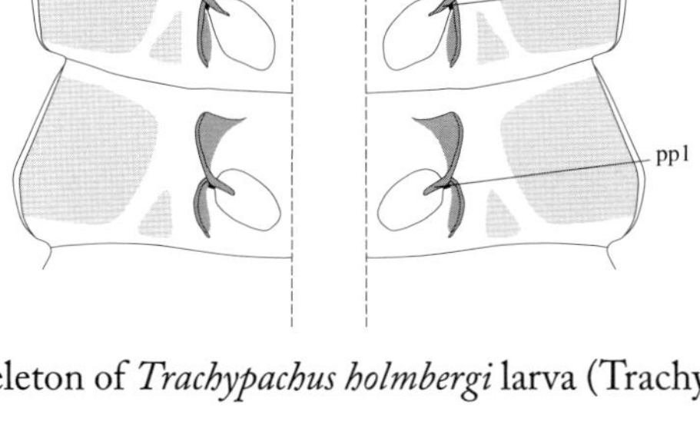

Fig. 20. Thoracic endoskeleton of *Trachypachus holmbergi* larva (Trachypachidae). Damaged part is shown by dotted line.

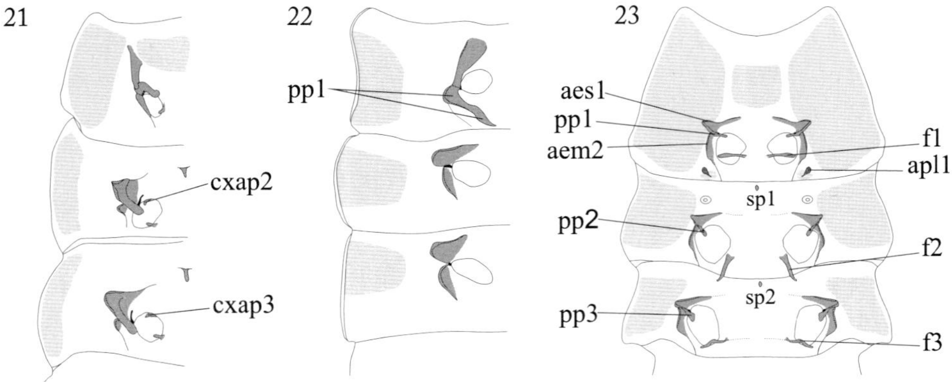

Figs 21-23. Thoracic endoskeleton of larvae of other Adephaga families (see abbr. in text): 21 – Gyrinidae; 22 – *Noterus crassicornis* (Noteridae); 23 – Dytiscidae.

endoskeleton also reduced, *aes2-3* uniformly widened anteriad, *pp2-3* not distinct; *f2-3*, *ais2-3*, *sp2-3* absent.

Haliplidae: Larvae of *Haliplus* and *Peltodytes* were examined. Rather unexpectedly, the structure of the endoskeleton of these two genera differs greatly from each other (Figs 24-25).

Prothoracic endoskeleton of *Haliplus* sp. consists of poorly developed *aes1* and *aem1*, and narrow, elongated *pp1*. *F1* and *ais1* absent. ***Meso-and metathorax*** also without *f2-3*, *ais2-3*, and *sp2-3*. *Aes2-3* widen near the border with *aem2-3*. *Pp2-3* not distinct.

Thoracic endoskeleton of *Peltodytes caesus* resembles that of some Carabidae. ***Prothoracic endoskeleton***: the characteristic features are large broad *pp1*, practically equal to episternite, poorly developed *aem1*; *f1* almost two times longer than *f2-3*, *ais1* absent. ***Meso-and metathorax*** consist of and *aes2-3* sharply extended in middle part, relatively small *f2-3* and rounded *sp2-3*. *Ais2-3* absent.

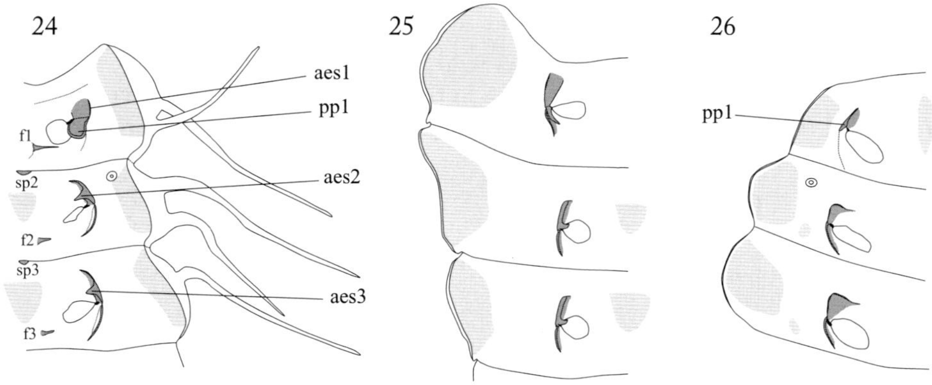

Figs 24-26. Thoracic endoskeleton of larvae of other Adephaga families (see abbr. in text): 24 – *Peltodytes caesus*; 25 – *Haliplus* sp (both Haliplidae); 26 – *Rhysodes comes* (Rhysodidae).

Rhysodidae: have a reduced thoracic endoskeleton (Fig. 26). ***Prothoracic endoskeleton***: relative length of *pp1* is the least in Adephaga larvae examined, *aem1* poorly developed; *f1* and *ais1* absent. ***Meso-and metathorax***: *aes2-3* and *aem2-3* well developed and resembles those of Carabidae. *Pp2-3* not distinct. *F2-3*, *ais2-3* and *sp2-3* absent.

DISCUSSION

Study of the thoracic endoskeleton of Carabidae larvae revealed the relative constancy of its structure within the family except for highly specialized ambush predators that live in holes (Cicindelini and Orthogonini). The only character that varies significantly in the majority of other examined Carabidae larvae is the pleural process of the prothoracic episternite *(pp1)*. Other elements of the endoskeleton vary mostly in relatively size (for example, *ais2-3*, *sp2-3*, often *f1-3*) and have little use in systematics. Thus, several distinct modifications of *pp1* were observed, and this character was the most valuable in identification of possible variants of thoracic endoskeleton of Carabidae larvae (Fig. 27). Unique modifications of *pp1* were found in Loricerini, Broscini and Scaritini. The prothoracic endoskeleton of *Loricera pilicornis* is characterized by bifurcated pp1, which was not observed in other examined Carabidae. At the same time, the structure of other endoskeletal elements remains typical for carabid larvae. A close relationship between Cicindelini and Loricerini as proposed by Arndt (1993), was not supported during the present study of the larval endoskeleton of these two tribes. The thoracic endoskeleton of Scaritini larvae differs from those of Broscini in the triangular shape of *pp1* and slightly bifurcated *f2-3*. The only character which is similar in larvae of these two tribes is the massive *f2-3*. However, it is more likely a case of convergence, rather than evidence of close relationship between these taxa. Both Scaritini and Broscini larvae are active soil-burrowers, so it is possible that broad furcas in combination with enlarged *ais2-3* provide the additional area for thoracic legs musculature.

Modification of *pp1* in examined Pelophilini, Carabini and Cychrini larvae is similar – it is massive and triangular, sharply narrowed anteriad. Moreover, endoskeletal structure in general is also almost identical, except of Cychrini (*f1-3* lesser developed, *p2-3* distinct). Platynini, Zabrini and Harpalini larvae possess practically identical thoracic endoskeleton, *pp1* flat and broad; there is a tendency of increasing in width of *pp1* from Platynini to Harpalini, but it is not significant for distinguishing different types of endoskeleton for these tribes. Thus, based on examined endoskeletal preparations, the relatively close relationships among Pelophilini-Carabini-Cychrini, and Platynini-Zabrini-Harpalini (Figs 27-28) was confirmed, which corresponds with previous studies on their larvae (Arndt, 1993, 1998). Thus, it is possible to use this feature in a systematic study at the tribal or family level of Carabidae and other Coleoptera, but further research is still needed. Nevertheless, the possibility of parallel development also should be taken into consideration.

Extensive enlargement of *pp1-2* and *sp3* in Cicindelini can be explained by the fact that the first thoracic segment does not participate in movement, but forms the united

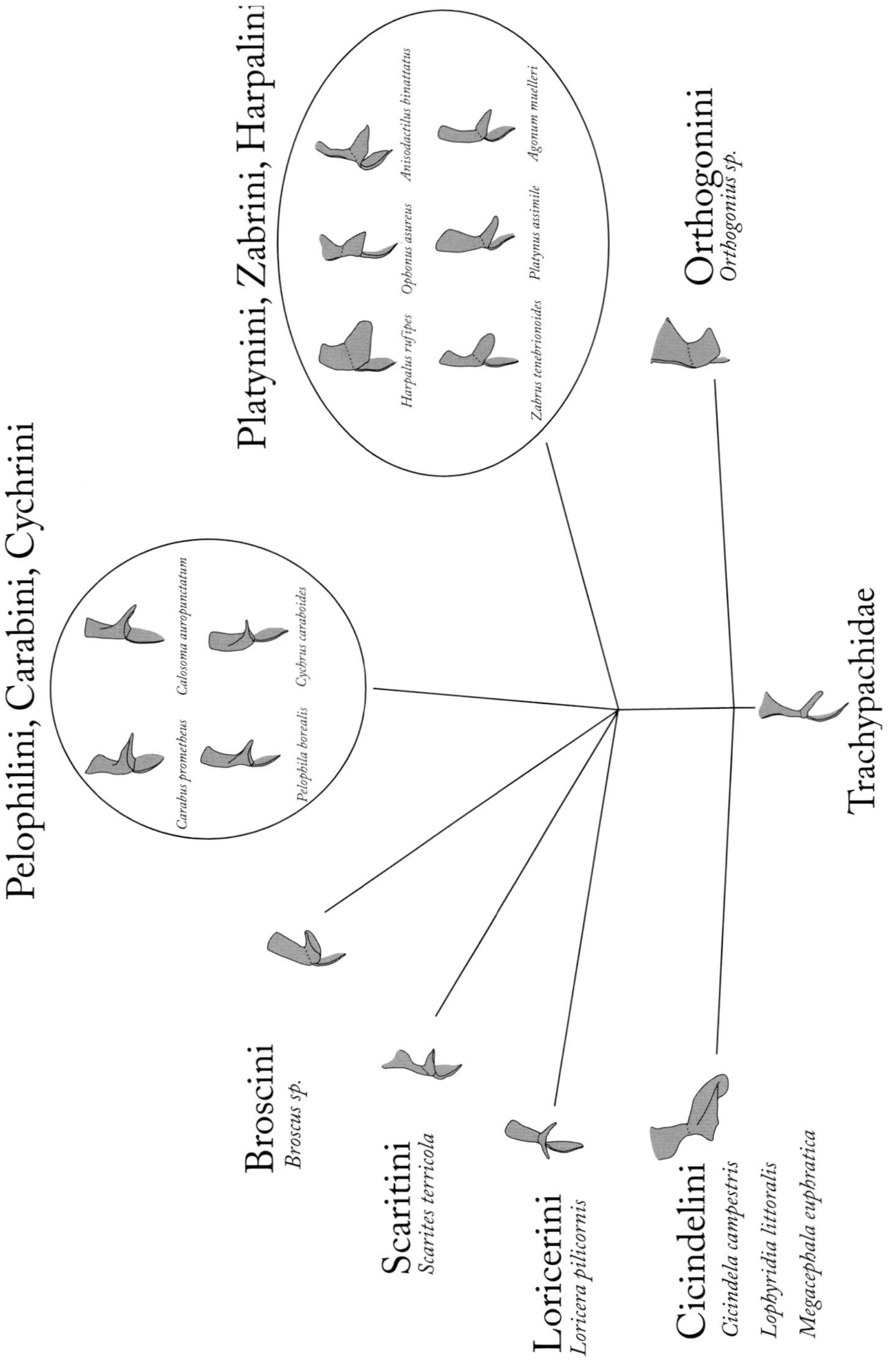

Fig. 27. Variety of pleural processes of episternite (pp1) in Carabidae larvae.

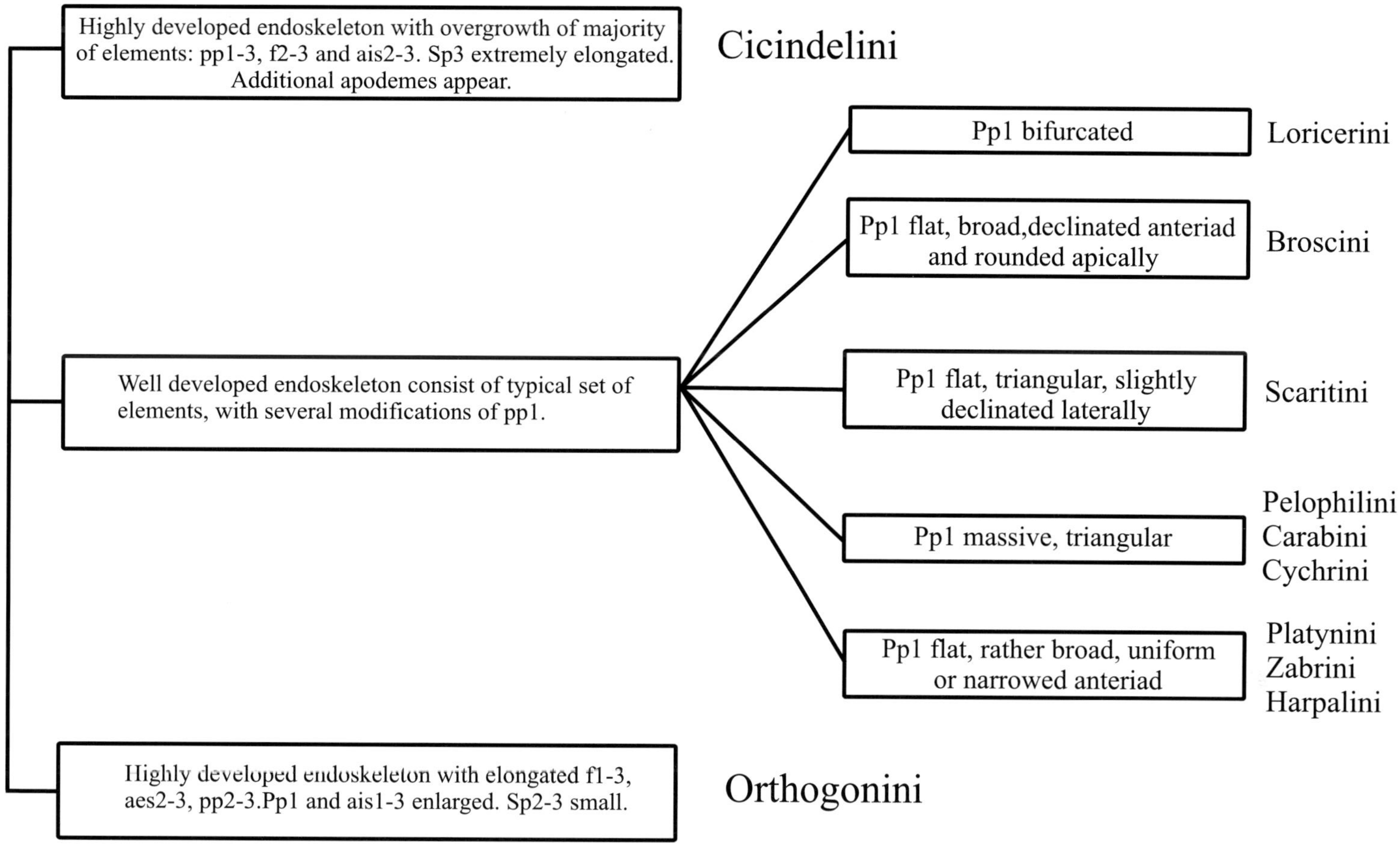

Fig. 28. Possible variants of thoracic endoskeleton structure in larvae of Carabidae

functional block with the head. This structure is supported by well developed musculature. The meso- and metathorax, on the other hand, are associated with locomotion.

The thoracic endoskeleton of *Orthogonius* is also characterized by elongation of apodemes (especially well seen in structure of *f2-3*, *sp2-3*. and *ais1-3*).

Gyrinidae and Dytiscidae larvae possess a well developed endoskeleton with enlarged *pp2-3*. Gyrinidae larvae crawl about submerged objects, using their characteristic apical abdominal hooks; they also swim in an undulating fashion by using the abdominal gills. Dytiscidae larvae are actively swimming predators. Well developed endoskeletal structure in dytiscid larvae can be correlated with active locomotion in the water column, or with specific locomotion of benthic Gyrinidae larvae. However, within the larvae of these two families were found specific elements, which made them quite distinguishable.

Reduction of the endoskeleton is observed in Rhysodidae and Noteridae larvae. Slow moving wood-boring *Rhysodes comes* larvae are characterized by a reduced endoskeleton with very small *pp1*. Larvae of *Noterus crassicornis* possess a well developed, enlarged *pp1*, which is possibly correlated with silt-burrowing life style. The short and massive prothoracic legs are used for separation of silt particles, and are supported by well developed muscles, which are attached to *pp1*. Close relationship between Noteridae and Dytiscidae proposed by some authors (e.g. Franciscolo, 1979; Miller, 2001) was not supported on endoskeletal characters in the present study.

With regard to similar endoskeleton structure at the generic level, it was unexpected to observe the striking differences in examined larvae of *Haliplus* and *Peltodytes*. Any conclusions can be made only after other larval Haliplidae are examined. The similarity of *Noterus* and *Haliplus* endoskeletal structure may support a close relationship between Noteridae and Haliplidae as was suggested by Ruhnau (1986).

Summarizing the facts mentioned above, it can be stated that well defined thoracic endoskeleton is observed in actively-moving larvae (Carabidae excl. Cicindelini and Orthogonini, Dytiscidae, Haliplidae, in part), or in taxa with specialized locomotion (e.g., Cicindelini, Orthogonini, Gyrinidae). Reduction of the endoskeleton was observed in slow-moving larvae of Rhysodidae and Noteridae; this may be considered a specialization during development in specific habitats.

As a result of accumulated data on Carabidae and other families, it is possible to state that the endoskeletal characters can be used in the taxonomy at family or subfamily level (with exception of possible cases of reduction), though it is unsuitable for the systematics of lower rank taxa due to the numerous cases of parallel development.

In conclusion it is necessary to point out that this study does not pretend to provide a complete coverage of endoskeletal structure of Adephagan larvae. A number of families – Amphizoidae, Hygrobiidae and Trachypachidae – were not available for the present work or were observed based on damaged material; also, some important Carabidae tribes (Brachinini, Paussini, etc.) were not available for examination. Further research is needed, with special attention to thoracic muscles, for the study of the locomotory complex of Coleoptera larvae as a dynamic whole.

ACKNOWLEDGEMENTS

The author is greatly appreciated to Professor Dr K.V. Makarov, Dr A.V. Matalin (MPSU), Dr A. Prokin (VSU) and Dr D.A. Pollock (ENMU) for providing with larval material; Dr D.A. Pollock is also thanked for checking the English in the manuscript.

REFERENCES

Arndt, E. (1993). Phylogenetische Untersuchungen larvalmorphologischer Merkmale der Carabidae (Insecta: Coleoptera). – Stuttgarter Beiträge zur Naturkunde Serie A (Biologie) 488: 1-56.

Arndt, E. (1998). Phylogenetic investigation of Carabidae (Coleoptera) using larval characters. – In: Ball, G.E., Casale, A. & Vigna Taglianti, A., eds. Phylogeny and classification of Caraboidea symposium (28 August, 1996, Florence, Italy). 20 International Congress of Entomology. Museo Regionale di Scienze Naturali, Torino. 1-543. Chapter pagination: 171-190.

Barlet, J. (1992). Observations sur le thorax de larves de cicindeles (Insectes, Coléoptères). – Bulletin de la Société Royale des Sciences de Liege. 61(5): 339-349.

Beutel, R.G. (1993). Phylogenetic analysis of Adephaga (Coleoptera) based on characters of the larval head. – Systematic Entomology 18(2), 127-147.

Beutel, R.G., Ribera, I. & Bininda-Emonds, O.R.P. (2008). A genus-level supertree of Adephaga (Coleoptera). – Organisms, Diversity & Evolution 7: 255–269.

Franciscolo, M.E. (1979). Fauna d'Italia, Vol. XIV. Coleoptera. Haliplidae, Hygrobiidae, Gyrinidae, Dytiscidae. – Calderini, Bologna.

Miller, K.B. (2001). On the phylogeny of the Dytiscidae (Insecta: Coleoptera) with emphasis on the morphology of the female reproductive system. – Ins. Syst. Evol. 32: 45-92.

Paulian, R. (1944). L'endosquelette thoracique des larves d'insectes. – Mem. Mus. Hist. nat. Paris (N.S.) 18: 191-218.

Ruhnau, S. (1986). Phylogenetic relations within the Hydradephaga (Coleoptera) using larval and pupal characters. – Entomol. Basil. 11: 231–272.

Sharova, I.Kh. (1981). Life forms of carabids (Coleoptera, Carabidae). – Nauka, Moscow.

L. Penev, T. Erwin & T. Assmann (Eds) 2008
BACK TO THE ROOTS AND BACK TO THE FUTURE. TOWARDS A NEW SYNTHESIS AMONGST TAXONOMIC, ECOLOGICAL AND BIOGEOGRAPHICAL APPROACHES IN CARABIDOLOGY
Proceedings of the XIII European Carabidologists Meeting, Blagoevgrad, August 20-24, 2007, pp. 173-182.

Sofia–Moscow

Multilayer structural colours in *Poecilus lepidus* (Coleoptera, Carabidae)

Dietrich Mossakowski[1], Wilfried Paarmann[2], Wolfgang Rohe[2], Ingrid Lüchtrath[2] & Thorsten Assmann[3]

[1] University of Bremen, Institute for Ecology and Evolutionary Biology. P.O. Box 330440, D 28334 Bremen, Germany. E-mail: dmossa@uni-bremen.de
[2] HAWK, Fakultät Ressourcenmanagement, Büsgenweg 1A, D 37077 Göttingen, Germany
[3] Institute of Ecology and Environmental Chemistry, University of Lueneburg, Scharnhorststrasse 1, D 21335 Lüneburg, Germany

SUMMARY

The Carabid beetle *Poecilus lepidus* occurs in different colour forms, which are controlled by a gene with at least five alleles. Spectra of the colour forms were measured and the cuticle studied by electron microscopy.

The three colour forms (blue, green, and black) from Italy show simple Mendelian heredity with the genotypes bb, bs, and ss. They display reflection spectra, which are caused by a multilayer system of electron dense and less dense layers. The same system exists in German populations, but the colours are closer together (bluish green, yellowish green and red; genotypes gg, rg, and rr). In addition to this system, in German populations a violet form was found in the genetic analysis (Vr, Vg), the colour of which is also produced by a multilayer system. The colour measurements confirmed that a multilayer system exists not only in the heterozygous males and females, but also in the black homozygous VV females with a maximum in the near ultraviolet. At the other end of the spectrum, at least two black forms were detected by their reflectance peaks in the infrared range. These peaks are beyond the visible spectrum and are also due to a multilayer system. However, the heredity of these forms is unknown.

Keywords: Structural colours, multilayer, interference, Carabidae, *Poecilus lepidus*

INTRODUCTION

Colours are due to two principally different effects: (i) pigment absorbtion of a part of the visible spectrum and reflection of another part, which causes the colour impression, and (ii) physical effects at thin structures producing structural colours. Besides other types, structural colours include interference colours of multilayer systems in the outer part of the insect cuticle. The cuticle must be transparent to allow the incident and reflected beams to transmit light to and from the layers.

The surface colours of *Poecilus lepidus* are mostly structural colours of a multilayer system. These layers require a thickness much smaller than the wave length of the produced colour. This is the reason why a study of such structures was one of the first applications of electron microscopy (Anderson & Richards, 1942). In Carabid beetles, multilayer structural colours were described in *Cicindela* species by Mossakowski (1980), and Schultz & Rankin (1985); in *Laemostenus* species by Mossakowski (1982).

Studies on the heredity of colour morphs in ground beetles are still very rare and limited to European *Carabus* species (Puisségur, 1964) and the American *Agonum decorum* (Liebherr 1983). But no study of colour morphs has been done combining genetics, electron microscopy, and colour measurement analysis of these morphs. The scope of this paper is to describe the cuticle structure and reflectance spectra of defined probes, the heredity of which are known (Paarmann et al., 2008).

MATERIAL AND METHODS

Specimens of *Poecilus lepidus* out of the crossbreeding experiments of Paarmann et al. (2008) were used for colour measurements. The Italian specimens were collected at Lago Maggiore, the German ones came from the Lüneburger Heide. Parts of the elytra of the same specimens were prepared for electron microscopy. Thus, the genetic background of the analyzed colour morphs was known. These are indicated in Figs 2-7. Unless mentioned in the text, all data and statements given in the text are restricted to the more or less plane surface of the elytral cuticle.

Colour measurement was done using a diode array spectrometer (MSC 500, Carl Zeiss, Jena). The measure geometry was 0°/0°. Thus, fibres of the light cable led the light from the source to the probe, and other fibres received the reflected light. The diameter of the measuring spot was about one millimetre. A measurement was taken every four nanometres. $BaSO_4$ was used as a standard. The reflectance curves were calculated and managed with the program Aspect Plus (Carl Zeiss).

EM technique

Fresh pieces of cuticle were fixed in 3% glutaraldehyde in 0.1 M Na-cacodylate buffer and embedded after Spurr (1969) (hard). An ultra microtome, ULTRACUT E, was used to make sections of about 60 nm thickness, which were contrast stained in 2% uranylacetate and lead citrat after "Reynolds" (30 min). An electron microscope ZEISS EM 10 A was used and documentation was done by MACO EM-Film EMS. Sections of all colour morphs were documented at a magnification of 40,000.

RESULTS

Electron microscopy

The elytral cuticle of *Poecilus lepidus* is variable in thickness, being 50 to 70 μm thick at the intervals and more than half as thick at the elytral striae (Fig. 1). The exocuticle is strongly pigmented; and underlies the multilayer system. In addition, the thickness of this system varies due to the number of layers and their thicknesses.

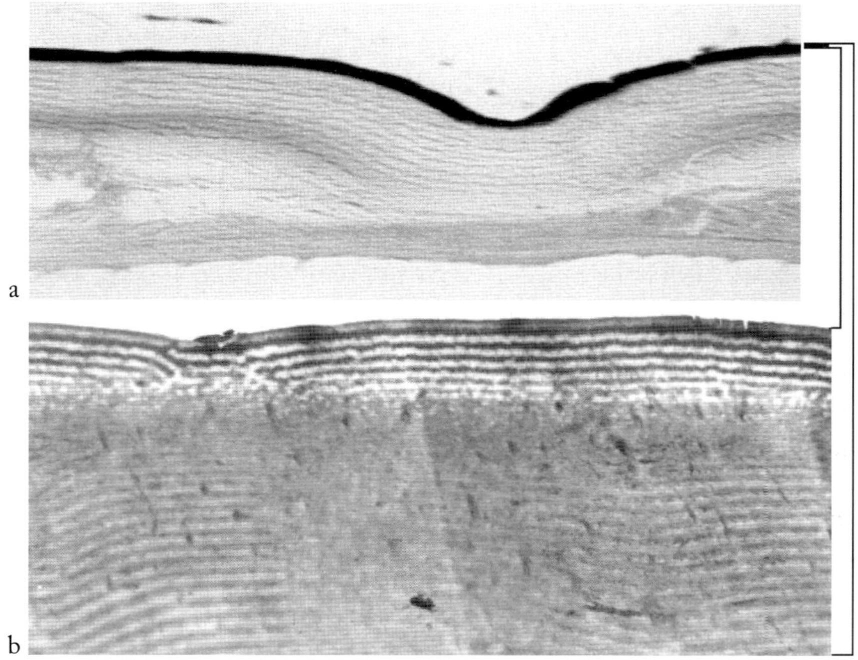

Fig. 1. Cross sections of the elytral cuticle of *Poecilus lepidus*.
a: Micrograph of a semi-thin section (light microscope). At the left and right sides, an elytral interval (50 μm thick) may be seen, in between an elytral stria (about 35 μm).
b: Micrograph of the outer part of cuticle (Transmission Electron Microscope). Alternating electron dense and less dense layers build up the multilayer system which is about 0.75 μm thick.

The colour producing stack consists of two alternating layers: an electron dense and another less dense layer, which appears nearly white in the electron micrographs. The whole stack of layers is crossed by wax canals. In some of the micrographs, thin layers may be seen, which are of the same dimension as in the pigmented part below (10-20 nm thick, Fig. 1, below).

According to the order of SEM micrographs in Figs 2 and 3, the thicknesses of the layers decreases from the IR to the UV range corresponding with the change in colour

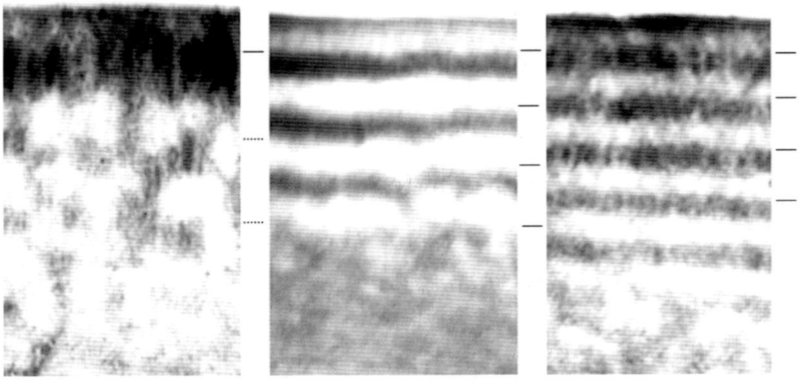

Fig. 2. Multilayer systems in the cuticle of Italian *Poecilus lepidus* colour forms.
Bars at the right side of micrographs indicate the thickness of the outer most three alternating layer pairs. Genotypes in brackets. The electron dense layers are mainly dissolved in the black form.

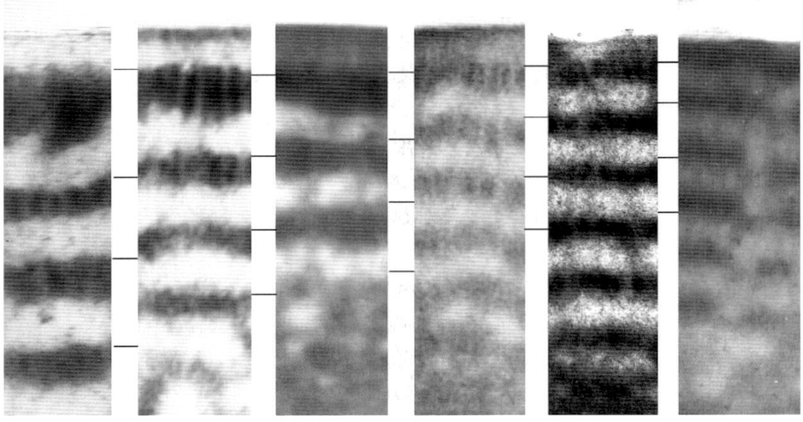

Fig. 3. Multilayer systems in the cuticle of German *Poecilus lepidus* colour forms.
Bars at the right side of micrographs indicate the thickness of the outer most three alternating layer pairs. Genotypes in brackets; Vx represents Vr and Vg. Black specimens exist at both ends of the spectrum.

of the cuticle of the particular colour form. In correspondence to the colour forms found, in the Italian populations specimens occur with three different dimensions of the layers. The Italian black form was the only one in which we found a reduced multilayer system. One electron dense layer was present at most places in the elytral cuticle (Fig. 2). The dense layers below are more or less irregular: they appear under dissolution.

At German sites, a similar system of three genotypes was found (Fig. 3), but additionally, a violet form could be differentiated and black specimens with a well developed multilayer system exist. These are equally near or within the IR as in the UV range.

Colour measurements

Italian *Poecilus lepidus* differ strikingly in colour. Three colour morphs are known (Fig. 4; genotypes bb: blue, bs: green, ss: black). The reflectance curves display the form characteristic for structural colours produced by a multilayer system. Blue and green cuticles show relatively high peaks between 35 and 40% reflectance, while black cuticle reflected only about 17% of the incident light. The curve of the black form has an additional peak near the ultraviolet range.

A similar system of three colour forms exists in German *Poelicus lepidus* populations, but the colour peaks are arranged more closely together than in the Italian ones (Fig. 5; gg: bluish green, gr: yellowish green, rr: red). This fits with the lower thickness of the colour producing multiple layers (Fig. 3). In addition to this basic status, other kinds of colour

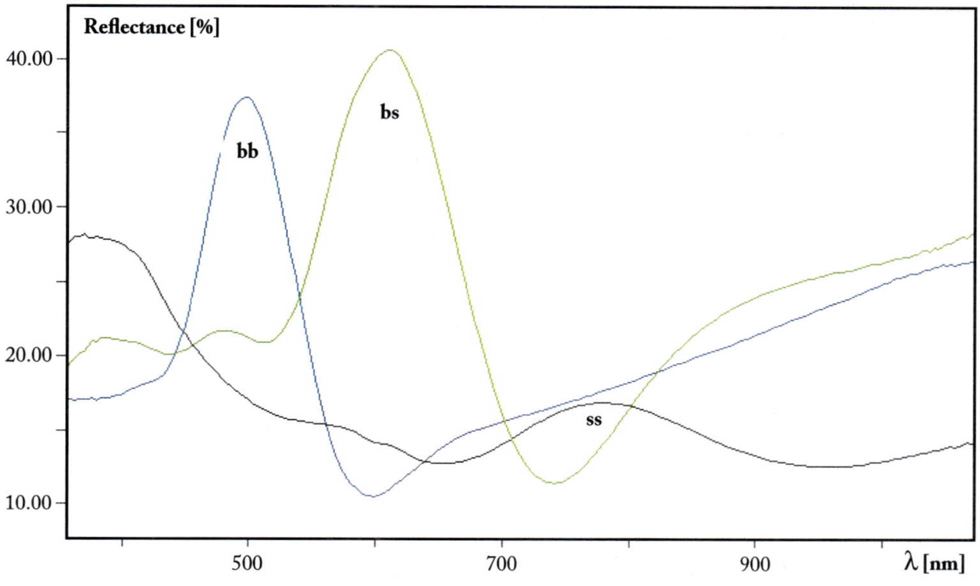

Fig. 4. Reflectance curves of Italian *Poecilus lepidus* colour forms.
Letters below the peaks indicate the genetic status of the respective forms. The black form (ss) displays an extra peak at λ/2.

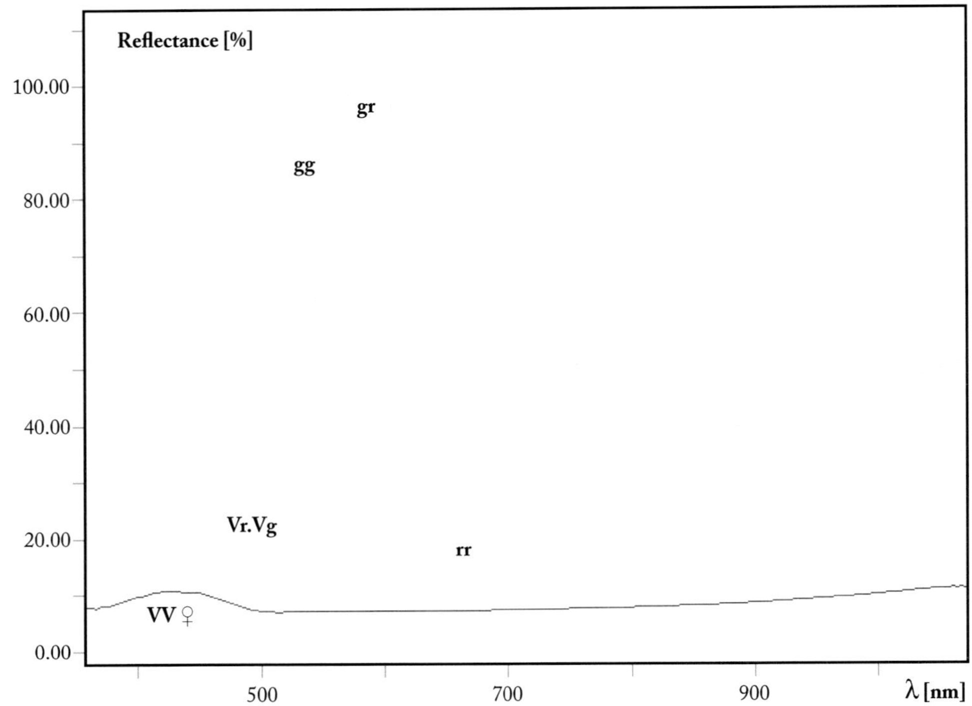

Fig. 5. Reflectance curves of German *Poecilus lepidus* colour forms.
The genetic status is indicated by letters. Homozygous beetles of the 'violet' allele could be only found in females, they differ from heterozygous ones in their black appearance.

morphs are found in German *Poecilus lepidus*: violet and different black forms. The 'violet' form is due to the allele V, which is dominant over all other alleles tested, and displays a different colour between heterozygous versus homozygous specimens (Paarmann et al., 2008; compare Fig. 5 and 6: VV versus Vr, Vg). The heterozygous specimens show a peak in the visible range of the spectrum (violet). The homozygous form was only found in females, which have a black elytral surface. Only the lateral edge of elytra is violet in these females. Although of relatively low amplitude, this peak of the black cuticle surface is also produced by a multilayer system as shown in Fig. 3.

Different black forms occur in Germany (Fig. 6). The curve of VV and Vr specimens of Fig. 5 was included as a standard of comparison for the enlarged scale of the ordinate axis. The reflectance spectra a and b represent *Poecilus lepidus* forms, which are black, but display reflectance peaks near or in the IR range, respectively, due to a multilayer system (Fig. 3, left most EM micrograph from the same specimen as curve b). The genetics of these two forms remain unknown. The last curve (c) represents the reflectance of a black sternum and was not analysed in more detail.

The wavelengths of reflectance peaks we measured seem to be greater than those of the colour visible to the naked eye (see discussion).

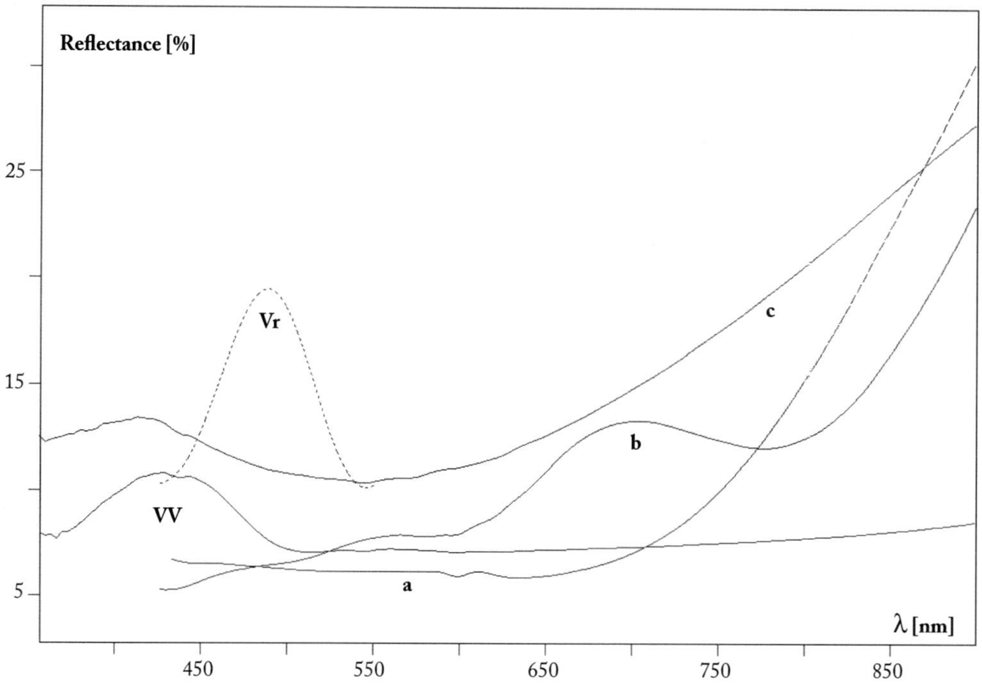

Fig. 6. Reflectance curves of German black *Poecilus lepidus*.
a: black specimen with a multilayer system shown in Fig. 3, left side; **b**: black specimen with a peak at 704 nm due to a multilayer system; **c**: black cuticle of a sternite. VV and Vr from Fig. 5.

DISCUSSION

Multilayer system

Two general types of multilayer colour systems have been described in insects. Lamellicorn beetles display a helocoidal multilayer system, which is characterised by the helicoidal structure of the chitin-protein micelles and its optical activity (Neville & Caveny, 1969). Many other families of beetles have a system of alternative layers without optical activity.

The structural colours of *Poecilus lepidus* are produced by a non-ideal type of such a multilayer system, because there is no evidence for a helicoidal system and the thickness of the alternating layers is different. Due to the extreme differences in electron density, the two layers should have a different refractive index. We found deviations of different kinds in the SEM micrographs. On the one hand, there are irregularities in the thickness of a single layer, which we interpret as an artefact established by cutting such a brittle material. At the other hand, the number of layers is not constant over the whole surface. In some places, a layer vanishes by dissolution at the pigmented background. But this phenomenon results only in a low effect on the intensity of the produced colour.

Most multilayer systems described, produce effects of constructive interference in the visible part of the spectrum. The only exception was the Carabid beetle *Laemostenus algerinus,* displaying a black colour due to a multilayer system. The thickness of the layers is very narrow, producing an invisible peak in the ultraviolet range of the spectrum (Mossakowski, 1982). In *Poecilus lepidus* we found a comparable effect: black specimens with an invisible peak in the infrared range.

Reflectance curve – visible impression

There seems to be a marked difference between the colour of visible impressions and the wavelength peaks of some of the measured curves. This effect may be due to different reasons: (i) the colour impression is due not only to the peak rather than to the other parts of the spectrum and (ii) the storage of the elytra in 60% ethanol. The latter may have resulted in a moderate swelling of the layers. To get a colour value for the measurement curve, we calculated the x and y values in the CIE system and put them into the CIE colour triangle. The projection of the x, y values onto the border of the triangle gave wavelengths nearer to the colour impression by the naked eye.

The height of a peak should not be overestimated, because it depends on the surface structure of the cuticle and the exact measurement geometry. The elytral surface of *Poecilus lepidus* cuticle is structured by the matrix of the epidermal cells, which is smoother in males than in females. Therefore, the colour measurements were performed with male specimens except the genotype VV, only found in females. Furthermore, the cuticle is not absolutely plane. The height of the peak also depends on the exact position of the cuticle probe to incident beam: a low deviation from the 0°/0° geometry results in a drastic decrease of the peak.

The curve of the red form shows an additional peak near the ultraviolet range (Fig. 5). This may be the secondary peak at $\lambda/2$. This peak is at the end of the visible spectrum, therefore we will not remark on it. The same situation is present in the cuticles with a black colour impression. Both peaks lie in or near that part of the visible spectrum where the resolution of the human eye is low. Therefore, the appearance is black for the naked eye (compare Fig. 4, form ss).

Heredity, ultrastructure, and reflectance

Fig. 7 gives a synoptic overview over the position of *Poecilus lepidus* colour forms in relation to the sunlight spectrum. The arrangement of colour bars representing colour forms was done according to the colour impression of the naked eye. It is obvious that a similar system of two alleles exists in Italian (s, b) and in German (r, g) populations, but the German colour forms are closer together. All colour forms studied so far display multilayer systems producing these spectacular colours. Only one exception was found in the black Italian form: In these beetles the multilayer system is reduced. The electron dense layers

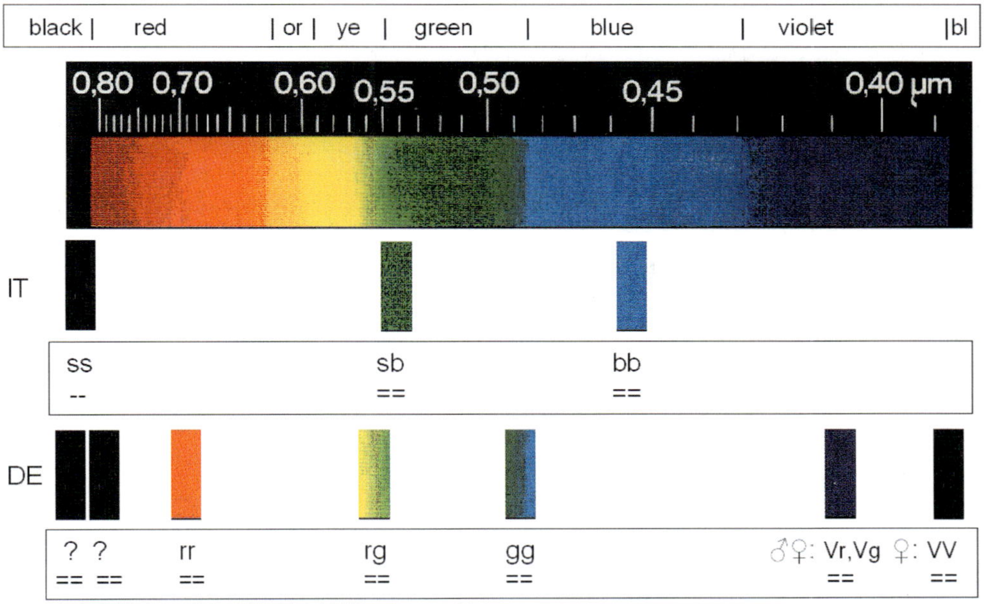

Fig. 7. Schematic arrangement of the colour morphs of *Poecilus lepidus* in a prismatic sunlight spectrum. The bars represent a colour morph placed at that wavelength of colour impression by the naked eye.
Colour names bl: black, or: orange, ye: yellow; IT: Italian colour morphs and their genotypes.; DE: German colour morphs and their genotypes; Letters below colour bars indicate genotypes; = =: multilayer system present; -: number of layers reduced because of their dissolution; ?: genetics unknown.

appear dissolved. At some places of the cuticle, we did not find any intact layer; therefore, the reflected colour is due to black background pigment at those positions.

5. ACKNOWLEDGMENTS

We would like to express cordially thanks to Anke Toltz for embedding the cuticle probes and performing the electron microscopy procedures, to Martin Patzwald and Felix Kerstan (Carl Zeiss, Jena) for their kind help in colour measuring, to Michael Vicker (University Bremen) who corrected the English, and to Hartmut Greven (University Düsseldorf) for helpful comments.

REFERENCES

Anderson, T.F. & Richards, A.G. (1942). An electron microscopical study of some structural colors in insects – J. appl. Phys. 3: 748-758.

Liebherr, J.K. (1983). The genetic basis for polymorphism in the ground beetle *Agonum decorum* (Coleoptera: Carabidae). – Annals of the Entomological Society of America 76: 349-358.

Mossakowski, D. (1980). Reflection measurements used in the analysis of structural colours of beetles. – Journal of Microscopy 116: 351-364.

Mossakowski, D. (1982). Pigmentation as a character for the reconstruction of evolution in cave beetles. – In: Environmental Adaptation and Evolution (Mossakowski, D. & Roth, G., eds). Gustav Fischer, Stuttgart, New York, p. 195-207.

Neville, A.C. & Caveny, S. (1969). Scarabaeid beetle exocuticle as an optical analogue of cholesteric liquid crystals. – Biol. Rev. 44: 531.

Paarmann, W., Aßmann T., Rohe, W., Lüchtrath, I. & Mosakowski, D. (2008). Heredity of the elytral colour in adults of *Poecilus lepidus* Leske (Col., Carabidae). – In: Back to the Roots and Back to the Future. Towards a New Synthesis amongst taxonomical, ecological and biogeographical approaches in carabidology (Penev, L., Erwin, T. & Assmann, T., eds). Pensoft, Sofia-Moscow, p. 183-194.

Puissegur, C. (1964). Recherches sur la génetique des Carabes. – Vie et milieu. Supplement 18: 1-288.

Schultz, T.D. & Rankin, M.A. (1985). The ultrastructure of the epicuticular interference reflectors of tiger beetles (*Cicindela*). – Journal of experimental biology 117: 87-110.

Spurr, R. (1969). A low-viscosity epoxy resin embedding medium for electron microscopy. – J. Ultrastructure Res. 26, 31.

L. Penev, T. Erwin & T. Assmann (Eds) 2008
Back to the Roots and Back to the Future. Towards a New Synthesis amongst Taxonomic, Ecological and Biogeographical Approaches in Carabidology
Proceedings of the XIII European Carabidologists Meeting, Blagoevgrad, August 20-24, 2007, pp. 183-194.

Heredity of the elytral colour in adults of *Poecilus lepidus* Leske (Coleoptera, Carabidae)

Wilfried Paarmann[1], Wolfgang Rohe[1], Ingrid Lüchtrath[1], Thorsten Assmann[2] & Dietrich Mossakowski[3]

[1] HAWK, Fakultät Ressourcenmanagement, Büsgenweg 1 A, D 37077 Göttingen. E-mail: rohe@hawk-hhg.de
[2] Institute of Ecology and Environmental Chemistry, Leuphana University Lüneburg, Scharnhorststr. 1, D 21332 Lüneburg. E-mail: assmann@uni.leuphana.de
[3] Institute for Ecology & Evolutionary Biology, University of Bremen, P. O. Box 330440, D 28334 Bremen

SUMMARY

The heredity of altogether seven colour morphs of the ground beetle species *Poecilus lepidus* was studied by crossbreeding experiments: black, dark green and blue morphs found in Italy (Lago Maggiore, Apennine), and red, yellowish green, bluish green and violet from a population in Germany (Lüneburger Heide). In their natural populations the colour morphs black, red, bluish green and blue are homozygous: ss (black), rr (red), gg (bluish green), bb (blue). Other green colour morphs are heterozygous: dark green (Lago Maggiore population) bs, yellowish green (Lüneburger Heide) gr. Crossbreeding experiments between beetles of the 'German' and the 'Italien' strains showed the red allele (R) is dominant over black (s), and bluish green (G) dominant over blue (b). Heterozygous beetles with the co-dominant alleles gs and br have a green colour. The dominant allele V (violet) was only found in the 'Lüneburger Heide' population, where it seems to be very rare.

Violet colour morphs in the 'Lüneburger Heide' population should mainly have the allele combination Vr, because the red colour morph is dominant there (82%). Beetles with combination Vg are also violet, but the bluish green colour morph (gg) is comparable rare (3%). Beetles homozygous for the allele V should be extremely rare in the 'Lüneburger Heide' population. Only in the crossbreeding experiments homozygous females were achieved. They were almost black with a small violet brim of the elytra.

The ecological significance of colour forms in *Peocilus lepidus* may be interpreted partly as mimetic effects.

Keywords: *Poecilus lepidus*, colour forms, heredity

INTRODUCTION

Poecilus lepidus has a Palaearctic distribution ranging from the Pyrenees to the Amur region in Siberia and from Trondheim region in Norway to central Italy (Lindroth, 1949; Turin, 2000).

A number of colour morphs from *P. lepidus* of different regions are known: blue, green, and copper-coloured (Erichson et al., 1860; Lindroth, 1949; Turin, 2000; for an overview of the colour morphs of *P. lepidus* s. str, and *P. lepidus gressorius* see Schatzmayr, 1942-43). Erichson et al. (1860) and Turin (2000) additionally mention black morphs. From Kirschenhofer (personal communication 1984) we have the information that in the Pyrenees uniformly black populations exist (in the Bigorre-massive, Assmann found exclusively black *Poecilus lepidus*, too). The Italian populations in the Apennine are uniformly blue (cf. Schatzmayr, 1942-43). This subspecies, *P.l. gressorius,* has females with shining elytra while the elytra of the subspecies *lepidus* is dull. Kirschenhofer mentions a violet specimen in his collection. Violet specimens were also recorded by Horion (1941). In Western Siberia also black, green and cupper coloured beetles appear (Zinovyev, personal communication 2007).

P. lepidus is a species that can easily bred under laboratory conditions (Kegel, 1989; Paarmann, 1990; De Vries, 2000). Laboratory studies on the environmental control of the life cycle of *P. lepidus* (Paarmann, 1990) led to the idea to study the heredity of colour morphs.

Up to now only very few studies on the potential value of these different colour morphs for an adaptation to different habitat conditions are already done. Mossakowski (1980) stated differences in reduction of solar radiation absorbance and heat gain for green versus red morphs of *Cicindela campestris*. Schultz (1986) demonstrated the role of structural colours in predator avoidance by tiger beetles of the genus *Cicindela*. Van Natto & Freitag (1986) got differences in the amount of reflectance in a comparison of carabids with structural colours and black pigmentation. Terrell-Nield (1990) found a positive relationship between ground temperature and percentage of black colour morphs of the legs in *Pterostichus madidus.*

For most of these studies the heredity of the colours is a precondition for further purpose (e.g. to determine selection coefficients). *Poecilus lepidus* is a suitable model organism because of its different habitats (see discussion), many colour morphs and the possibility to rear it under laboratory conditions, which is essential for crossbreeding experiments.

MATERIAL AND METHODS

For our crossbreeding experiments we used three strains of different origin:

1. North Italy – Lago Maggiore (Fig. 1, I). This strain was founded in 1979 with only three specimens of the colour morphs black, blue and green each. It is still in culture in the 34th generation now.

2. Central Italy – Apennine (Fig. 1, Plg). Only one pregnant blue female founded this strain.
3. Germany – Lüneburger Heide (Fig. 1, D). Founded by many specimens of the colour morphs red (copper coloured), yellowish green, bluish green and violet.

Some of the data were collected during the studies dealing with the life cycle control. For breeding conditions see Paarmann (1990). During the following studies we simulated winter conditions by keeping the adults in 8 °C and short day conditions (8h light) or in continuous darkness at 4 °C. To reach maturation, we transferred them into 18 or 20 °C and long day conditions (16h light). Larvae were either collected directly from the culture vessel or the females were separated on fine moisten sand. After a few days the sand was washed through a sieve to extract the eggs. The eggs were kept on moisten sand, separated by pieces of plastic drinking pipes, pressed into the sand. After hatching, the larvae were transferred into glass tubes of a height of 7.5 cm and a diameter of 2.5 cm. They were filled with moist peat moss up to a height of 5 cm. The larvae were fed with pieces of mealworms during their whole development.

The determination of the colour morph was done with beetles older than one month (hardened beetles). In some beetles a colour change was observed during the first four

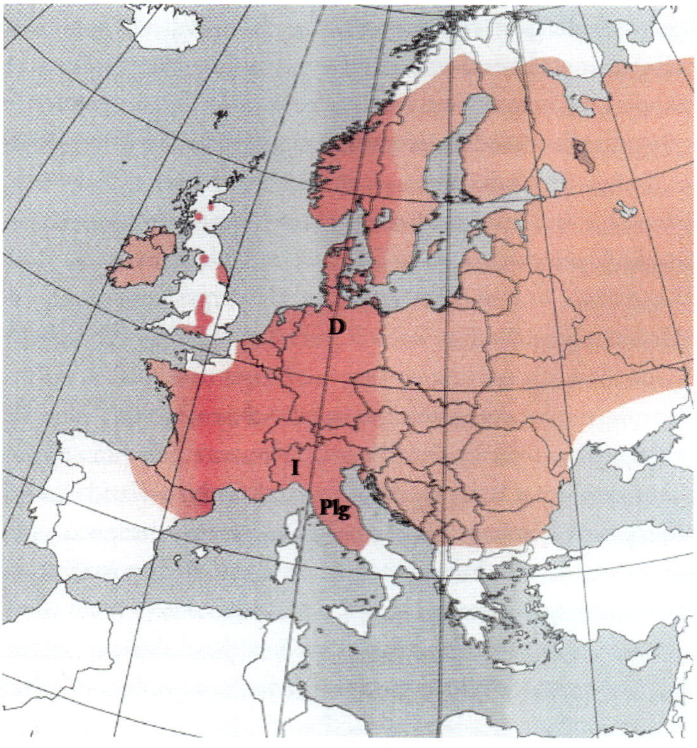

Fig. 1. Distribution of *Poecilus lepidus* in Europe. Map from Turin (2000). Places were the parental generations of our laboratory strains were collected are marked: D = Lüneburger Heide (Germany), I = Lago Maggiore, Plg = Apennine (subspecies *gressorius*), both places in Italy.

weeks of their life. Especially in the red (copper-coloured) morph some beetles appeared to be green during the first weeks of their life, but then at least the elytra changed to red. Such colour changes after moulting are due to the post-ecdysal development of the colour producing layers as shown for Cicindelids by Schultz & Rankin (1985), who described a gradual increase of the thickness of the layers.

A series of 176 *P. lepidus* beetles, collected with pit fall traps in the 'Lüneburger Heide' during a period from 11.04. to 14.10.81 (see Mossakowski et al., 1990), were sorted for their colour morphs.

All crossing experiments were done with virgin females, mostly in groups between four and ten females and a similar number of males. Females of *P. lepidus* have a spermatheca and are able to store sperm from multiple mating (De Vries, 2000). In consequence, the numbers of mating could not be counted and the offspring cold not be separated in our breeding experiments. There is only one exception: the females, homozygous for the allele violet (V) (see Tab. 4).

We use the chi-square test of Exel (Chitest) to determine whether there was a significant difference between observed and expected frequencies (95% confidence level).

RESULTS

Heredity of the colour morphs from the Italian strain

The colour morphs blue and black are homozygous while the colour morph green is heterozygous: carrying one allele of the blue and one of the black morph. In the heterozygous type the gene expression is intermediate (green). The heredity follows the Mendelian rules. The deviations from the expected values are not significant.

Table 1. Heredity of the colour morphs blue (bb), black (ss), and green (bs) from the Italian strain of *Poecilus lepidus*. Genotypes in brackets.

coupled colour morphs	resulting colour morphs	F1 (n)	F1 (%)
blue (bb) x blue (bb)	blue (bb)	429*	100
black (ss) x black (ss)	black (ss)	50	100
blue (bb) x black (ss)	green (bs)	47	100
blue (bb) x green (bs)	green (bs)	36	54
	blue (bb)	32*	46
black (ss) x green (bs)	green (bs)	20	61
	black (ss)	13	39
green (bs) x green (bs)	green (bs)	60	49
	blue (bb)	29	24
	black (ss)	34	28

* including *P. l. gressorius*

Heredity of the colour morphs from the German strain

The following distribution of colour morphs was found in the catch series from 1981 in the 'Lüneburger Heide' (n=176):

red	144 (82%)
yellowish green	17 (10%)
bluish green	6 (3%)
black	9 (5%)

No specimen of the colour morph violet was among them.

Red (copper coloured), yellowish green, bluish green

The heredity of the colour morphs red, yellowish green and bluish green is summarized in Table 2. It is similar to the heredity of the Italian colour morphs black, green, and blue. Bluish green and red are homozygous while yellowish green is heterozygous for one red and one bluish green allele with an intermediate gene expression. No specimens of the German black colour morph were available for the crossbreeding experiments.

Table 2. Heredity of the colour morphs bluish green (gg), red = copper coloured (rr), and yellowish green (gr) from a German strain of *Poecilus lepidus*. Genotypes in brackets.

coupled colour morphs	resulting colour morphs	F1 (n)	F1 (%)
red (rr) x red (rr)	red = copper coloured (rr)	49	100
bluish green (gg) x bluish green (gg)	bluish green (gg)	17	100
bluish green (gg) x red (rr)	yellowish green (gr)	46	100
red (rr) x yellowish green (gr)	yellowish green (gr)	23	53
	red (rr)	20	47
bluish green (gg) x yellowish green (gr)	yellowish green (gr)	24	44
	bluish green (gg)	31	56
yellowish green (gr) x yellowish green (gr)	yellowish green (gr)	19	44
	bluish green (gg)	12	28
	red (rr)	12	28

The devitations from the expected values are not significant.

Violet

During the crossbreeding experiments some unexpected results occurred which made it necessary to differentiate an additional allele V for the colour violet.

Violet beetles are heterozygous for the dominant allele V and a recessive allele: either red (r) or bluish green (g) (Tab. 3)

Table 3. Heredity of the colour morph violet (V) in the 'Lüneburger Heide' strain (Germany). Genotypes in brackets

coupled colour morphs	resulting colour morphs	F1 (n)	F1 (%)	expected
violet (Vr) x red (rr)	violet (Vr)	21	36	1:1
	red (rr)	38	64	
violet (Vr) x violet (Vr)	violet (Vr + VV?)	28	70	3:1
	red (rr)	12	30	
violet (Vr) x violet (Vr)	violet (Vr +VV?)	40	67	3:1
(F1 see Tab. 4)	red (rr)	20	33	
violet (Vr) x bluish green (gg)	violet (Vg)	14	38	1:1
	yellowish green (gr)	23	62	
violet (Vr) x yellowish green (gr)	violet (Vr + Vg)	36	77	2:1:1
	yellowish green (gr)	8	17	
	red (rr)	3	6	

The deviations from the expected values (Vr x rr: p-value = 0.023, Vr x gr: p-value = 0,01) may be partly due to the relative low numbers. The occurrence of homozygote violet specimens could not be verified. The relationship of about 2:1 in the results instead of the expected value 3:1 may be due to a deficiency of this genotype.

In the 'Lüneburger Heide' population V is the most rare allele. That means that under natural conditions violet beetles are mostly heterozygotes. Under breeding conditions in the laboratory we got some females with nearly black elytra. Only the brim of the elytra was violet. When we crossbred them with red males (rr) all resulting beetles of the F1 were violet (n = 28). In this experiment 4 females were included and we did not know if we got the offspring from one or more females. Therefore we separated them in a second breeding experiment but the offspring of all females was violet (female 1 – n = 26, female 2 – n = 42, female 3 – n = 50, female 4 – n = 31). The results of both experiments are summarized in Table 4.

Table 4. Heredity of the colour morph black with a violet brim of the elytra in the 'Lüneburger Heide' strain (Germany). Genotypes in brackets.

coupled colour morphs	resulting colour morphs	F1 (n)	F1 (%)
black with a violet brim(VV) x red (rr)	violet (Vr)	177	100

Homozygous females therefore do not show the typical violet colour. They look almost black with only a small violet brim around the elytra. Up to now we did not detected any homozygous males for the variant V.

Crossbreeding of colour morphs from the German and Italian strains

From the crossbreeding experiment we can conclude that the red allele (R – German strain) is dominant over the allele black (s – Italian strain) in heterozygous beetles. Similarly bluish green (G – German strain) is dominant over blue (b – Italian strain). Heterozygous beetles of the allele combinations gs and br are green like the combinations bs (Italian strain) and gr (German strain). They show a similar intermediate colour and therefore a co-dominant allele expression. We did not try to divide the green colour morphs of these crossbreeding experiments into a German type (yellow green) or an Italian type (darker green) (see Figs 2-3) due to the fact that these phenotypes are sometimes difficult to distinguish in given specimens (the subsuming of similar alleles is not an argument against the heredity of the colour morphs). The single crossing experiments are summarized in Table 5.

a b c

Fig. 2. Colour morphs of the Lago Maggiore strain: a) black (ss), b) blue (bb) and c) dark green (bs). Alleles in brackets.

a b c d

Fig. 3. Colour morphs of the Lüneburger Heide strain: a) red (rr), b) bluish green (gg), c) yellowish green (gr) and d) violet (Vr).

Table 5. Crossbreeding experiments of German and Italian colour morphs of *Poecilus lepidus*. Allele combinations in brackets, I – Italian, G – German colour morph; hybrids are not marked with I or G.

coupled colour morphs	resulting colour morphs	F1 (n)	F1 %
black (ss) I x red (RR) G	red (Rs)	37	100
red (Rs) x red (Rs)	red (RR, Rs)	30	79
	black (ss)	8	21
blue (bb) I x red (rr) G	green (br)	57	100
black (ss) I x yellowish green (gr) G	green (gs)	19	54
	red (Rs)	16	46
green (bs) I x bluish green (gg) G	green (gs)	13	43
	bluish green (Gb)	17	57
bluish green (Gb) x bluish green (Gb)	bluish green (GG, Gb)	8	62
	blue (bb)	5	38
bluish green (Gb) x blue (bb)I	bluish green (Gb)	7	54
	blue (bb)	6	46
green (bs) I x yellowish green (gr) G	green (gs + br)	21	46
	bluish green (Gb)	16	35
	red (Rs)	9	20

The deviations from the expected values are not significant.We did not yet crossbred the colour morph violet from the German strain with the colour morphs from the Italian strains.

DISCUSSION

We have to conclude that there are at least two alleles (on one locus) from the Italian strain and three alleles (on one locus) from the German strain, which are controlling in a dominant, co-dominant or recessive way the colour of the elytra. Few rare phenotypes, e.g., black from the German strain could not be incorporated into the experiments. Crossbreeding between specimens with alleles from both origins proof, that some of the alleles are different (despite the fact that the phenotypes are very similar) and that the alleles controlling the colour are localized on the same locus. The high number of colour morphs of *Poecilus lepidus* is therefore controlled by one gene locus with multiple alleles.

Besides *Poecilus lepidus*, some other polychromous carabid beetles occur in heath lands, peat bogs and other open habitats, e.g., *P. versicolor, P. kugelanni, Carabus nitens, Cicindela campestris* (cf. Schatzmayr, 1942-43; Mossakowski, 1980; Assmann & Forman, 1981), Agonum *ericeti* (Främbs et al., 2002), and *Agonum sexpunctatum* (own observation).

The heredity of colour morphs of polychromous species has been only rarely studied in carabid beetles. Puisségur (1964) used crossbreeding experiments to show that blue and green colour morphs of *Carabus solieri* are controlled by two co-dominant alleles.

The blue or green coloured elytra in *Carabus auronitens* specimens is controlled by two alleles (the genetic variant for green is dominant over the one for the blue elytra) (Puisségur, 1964). Liebherr (1983) found in his study on the morphs of *Agonum decorum* independent alleles producing the red and hirsute phenotype which were dominant to those producing the green and glabrous conditions.

Moreover, Puisségur (1964) was able to show that the genetic basis of the colouration of elytra and pronotum are independent from each other (interspecific crossbreeding with *C. rutilans* and *C. hispanus*). From our study on *Poecilus lepidus* we believe that the genetic basis for the colour of the pronotum and the elytra seems to be controlled by two different loci. This aspect should be kept into mind if studying the selection pressure on colour morphs in the field.

The pigment within the cuticle of all colour morphs of *P. lepidus* is black. This is a very common pigment colour among ground dwelling beetles like Carabidae and especially darkling beetles (Tenebrionidae). Black colour might be useful for body temperature regulation. Especially beetles in arid environments raise their body temperature above air temperature by sun basking. Earlier activity in the morning helps to save body water (less evaporation). It can also be used to reach maturity earlier in the season, as shown by the desert dwelling ground beetle *Thermophilum sexmaculatum* (Erbeling & Paarmann, 1985, 1986). Forest dwelling and nocturnal ground beetle of the temperate zone are more often black than beetles in open landscape and with preferred diurnal activity (Lauterbach, 1964; Thiele & Weber, 1968; Luff, 1978; Löser, 1980; Desender et al., 1984; Kegel, 1990). Dark colour may be a good camouflage in darker environments like leaf litter in the forest. For night active beetles there is no need to have other colours than black, while it may be of selective advantage to have other colours if the beetles are day active: for example green in a green environment of grasses and herbs at a meadow. It could lower the chances to be caught by a day active bird, compared to other colour morphs living in the same habitat. Such diurnal bird species seem to play a decisive roll as ground beetle predators (Thiele, 1977). Red colour morphs, dominant in heath land, may be better camouflaged in this habitat than the other colour morphs.

Schultz (1986) pointed out that background matching of colour patterns is present in many American *Cicindela* species. He also described the occurrence of a few other species with highly reflective colours. When they fly into deep shadow the beetles disappear abruptly.

Along riversides of the Apennine only the blue morph of *P. l. gressorius* appears. The whole population is homozygous for the allele blue (b). In other habitats with differing surface colours or denser green vegetation, other morphs might be favoured by selection. Deviations from Hardy-Weinberg-equilibrium (HWE) can give first indications for a selection against any genotypes. The data from the Lüneburger Heide (cf. above) give only weak (and not highly significant) evidence because of more homozygotes of the variant g than expected after HWE. In habitats were the heterozygous morph green has an higher adaptive value, the homozygous morphs blue (bb) or black (ss) as well as red (rr) or bluish green (gg) cannot vanish. One forth of the offspring of green parents will be black or red, and one fourth blue or bluish green.

Reasons for rarity of the violet morph, caused by the dominant allele V, may be due to a developmental depression of the homozygote. Field studies are necessary to get more information about the percentage of beetles of this colour in the 'Lüneburger Heide' population. No specimen was found in a series of 176 beetles.

The studied colours, with the exception of black, are structural colours, caused through light reflection and interference by thin layers in the upper cuticle in front of the black background. So our study of the heredity of colour morphs was a study of the heredity of the thickness of light reflecting layers. For details see Mossakowski et al. (2008).

Wilmer & Unwin (1981) studied heat gain and loss in relation to weight and reflectance of insects. They emphasized size and reflectance as an important feature, which regulate body temperature. For moderate or large sized diurnal species it should be a great advantage to be highly reflective. They can thus avoid overheating by radiation. To be coloured may have two advantages to diurnal Carabidae: protection against birds and overheating (Mossakowski, 1980; Schultz, 1986).

It is still not known if there are differences in the daily activity of the different colour morphs. May be, that darker morphs tend to be more nocturnal than brightly coloured ones. Life span of colour morphs of the *P. lepidus* Lago Magiore strain differs under long day (16h light) and short day (8h light) at 20 °C (Paarmann, 1990). In the first case life span is significantly shorter for green males compared to blue and black males, while under short day condition it is significantly shorter in black males than in blue or green.

ACKNOWLEDGEMENTS

Collecting the P generation of the *P. lepidus gressorius* strain was possible only because of a detailed description of the population site in the Apennine mountains by E. Contarini (Bagnacavallo Ravenna) and an invitation to Italy by P. Brandmayr (Universita della Callabria, Arcavacata di Rende). S. Albrecht bred and crossed *P. lepidus* beetles with great enthusiasm.

REFERENCES

Assmann, T. & Forman, F. (1981). Die Carabidenfauna des Naturschutzgebietes Venner Moor (Landkreis Osnabrück), 1. Teil: Die *Cicindela*-Arten des Naturschutzgebietes und Bemerkungen über eine *Cicindela campestris*-Population im Schweger Moor. – Osnabrücker naturwiss. Mitt. 8: 173-176.

Desender, K., Mertens, J., D'Hulster, M. & Berbiers, P. (1984). Diel activity patterns of Carabidae (Coloeptera), Staphylinidae (Coleoptera), and Collembola in a heavily grazed pasture. – Review of Ecology and Soil Biology, 2121: 347-362.

De Vries, H. (2000). Multiple paternity in ground beetles. – Mitt. Dtsch. Ges. Allg. Angew. Ent. 12: 441-445.

Erbeling, L. & Paarmann, W. (1985). Diel activity patterns of the desert carabid beetle *Thermophilum* (=*Anthia*) *sexmaculatum* F. (Coleoptera: Carabidae). – Journal of Arid Environment, 8: 141-155.

Erbeling, L. & Paarmann, W. (1986). The role of a circannual rhythm of thermoregulation in the control of the reproductive cycle of the desert carabid beetle *Thermophilum sexmaculatum* F. – In: Carabid beetles. Their Adaptations and Dynamics (Den Boer, P.J., Luff, M.L., Mosskowski, D. & Weber, F., eds). Gustav Fischer, Stuttgart, New York, p. 125-146.

Erichson, W.F., Schaum, H., Kraatz, G. & Kiesenwetter, H.v. (1860). Naturgeschichte der Insekten Deutschlands, Berlin, 791 pp.

Främbs, H., Dormann, W. & Mossakowski, D. (2002). Spatial Distribution of Carabid beetles on Zehlau Bog. – Baltic J. Coleopterol. 2 (1): 7-15.

Horion, A. (1941). Faunistik der deutschen Käfer. 1. Adephaga – Caraboidea. – Goecke & Evers, Krefeld.

Kegel, B. (1989). Laboratory experiments on the side effects of selected herbicides and insecticides on the larvae of three sympatric *Poecilus* species. – Journal of applied Entomology, 108: 144-155.

Kegel, B. (1990). Diurnal activity of carabid beetles living on arable land. – In: The role of ground beetles in ecological and environmental studies (Stork, N.E., ed.). Andower, Hampshire, 65-76.

Lauterbach, A.W. (1964). Verbreitungs- und aktivitätsbestimmende Faktoren bei Carabiden in sauerländischen Wäldern. – Abhandlungen aus dem Landesmuseum für Naturkunde zu Münster in Westfalen, 26: 1-100.

Liebherr, J.K. (1983). Genetic basis for polymorphism in the ground beetle Agonum decorum (Coleoptera: Carabidae). – Annals of the Entomological Society of America 76: 349-358.

Lindroth, C.H. (1949). Die Fennoskandischen Carabidae. – Kungl. Vetensk. Vitterh. Samh. Handl. (Ser. B4), 3. Allgemeiner Teil, 911 pp.

Löser, S. (1980). Zur tageszeitlichen Aktivitätsverteilung von Arthropoden der Bodenstreu (Coleoptera, Diplopoda, Isopoda, Opiliones, Aranea) eines Buchen-Eichen Waldes (Fago-Quercetum). – Entomologia Generalis, 6: 169-180.

Luff, M.L. (1978). Diel activity patterns of some field Carabidae. – Ecological Entomology, 3: 53-62.

Mossakowski, D. (1980). Reflection measurements used in the analysis of structural colours of beetles. – Journal of microscopy 116: 351-364.

Mossakowski, D., Främbs, H. & Baro, A. (1990). Carabid beetles as indicators of habitat destruction caused by military tanks. – In: The role of ground beetles in ecological and environmental studies (Stork, N. E., ed.). Intercept, Andover, Hampshire, p. 237-243.

Mossakowski, D., Paarmann, W., Rohe, W., Lüchtrath, I. & Aßmann, T. (2008). Multi-layer structural colours in *Poecilus lepidus* (Col., Carabidae). – In: Back to the Roots and Back to the Future. Towards a New Synthesis amongst taxonomical, ecological and biogeographical approaches in carabidology (Penev, L., Erwin, T. & Assmann, T., eds). Pensoft, Sofia-Moscow, p. 173-182.

Paarmann, W. (1990). *Poecilus lepidus* Leske (Carabidae, Coleoptera), a species with the ability to be a spring and autumn breeder. – In: The role of ground beetles in ecological and environmental studies (Stork, N.E., ed.). Andover, Hampshire. p. 259-267.

Puisségur, C. (1964). Recherches sur la génétique des carabes. – Vie et milieu. Supplement 18: 1-288.

Schatzmayr, A. (1942-43). Bestimmunsgtabellen der europäischen und nordafrikanischen *Pterostichus*- und *Tapinopterus*-Arten. – Koleopterologische Rundschau 27: 1-144.
Schultz, T.D. (1986). Role of structural colors in predator avoidance by tiger beetles of the genus Cicindela (Coleoptera: Cicindelidae). – Bulletin of the Entomological Society of America 32: 142-146.
Schultz, T.D. & Rankin, M.A. (1985). Developmental changes in the interference reflectors and colorations of tiger beetles (Cicindela). – Journal of experimental biology 117: 111-117.
Terrell-Nield, C.E. (1990). Distribution of leg colour morphs of *Pterostichus madidus* (F.) in relation to climate. – In: The Role of Ground Beetles in ecological and environmental studies. Intercept Publishers (Stork, N.E., ed.). Andover, Hampshire. p. 39-51.
Thiele, H.U. (1977). Carabid beetles in their environments. – Zoophysiology and Ecology 10. Springer, Berlin, Heidelberg, New York, 369 pp.
Thiele, H.U. & Weber, F. (1968). Tagesrhythmen und Aktivität bei Carabiden. – Oecologia, 1: 315-355.
Turin, H. (2000). De neederlandse loopkeepers – verspreiding en oecologie (Coleoptera: Carabidae). – Nederlandse Fauna 3, National Naturhistorisch Museum Naturalis, KNNV Uitgeverij, European invertebrate survey – Nederland, 666 pp.
Van Natto, C. & Freitag, R. (1986). Solar radiation reflectivity of *Cicindela repanda* and *Agonum decentis* (Coleoptera, Carabidae). – Canadian entomologist 118: 89-95.
Wilmer, P.G. & Unwin, D.M. (1981). Field analysis of insect heat budgets: reflectance, size and heating rates. – Oecologia, 50: 250-255.

L. Penev, T. Erwin & T. Assmann (Eds) 2008
BACK TO THE ROOTS AND BACK TO THE FUTURE. TOWARDS A NEW SYNTHESIS AMONGST TAXONOMIC, ECOLOGICAL AND BIOGEOGRAPHICAL APPROACHES IN CARABIDOLOGY
Proceedings of the XIII European Carabidologists Meeting, Blagoevgrad, August 20-24, 2007, pp. 195-206.

Patterns of molecular variability in Carabid beetles mostly from the Baltic Sea coast

Nordfried Kamer, Wolfgang Dormann & Dietrich Mossakowski

University of Bremen, Institute for Ecology and Evolutionary Biology. P.O. Box 330440, D 28334 Bremen, Germany. E-mail: dmossa@uni-bremen.de

SUMMARY

Sequence variability was studied in six halobiontic or halophilic salt marsh species, which represent animals with high dispersal power. In contrast, *Carabus clatratus* was included in the study as a flightless species common in the same habitats.

Four of the salt marsh specialists show a pattern of molecular variability congruent with the hypothesis that high dispersal power results in low variability (*Bembidion fumigatum*) or in high variability, which is not correlated with geographical patterns (*Bembidion pallidipenne, B. tenellum*, and *Dyschirius salinus)*.

Two other specialists (*B. minimum, Anisodactylus poeciloides*) have a distribution of haplotypes that form geographical clusters. In particular, *B. minimum* has one haplotype restricted to coastal habitats and a different one inland. This distribution seems to contradict the wide distribution area and high mobility of this species.

Carabus clatratus is the only species under study which is not able to disperse by flight. The populations of this beetle show numerous ND5 haplotypes. Three continental ones are distributed mainly in the west, the east or at the coast, respectively (main haplotypes). The sequences of animals from Ireland are quite different from those from the continent (ND5: p-distance = 3.5 - 4.3%). These findings were interpreted by postulating a south-western glacial refuge, which is supported by 12S RNA data: the Irish *C. clatratus* cluster with specimens from the South of France and the Tuscany.

Keywords: Molecular variability, mt DNA sequences, salt marsh, Carabidae

INTRODUCTION

Genetic variability within species and populations is a measure of biodiversity and may be used for other purposes like reconstruction of refuges, pathways of postglacial colonisation, metapopulation studies, questions of species borders, etc. In a project on biodiversity of salt meadows at the Baltic Sea in Mecklenburg-Western Pomerania we studied the within species diversity of six halobiontic or halophilic Carabid species, respectively, by analysing DNA sequences.

Salt habitats are more or less isolated along the German Baltic Sea coast. Therefore, populations of Carabid species dwelling in such habitats may be isolated to some degree. A more strict isolation may be given for populations at inland salt localities.

It is well known that most salt marsh species in general are able to fly and that they use this capacity. The only exception is *Pogonus chalceus,* with specimens able to fly and others with reduced flight muscles or reduced hind wings (Desender, 1989). But this species does not occur at the German Baltic coast.

Additionally, we studied *Carabus clatratus* as a representative of species with low dispersal power due to reduced hind wings. *Carabus clatratus* is halotolerant and very abundant at the peaty salt meadows of the Baltic Sea coast. In general, it is not able to fly but some few specimens with well developed hind wings are known (Lindroth, 1949). The first results on this species were published by Kamer et al. (2005).

Genetic studies on salt marsh Carabid beetles are rare. Desender et al. (1998) described a higher genetic variability but a smaller genetic differentiation between populations in the full winged *Dicheirotrichus gustavii* in comparison with the polymorphic *Pogonus chalceus* in populations varying in size and isolation. The variability found displayed patterns at a more local scale, but not at a geographical scale. Dhuyvetter et al. (2005) found no significant correlation between genetic variability (allozyme and microsatellite markers) and population size and habitat area in fragmented populations of *Pogonus chalceus* in Belgian salt marshes. The variability of Atlantic and Mediterranean populations of the same species displayed geographical differences on a large scale and a higher dispersal power combined with lower genetic differentiation of the Mediterranean populations than those of the Atlantic coast (Desender & Serrano, 1999).

The scope of this paper is to describe patterns of molecular variability of some selected species of Carabid beetles of the Baltic Sea coast dwelling in salt marsh habitats. We expected that the halobiontic and halophilic species display a lower degree of variability, because of their high dispersal power than the halotolerant *Carabus clatratus* living in high abundances in the same habitat, but unable to fly.

STUDY AREA AT THE BALTIC SEA COAST

The study area of this project is marked in Fig. 1 and more details are shown in the maps of Figs 2-7. Three scales of geographical dimensions were studied: (i) the Baltic Sea coast of Mecklenburg-Western Pomerania (project area, rectangle in Fig. 1). The interpretation of the variability at a local scale needs some knowledge of the situation at a larger scale. Therefore, we additionally analysed (ii) specimens from inland salt habitats and the North Sea coast and (iii), in particular, in *Carabus clatratus* material from populations of its European distribution area (Fig. 7).

MATERIAL AND METHODS

Because a low amount of variability was expected, mitochondrial DNA sequences as rapidly evolving characters were used. In Carabid beetles, lower rates of molecular evolution than the usual 2 % of mitochondrial DNA have been published (below 1 % per myr: Su et al., 1998; Prüser & Mossakowski, 1998). A part of the CO2 gene (751 base pairs) and ND5 (1083 bp) were used to analyse variability of salt marsh species. CO2 data were presented for *Anisodactylus poeciloides* (5 populations, 9 specimens), *Bembidion fumigatum* (5, 10), *B. minimum* (12, 12), *B. pallidpenne* (5, 9), *B. tenellum* (4, 5), and *Dyschirius salinus* (6, 6). For *Carabus clatratus,* we present data mainly of the ND5 gene (43, 121). In order to be successful with dry collection material, we sequenced a part of the 12S RNA gene (311 bp) in *C. clatratus.*

PCR and sequencing was performed by using the primer pairs 12Sai and 12Sbi for 12S RNA (Simon et al., 1994), His and Phe for ND5 (Su et al., 1996), and A-tLEU and B-tLYS for CO2 (Liu & Beckenbach, 1992). In some cases intermediate primers were used for ND5 (270 and 270r, 850 and 850r; Düring & Brückner, 2000).

The ClustalX program (Thompson et al., 1997) was used for alignment (default option). Phylogenetic analysis was performed using the Maximum Likelihood Method (DNAML in PHYLIP 3.6, Felsenstein, 2004; base freqency = empirical, tranversions/transitions = 2:1, global rearrangements). Also calculations were done using Maximum Parsimony (PAUP* 4.10, Swofford, 1998) and Minimum Evolution (MEGA 2.1, Kumar et al., 2001). Only ML results are shown because results differed only in insignificant details from the other methods. Branch support was checked by bootstrap analysis (1,000 pseudo replicates).

Additional sequences were included from Genbank in the analysis: *Carabus maacki* (D50358 ND5, Su et al., 1996) as a sister species of *C. clatratus,* formerly stated as its subspecies; *C. nodulosus* (AF231700: ND5, Arndt et al., 2003), *C. granulatus* (AF 219473: ND5, Sota & Vogler, 2001), *C. nemoralis* (AB047265: ND5, Imura et al., 2000), *C. guerini* (AB047277: ND5, Imura et al., 2000) and *C. splendens* (AF190030: 12S, Duering & Brueckner, 2000) were chosen as outgroup species.

The sequence of *C. clatratus* Geestmoor 1 was taken from Arndt et al. (2003; AF231688: ND5)

All other *C. clatratus, C. arcadicus* and the salt marsh specialists were analysed for this paper „(gene bank accession numbers: EU790645-EU790667 ND5, EU798727-EU798747 12S, EU839504-EU839553 CO_2).

RESULTS

We found quite different patterns of molecular variation, which depend on the species.

- Species without any variability

In *Bembidion fumigatum* populations, we found only one haplotype (Fig. 1). All the other species showed variability, some without, others in a geographical context.

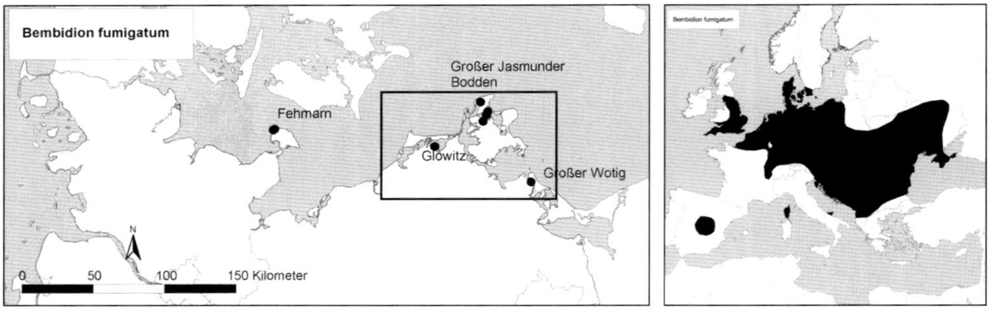

Fig. 1. Distribution of haplotypes in *Bembidion fumigatum*. The rectangle indicates the main study area. Distribution map (right) modified from Turin (2000).

- Species variable - no geographical context

The species *Bembidion tenellum* (Fig. 2), *Dyschirius salinus* (Fig. 3) and *Bembidion pallidipenne* (Fig. 4) displayed variability, but no correlations with geographical patterns are obvious.

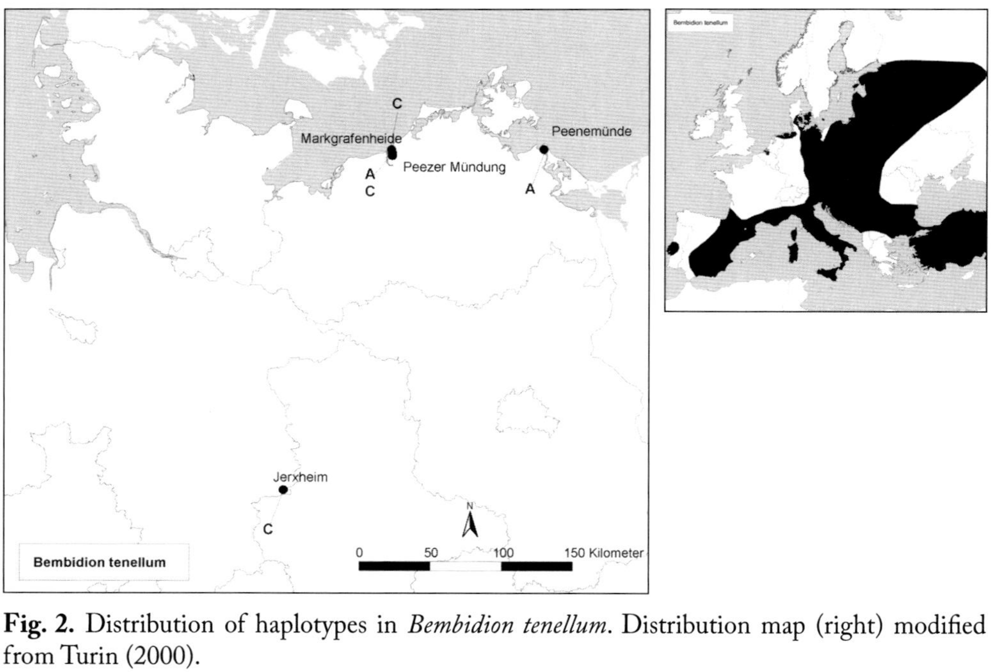

Fig. 2. Distribution of haplotypes in *Bembidion tenellum*. Distribution map (right) modified from Turin (2000).

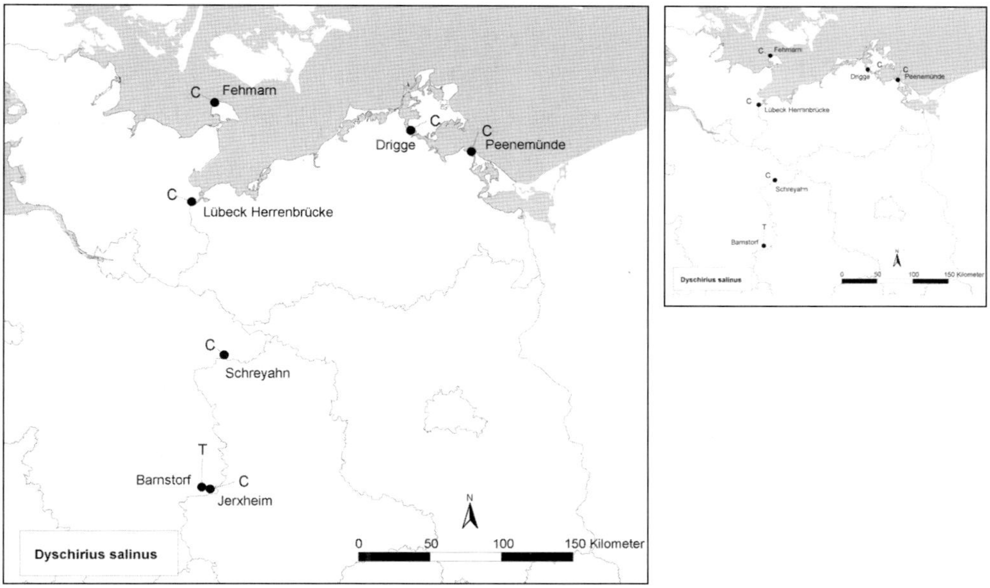

Fig. 3. Distribution of haplotypes in *Dyschirius salinus.* Distribution map (right) modified from Turin (2000).

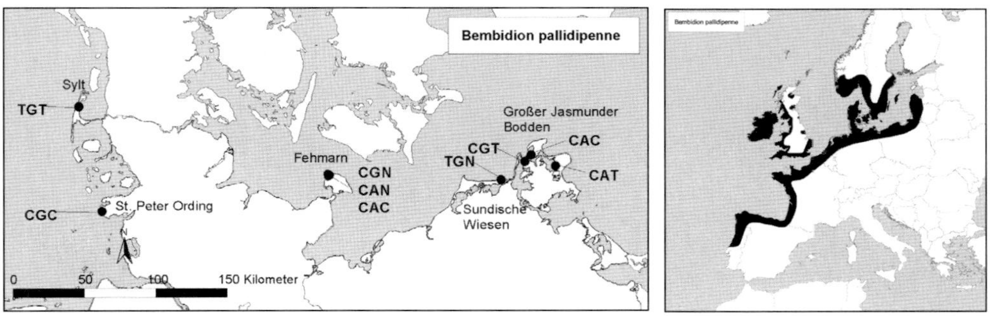

Fig. 4. Distribution of haplotypes in *Bembidion pallidipenne.* Distribution map (right) modified from Turin (2000).

- Species variable - with geographical context
  - Coastal versus inland populations

In *Bembidion minimum*, all the inland populations under study have the identical haplotype, the same was found for the coastal ones. But they are different (Fig. 5). We obtained a low but stable p-distance of max. 0.28%.

  - North-West versus South-East?

In *Anisodactylus poeciloides* we found differences, which might be interpreted in a geographical context (Fig. 6). This species occurs only very locally and is not known from the German North Sea coast. We included two *Anisodactylus* specimens from the South of France (Camargue).

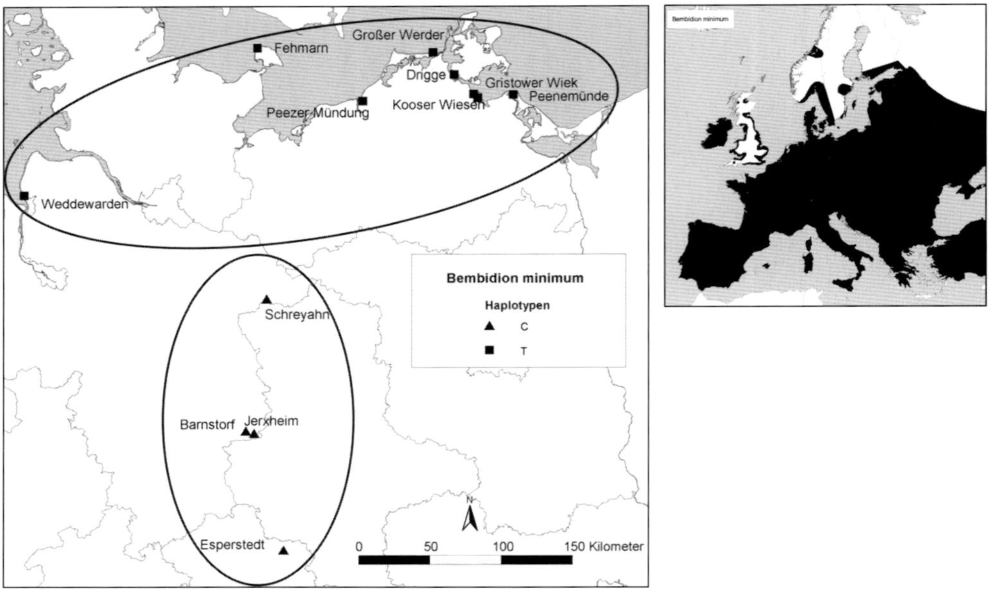

Fig. 5. Distribution of haplotypes in *Bembidion minimum.* Distribution map (right) modified from Turin (2000).

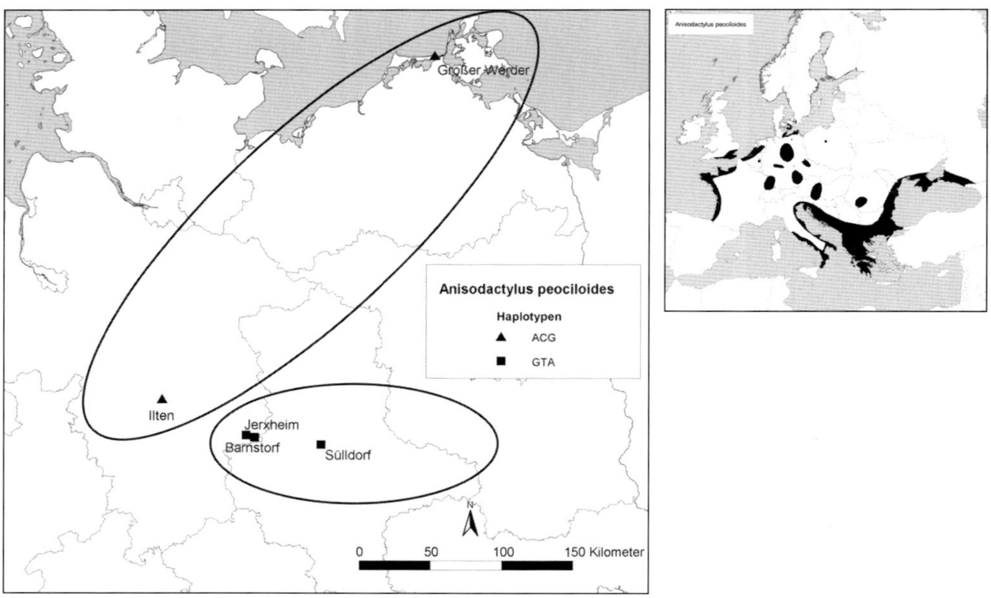

Fig. 6. Distribution of haplotypes in *Anisodactylus poeciloides.* Distribution map (right) modified from Turin (2000).

Table 1. Sequences differences in *Anisodatylus poeciloides* and *A. virens* (Camargue). Number in brackets: total number of positions; other numbers: sequence position.

	ND5 (1083)				CO2 (751)			
	24	460	468	513	89	350	418	551
Großer Werder					T	C	C	G
Großer Werder					T	C	**T**	G
Ilten					T	C	C	G
Jerxheim					C	T	C	A
Sülldorf	A	G	A	C	C	T	C	A
Barnstorf	A	G	A	C	C	T	C	A
Camargue 1	A	**A**	A	C	C	T	C	A
Camargue 2	**G**	G	**G**	**T**	C	T	C	A

Six out of 1834 positions are variable (Table 1). The German populations under study display three haplotypes in CO2 and no difference between two populations in ND5. The French specimens belong to *A. virens*, the sister species of *A. poeciloides*. The genetic distance between the two Camargue specimens is lower than 0.4 %, between French and German haplotypes lower than 0.1 and 0.3, respectively. Because of the very low genetic difference between both species, their status should be tested using more material.

- A species with low dispersal power and a complex geographical pattern

Three main ND5 haplotypes were found in most specimens of *Carabus clatratus*. Additionally, there was a lot of singleton variability (121 specimens of continental populations with 28 different haplotypes).

Continental *Carabus clatratus* cluster together with high or good bootstrap support (Figs 8-9). The same picture results when many additional specimens were included (see Kamer et al., 2005). One specimen from Hiddensee (Karkensee 2) has the same haplotype as the uppermost cluster in Fig. 8. It is placed at the base of the continental cluster, because of its differences in other positions.

The Irish specimens show differences in three positions in ND5 and build their own cluster in Fig. 8. They differ largely from the continental animals (p-distance about 3-4 %; Table 2). Only *C. maacki*, the closest relative to *C. clatratus*, shows higher differences to the continental forms (4.5-5 %) in this species complex. The specimens from the south of France (Camargue) and Italy (Tuscany) cluster together with the Irish specimens in the 12S RNA tree (Fig. 9). Interestingly, the Camargue specimen was placed with the specimen from Galway, in the north-western part of the Republic of Ireland, south of the other locality (compare Fig. 7).

Table 2. Substitutions of ND5 in *Carabus clatratus* populations and some outgroup species.

	N	p	K2P
Within Ireland	0 - 2	0.0 - 0.4	0 - 1.0
Within Continent	0 - 5	0.0 - 1.0	0 - 0.4

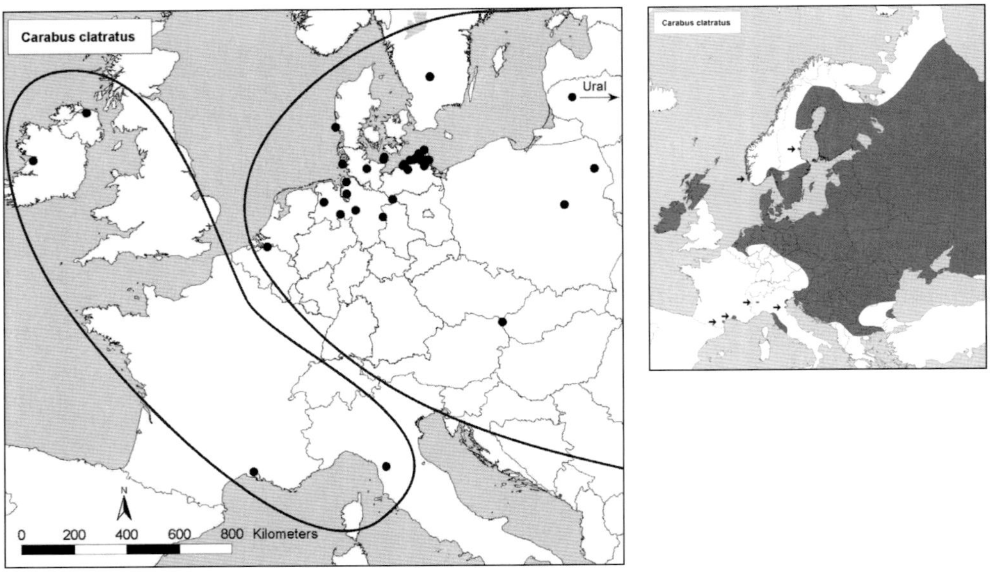

Fig. 7. European distribution area of *Carabus clatratus*.
Left: Western and 'continental' clade of *C. clatratus*. Dots indicate localities from which material was included in this paper. Right: distribution area of *C. clatratus* redrawn and modified from Turin (2000).

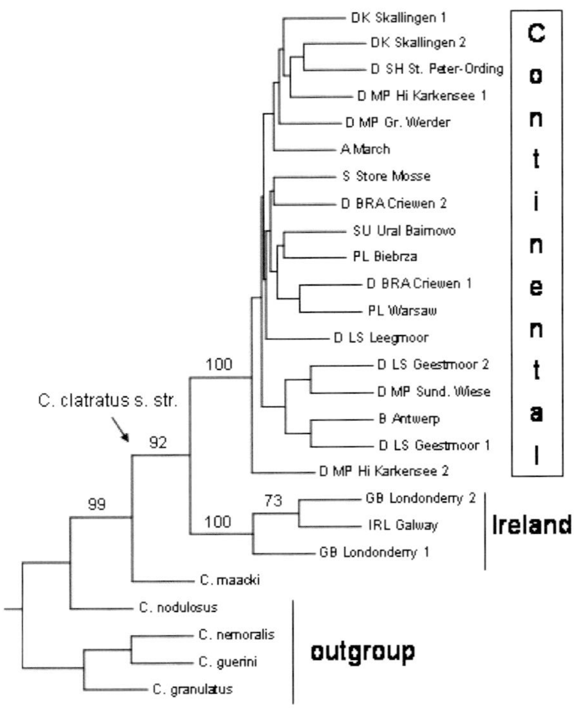

Fig. 8. Maximum Likelihood molecular tree (ND5) of *Carabus clatratus*.
Numbers above the branches are bootstrap values >50%.

	N	p	K2P
Continent – Ireland	**16 - 20**	**3.5 - 4.3**	**3.5 - 4.7**
Ireland – *C. maacki*	**19 - 20**	**3.9 - 4.1**	**4.2 - 4.4**
C. maacki – Continent	**24 - 28**	**4.9 - 5.5**	**5.4 - 6.3**
Continent – outgroup	40 - 52	8.2 -10.9	9.5 - 12.7
Ireland – outgroup	35 - 44	7.2 - 7.4	8.0 - 10.3
C. maacki – outgroup	41 - 57	8.4 -10.7	9.5 - 12.7
Outgroup - outgroup	33 - 57	6.8 -11.7	7.5 - 14.0

N: number of pairwise substitutions; p: p-distance (n in %); K2P: distance with Kimura-2-Parameter (Gamma). Number of specimens/taxa: Continent: 118, Ireland 3; *C. maacki* 1, outgoup 4 (cf. Fig. 8).

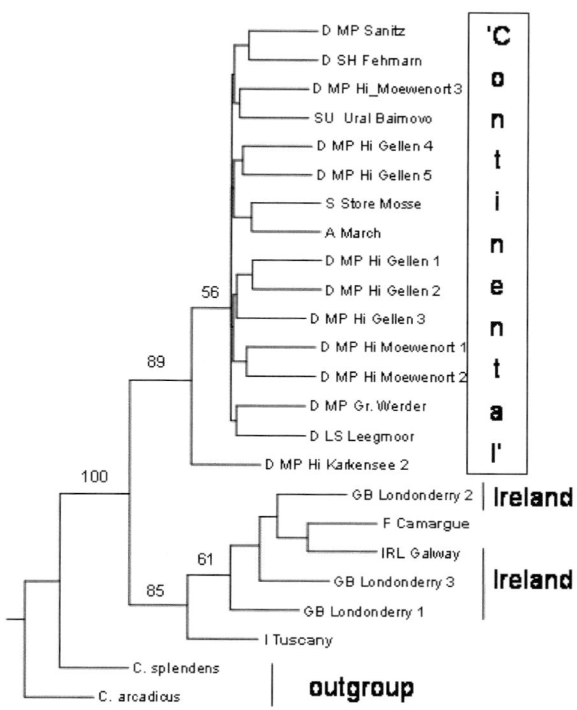

Fig. 9. Maximum Likelihood molecular tree (12S RNA) of *Carabus clatratus*. Numbers above the branches are bootstrap values >50%.

DISCUSSION

Mobility

We found four salt marsh specialists with a pattern of molecular variability, which is congruent with their high dispersal power. This may result in a low variability (*Bem-*

bidion fumigatum) or in high variability and no correlation with geographical conditions (*Bembidion palliipenne, B. tenellum*, and *Dyschirius salinus)*. Recently it appears that *B. fumigatum* is in a process of widening its distribution area; *B. pallidipenne* is a very flight-active species, therefore the high variability without geographical pattern is no surprise.

In two specialists of salt habitats (*B. minimum, Anisodactylus poeciloides*), the distribution of haplotypes build geographical clusters. The distribution of haplotypes in *B. minimum* seems to contradict the wide distribution area and the high mobility of this species.

In *Bembidion minimum*, all the inland populations under study have the identical haplotype and the coastal specimens have a different one. This was the most surprising result, because of the occurrence of this species far from salt habitats inland. *B. minimum* is a halophilic species and quite common in salty habitats. As far as we know, specimens are rare in salt free habitats. One interpretation may be that this is due to their high dispersal power. But this seems to contradict the pattern we found. The observed pattern in *Bembidion minimum* may be the result of colonisation from two different regions, which are not very distant, because of the relative low genetic distance.

Carabus clatratus is the only species under study, which is not able to disperse by flight. The populations of this beetle show numerous haplotypes, three of the continental ones are distributed mainly in the west, the east or at the coast, respectively (main haplotypes, see Kamer et al., 2005).

Differentiation by distance

No support for this hypothesis was found. In *C. clatratus* a higher variability was found in the central European populations in comparison with western and eastern ones on the continent. This phenomenon might be interpreted as a hint on a glacial refuge in the middle European region (Pawlowski, 1986: *C. menetriesi* in Poland, 25,000 years BP).

Human transport

As we pointed out earlier (Kamer et al., 2005), the high variability at Karkensee on the isle of Hiddensee may be caused by human transport due to the special environmental and historical conditions there. Only one haplotype was found on the whole island except at one small place with the three main haplotypes. Additionally, one specimen from there did not cluster with the specimens of its main haplotype due to specific substitutions (cf. Figs 8-9: 'Karkensee 2'). The position of this specimen in the trees is an additional hint for the special situation at Karkensee.

Large scale differentiation in *Carabus clatratus*

The Irish specimens differ largely from the continental animals (p-distance in ND5 about 3-4 %; Table 2). The respective values for *C. maacki* range from 4.5 to 5 %, a form formerly included into *C. clatratus*, but now established as a separate species (Imura et al., 1998; Deuve, 2004). The specimens from southern France and Italy build a cluster together with the Irish ones in the 12S RNA tree (Fig. 9). We suggest the existence of a separate refuge during the glaciation period in the south-west of the extant distribution. Interestingly, the Camargue specimen forms a cluster with a single specimen from the north-western part of the Republic of Ireland. Although not supported by high bootstrap values, this fits with a scenario of a separate refuge.

ACKNOWLEDGMENTS

Our cordially thanks are due to many colleagues who collected some *C. clatratus* for us or helped us to get material: Roy Anderson, Wulf Carius, Achille Casale, Konjev Desender, Alain Drumont, Ron Felix, Herbert Fraembs, Konstatin B. Gongalsky, Björn A. Hatteland, Axel Hochkirch, Jarmo Holopaien, Jacques Leplat, Martin Luff, Tibor Magura, Gerd Mathiak, Gerd Müller-Motzfeld, Wolfgang Paill, Artur Rutkiewicz, Pawel Sienkewicz, Jarek Sklodowski, Ivailo Stoyanov, Gyozo Szel, Vytautas Tamutis, Hans Turin, Friederike Zinner and Christoph Zöckler; to an unknown reviewer for helpful comments and to Michael Vicker who corrected the English.

REFERENCES

Arndt, E., Brueckner, M., Marciniak,M., Mossakowski, D. & Prueser, F. (2003). Phylogeny. – In: The Genus *Carabus* L. in Europe. A Synthes (Turin, H., Penev, L. & Casale, A., eds). – Pensoft, Sofia-Moscow, 511 p.

Desender, K. (1989). Heritability of wing development and body size in a carabid beetle, *Pogonus chalceus* Marsham, and its evolutionary significance. – Oecologia 78: 513-520.

Desender, K., Backeljau, T. Delahaye, K. & De Meester, L. (1998). Age and size of European saltmarshes and the population genetic consequences for ground beetles. – Oecologia 114: 503-513.

Desender, K. & J. Serrano (1999): A genetic comparison of Atlantic and Mediterranean populations of a saltmarsh beetle. – Belg. J. Zool. 129 (1): 83-94.

Deuve, T. (2004). Illustrated Catalogue of the Genus *Carabus* of the World (Coleoptera: Carabidae). –Pensoft, Sofia-Moscow. 461 pp.

Dhuyvetter, H., Gaublomme, E. & Desender, K. (2005). Bottlenecks, drift and differentiation: the fragmented population structure of the salt marsh beetle *Pogonus chalceus*. – Genetica 124 (2/3): 167-177.

Duering, A. & Brueckner, M. (2000). The Evolutionary History of the Tribe Molopini: A First Molecular Approach. – In: Natural History and applied Ecology of Carabid Beetles.

(Brandmayr, P., Lövei, G., Zetto-Brandmayr T., Casale, A. & Vigna Taglianti A., eds). Pensoft, Sofia-Moscow, p. 1-4.

Felsenstein, J. (2004). PHYLIP (Phylogeny Inference Package) version 3.6. Distributed by the author. Department of Genome Sciences, University of Washington, Seattle.

Imura, Y., Kim, C.-G., Su, Z.-H. & Osawa, S. (1998). An attempt at the higher classification of the Carabina (Coleoptera, Carabidae) based on morphology and molecular phylogeny, with special reference to *Apotomopterus*, *Limnocarabus* and *Euleptocarabus*. – Elytra 26: 17-35.

Imura, Y., Su, Z. H. & Osawa, S. (2000). Phylogeny in the division Archicarabomorphi (Coleoptera, Carabidae) viewed from mitochondrial ND5 gene sequences. – Elytra Tokyo 28, 223-228.

Kamer, N., Mossakowski, D. & Dormann, W. (2005). Carabids of salt meadows at the Baltic Sea coast in Mecklenburg-Western Pomerania (Germany) and their variability in mitochondrial genes – Proceedings of the 11th European Carabidologist Meeting. DIAS Report, 114: 145-150.

Kumar, S., Tamura, K., Jakobsen, I.B. & Nei, M. (2001). MEGA2: Molecular Evolutionary Genetics Analysis software. – Arizona State University, Tempe, Arizona, USA.

Lindroth, C. H. (1949). Die fennoskandischen Carabidae. Eine tiergeographische Studie. III. Allgemeiner Teil. – Göteborgs kungl. Vetensk. Vitterh. Samh. Handl. Ser. B 4, 3. 911 p.

Liu, H.L. & Beckenbach, A.T. (1992). Evolution of the mitochondrial cytochrome oxidase II gene among ten orders of insects. – Mol. Phylogenet. Evol. 1 (1): 41-52.

Pawłowski J. (1986). Próba wyznaczenia gatunków chrząszczy (Coleoptera) przewodnich dla granicy plejstocenu i holocenu w południowej Polsce. – Spraw. Pos. Kom. Nauk. 27(2): 368-370.

Prüser, F. & Mossakowski, D. (1998): Low substitution rates in mitochondrial DNA in mediterranean carabid beetles. – Insect Molecular Biology 7(2): 121-128.

Prüser, F. (1996). Variabilität mitochondrialer DNA-Sequenzen und die Phylogenie der Gattung *Carabus* Linné 1758 (Coleoptera, Carabidae). – Thesis, University of Bremen. 173 pp.

Simon, C., Frati, F., Beckenbach, A., Crespi, B., Liu, H. & Flook, P. (1994). Evolution, weighting, and phylogenetic utility of mitochondrial gene sequences and a compilation of conserved polymerase chain primers. – Ann. Ent. Soc. America 87 (6): 651-701.

Sota, T. & Vogler, A. P. (2001). Incongruence of mitochondrial and nuclear gene trees in the Carabid beetles *Ohomopterus*. – Syst. Biol. 50 (1), 39-59.

Su, Z.H., Ohama, T., Okada, T.S., Nakamura, K., Ishikawa, R. & Osawa, S. (1996). Phylogenetic relationships and evolution of the Japanese Carabinae ground beetles based on mitochondrial ND5 gene sequences. – J. Mol. Evol. 42 (2), 124-129.

Su, Z.H., Tominaga, O., Okamoto, M. & Osawa, S. (1998). Origin and diversification of hindwingless *Damaster* ground beetles within the Japanese islands as deduced from mitochondrial ND5 gene sequences (Coleoptera, Carabidae). – Mol. Biol. Evol. 15(8): 1026-1039.

Swofford, D.L. (1998). PAUP*. Phylogenetic Analysis Using Parsimony (*and Other Methods). Version 4.10. – Sinauer Associates, Sunderland, Massachusetts.

Thompson, J.D., Gibson, T. J., Plewniak, F., Jeanmougin, F. & Higgins, D.G. (1997). The ClustalX windows interface: flexible strategies for multiple sequence alignment aided by quality analysis tools. – Nucleic Acids Research, 24: 4876-4882.

Turin, H. (2000). De Nederlandse Loopkevers: Verspreiding En Oecologie (Coleoptera Carabidae). – De Nederlandse Fauna 3. Nationaal Natuurhistorisch Museum Naturalis, Leiden. 666 pp.

L. Penev, T. Erwin & T. Assmann (Eds) 2008
BACK TO THE ROOTS AND BACK TO THE FUTURE. TOWARDS A NEW SYNTHESIS AMONGST TAXONOMIC, ECOLOGICAL AND BIOGEOGRAPHICAL APPROACHES IN CARABIDOLOGY
Proceedings of the XIII European Carabidologists Meeting, Blagoevgrad, August 20-24, 2007, pp. 207-240.

Sofia–Moscow

Beetles (Coleoptera) in High Arctic

Yurii I. Chernov & Olga L. Makarova

Severtsov Institute of Ecology and Evolution, Russian Academy of Sciences, Leninsky pr., 33, Moscow 119071, Russia. E-mail: lsdc@mail.ru

SUMMARY

Beetles of the suborder Adephaga and the series Staphyliniformia (suborder Polyphaga), are disproportionately well represented in the Arctic. The High Arctic, i.e. the territory encompassing both the subzone of arctic tundra and the zone of polar deserts, is populated by at least 71 coleopteran species belonging to ten families. The set of beetle families in the High Arctic is stable enough in different sectors of the Arctic as well as in the subarctic highlands. The most common beetle families are: Staphylinidae, Carabidae, Dytiscidae, Chrysomelidae, and Curculionidae. Rove beetles clearly demonstrate superiority over ground beetles in the High Arctic. The beetle fauna of the polar deserts consists of relatively small, wingless, polytopic and polyphagous species with vast distributions (*Micralymma brevilingue*, Staphylinidae; *Chrysolina subsulcata, Ch. septentrionalis,* Chrysomelidae; *Dienerella filum, D. elegans*, Latridiidae); all the species recorded from polar deserts prefer intrazonal, often zoogenic, habitats where the snowless season is longer than in the surrounding habitats. The coleopteran assemblages of Greenland and the Canadian High Arctic are greatly impoverished probably due to the relatively young ages of those territories. Among the High Arctic beetle faunas, two polar types are recognizable, i.e. migratory and continual. The relatively young migratory faunas have mainly developed on islands after the last glaciation and contain large portions of macropterous species; such typical wingless arctic species as those in the genera *Pterostichus* and *Chrysolina* are absent. The continual faunas comprise mainly wingless, brachypterous or dimorphic species. Apparently, *M. brevilingue* is to be considered as the most cold-tolerant beetle species in the Northern Hemisphere.

Keywords: Coleoptera, High Arctic, polar desert, tundra, distribution, flight capability, faunogenesis

INTRODUCTION

Beetles (Coleoptera), especially ground beetles (family Carabidae), have long been used as an important model group in zoogeographical investigations. The coleopteran fauna of the Arctic is of special interest because this area, its northern belt in particular, is populated by a multitude of species showing the most pronounced adaptations to the adverse conditions of the globe's thermal minimum. At the same time, although there are numerous faunistic records from high-latitude landscapes scattered in the literature, no attempts at analyzing the general taxonomic structure of the order in the Arctic and its regional or zonal peculiarities have hitherto been made.

Based on published evidence (Poppius, 1910; Yakobson, 1905-1916; Økland, 1928; Danks, 1981; Böcher, 1988; Khruleva, 1991, 2007; Ryabitsev, 1995; Khruleva & Korotyaev, 1999; Chernov et al., 2000; Anderson R.S., 1997; Ball & Currie, 1997; Ryabukhin, 1999; Larson et al., 2000; Zinoviev & Olshwang, 2003; Andreyeva & Petrov, 2004; Coulson & Refseth, 2004 etc.), faunistic collections and expert evaluations of authors, the Arctic coleopteran fauna, i.e. from the territory covering the zones of tundra and polar deserts, totals up to some 600 species belonging to 23 families. Five families clearly dominate in species diversity and in some other features of adaptive success in tundra landscapes: ground beetles (Carabidae), rove beetles (Staphylinidae), diving beetles (Dytiscidae), leaf beetles (Chrysomelidae) and weevils (Curculionidae). More than half of the Arctic coleopteran fauna are represented by Carabidae and Staphylinidae, with the bulk of characteristic arctic species (up to 100) belonging to these two families. Representatives of Dytiscidae, Chrysomelidae and Curculionidae compose altogether about 180 species in the Arctic. Yet among the leaf beetles only seven species can be considered as truly arctic (Danks, 1981; LeSage, 1991; Bieńkowski, 2001, 2004, 2007). Among the diving beetles found in the tundra, arcto-boreal forms appear to predominate while only two species can be termed as arctic, that is not occurring in the taiga zone (Larson et al., 2000). At present, among the weevils we see up to nine truly arctic species (Anderson R.S., 1989, 1997; Khruleva & Korotyaev, 1999) but, as our knowledge of this family in the Arctic is still insufficient, this number could prove to be larger.

All other families are notably inferior in species diversity to the above five, many of them being represented only by a few species. Specialized arctic species are absent from most of these families, only five contain a single species whose distribution can be defined as more or less arctic. Nevertheless, some of these families are remarkable components of tundra faunas, especially of their southern variants. Such families are crawling water beetles (Haliplidae), water scavenger beetles (Hydrophilidae), burying beetles (Silphidae), pill beetles (Byrrhidae), click beetles (Elateridae), ladybirds (Coccinellidae), and catopids (Catopidae). Leather-winged beetles (Cantharidae) and soft-winged flower beetles (Melyridae) are represented by a few characteristic species only in some districts of the Arctic. Some more families, namely jewel beetles

(Buprestidae), longhorn beetles (Cerambycidae), dermestid beetles (Dermestidae), brown scavenger beetles (Latridiidae) and flat bark beetles (Cucujidae), are found in different parts of the tundra zone, but mainly either as local invaders from the south or adventitious and synanthropic elements.

As a "negative" feature of the taxonomic composition of the Arctic coleopteran fauna, a total or nearly so absence of such important groups of Polyphaga as the superfamily Scarabaeoidea and the enormous complexes of Cucujoidea and Tenebrioidea must be mentioned first of all. The absence of most of the generally dendrophilic (in the broad sense) families seems to be the most important ecological peculiarity of the arctic beetle fauna. This lack largely concerns the truly xylophages like bark beetles, buprestids, and cerambycids. The families Buprestidae and Cerambycidae, apart from obligate dendrobionts, comprise a diversity of species and groups associated with grass roots and stems which are also widespread in herbaceous habitats in arid and semi-arid landscapes. So their absence in the Arctic is caused not only by trophic specializations, but also by the general adaptive potential of these taxa, probably due to some delicate eco-physiological peculiarities determining the processes of growth and development. This observation is further supported by the absence of middle-sized and small families considered primarily or basically as dendrophilous in the broad sense and comprising different trophic groups, including phyto-, sapro-, saproxylo-, mycetophages and predators. These are Anobiidae, Ostomatidae, Cleridae, Nitidulidae, Rhizophagidae, Erotylidae, Melandryidae, Mycetophagidae etc. Their representatives are widely distributed in forests, including the taiga zone. Clearly, tundra communities cannot be entirely sheltered from colonization by these groups. Nonetheless, it seems that a successful adaptation to arctic environments is inconsistent not only with a xylobiotic life-style, but even with an inclination to dendrophily in general. Thus, the decline of numerous taxa in the Arctic results primarily from climatic factors while the tundra simply lies beyond their latitudinal range. Thus, tenebrionids are excluded from the beetle fauna already in the subzone of northern taiga while lamellicorn beetles show a sharply reduced diversity at the northern tree line, even though *Aphodius* species do occur up to the shrub tundra northwards.

Biological progress in different beetle taxa in the Arctic seems to be somehow related to their phylogenetic position. The greatest adaptive capacities in high latitudes are shown by a rather primitive complex of primarily (or mainly?) carnivorous beetles of the suborder Adephaga (Figs 1-2). This suborder comprises nine families (Lawrence & Newton, 1995), four of which inhabit the tundra zone: Carabidae, Dytiscidae, Haliplidae and Gyrinidae. The proportion of Adephaga in the Arctic is much greater than beyond the cold belt (Table 1), this possibly resulting from carnivory. Within the huge suborder Polyphaga, the staphyliniform complex of families (Hydrophilidae, Hysteridae, Staphylinidae, Silphidae) is represented well enough. These moderately advanced beetles that show certain plesiomorphic features are also carnivorous to this or that extent. The phylogenetically most advanced branch of Coleoptera, including the bulk of phytophagous taxa (series Cucujiformia), is represented in the Arctic by

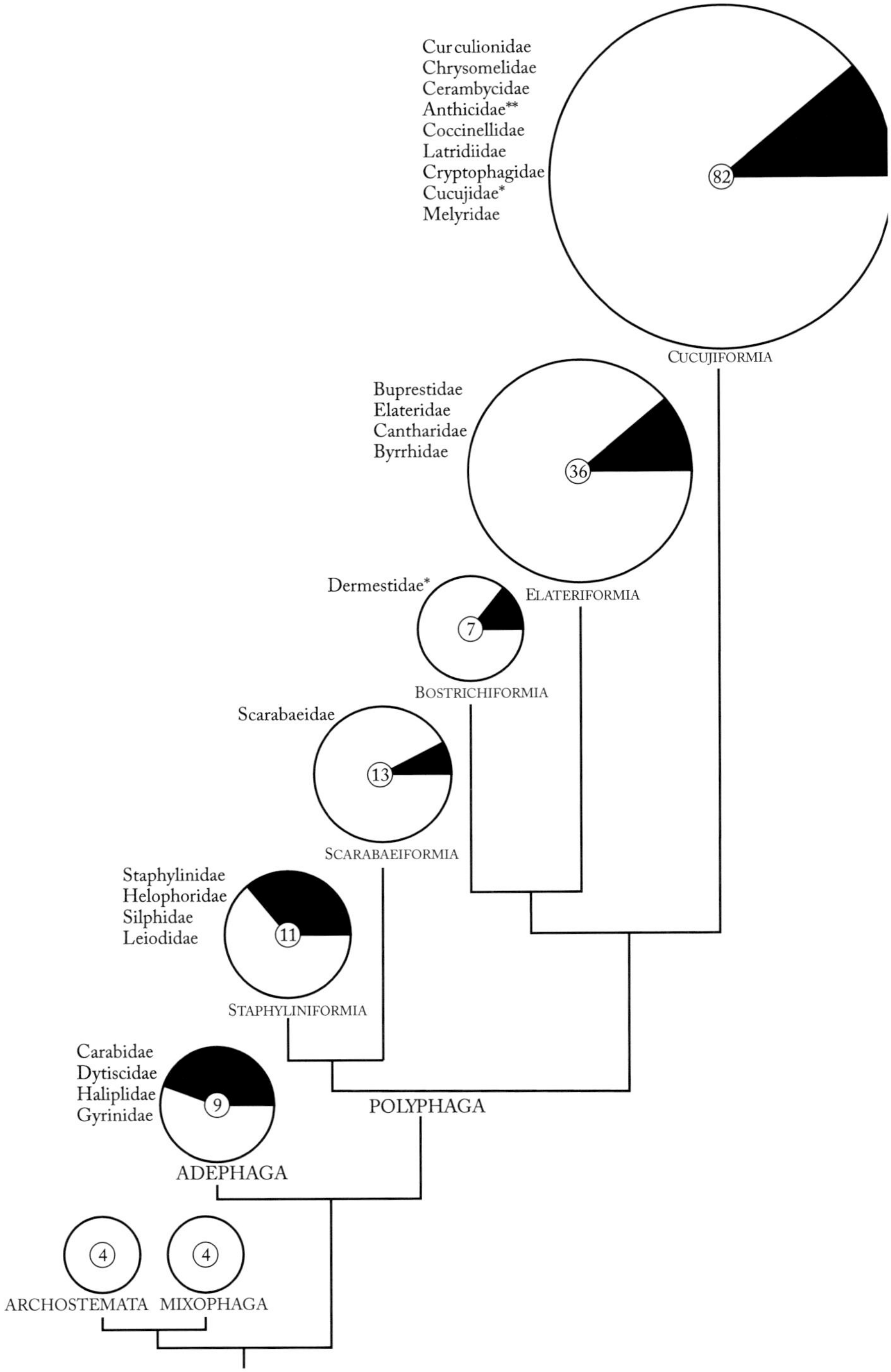

Fig. 1. Proportions of the beetle families occurring in the Arctic (black color) within the major taxa. The square of circles corresponding to number of families (mentioned in the circle centre). * - regular anthropogenic introductions, ** - probably occasional finds.

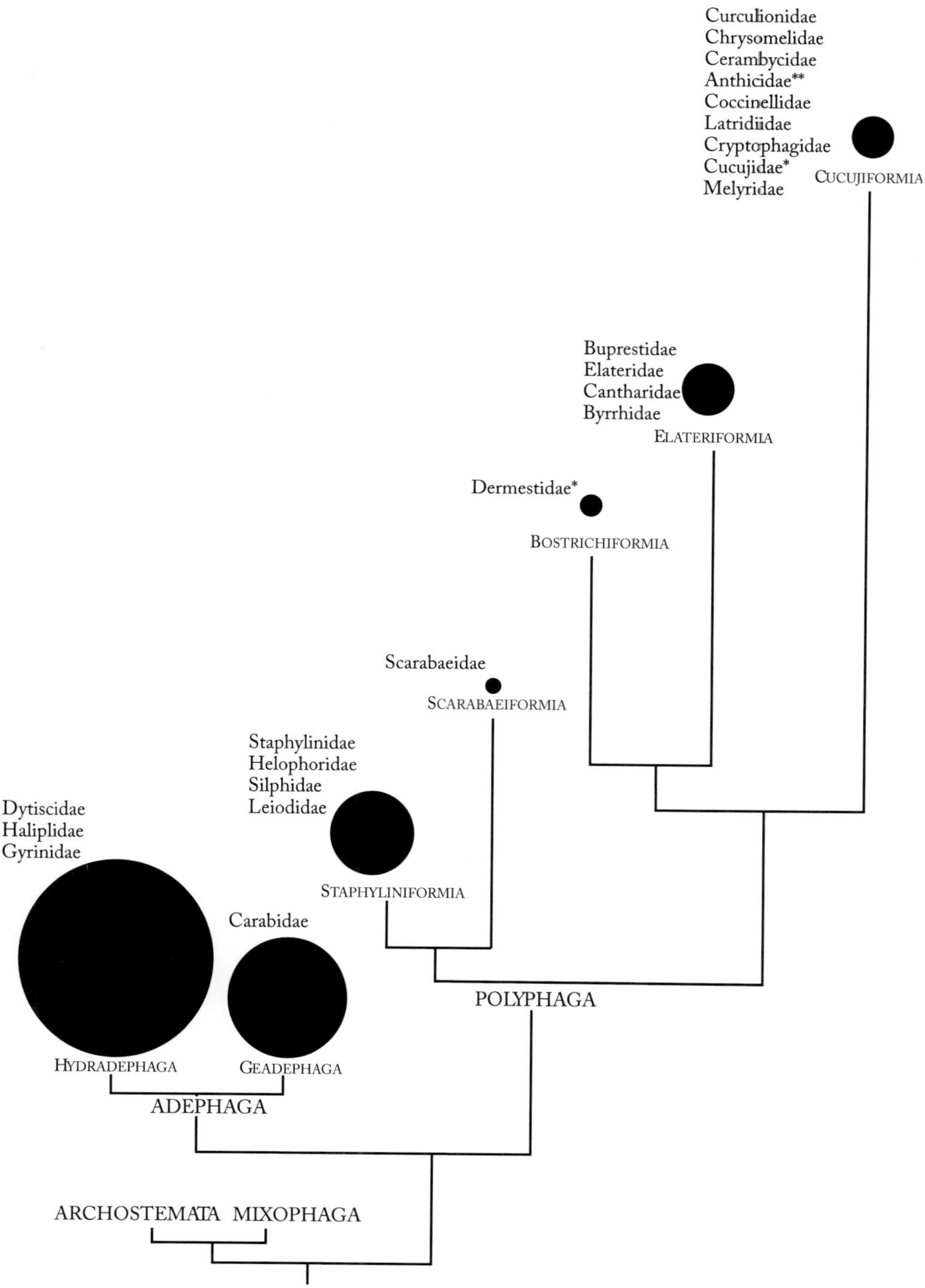

Fig. 2. Species representativeness in the Arctic (% of World fauna) of the major beetle taxa. The square of circles corresponding to proportion size. * - regular anthropogenic introductions, ** - probably occasional finds.

the largest beetle superfamilies Chrysomeloidea and Curculionoidea, but still these taxa are inferior to Carabidae, Staphylinidae and even Dytiscidae by both absolute and relative parameters of diversity (Table 1), especially in the northern part of the tundra zone. Thus, the taxonomic structure of the Arctic coleopteran fauna repeats the general pattern characteristic of other components of the high-latitude biota, namely the clear prevalence of rather primitive groups (members of the lower branches of the phylogenetic tree of a given macrotaxon) and the moderate proportion of a few phylogenetically most advanced branches (Chernov, 1988, 1995, 2002). Another characteristic feature of the arctic coleopteran fauna is a clearly more pronounced decrease in the diversity of typical phytophages in comparison with carnivorous forms towards the higher latitudes (Chernov, 1992).

The global distribution of the biota is well-known to show distinct and regular latitudinal changes (Chernov & Penev, 1993; Chernov, 1995). These alterations are especially formidable in the cold climatic zones and can easily be seen using insect faunas taken as examples. Insect communities formed at the margins of climatic gradients are of great interest for elucidating the adaptive potential of individual taxa, as well as for an analysis of the general principles of global biodiversity distribution. The objective of our work was to analyze the taxonomic structure of the order Coleoptera at the northern border of its range as based on new evidence.

Table 1. Some parameters describing the diversity and representativeness of the main beetle taxa in the Arctic.

Suborder, group, series	Number of families in the World	Representativeness in the Arctic, family level, %	Number of species in the World	Number of species in the Arctic	Representativeness in the Arctic, species level, %	Percentage in the World beetle fauna, %	Percentage in the Arctic beetle fauna, %
Suborder Archostemata	4	0	40	0	0	0.01	0
Suborder Myxophaga	4	0	94	0	0	0.02	0
Suborder Adephaga							
Hydradephaga	6	50.0	5,560	85	1.53	1.47	12.89
Geadephaga	3	33.3	35,000	200	0.57	9.26	30.35
Suborder Polyphaga							
Staphyliniformia	11	36.0	64,542	183	0.28	17.07	27.77
Scarabaeiformia	13	8.0	35,000	2	0.01	9.23	0.30
Elateriformia	36	11.0	41,780	46	0.11	11.06	6.98
Bostrichiformia	7	14.0	4,399	1	0.02	1.16	0.15
Cuccujiformia	82	11.0	191,616	142	0.07	50.69	21.55

Note. Calculated from Lawrence & Newton (1995), Hunt et al. (2007) and from our data and estimations.

METHODS AND APPROACHES

We have analyzed the beetle species assemblages inhabiting the coldest parts of the Holarctic, such as arctic tundra, polar deserts and their altitudinal analogs, viz. nival deserts. These landscapes are characterized by a very low heat supply, low average annual and average summer temperatures, scarce vegetation, frost boiled soils and an extremely short vegetation season (Chernov & Matveyeva, 1997).

There are several approaches to dividing the Arctic into latitudinal zones (Fig. 3). We usually use the partition design suggested by B. Gorodkov (1935) and followed since then in the Russian literature. A different pattern is mostly accepted in the West, proposing a division of the Arctic into two large belts, namely, the Low Arctic, where the average temperatures during the warmest month range between +5°C and +10°C, and the High Arctic, where the average temperatures during the warmest month lie below +5°C. A third approach follows Elvebakk et al. (1999), currently being intensely elaborated by Walker et al. (2003).

Below we shall restrict our analyses to the beetle groupings of the High Arctic, namely, the zone of polar deserts and the subzone of arctic tundra (zones A and B, Fig. 4), where the mean July temperatures do not exceed +5 to +6°C, as a rule, and the vegetation cover is dominated by lichens, mosses, as well as by a few species of herbs and prostrate dwarf shrubs*. Naturally, similar conditions also exist in subarctic mountains. Fig. 5 shows the locations of the faunas analyzed. Much of the new information is original (marked

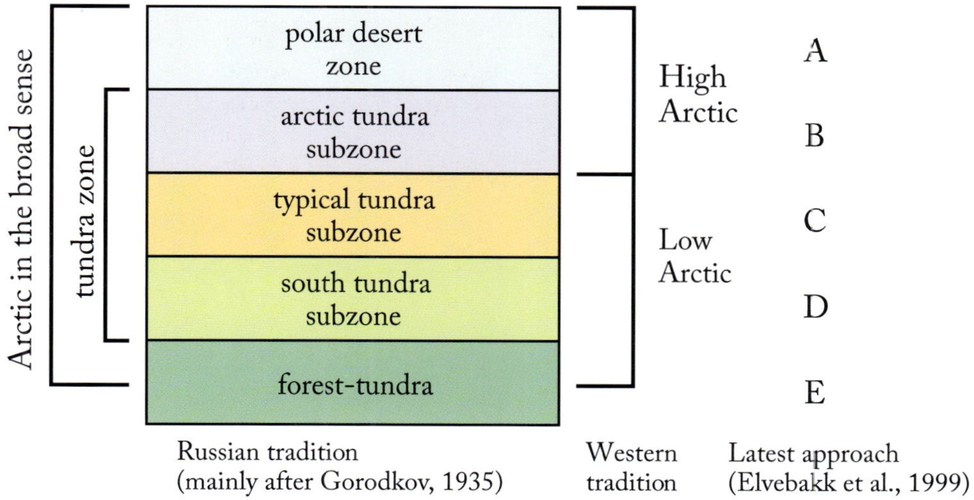

Fig. 3. Zonal division of the Arctic based on different approaches.

* The actual mean July temperature is regarded as the most meaningful, that is why the Wrangel Island (+8...+10°C in the central part) as a whole is excluded from the consideration of the High Arctic fauna, but Cape Barrow (+3.8°C) on the contrary is included.

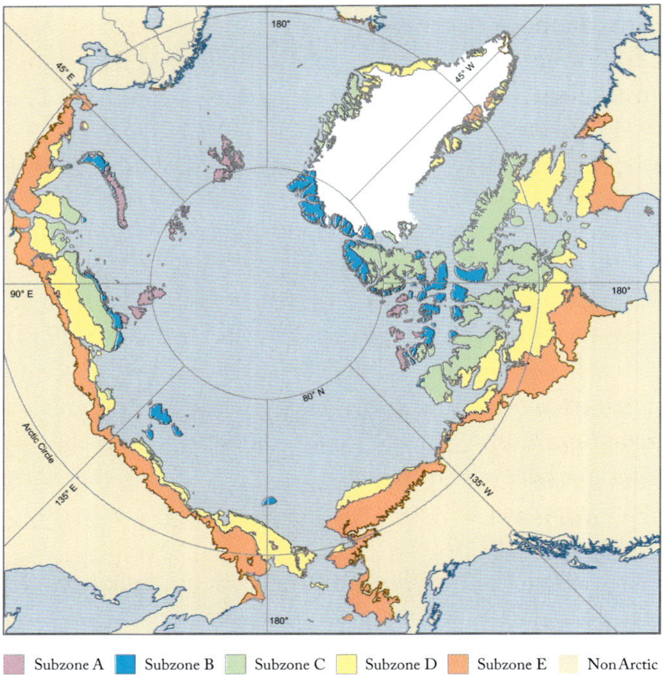

Fig. 4. Vegetation map of the circumpolar Arctic (after Walker et al., 2003).

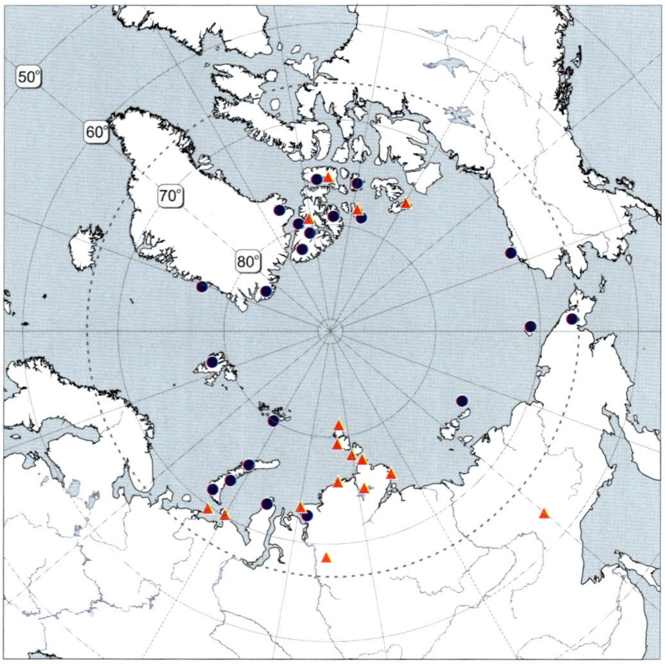

Fig. 5. Map showing the coldest sites whence local beetle faunas were analyzed. Circles: literature data; triangles: original data.

by triangles), the remaining evidence derives from published sources (marked by circles). Various field collection techniques were utilized: pitfall trapping, water trapping with bait for Dytiscidae, Tullgren funnel extraction for the smallest beetles, and hand-sorting. The flight ability of separate species was analyzed with the use of published information, that provided by specialists, as well as that revealed through beetle dissections.

RESULTS AND DISCUSSIONS

Structure of the High Arctic fauna

The enormous order of beetles, totaling about 378,000 known species (Hunt et al., 2007) belonging to 166 families (Lawrence & Newton, 1995), is represented in the High Arctic by only a few dozen species from ten families. Beetles, comprising about one-third of all insects, reduce their diversity successively and sharply towards the high altitudes, down to 3-4% (Böcher, 1988; Danks, 1990; Chernov, 1995, 2002). Only a few species live in the arctic polar deserts (Makarova et al., 2007; unpublished data). In the polar deserts of the Antarctic, beetles are totally absent (see references in: Block, 1992). This extreme suppression of Coleoptera, both numerical and proportionate, near the poles of both hemispheres reflects a trend in a global biodiversity usually decreasing from the tropics to the high latitudes. This pattern strongly contrasts that of the similarly diverse Diptera, as their percentage appears to be the largest amongst the insects in the cold belts (Table 2). Among beetles, there are truly arctic species, some even high-arctic ones. Local beetle faunas within the arctic tundra subzone usually fail to exceed 20 species, ranging up to 4 species only within polar deserts (Table 3).

In the coldest arctic landscapes of the Holarctic, comprising the subzone of arctic tundra and the zone of polar deserts (High Arctic), 71 beetle species belonging to ten families have been registered (Table 4). Members of five families are especially characteristic: Carabidae, Dytiscidae, Staphylinidae, Chrysomelidae and Curculionidae. The set of

Table 2. Shares of the main insect orders (%) in the faunas of the world, the Arctic and the High Arctic (after Chernov, 2002).

Order	World	Arctic	High Arctic
Homoptera and Heteroptera	8	5	2
Coleoptera	33	13	4
Lepidoptera	17	11	6
Hymenoptera	17	14	20
Diptera	18	49	57
Others, including parasites	7	8	11

Sources: Danks (1981, 1989, 1990); Historical Development of the Class Insecta (1980), multivolume identification keys to the insects of the USSR European part and of the USSR Far East, numerous regional publications and authors' data.

Table 3. Beetle species diversity, family composition, and flight capabilities in the coldest areas of the northern Holarctic.

Island/District	Latitude, °N	Mean July temperature °C[1]	Number of species[2]	Diversity of separate families	Sources of species records	Flight capabilities of the fauna, %		
						+	±	–
Arctic tundra subzone								
1. Spitsbergen, Svalbard	76.6-80.1	+4.5	19(17)	Ca(1), Dy(1), St(9), By(1), Cr(2), An(1), La(1), Ch(1), Cu(2)	Fjellberg, 1983; Coulson, 2000; Coulson, Refseth, 2004	76	18	6
2. Severnyi Island, Novaya Zemlya	73.2-77.0	+4.0	12(10)	Ca(2), St(8), Ch(2)	Jacobson, 1898; Økland, 1928	50	10	40
3. Meduza Bay, NW Taimyr	73.4	+5.0	17(15)	Ca(3), Dy(1), St(9), Ch(2), Cu(2)	Khruleva, 1999 with specification	20	20	60
4. Uboynaya River (lower reaches), NW Taimyr	73.6	+4.5	14(13)	Ca(5), St(5), Ch(3), Cu(1)	Chernov, 1992; new data	8	15	77
5. Taimyr Lake (northern coast), N Taimyr[3]	74.6	+6.5	19(19)	Ca(8), St(7), Ch(2), Cu(2)	New data	26	11	63
6. M. Pronchishcheva Bay, NE Taimyr	75.9	+4.0	9(9)	Ca(3), St(3),Ch(3)	Chernov, 1978; Chernov et al., 2001; new data	11	11	78
7. Cape Barrow, Alaska	71.2	+3.8	15(12)	Ca(4),Dy(1), St(6), Ch(4)	Hurd,Lindquist,1958;Brown,1962; Campbell,1988; Nelson,2001	0	17	83
8. Ellesmere Island, Canadian Arctic Archipelago	76.1-83.2	+6.1	10(8)	Dy(3), St(4), Cr (1), La(1), Cu(1)	Brown, 1937; Downes, 1988; Brodo, 2000; Larson et al., 2000	25	37	38
9. Devon Island, Canadian Arctic Archipelago	74.4-77.1	+5.2	3(1)	Ca(1), Dy(1), St(1)	Lindroth, 1968; Ryan, 1977; Larson at al., 2000	?	?	?
10. NW Greenland	70.0-80.0	+4.8	5(5)	Dy(2), St(1), By(1), Cc(1)	Böcher, 1988	60	20	20
11. NE Greenland	69.0-80.0	+3.9	6(6)	Ca(1), Dy(1), St(2), By(1), Cc(1)	Böcher, 1988	66	17	17
12. N Greenland	80.0-83.0	+3.7	2(2)	Dy(1), St(1)	Böcher, 1988	50	50	0
Nival deserts								
13. Putorana Plateau (Yt-Kueol Lake environs), 700-900 m a.s.l., SW Taimyr	69.13	+2.0+6.0[4]	14(14)	Ca(5), St(7), Ch(2)	New data	64	7	29

14. Suntar-Khayata Ridge, Yakutia, 2200 m a.s.l.	63.2	+2.0+3.0[4]	9(7)	Ca(2), St(1), By(1), Cr(1), La(3), Cu(1)[5]	New data	28	29	43
15. Devon Plateau, 400 m a.s.l., Canadian Arctic Archipelago	75.7	+3.8	0	0	Chernov, 2004	0	0	0
Polar deserts								
16. Tundra Akademii (lower reaches of Hydrographs' River, Wrangel Island[6])	71.5	+1.0	4(4)	Ca(1), St(2), Ch(1)	Khruleva, 1991; Zerche, 1993	0	0	100
17. Bolshevik Island, Severnaya Zemlya	77.9-79.4	+2.2	4(4)	St(1), La(1), Ch(2)	Makarova et al., 2007	0	25	75
18. Komsomolets Island, Severnaya Zemlya	80.0-81.3	? +1.0	1(1)	St(1)	Makarova et al., 2007	0	0	100
19. Bathurst Island, Canadian Arctic Archipelago	75.0-76.7	+4.5	2(2)	Dy(2)	Danks, 1980; Larson et al., 2000	50	50	0
20. Melvill Island, Canadian Arctic Archipelago	74.4-76.8	+4.0	1?	Cu(1)	Anderson R.S., 1989	0	0	100
21. Ellef Ringnes Island, Canadian Arctic Archipelago	78.8	+3.2	1	La(1)	McAlpine, 1965; new data	0	0	100
22. Meigen Island, Canadian Arctic Archipelago	79.7-80.2	+3.0	0	0	McAlpine, 1965	0	0	0
23. Cape Chelyuskin, N Taimyr	77.7	+1.5	0	0	Chernov et al., 1979	0	0	0
24. Hocker Island, Franz Josef Land	80.1-80.4	+1.3	0	0	Bulavintsev, Babenko, 1983	0	0	0

Abbreviations: Ca - Carabidae, Dy – Dytiscidae, St – Staphylinidae, By – Byrrhidae, Cr – Cryptophagidae, La – Latridiidae, An – Anthicidae, Ch – Chrysomelidae, Cu – Curculionidae, Cc – Coccinellidae.

[1] – the highest value per district is given; [2] – in parentheses, the number of species with known flight capacities is given; [3] - district situated at the border between the typical tundra and arctic tundra subzones; [4] – the value was calculated based of numerous measures of temperature at different altitudes as compared to data from the nearest weather stations at the foothills; [5] – leaf beetles were not captured in pitfall traps at 2200 m a.s.l., but *Chrysolina septentrionlis* and *Chrysolina cavigera tolli* inhabit small meadows among rockeries of the same slope at 1,850 m a.s.l.; [6] – district located outside of the polar desert zone and the formation of the desert-like landscape here (Khruleva, 1991) is apparently caused by the sea cooling effect; "+" – flying species, "±" - dimorphic species (some specimens with reduced hind wings and/or wing muscles), "-" – hind wings and/or wing muscles always reduced.

Table 4. List of the Coleoptera encountered in the coldest areas of the northern Holarctic.

Species, flight capability[1)]	Arctic tundra subzone												Nival deserts		Polar deserts					
	1	2	3	4	5	6	7	8	9	10	11	12	13	14	16	17	18	19	20	21
Carabidae																				
Nebria nivalis (Payk.) (+)													+							
Notiophilus aquaticus (L.)(±)				+																
Bembidion aeuriginosum (Gebler) (?)													+							
Bembidion grapii (Gyll.) (±)											+									
Bembidion hasti C. Sahlb. (+)					+								+							
Pterostichus (Cryobius) ventricosus (Esch.) (-)		+			+															
Pterostichus (Cryobius) brevicornis (Kirby) (-)			+	+	+	+							+							
Pterostichus (Cryobius) pinguedineus (Esch.) (-)			+	+	+	+	+						+		+					
Pterostichus (Cryobius) nivalis (R. Sahlb.) (-)							+													
Pterostichus (Cryobius) cf. longipes (Popp.) (-)					+									+						
Pterostichus (Cryobius) tareumiut Ball (-)							+													
Pterostichus (Lenapterus) agonus Horn. (-)				+	+															
Pterostichus (Tundraphilus) pfitzenmayeri Popp. (-)														+						
Curtonotus alpinus (Payk.) (±)		+	+	+	+	+	+		+											
Amara quenseli (Schönh.) (±)	+																			
Amara glacialis (Mnnh.) (+)					+															
Dytiscidae																				
Hydroporus striola Gyll. (±)	+																			
Hydroporus lapponum (Gyll.) (±)								+												
Hydroporus polaris Fall (?)								+	+											
Hydroporus morio Aubé (±)								+		+		+						+		
Agabus moestus (Curtis) (±)			+				+													
Stictotarsus griseostriatus (DeGeer) (+)																		+		
Colymbetes dolabratus (Payk.) [= *C. groenlandicus* Aubé] (+)										+	+									

Species, flight capability[1)]	Arctic tundra subzone												Nival deserts		Polar deserts					
	1	2	3	4	5	6	7	8	9	10	11	12	13	14	16	17	18	19	20	21
Staphylinidae																				
Micralymma brevilingue Schiødte (-)			+	+	+	+	+	+		+	+				+	+	+			
Micralymma marinum (Strøm) (-)	+																			
Micralymma sp.							+													
Phyllodrepa angustata (Mäklin) (+)		+	+		+															
Coryphiomorphus hyperboreus (Mäklin) (+)		+											+							
Coryphiomorphus sp. (?)			+																	
Eudectus whitei Sharp (+)		+	+																	
Eudectus reductus Zerche (-)															+					
Olophrum boreale (Payk.) (+)	+												+							
Omalium caesum Grav. (+)	+																			
Omalium septentrionis Thoms. (?)	+																			
Eucnecosum brachypterum (Grav.) (±)	+																			
Acidota crenata (Fabr.) (+)													+							
Holoboreaphilus nordenskioldi (Mäklin) (-)		+					+[2]													
Bryophacis punctipennis (Thoms.) (+)													+							
Tachinus arcticus (Motsch.) (+)			+	+	+	+														
Tachinus brevipennis J. Sahlb. [= *T. apterus* Mots.] (-)			+	+	+		+													
Tachinus instabilis Mäklin (+)							+[3]													
Atheta (Atheta) holtedahli Mnst. (?)		+																		
Atheta (Alaobia) trinotata (Kr.) (?)		+																		
Atheta (Coproceramius) graminicola (Grav.) (+)	+												+							
Atheta (Oreostiba) lenensis Popp. (-)					+															
Atheta (Boreophilia) subplana (J. Sahlb.) (+)	+				+															
Atheta (Boreostiba) sibirica Mäklin [= *A. frigida* (J. Sahlb.)] (+)		+											+							
Atheta sp. 1 (?)	+																			

Species, flight capability[1)]	**Arctic tundra subzone**												**Nival deserts**		**Polar deserts**					
	1	2	3	4	5	6	7	8	9	10	11	12	13	14	16	17	18	19	20	21
Atheta sp. 2 (+)														+						
Gnypeta brincki Palm [= *G. canaliculata* J. Sahlb.] (?+)			+																	
Gnypeta cavicollis J. Sahlb. (+)		+						?			+	+								
Gnypeta sp.1 (?+)								+	+											
Gnypeta sp.2 (?+)								+												
Stenus frigidus (J. Sahlb.) (-)			+																	
Stenus lagopodis Ryvkin (+)													+							
Lathrobium poljarnis tchernovi Tikh. (-)			+	+	+	+														
Quedius mesomelinus (Marsch.) (-)								+												
Philonthus sp. (?)	+																			
Staphylinidae gen. sp. (?)				+			+													
Byrrhidae																				
Byrrhus fasciatus Förster (+)										+	+									
Morychus cf. *aeneus* (Fabr.) (+)														+						
Simplocaria metallica (Sturm) (+)	+																			
Cryptophagidae																				
Cryptophagus corticinus Thoms. (+)														+						
Gryptophagus acutangulus Gyll. (+)								+												
Atomaria atricapilla angulicollis Kangas (+)	+																			
Atomaria lewisi Reitt. (+)	+																			
Coccinellidae																				
Coccinella transversoguttata Falderman (+)										+	+									
Latridiidae																				
Latridius minutus (L.) (+)	+							+												
Dienerella filum Aubé (±)																+				
Dienerella cf. *filiformis* Gyll. (±)														+						
Dienerella elegans Aubé (-)																				+

Species, flight capability[1]	Arctic tundra subzone												Nival deserts		Polar deserts					
	1	2	3	4	5	6	7	8	9	10	11	12	13	14	16	17	18	19	20	21
Corticaria rubripes Mann. [= *C. linearis* Payk.] (+)														+						
Enicmus sp. (?)														+						
Anthicidae																				
Anthicus flavipes (Panz.) (+)	+																			
Chrysomelidae																				
Chrysolina (Arctolina) septentrionalis (Mén.) (-)		+	+	+	+	+	+						+			+				
Chrysolina (Arctolina) subsulcata (Mannh.) (-)				+		+	+									+				
Chrysolina (Arctolina) magniceps (J. Sahlb.) (-)							+								+					
Chrysolina (Pleurosticha) cavigera cavigera (J. Sahlb.) (-)							+													
Chrysolina (Pleurosticha) cavigera tolli (Jac.) (-)													+							
Hydrothassa hannoverana F. (-)		+	+	+	+	+														
Gonioctena sp. (+)	+																			
Curculionidae																				
Hypera diversipunctata (Schrank) (<u>+</u>)			+		+															
Hypera ornata (Cap.) (-)				+										+						
Isochnus flagellum (Erics.) (+)	+																			
Isochnus arcticus (Korot.) (-)			+		+			+											+	
Orchestris saliceti Fabr. (?)	+																			

Notes. Numbers 1-21 in the banner headline match the local faunas quoted in Table 3. Sources of data: see Table 3.

[1] Flight capabilities: (+) – flying species, (<u>+</u>) - dimorphic species (some specimens with reduced hind wings and/or wing muscles), (-) – hind wings and/or; wing muscles reduced, (?) – no data; [2] – ? mentioned as Omaliinae gen. sp. (Hurd & Lindquist, 1958), specification was suggested on the base of nearest records (Campbell, 1978); [3] – ? mentioned as Tachyporinae gen. sp. (Hurd & Lindquist, 1958), specification was suggested on the base of nearest records (Campbell, 1988).

these families is rather stable in different sectors of the Arctic, as well as in the subarctic highlands (Table 3). Besides this, species of Byrrhidae, Latridiidae and Cryptophagidae are found regularly. The presence of Coccinellidae and Anthicidae varies with the regions of study. Most of the species living in the High Arctic show Holarctic distributions (67%), with Palaearctic species (26%) clearly prevailing over cosmopolitan (5%) and Nearctic ones (2%). The landscape-zonal structure of the fauna is rather diverse: arcto-boreal distribution patterns are common (34% of the species), but truly arctic components are also significant (31%). Arcto-boreomontane species contribute to only 9% of the fauna, but polyzonal forms are numerous (24%).

In the High Arctic, as in the arctic fauna as a whole, the absolute prevalence of Carabidae and Staphylinidae is apparent (13 and 29 species, respectively). Dytiscidae, Chrysomelidae and Curculionidae are represented by 5-7 species each. So, the two main families contribute to 60% of total beetle fauna, and the first five families up to 86%. Such a taxonomic composition, being constant and distinct from that of the southern areas in different sectors of the Arctic, argues the peculiarity of the High Arctic coleopteran fauna.

Ground beetles (Carabidae) have traditionally been considered as the most common among the arctic beetles, constituting about one-third of the total (Chernov et al., 2000). In the Arctic fauna in the broad sense, the family accounts for about 200 species. The number of carabid species in local faunas is closely related to summer temperatures (Chernov & Penev, 1993), the correlation coefficient being 0.87 (Chernov et al., 2000). The most cold-resistant ground beetles are listed in Table 4. The wingless carabids belonging to *Pterostichus* Bon. are especially common, comprising up to 5 species in particular local faunas. Among the cryotolerant ground beetles there are also representatives of the genera *Bembidion* Latr., *Notiophilus* Dum., *Nebria* Latr. and *Amara* Bon., as well as the very typical *Curtonotus alpinus* (Table 4). Half of these species, including *C. alpinus*, are well-known as dimorphic with respect to development of the hind wings. The poverty of carabids in the North American High Arctic (Danks, 1981; Böcher, 1988) is due probably to limited dispersion abilities of the truly arctic species. The development time of the biota of Queen Elizabeth Islands and Greenland since the last glaciation has evidently been not long enough for their recovery. This is strongly supported by subfossil fragments of *C. alpinus* found in the last interstadial deposits in different parts of Greenland (Böcher, 1989; Bennike et al., 2000) and Ellesmere Island (Blake & Matthews, 1979, cited in: Brodo, 2000). Being so common in the arctic tundra subzone, carabids are absent from polar deserts proper* (Table 3). Thus, the isotherm of the mean July temperature of +2°C is an insurmountable climatic barrier for ground beetles (Chernov et al., 2000). Interestingly, it is to the genera *Pterostichus, Bembidion* or *Nebria* that most of the carabids known as winter-active species or dwelling under or above the snow cover appear to belong (Renken, 1956; Evans, 1969; Kaufman, 1971; Aitchison, 1979; Lindroth, 1992 etc.).

* The occurrence of *Pterostichus pinguedineus* in the coastal part of Tundra Akademii, Wrangel Island (lower reaches of Hydrographs' River), with very harsh desert-like environments (Khruleva, 1991), is probably due to ongoing downstream dispersal from the warmer inland areas of the island.

The diversity of rove beetles, Staphylinidae, in the high latitudes is most significant. At least 29 species inhabit High Arctic landscapes (Table 4). The most common staphylinids there belong to the genera *Micralymma* Westwood, *Phyllodrepa* Thompson, *Coryphiomorphus* Zerche, *Eudectus* Redtenbacher, *Tachinus* Gravenhorst, *Gnypeta* Thompson, *Atheta* Thompson or *Lathrobium* Gravenhorst. The prevalence of the subfamilies Omaliinae (12 species) and Aleocharinae (10) is strongly pronounced, altogether accounting for 76% of the total fauna. It is members of these subfamilies that have been mentioned most often as cryophiles, chionophiles or winter-active beetles in temperate climate areas (Chapman, 1954; Renken, 1956; Heydemann, 1956; Tikhomirova, 1973; Aitchison, 1979 etc.). The latitudinal maxima are also associated with these subfamilies. Thus, the northernmost records of the family belong to *Micralymma brevilingue*, Omaliinae (81.0°N), in the Far North of the Siberian sector (Makarova et al., 2007) and to *Gnypeta* spp., Aleocharinae (up to 82.3°N), in the Nearctic (Böcher, 1988, Brodo, 2000)*. Species of *Micralymma* are wingless, whereas members of *Gnypeta* fly (V. Semenov, personal communication). The very small (3-3.5 mm), flying *Gnypeta cavicollis* seems to be a High Arctic specialist.

The diving beetles, Dytiscidae, occur with seven species in the High Arctic, but they are obviously underestimated in the Eastern Hemisphere. Among the northernmost diving beetles, only *Hydroporus polaris* and *Agabus moestus* are truly tundra species (Larson et al., 2000), the others are widely distributed arcto-boreal forms. Very small *Hydroporus* spp. are especially common. A temporarily reduced flight musculature is often observed (Eriksson, 1972; Larson et al., 2000).

Among six species of the family Chrysomelidae recorded in the High Arctic, strictly arctic species predominate (Table 4). All of them belong to the wingless** subgenera of *Chrysolina*, whose members are very typical of the Arctic, and often coexist. The remarkable present-day absence of *Chrysolina* species on Canadian arctic islands, in Greenland and on Svalbard is very likely due to Pleistocene glacial events, coupled with poor dispersal capacities of these beetles. The richness of local *Chrysolina* faunas in Beringia (Brown, 1962; Bieńkowski, 2004, 2007; Khruleva, 2007) with relatively stable environments during the Quaternary supports this conclusion.

Among the northernmost weevils, only some species of the genera *Hypera* and *Isochnus* are regularly recorded in the Artcic. The wing-dimorphic *H. diversipunctata* and the brachypterous *H. ornata* inhabit the entire tundra zone and are often common, but their ranges extend far beyond the Arctic. The small (2 mm) wingless *Isochnus arcticus*, developing on *Salix arctica* (adults found under different *Salix* species), is evidently a truly Arctic specialist.

The remaining beetle families found in the High Arctic (Byrrhidae, Cryptophagidae, Coccinellidae, Latridiidae, Anthicidae) are represented by a few species each (Table 4), all usually polyzonal and capable of flight.

* It is noteworthy that the altitudinal record of the order Coleoptera (5,600 m a.s.l. in the Himalaya) also belongs to an aleocharine beetle, *Atheta hutchinsoni* Gam. (Mani & Giddings, 1980).

** In some species, vestigial hind wings are present (Bieńkowski, 2004, 2007).

The Carabidae to Staphylinidae ratio

In northern boreal areas and southern tundra sites, the family Staphylinidae often contains 1.5-3.0 times more species than Carabidae do, this trend growing stronger to the North up to a total absence of ground beetles in polar deserts (Table 5). It also seems that rove beetles are generally more psychrophilic, a lot of them in temperate regions being associated with moist coasts and banks. Obviously because staphylinids on the average seem to be smaller and more diverse in trophic ecology (mycophagy, parasitoidism and saprophagy being common), they demonstrate clear superiority over carabids in the Arctic.

Like Carabidae, the family Staphylinidae contains cold-hardened, including freeze-tolerant (Luff, 1966; Miller, 1982; Rossolimo, 1995; Slabber & Chown, 2005), as well as chionophilous species (Chapman, 1954; Heydemann, 1956; Smetana, 1958; Lindroth, 1992). The range of preferred temperatures and its geographical and seasonal changes are also comparable in these two families (Tikhomirova, 1973; Rossolimo & Rybalov, 1995; Rossolimo, 1997). Some of them can move and feed under zero and subzero temperatures (Chapman, 1954; Heydemann, 1956; Kaufmann, 1971; Aitchison, 1979). *Pterostichus brevicornis* normally walks at down to -12°C (Baust, 1972). A number of temperate carabid and staphylinid species breed in winter (Heydemann, 1956; Evans, 1969; Topp & Smetana, 1998 etc.), using the so-called "winter biocoenotic vacuum", when countless inactive preys are available. When sharing the same habitat, species from these families often show similar supercooling points (Luff, 1966; Block & Sømme, 1983; Rossolimo, 1995 etc.). In the Arctic, both carabids and staphylinids demonstrate elevated metabolic rates as compared to their counterparts in other climatic zones (Aunaas et al., 1983; Strømme et al., 1986).

Extended life cycles and a long adult life are considered as characteristic features of arctic carabid biology (Chernov et al., 2000). Within both families there are numerous species that are capable of hibernation in both adult and larval stage (Heydemann, 1956; Steel, 1970; Kaufmann, 1971; Andersen, 1983; Thayer, 1985; De Zordo, 1979; Korobeinikov, 1990; Lindroth, 1992 etc.). This can be treated as a prerequisite to life in the Arctic (Paarmann, 1979; Korobeinikov, 1990; Sota, 1994). The life cycles of most of the arctic carabids last about two years (Kaufmann, 1971; Korobeinikov, 1990; Lindroth, 1992; Filippov, 2007). In this case, the first hibernation is in the larval stage, the second one as adults. However, the sum of the positive temperatures in the polar desert zone does not allow even such an extended cycle to be realized. There is no evidence of a 3-year long development, although such suggestions exist (De Zordo, 1979). In the Arctic, a simultaneous hibernation of different instar larvae and of adults from different generations is fairly well documented only for large *Carabus* species (Korobeinikov, 1990a; Ryabitsev, 1998). All larval instars hibernate together with adults in alpine populations of *Amara quenseli* (De Zordo, 1979). Yet there is no proof of any ground beetle repeatedly overwintering as larvae, moreover if the larva hibernates, the pupation occurs just in spring (K. Makarov, personal communication). So, the carabid life cycles, unlike that of, e.g. some leaf beetles, seem to be not longer than two years in the Arctic. That is why the deficit of heat in polar deserts strongly limits carabid immigration. It is an annual life

Table 5. Number of species belonging to the families Staphylinidae and Carabidae in the world fauna and in different regions/districts of the northern Holarctic.

Region/district	Staphylinidae	Carabidae	Sources
World fauna	46,200	32,561	Lorenz, 1998; Newton et al., 2001
Some large northern regions			
1. Canada and Alaska	1,129	946	Bousquet, 1991
2. Alaska	308	237	Bousquet, 1991
3. Yukon	179	209	Anderson R.S., 1997a
4. Fennoscandia	1,253	391	Silfverberg, 2004
5. Moscow Region	753	260	Tikhomirova, 1982; Fedorenko, 1988, V. Semenov, pers. comm.
6. White Sea islands	82	48	Byzova et al., 1986
7. Greenland	11	4	Böcher, 1988
South tundra subzone			
1. Near Shchuchye, SW Yamal	52	32	A. Sokolov & K. Makarov, pers. comm.
2. Upper reaches of Lower Agapa R., SW Taimyr	43	28	Sokolov, 2003; K. Makarov, pers. comm.
3. Dolgiy Island, Barents Sea	30	18	Unpublished data
4. Basin of Amguema R., Chukotka	64	21	Marusik, 1993
Typical tundra subzone			
1. Cape Blizhniy, Taimyr Lake, N Taimyr	7	8	Unpublished data*
2. Yuzhnyi Island, Novaya Zemlya	13	8	Økland, 1928
Arctic tundra subzone			
1. Svalbard	9	1	Coulson & Refseth, 2004
2. Severnyi Island, Novaya Zemlya	8	1	Økland, 1928
3. Meduza Bay, NW Taimyr	9	3	Khruleva, 1999
4. Cape Barrow, Alaska	6	3	Hurd & Lindquist, 1958; Nelson, 2001
5. Ellesmere Island, Canadian Arctic Archipelago	4	0	Brown, 1937; Downes, 1988; Brodo, 2000
6. Devon Island, Canadian Arctic Archipelago	1	1	Lindroth, 1968; Ryan, 1977
7. Peary Land, N Greenland	2	0	Böcher, 1988
Polar deserts			
1. Eastern part of Tundra Akademii, Wrangel Island	3	1	Khruleva, 1991; Zerche, 1993
2. Bolshevik Island, Severnaya Zemlya Archipelago	1	0	Makarova et al., 2007

* - The diversity of rove beetles was possibly underestimated.

cycle that is typical of species from the tribes Notiophilini and Bembidiini (Andersen, 1969, 1983; Filippov, 2007), both of which could have some advantages in the High Arctic due to their small size. Unsurprisingly, their expansion to the North is confined to the subzone of typical tundra, as a rule*. Thus, the total absence of carabids in polar deserts is very probably due to insurmountable temperature limitations.

Some rove beetles show a very low thermal threshold of development (+1.8°C) (Topp & Smetana, 1998). Numerous staphylinids, in contrast to carabids, inhabit different enriched temporal substrates (particularly members of Aleocharinae) and have a shortened generation time, up to 13 days in favorite conditions (Miller & Williams, 1983). The least duration per generation for carabids (*Bembidion* spp.) at high temperatures is probably not less than a month (Andersen, 1969). In natural conditions, where a small ground beetle (*Bembidion*) and a small rove beetle (*Tachyporus*) had similar life cycles and development time, it is the latter species that showed a much higher (by one order of magnitude!) effectiveness of fecundity (Petersen, 1998). The physiological capability of forcing the development and reproducing during the short active season could appear as an important advantage of staphylinids in the arctic conditions.

Beetles in polar deserts

The composition of beetle families at the northern limit of distribution of the order is of a particular interest. Earlier, beetles were considered as absent from polar deserts (McAlpine, 1965; Korotkevitch, 1972; Chernov et al., 1979; Bulavintsev & Babenko, 1983)**. We conducted four expeditions to the Severnaya Zemlya Archipelago and also worked on Cape Chelyuskin (Northern Taimyr), Ellef Ringnes Island and Plateau Devon, Princess Elizabeth Islands (Fig. 5). Our colleagues collected insects on different islands of Franz Josef Land, on De Longa Islands (V.I. Bulavintsev) and in the desert-like landscapes of Wrangel Island (Khruleva, 1991). Most of polar desert areas are indeed devoid of beetles. However, on the islands of Severnaya Zemlya Archipelago and on Ellef Ringnes Island five beetle species have been revealed (Makarova et al., 2007; unpublished data from O. Makarova's sampling in 2005). The beetle assemblage of Severnaya Zemlya's polar deserts should probably be referred to as the most cold-resistant ("last one") coleopteran fauna near the Pole in the Northern Hemisphere, demonstrating the limits of adaptive potentials of Coleoptera. The composition of taxa encountered is rather unexpected (Table 3): carabids and dytiscids are absent, chrysomelids and latridiids are present. The polar deserts are populated only by wingless rove beetles (*Mycralymma*) and leaf beetles (*Chrysolina*), as well as by the dimorphic latridiids (*Dienerella*)***.

* The single *Bembidion* species occurring in the High Arctic, *B. hasti*, possibly hibernates in any stage and has a biennial cycle of development (Andersen, 1969, 1983).

** We accept the narrow treatment of the polar desert biome as suggested by Korotkevitch (1972) and Alexandrova (1983).

*** All specimens of brown scavenger beetles found by us in the High Arctic were brachypterous.

Among the beetles in the polar deserts, there is only one common species, the rove beetle *Micralymma brevilingue*. On the Severnaya Zemlya Archipelago (Makarova et al., 2007) and in similar landscapes on the Northern Plain of Wrangel Island (Khruleva, 1991), this predatory beetle occurs in a wide range of habitats. It exists even in the Cape Arktichesky area on Komsomolets Island, Severnaya Zemlya (81°09'N) at a mean July temperature of about +1°C. We assume it is the most cold-tolerant beetle species of the Northern Hemisphere. This may be facilitated by a total absence of direct competitors and a high abundance of springtails as potential pray, since the population numbers of springtails exceed there those in any other natural zone (Babenko & Bulavintsev, 1997). The abundance of the collembolans *Folsomia binoculata* (Walgren) and *F. taimyrica* Martynova, which are consumed by the beetles in laboratory (Makarova et al., 2007), can reach 2,000 ind./dm^2 in zonal communities (Babenko, 2000), i.e. exceeding the population density of the predator by three orders of magnitude. It seems very likely that the beetle's diet is actually more diverse. Its nearest relative, *Micralymma marinum* (Ström, 1783), consumes not only collembolans (Steel, 1958), but also littoral mites (Anderson R., 1997) and dead dipteran larvae (Thayer, 1985). Our quantitative surveys in the polar deserts of the Bolshevik Island, Severnaya Zemlya show this polytopic beetle, a typical dweller of coasts and islands, to clearly prefer dry, well-drained habitats with an extended vegetation season (Fig. 6, *2-4*). This probably shows that the thermal conditions in this region correspond to the threshold values necessary for the beetle development. Samples taken on the Bolshevik Island during the entire summer season and in September from under snow usually include larvae of all instars and adults, indicating a heterochronous development that appears to take more than one year. In Greenland, overwintering larvae and adults have also been found in *M. brevilingue* and *M. marinum*, this being uncharacteristic of most of the Greenland staphylinids (Böcher, 1988).

The remaining beetles found in polar deserts are only represented by a few species and specimens. Our finds of *Chrysolina septentrionalis* (Fig. 6-*1*) and *C. subsulcata* on two islands of the Severnaya Zemlya Archipelago confirm their status of euarctic species (Chernov, 1978) and shift the northern limit of the family distribution by more then 2° to the North, up to 79°35'N (Makarova et al., 2007). Larvae of these species in Taimyr tundras hibernate not less than two times (Chernov, 1978a). On the Wrangel Island, larvae of *C. magniceps* (Sahlberg, 1887)*, a species close to *C. subsulcata*, overwinter up to 4 times (Khruleva, 1994), the whole cycle of these beetles taking 5-6 years (Chernov et al., 1994). The most important adaptations of arctic *Chrysolina* that ensure their survival in such high latitudes are polytopy, polyphagy, ovoviviparity, and the capability of developing many years without a seasonal fixation of stages (Chernov, 1978; Chernov et al., 1994). The absence of *Chrysolina* from North American polar deserts is highly remarkable (see above).

Among Lathridiidae of the High Arctic, only members of *Corticaria* and *Lathridius* have hitherto been recorded (Brown, 1937; Danks, 1981; Khruleva, 2007). Earlier, they

* Originally referred to as *C. subsulcata* (see Khruleva, 1994).

Fig. 6. Preferred and avoided habitats of beetles in polar deserts. *1–4* Bolshevik Isl., Severnaya Zemlya Archipelago; *1* - *Chrysolina septentrionalis* female on the buttercap *Ranunculus sulphureus*; *2* – colony of collared lemmings on a sandy slope, the typical habitat of *C. septentrionalis*; *3, 4* – river benches with high abundance of *Micralymma brevilingue* (*3* – Golysheva R., *4* – Lagernaya R.); *5, 6* – Ellef Ringnes Isl., Princess Elizabeth Islands; *5* - colony of collared lemmings, the place where *Dienerella elegans* was found; *6* – zonal community on interfluvial plains where the beetles are absent.

were accepted as occasional invaders in the Arctic (mostly with goods), incapable of overcoming the tree line (Campbell et al., 1979; Danks, 1981; Bousquet, 1991a etc.). However, our pitfall trapping in the mountain tundra of Yakutia (Sakha), up to the nival deserts, as well as in the southern tundras of Europe, Yamal Peninsula and Taimyr show that species of Latridiidae are a common and sometimes diverse component of the soil-bound fauna. Thus, Lathridiidae can be considered permanent residents of high latitudes. Moreover, the ranges of many lathridiids, especially *Dienerella* species, are often cosmopolitan. The key features ensuring their occupation of polar and nival deserts seems to be the very small size (usually less than 3 mm), euryoky, and broad mycophagy (see: Hinton, 1945). In addition, numerous species possess wing dimorphism up to a full absence of the hind wings and the unity of elytra.

Therefore, one can see that individual adaptations of separate genera and species of non-carabid beetles surpass those of carabids in the polar desert zone, in spite of their remarkable commonness in the Arctic as a whole. All beetles found in polar deserts are small-sized and show vast distributions, albeit a lack in wings.

Despite intense collecting efforts, four beetle species of the five found in polar deserts are only represented by occasional findings, mostly from well-drained slopes, river terraces, elevations and steep margins. These habitats are characterized by a thinner snow cover which is blown off with the wind, faster snowmelt, and a deeper thawing of the permafrost in summer. All this extends the local vegetation season with 2-4 weeks in comparison to the neighbouring plainy zonal environment (Makarova, 2002). It is here that sea-bird rookeries, lemming colonies and look-outs of the birds of prey are situated. Lemming impact is especially significant in polar deserts. They really occupy key positions there (McAlpine, 1965). Many species of mites, collembolans and insects are confined to these habitats with heat benefits (Babenko, 2000, 2003; Makarova, 2002). *Chrysolina septentrionalis* on two islands of the Severnaya Zemlya Archipelago and *Dienerella elegans* on Ellef Ringnes Island are found only in the turf of lemming hills (Fig. 6-*2,5*). The abundance of *Micralymma brevilingue* is the highest also in these sites (Makarova et al., 2007). The distribution of beetles on the Severnaya Zemlya must have been more uniform both during the glaciations, when the summer temperatures in northern Asia were 5-7°C higher than the present-day values (Alfimov & Berman, 2004), and during the Holocene thermal maximum (10.0-8.8 thousands years ago) with the summer temperatures 4-6°C higher than at present (Bolshiyanov & Makeyev, 1995). It is quite possible that the significance of thermal conditions for beetle development is indirectly confirmed by the relatively small size of the species found (less than 8 mm; in three species, less than 3 mm).

Sectoral peculiarities of the High Arctic beetle faunas

A strong positive correlation between species richness and summer temperature is well-known for arctic landscapes (Rannie, 1986; Chernov, 1989, 1995, 2002; Chernov &

Penev, 1993; Matveyeva 1998; Chernov et al., 2000). This appears to hold true for the High Arctic beetles as well (Fig. 7). However, the relationships between beetle diversity and temperature in the different sectors of the Arctic are remarkably distinguishable. In the Greenland and Canada sectors, beetle diversity is much lower than in the other

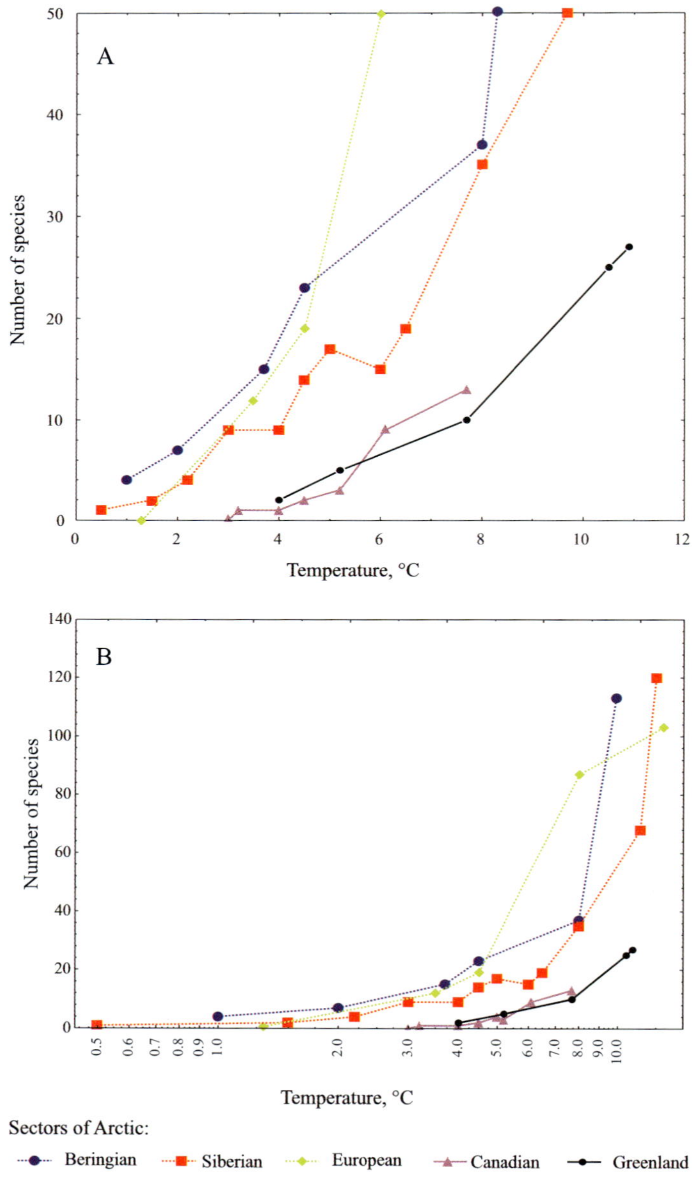

Fig. 7. Local beetle diversity change in relation to mean July temperatures in different sectors of the Arctic: A: non-transformed data; B: log-transformed data. Besides of the sources mentioned in the Tables 3 and 5, the following publications were used as well: Chernov & Matveyeva, 1997; Kolesnikova, 2000; Leshko, 2000; Khruleva, 2001.

sectors at the same temperature levels. It is these areas that are almost fully glaciated still recently enough and their beetle assemblages are strongly impoverished, being members of a young biota (Matthews, 1979; Bennike, 1999; Sadler, 1999). The small areas free of ice (nunataks and the like) could not support most of the beetle species because of too low ambient temperatures (Scudder, 1979; Böcher, 1988, 1997; Sadler, 1998). Thus, the Greenland insect fauna contains less than 14% as compared to a climatically analogous continental fauna (Downes, 1988); the same parameter for carabids was 17% (Coope, 1986). Being relatively young, these beetle assemblages support great proportions of vagile, flying species (Table 3, Fig. 8), though this active way of repopulation was not considered to be the main one (Coope, 1986; Böcher, 1988). The beetle fauna of Svalbard is comparatively rich (Coulson & Refseth, 2004) in spite of an almost total glaciation in the late Pleistocene. This is possibly related to favourable sea and air currents and rather mild climate facilitating repopulation (Coulson, 2000). In the Svalbard beetle fauna, "good" flyers also strongly predominate (Fig. 8).* The most characteristic arctic species belonging

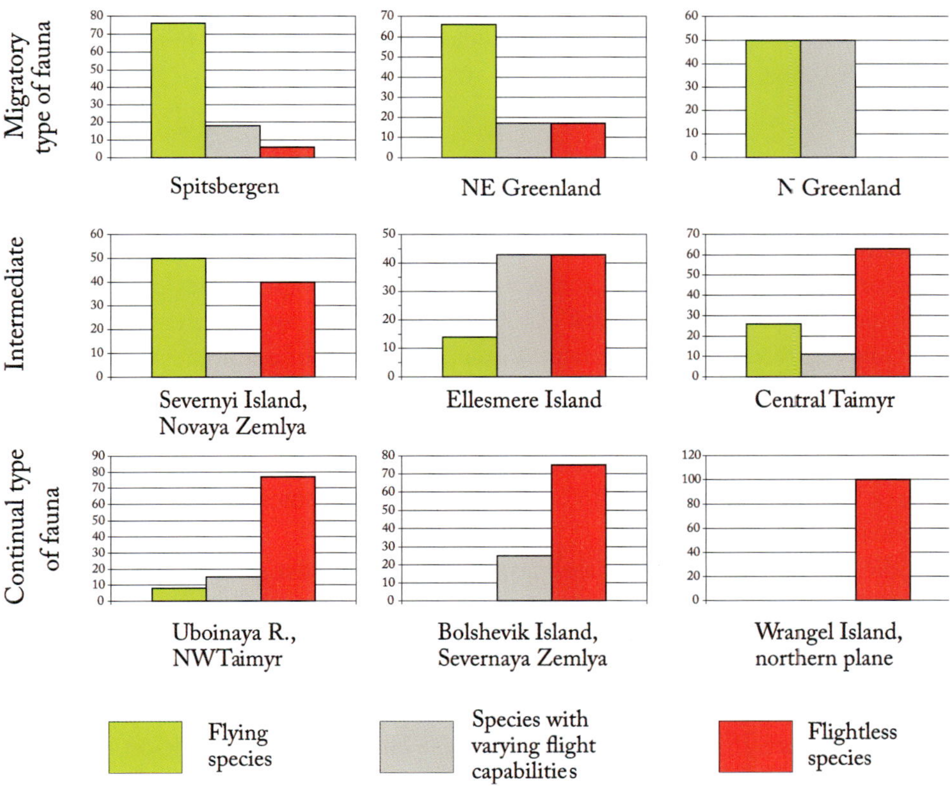

Fig. 8. Flight capability in different beetle faunas in the High Arctic. Sources: as in Table 3.

* The similar situation was found in the Scandinavian alpine carabid assemblages (Nilsson et al., 1993). Against the expectations, all altitudinal specialists appeared to be macropterous, that can be caused by the relative recentness of the deglaciation process and the lower dispersal capability of brachypterous arctic species.

to wingless subgenera of *Pterostichus* and *Chrysolina* are totally absent from the Canadian High Arctic, Greenland and Svalbard (Danks, 1981; Böcher, 1988; Coulson & Refseth, 2004), but are especially diverse in arctic areas with a successively developed biota (Ball & Currie, 1997; Bieńkowski, 2004, 2007; Khruleva, 2007). The share of non-flying beetles is generally increasing northwards (Lindroth, 1957; Downes, 1965; Roff, 1990). This trend reflects both a greater stability of habitats in a colder climate and the evident advantages gained through reduction of the hindwings and/or wing musculature. Flight deprivation vastly economizes the energy and body space essential for reproduction and development (Southwood, 1962; Roff, 1990). This is why the arctic beetle faunas developing continually comprise mainly brachypterous or dimorphic species (Ball & Currie, 1997; Bieńkowski, 2004, 2007; Khruleva, 2007; Makarova et al., 2007). Thus, two opposite types of High Arctic beetle faunas can be recognized, migratory and continual ones. Migratory faunas are relatively young and consist mainly of actively dispersing species, in most cases able to fly. To the contrary, continual faunas consist mostly of dimorphic, brachypterous or entirely wingless forms. It is supposed that continual faunas are more or less stabilized having been developing *in situ* over longer period of time than migratory ones. Thus, the general flight capability of the species of a particular local beetle fauna in the Arctic reflects its origins, and to a certain extent - its age, as well.

CONCLUSIONS

The composition of beetle families in the High Arctic is stable enough in the different sectors of the Arctic, as well as in subarctic highlands. Throughout the Arctic, Carabidae and Staphylinidae are reigning, but at the northern cold limit of Coleoptera range as a whole, only apparently deficient faunas occur, consisting of a few members of Chrysomellidae, Lathridiidae and Staphylinidae.

Rove beetles demonstrate clear superiority over ground beetles all over the Arctic, this possibly being rooted in the smaller average size and more diverse ecologies. Their conquest of polar deserts seems to be related to a higher potential to flex and force the individual development.

The beetle faunas of polar deserts consist of a few, rather small, wingless, polytopic and polyphagous species with vast distributions. It is a complex of adaptations, not a single adaptation alone, which is responsible for beetle survival in the extreme environments (see Hodkinson, 2005).

In polar desert landscapes, beetles tend to prefer intrazonal, relatively warm, often zoogenic habitats. If the climatic change of the last few decades goes on, predictably rapid alterations in local beetle compositions are most probably to be due to dispersal and extinction (Ashworth, 2001). Lemming hills representing the main micro-hotspots of biodiversity in the High Arctic (McAlpine, 1965; Korotkevich, 1972; Chernov, 1978; Babenko, 2000; Makarova, 2002) can serve as suitable sites for monitoring the insect fauna fluctuations.

Two opposite types of High Arctic beetle faunas can be recognized: migratory and continual (see above). Examples of migratory faunas are the beetle lists of Greenland and Svalbard, where postglacial flying migrants clearly prevail. The beetle faunas developing continually *in situ* over a long period mainly comprise brachypterous and/or dimorphic species.

ACKNOWLEDGMENTS

The authors are grateful to all specialists who provided beetle identification: K.V. Makarov, A.O. Bieńkowski, A.V. Sokolov, P.N. Petrov, B.A. Korotyaev, S.V. Saluk, V.G. Grachev (Russia), V.I. Gusarov (Norway), L. Zerche (Germany). The unfailing help of A.G. Listkov, V.V. Larin, V.A. Ishkov, N.V. Matveyeva, I.A. Lavrenenko, O.V. Lavrenenko, D.N. Krasavin, R.A. Krasavina (Russia), D. Walker, W. Gould, G. Gonzales (USA), L.C. Bliss (Canada) in the organization of field work is greatly acknowledged. The valuable help of K.V. Makarov in preparing the illustrations, of S.I. Golovatch in improving the English text, and of L. Penev in discussion of the results are highly appreciated. We are very grateful to E.E. Lindquist who provided data on beetle assempbalges at Cape Barrow (mimeo of 1958). The work was financially supported by the Russian Foundation for Basic Research, through the Support Programme for Leading Academic Schools of Russia and a RAS grant within the Programme "The Origin and Evolution of the Biosphere".

REFERENCES

Aitchison, C.W. (1979). Winter-active subnivean invertebrates in Southern Canada. II. Coleoptera. – Pedobiologia 19: 121-128.

Alexandrova, V.D. (1983). [Vegetation of the USSR Polar Deserts]. – Nauka, Leningrad. (In Russian).

Alfimov, A V. & Berman, D.I. (2004). [The thermophytic tundra-steppe expansion and the climate of the last glacial maximum on the northeast of Asia]. – Kriosphera Zemli, Novosibirsk 8(4): 78-87. (In Russian).

Andersen, J. (1969). Habitat choice and life history of Bembidiini (Col., Carabidae) on river banks in Central and Northern Norway. – Norsk ent. Tidsskr. 17: 17-65.

Andersen, J. (1983). The life cycles of the riparian species of *Bembidion* (Coleoptera, Carabidae) in northern Norway. – Notulae Entom. 63: 195-202.

Anderson, R. (1997). Rove beetles (Coleoptera: Staphylinidae). – In: Northern Ireland Species Inventories. Recorder database of CEDaR, The Northern Ireland environmental records centre, p. 1-108.

Anderson, R.S. (1989). Revision of the subfamily Rhynchaeninae in North America (Coleoptera: Curculionidae). – Trans. Am. Entom. Soc. 115: 207-312.

Anderson, R.S. (1997). Weevils (Coleoptera: Curculionoidea, excluding Scolytinae and Platypodinae) of the Yukon. – In: Insects of the Yukon (Danks, H.V. & Downes, J.A., eds). Biological Survey of Canada (Terrestrial Arthropods), Ottawa, p. 523-562.

Anderson, R.S. (1997a). An overview of the beetles (Coleoptera) of the Yukon. – In: Insects of the Yukon (Danks, H.V. & Downes, J.A., eds). Biological Survey of Canada (Terrestrial Arthropods), Ottawa, p. 405-444.

Andreyeva, T.R. & Petrov, P.N. (2004). [The aquatic Adephaga beetles (Coleoptera) of Southern Yamal Peninsula and the Polar Urals]. – Byulleten' moskowskogo obshchestva ispytateley prirody, Moskva. Otdel biologicheskii 109(3): 9-20. (In Russian).

Ashworth, A.C. (2001). Chapter 8. Perspectives on beetles and climate change. – In: Geological Perspectives of Global Climate Change (Gerhard, L.C., Harrison, W.E. & Hanson, B.M., eds). American Association of Petroleum Geologists Studies in Geology, 47. Tulsa, Oklahoma, p. 153-168.

Aunaas, T., Baust, J.G. & Zachariassen, K.E. (1983). Ecophysiological studies on arthropods from Spitsbergen. – Polar Research 1(3): 235-240.

Babenko, A.B. (2000). Collembolan assemblages of polar deserts and subarctic nival communities. – Pedobiologia 44: 421-429.

Babenko, A.B. (2003). Landscape chorology of springtails in the Taimyr Peninsula. 1. Biotopic preferences of species. – Entom. Rev. 83(8): 1003-1021.

Babenko, A.B. & Bulavintsev, V.I. (1997). Springtails (Collembola) of Eurasian polar deserts. – Russian J. Zool. 1(2) 177-184.

Ball, G.E. & Currie, D.C. (1997). Ground beetle (Coleoptera: Trachypachidae and Carabidae) of the Yukon: geographical distribution, ecological aspects, and origin of the extant fauna. – In: Insects of the Yukon (Danks, H.V. & Downes, J.A., eds). Biological Survey of Canada (Terrestrial Arthropods), Ottawa, p. 445-489.

Baust, J.G. (1972). Temperature-induced neural adaptation: motoneuron discharge in the Alaskan beetle *Pterostichus brevicornis* (Carabidae). – Comp. Biochem. Physiol. A41: 205-213.

Bennike, O. (1999). Colonization of Greenland by plants and animals after the last ice age: a review. – Polar Record 35: 323-336.

Bennike, O., Björck, S., Böcher, J. & Walker, I. (2000). The Quaternary arthropod fauna of Greenland: a review with new data. – Bull. Geol. Soc. Denmark 47: 111-134.

Bieńkowski, A.O. (2001). A study on the genus *Chrysolina* Motschulsky, 1860, with the checklist of all the described subgenera, species, subspecies, and synonyms (Coleoptera: Chrysomelidae: Chrysomelinae). – Genus 12(2): 1-236.

Bieńkowski, A.O. (2004). A review of the subgenus *Arctolina* Kontkanen, 1959 of the genus *Chrysolina* Motschulsky, 1860 (Coleoptera: Chrysomelidae: Chrysomelinae). – Genus 15(2): 187-233.

Bieńkowski, A.O. (2007). A revision of the subgenus *Pleurosticha* Motschulsky, 1860 of the genus *Chrysolina* Motschulsky, 1860 (Coleoptera: Chrysomelidae: Chrysomelinae). – Russian Entom. J. 16(2): 219-235.

Blake, W. & Matthews, J.V. (1979). New data on an interglacial peat deposit near Makinson Inlet, Ellesmere Island, District of Franklin. – Geological Survey of Canada. Paper 79-1A: 157-164.

Block, W. (1992). An Annotated Bibliography of Antarctic Invertebrates (Terrestrial and Freshwater). – British Antarctic Survey, Cambridge.

Block, W. & Sømme, L. (1983). Low temperature adaptation in beetles from the Sub-Antarctic Island of South Georgia. – Polar Biol. 2: 109-114.

Böcher, J.J. (1988). The Coleoptera of Greenland. – Meddr. Grønland, Biosci. 26: 1-100.

Böcher, J., (1989). First record of an interstadial insect from Greenland: *Amara alpina* (Paykull, 1790) (Coleoptera: Carabidae). – Boreas 18: 1-4.

Böcher, J., (1997). The history of the Greenland insect fauna – with emphasis on living and fossil beetles. – Zoological Museum of the Copenhagen University, Copenhagen, p. 1-20.
Bolshiyanov, D.Yu. & Makeyev, V.M. (1995). [The Severnaya Zemlya Archipelago: glaciation and Formation of the Environments]. – Gidrometeoizdat, St. Petersburg. (In Russian).
Bousquet, Y. (1991). Synoptic table. – In: Checklist of Beetles of Canada and Alaska (Bousquet, Y., ed). Biosyst. Research Centre, Research Branch, Agriculture Canada, Ottawa, Ontario, p. 1-3.
Bousquet, Y. (1991a). Family Lathridiidae, minute brown scavenger beetles. – In: Checklist of Beetles of Canada and Alaska (Bousquet, Y., ed). Biosyst. Research Centre, Research Branch, Agriculture Canada, Ottawa, Ontario, p. 239-241.
Brodo, F. (2000). The insects, mites and spiders of Hot Weather Creek, Ellesmere Island, Nunavut. – Geol. Survey of Canada. Bull. 529: 145-173.
Brown, W.J. (1937). The Coleoptera of Canada's eastern Arctic. – Can. Entom. 69(5): 106-111.
Brown, W.J. (1962). The American species of *Chrysolina* Mots. (Coleoptera: Chrysomelidae). – Can. Ent. 94(1): 58-74.
Bulavintsev, V.I. & Babenko, A.B. (1983). [Soil-bound invertebrates in the eastern sector of Franz Josef Land Archipelago]. – Zoologicheskii Zhurnal, Moskva 62(7): 1114-1116. (In Russian).
Byzova, Yu.B., Uvarov, A.V., Gubina, V.G., Zalesskaya, N.T., Zakharov, A.A., Petrova, A.D., Suvorov, A.A. & Vorobiyova, E.G. (1986). [Soil-bound Invertebrates of White Sea Islands of Kandalaksha Nature Reserve]. – Nauka, Moskwa. (In Russian).
Campbell, J.M. (1978). A revision of the North American Omaliinae (Coleoptera, Staphylinidae). 2. The tribe Coryphiini. – Mem. Entom. Soc. Canada 106: 20-87.
Campbell, J.M. (1988). New species and records of North American *Tachinus* Gravenhorst (Coleoptera: Staphylinidae). – Can. Entom. 120: 231-295.
Campbell, J.M., Ball, G.E., Becker, E.C., Bright, D.E., Helava, J., et al. (1979). 40. Coleoptera. – In: Canada and Its Insect Fauna (Danks, H.V., ed.). Mem. Entom. Soc. Canada 108: 357-387.
Chapman, J. (1954). Observation of snow insects in western Montana. – Can. Ent. 86: 357-363.
Chernov, Yu.I. (1978). [Structure of the Animal Populations in the Subarctic]. – Nauka, Moskva. (In Russian).
Chernov, Yu.I. (1978a). [Adaptive peculiarities of insect life cycles in the zone of tundra]. – Zhurnal obshchei biologii 39(3): 394-402. (In Russian).
Chernov, Yu.I. (1988). [Phylogenetic level and the geographical distribution of taxa]. – Zoologicheskii Zhurnal, Moskva 67(10): 1445-1458. (In Russian).
Chernov, Yu.I. (1989). [Heat condition and Arctic biota]. – Ekologia, Ekaterinburg 2: 49-57. (In Russian).
Chernov, Yu.I. (1992). [Who are more numerous in tundra – predators or phytophages?] – In: Tsenoticheskie Vzaimodeistviya v Tundrovykh Ekosistemakh (Chernov, Yu.I, ed.). Nauka, Moskva, p. 100-125. [In Russian].
Chernov, Yu.I. (1995). Diversity of the arctic terrestrial fauna. – In: Ecological Studies. V. 113. Arctic and Alpine Biodiversity (Chapin/Körner, eds). Springer-Verlag, Berlin, Heidelberg, p. 79-93.
Chernov, Yu.I. (2002). Arctic biota: taxonomic diversity. – Entom. Rev. 82 (Suppl. 1): S1-S23.
Chernov, Yu.I. (2004). The animal world of the polar desert on the Devon Island Plateau, Canadian Arctic Archipelago. – Entom. Rev. 84 (Suppl. 1): S15-S24.

Chernov, Yu.I., Makarov, K.V & Eremin, P.K. (2000). Family of ground beetles (Coleoptera, Carabidae) in the Arctic fauna: Communication 1. – Entom. Rev. 80(8): 1009-1021.

Chernov, Yu.I., Makarov, K.V & Eremin, P.K. (2001). Family of ground beetles (Coleoptera, Carabidae) in the Arctic fauna: Communication 2. – Entom. Rev. 81(1): 108-117.

Chernov, Yu.I. & Matveyeva, N.V. (1997). Chapter 16. Arctic ecosystems in Russia. – In: Polar and Alpine Tundra. Ecosystems of the World. 3 (Wielgolaski, F.E., ed.). Elsevier, Amsterdam, p. 361-507.

Chernov, Yu.I., Medvedev, L.N. & Khruleva, O.A. (1994). Leaf beetles (Coleoptera, Chrysomelidae) in the Arctic. – Entom. Rev. 73(2): 152-167.

Chernov, Yu.I. & Penev, L.D. (1993). [Biological diversity and climate]. – Uspekhi sovremennoi biologii 113(5): 515-531. (In Russian).

Chernov, Yu.I., Striganova, B.R., Ananieva, S.I. & Kuzmin, L.L. (1979). [Animal world of polar desert at Cape Chelyuskin.] – In: Arkticheskie Tundry i Polyarnye Pustyni Taimyra (Alexandrova, V.D. & Matveyeva, N.V., eds). Nauka, Leningrad, p. 35-49. (In Russian).

Coope, G.R. (1986). The invasion and colonization of the North Atlantic islands: a palaeoecological solution to a biogeographic problem. – Phil. Trans. R. Soc. Lond. B 314: 619-635.

Coulson, S.J. (2000). A review of the terrestrial and freshwater invertebrate fauna of the High Arctic archipelago of Svalbard. – Norw. J. Entom. 47: 41-63.

Coulson, S.J. & Refseth, D. (2004). Chapter 3. The terrestrial and freshwater invertebrate fauna of Svalbard (and Jan Mayen). – Skrifter, Norwegian Polar Institute 201: 57-122.

Danks, H.V. (1980). Arthropods of Polar Bear Pass, Bathurst Island, arctic Canada. – Syllogeus 25: 1-68.

Danks, H.V. (1981). Arctic Arthropods. – Tyrell Press Limited, Ottawa.

Danks, H.V. (1989). Insects of the boreal zone of Canada. – Can. Entom. 121(8): 626-690.

Danks, H.V. (1990). Arctic Insects: instructive diversity. – In: Canada's Missing Dimension: Science and History in the Canadian Arctic Islands. V. II (Harington, C.R., ed.). Canadian Museum of Nature, Ottawa, p. 444-470.

De Zordo, I. (1979). Phänologie von Carabiden im Hochgebirge Tirols (Obengurgl, Österreich) (Insecta: Coleoptera). – Ber. nat.-med. Ver. Innsbruck 66: 73-83.

Downes, J.A. (1965). Adaptation of insects in the arctic. – Ann. Rev. Ent. 10: 257-274.

Downes, J.A. (1988). The post-glacial colonization of the North Atlantic Islands. – Mem. ent. Soc. Can. 144: 55-92.

Elvebakk, A., Elven, R. & Razzhivin, V.Y. (1999). Delimitation, zonal and sectorial subdivision of the Arctic for the Panarctic Flora Project. – In: The Species Concept in the High North – A Panarctic Floral Initiative (Nordal, I. & Razzhivin, V.Y., eds). The Norwegian Academy of Science and Letters, Oslo, p. 375-386.

Eriksson, U. (1972). The invertebrate fauna of the Kilpisjärvi area, Finnish Lapland. 10. Dytiscidae. – Acta Soc. pro Fauna et Flora Fennica 80: 121-152.

Evans, M.E.G. (1969). The surface activity of beetles in a northern English wood. – Trans. Soc. Brit. Ent. 18(11): 247-262.

Fedorenko, D.N. (1988). [The fauna of ground-beetles (Coleoptera, Carabidae) of the Moskovskaya Area]. – In: Nasekomye Moskovskoi Oblasti: Problemy Kadastra i Okhrany. Nauka, Moscow, p. 20-46. [In Russian].

Fjellberg, A. (1983). Three species of Staphylinid Coleoptera new to Spitsbergen, Norway. – Fauna norv., Ser. B 30: 110-111.

Filippov, B.Yu. (2007). [Life cycles of some species of carabid beetles (Coleoptera: Carabidae) in the south tundra]. – Russian Entom. J. 16(4): 423-436. (In Russian).

Gorodkov, B.N. (1935). [Tundra Zone Vegetation of the USSR]. – Izdatelstvo Akademii nauk SSSR, Moscow-Leningrad. (In Russian).

Heydemann, V.B. (1956). Untersuchungen über die Winteraktivität von Staphylinidae auf Feldern. – Entom. Blätter 52: 138-150.

Hinton, H.E. (1945). A Monograph of the Beetles Associated with Stored Products. V. I. – Brit. Mus. Nat. Hist., London.

Historical Development of the Class Insecta. (1980). (Rodendorf, B.B. & Rasnitsyn, A.P., eds) – Nauka, Moscow.

Hodkinson, I.D. (2005). Adaptation of invertebrates to terrestrial Arctic environments. – Skrifter, Det Kongelige Norske Videnskabers Selskab: 1-45.

Hunt, T., Bergsten, J., Levkanicova, Z., Papadopoulou, A., John, O.S., Wild, R., Hammond, P.M., Ahrens, D., Balke, M., Caterino, M.S., Gómez-Zurita, J., Ribera, I., Barraclough, T.G., Bocakova, M., Bocak, L. & Vogler, A.P. (2007). A comprehensive phylogeny of beetles reveals the evolutionary origins of a superradiation. – Science 318 (21): 1913-1916.

Hurd, P.D. & Lindquist, E.E. (1958). Analysis of soil invertebrate samples from Barrow, Alaska. – Final Report, Arct. Inst. N. Am. Project ONR-173 and ONR-193, p. 1-24 (mimeo).

Kaufmann, T. (1971). Hibernation in the arctic beetle *Pterostichus brevicornis* in Alaska. – J. Kansas Entom. Soc. 44: 81-92.

Khruleva, O.A. (1991). [Spatial aspects of invertebrate population of the Wrangel Island]. – In: Populacii i Soobshchestva Zhivotnykh Ostrova Wrangelya. Gosudarstvennyi Komitet RSFSR po okhrane prirody, Moscow, p. 161-179. (In Russian).

Khruleva, O.A. (1994). Life cycle of the leaf cutting beetle *Chrysolina subsulcata* (Coleoptera, Chrysomelidae) on Wrangel Island. – Entom. Rev. 73(3): 29-38.

Khruleva, O.A. (1999). [Herpetobiotic arthropods (beetles and spiders) of the northeastern Taimyr]. – In: Materialy II (XII) Vserossiiskogo soveshchaniya po pochvennoi zoologii "Bioraznoobraziye i zhizn' pochvennoi systemy" (Striganova, B.R., ed.). KMK, Moscow, p. 135-136. (In Russian).

Khruleva, O.A. (2001). [Herpetobiotic arthropods of the Wrangel Island]. – PHD Thesis. Institut problem ekologii i evolyutsii imeni A.N. Severtsova, Moskva. (In Russian).

Khruleva, O.A. (2007). [Peculiarity of terrestrial invertebrate fauna of the Wrangel Island as a reflection of the originality of its environment]. – In: Priroda Ostrova Wrangelya: Sovremennyje Issledovaniya (Gruzdev, A.R., ed.). Asterion, St. Petersburg, p. 136-181. (In Russian).

Khruleva, O.A. & Korotyaev, B.A. (1999). Weevils (Coleoptera: Apionidae, Curculionidae) of Wrangel Island. – Entom. Rev. 79(9): 1119-1137.

Kolesnikova, A. (2000). 6.5. Beetles. – In: Pechora Delta. Structure and Dynamics of the Pechora Delta Ecosystems (1995-1999). Report No. 2000.037 (van Eerden M.R., ed). Inst. for Inland Water Management and Waste Water Treatment RIZA, Lelystad, p. 127-132.

Korobeinikov, Yu.I. (1990). [Seasonal dynamics of ground-beetle activities in the biocenoses of the Southern Yamal]. – In: Prostranstvenno-vremennaya Organizatsia Entomokompleksov Subarktiki. UrO AN SSSR, Sverdlovsk, p. 45-54. (In Russian).

Korobeinikov, Yu.I. (1990a). [Feeding and reproducing of carabids in the Subarktik conditions]. – In: Energetica Pitaniya i Rosta Zhivotnykh. UrO AN SSSR, Sverdlovsk, p. 65-72. (In Russian).

Korotkevich, E.S. (1972). [Polar Deserts]. – Hydrometeoizdat, Leningrad. (In Russian).
Laff, M.L. (1966). Cold hardiness of some beetles living in grass tussocks. – Entom. Exp. & Appl. 9: 191-199.
Larson, D.J., Alarie, Y. & Roughley, R.E. (2000). Predaceous Diving Beetles (Coleoptera: Dytiscidae) of the Nearctic Region, with Emphasis of the Fauna of Canada and Alaska. – NRC Research Press, Ottawa.
Lawrence, J.F. & Newton, A.F. Jr. (1995). Families and subfamilies of Coleoptera (with selected genera, notes, references and data on family-group names). – In: Biology, Phylogeny, and Classification of Coleoptera (Pakaluk, J. & Ślipiński, S.A., eds). Papers Celebrating the 80th birthday of Roy A. Crowson. Museum i Instytut Zoologii PAN, Warszawa. Reprinted with permission in: Publicaciones Especiales No. 3. Centro de Estudios en Zoologia, Universidas de Guadalajara, p. 779-1006.
LeSage, L. (1991). Family Chrysomelidae, leaf beetles. – In: Checklist of Beetles of Canada and Alaska (Bousquet, Y., ed). Biosyst. Research Centre, Research Branch, Agriculture Canada, Ottawa, Ontario, p. 301-323.
Leshko, Yu.V. (2000). 6.4. Zoobenthos and macrofauna. – In: Pechora Delta. Structure and Dynamics of the Pechora Delta Ecosystems (1995-1999). Report No 2000.037 (van Eerden M.R., ed). Inst. for Inland Water Management and Waste Water Treatment RIZA, Lelystad, p. 121-125
Lindroth, C.H., (1957). The Faunal Connections between Europe and North America. – Willey & Sons, Stockholm, N.-Y.
Lindroth, C.H. (1968). The ground-beetles (Carabidae, excl. Cicindelinae) of Canada and Alaska, Part 5. – Opusc. Entom., Supp. 33: 649-944.
Lindroth, C.H. (1992). Ground Beetles (Carabidae) of Fennoscandia. A Zoogeographic Study. Part I-III. – Smiths. Inst. Lib., Pauls Press, New Delhi.
Lorenz, W. (1998). Systematic List of Extant Ground Beetles of the World. – Germany, Tutzing.
Makarova, O.L. (2002). Acarocenoses (Acariformes, Parasitiformes) in polar deserts: 2. Cenotic relations, structure of communities, and the proportion of suborders. – Entom. Rev. 82(7): 857-875.
Makarova, O.L., Bieńkowski, A.O., Bulavintsev, V.I. & Sokolov, A.V. (2007). Beetles (Coleoptera) in polar deserts of the Severnaya Zemlya Archipelago. – Entom. Rev. 87(9): 1142-1154.
Mani, M.S. & Giddings, L.E. (1980). Ecology of Highlands. – Monogr. Biol., 40. The Hague Ac., Junk, Hague.
Marusik, Yu. M. (1993). [Terrestrial invertebrates]. – In: Ecologiya Basseina Reki Amguemy (Chukotka). Chast' 1 (Berman, D.I., ed.). Institut biologicheskikh problem severa, Vladivostok, p. 164-185. (In Russian).
Matthews, J.V. (1979). Tertiary and quaternary environments: historical background for an analysis of the Canadian insect fauna. – In: Canada and Its Insect Fauna (Danks, H.V., ed.). Mem. Entom. Soc. Canada 108: 31-86.
Matveyeva, N.V. (1998). [Zonation in Plant Cover of the Arctic]. – Proceedings of Komarov Botanical Institute 21, Russian Academy of Sciences, St. Petersburg. (In Russian).
McAlpine, J.F. (1965). Insect and related terrestrial invertebrates of Ellef Ringnes Island. – Arctic 18: 73-103.
Miller, K. (1982). Cold-hardiness strategies of some adult and immature insects overwintering in interior Alaska. – Comp. Biochem. Physiol. 73 A(4): 595-604.

Miller, K.V. & Williams, R.N. (1983). Biology and host preference of *Atheta coriari* (Coleoptera: Staphylinidae), an egg predator of Nitidulidae and Muscidae. – Ann. Entom. Soc. Am. 76: 158-161.

Nelson, R.E. (2001). Bioclimatic implications and distribution pattern of the modern ground beetle fauna (Insecta: Coleoptera: Carabidae) of the Arctic Slope of Alaska, U.S.A. – Arctic 54(4): 425-430.

Newton, A.F. Jr, Thayer, M.K., Ashe, J.S. & Chandler, D.S. (2001). Staphylinidae. – In: American Beetles: Vol. I Archostemata, Myxophaga, Adephaga, Polyphaga: Staphyliniformia (Arnett, R. & Thomas, M., eds). CRC Press, Boca Ration, p. 272-418.

Nilsson, A.N., Peterson, R.B. & Lemdahl, G. (1993). Macroptery in altitudinal specialists versus brachyptery in generalists – a paradox of alpine Scandinavian carabid beetles (Coleoptera: Carabidae). – J. Biogeogr. 20(2): 227-234.

Økland, F. (1928). Land- and Süsswasserfauna von Nowaja Semlja. – Report of the Scientific Results of the Norwegian Expedition to Nowaya Zemlya 1921. Kristiania 42: 1-125.

Paarmann, W. (1979). Ideas about the evolution of the annual reproduction rhythms in carabid beetles of the different climatic zones. – In: On the Evolution of Behavior in Carabid Beetles. Miscellaneous papers (den Boer, P.J. et al., eds). Vol. 18: 119-132.

Petersen, M.K. (1998). Fecundity and juvenile survival of *Bembidion lampros* and *Tachyporus hypnorum*. – Entom. Exper. & Appl. 87: 301-309.

Poppius, B. (1910). Die Coleopteren des arktischen Gebietes. – Fauna Arctica, Jena (Römer, F., Schaudinn, F. & Brauer, A., eds) 5(1): 289-447.

Rannie, W.F. (1986). Summer air temperature and number of vascular species in Arctic Canada. – Arctic 39(2): 133-137.

Renken, W. (1956). Untersuchungen über Winterlager der Insekten. – Z. Morph. Ökol. Tiere 45: 34-106.

Roff, D.A. (1990). The evolution of flightlessness in insects. – Ecol. Monogr. 60(4): 389-421.

Rossolimo, T.Ye. (1995). Cold resistance of Coleoptera of the Subarctic region (comparative analysis). – Entom. Rev. 74(2): 98-110.

Rossolimo, T. (1997). Temperature adaptations of Siberian *Pterostichus* species (Coleoptera: Carabidae). – Eur. J. Entomol. 94: 235-242.

Rossolimo, T.Ye. & Rybalov, L.B. (1995). Thermopreferenda of Coleoptera from Subarctic Region. – Entom. Rev. 74(8): 60-73.

Ryabitsev, A.V. (1995). [Species composition of beetles of the Northern Yamal and their seasonal activities]. – Mekhanismy Podderzhaniya Biologicheskogo Raznoobraziya. Ekaterinburg, Ekaterinburg: 127-128. (In Russian).

Ryabitsev, A.V. (1998). [Population and ecology of carabids of the Northern Yamal]. – Avtoreferat dissertatsii na soiskanie uchenoy stepeni kandidata biologicheskikh nauk. Labytnangi. (In Russian).

Ryabukhin, A.S. (1999). A Catalogue of Rove Beetles (Coleoptera, Staphylinidae, excluding Aleocharinae) of the Northeast of Asia. – Institute of Biological Problems of the North RAS, Pensoft Publishers, Sofia-Moscow – Magadan.

Ryan, J.K. (1977). Appendix 7. Invertebrates of Truelove Lowland. – In: Truelove Lowland, Devon Island, Canada: A High Arctic Ecosystem (Bliss, L.C., ed.). Univ. Alta Press, Edmonton, p. 699-703.

Sadler, J.P. (1998). "Is Greenland a zoogeographic unit?" A response to Bergersen. – J. Biogeogr. 25: 399-403.

Sadler, J.P. (1999). Biodiversity of oceanic islands: a paleoecological assessment. – J. Biogeogr. 26: 75-87.

Scudder, G.G.E. (1979). Present patterns in the fauna and flora of Canada. – In: Canada and Its Insect Fauna (Danks, H.V., ed.). Mem. Entom. Soc. Canada 108: 87-179.

Silfverberg, H. (2004). Enumeratio Coleoptratorum Fennoscandiae, Daniae et Baltiae. – Sahlbergia, Helsinki 9: 1-111.

Slabber, S. & Chown, S.L. (2005). Differential responses of thermal tolerance to acclimation in the sub-Antarctic rove beetle *Halmaeusa atriceps*. – Physiological Entomology 30: 195-204.

Smetana, A. (1958). Drabčikoviti – Staphylinidae. I Staphylininae. – Fauna ČSR, zv. 12, Praha.

Sokolov, A.V. (2003). Fauna of rove beetles (Coleoptera, Staphylinidae) in the southern tundra of Western Taimyr Peninsula. – Entom. Rev. 83(9): 1226-1230.

Sota, T. (1994). Variation of carabid life cycles along climatic gradients: an adaptive perspective for life-history evolution under adverse conditions. – Series Entomologica 52: 91-112.

Southwood, T.R.E. (1962). Migration of terrestrial arthropods in relation to habitat. – Biol. Reviews 37: 171-214.

Steel, W.O. (1958). Notes on the Omaliinae (Col., Staphylinidae). – Ent. mon. Mag. 94: 138-142.

Steel, W.O. (1970). The larvae of the genera of the Omaliinae (Coleoptera: Staphylinidae) with particular reference to the British fauna. – Trans. R. ent. Soc. London 122: 1-47.

Strømme, J.A., Ngari, T.W. & Zachariassen, K.E. (1986). Physiological adaptations in Coleoptera on Spitsbergen. – Polar Research 4 (n. s.): 199-204.

Thayer, M.K. (1985). *Micralymma marinum* (Stroem) in North America: biological notes and new distributional records (Coleoptera: Staphylinidae). – Psyche 92: 49-55.

Tikhomirova, A.L. (1973). [Morphological Peculiarities and Phylogenesis of Rove Beetles (with Catalogue of USSR Fauna)]. – Nauka, Moscow. (In Russian).

Tikhomirova, A.L. (1982). [Fauna and ecology of staphylinid beetles (Coleoptera, Staphylinidae) in the Moscow region]. – In: Pochvennye Bespozvonochnye Moskovskoi Oblasti. Nauka, Moskva, p. 201-222. (In Russian).

Topp, W. & Smetana, A. (1998). Distributinal pattern and development of the winter-active beetle *Quedius pellax* (Staphylinidae). – Global Ecology & Biogeography 7(3): 189-195.

Walker, D.A., Gould, W.A., Blliss, L.C. et al. (2003). Circumpolar Arctic Vegetation Map. Scale 1:7,500,000. – Conservation of Arctic Flora and Fauna (CAFF) Map No. 1. U.S. Fish and Wildlife Service, Anchorage, Alaska.

Yakobson, G.G. (1898). [Zoological investigations on the Novaya Zemlya in 1896. Insects of the Novaya Zemlya]. – Zapiski Imperatorskoi Akademii nauk, Ser. 8. 8(1): 171-244. (In Russian).

Yakobson, G.G. (1905-1916). [Beetles of Russia, Western Europe and Adjacent Countries]. – Izdatelstvo F.F. Devriena, St.-Petersburg. (In Russian).

Zerche, L. (1993). Monographie der paläarktischen Coryphiini (Coleoptera, Staphylinidae, Omaliinae). Supplement 1. – Beitr. Entom. 43(2): 319-374.

Zinoviev, E.V. & Olshwang, V.N. (2003). [Beetles of the north of the West Siberian Plain, Subpolar and Polar Urals]. – Biologicheskiye Resursy Polyarnogo Urala. Nauchnyi Vestnik, Salekhard 3(2): 37-60. (In Russian).

L. Penev, T. Erwin & T. Assmann (Eds) 2008
BACK TO THE ROOTS AND BACK TO THE FUTURE. TOWARDS A NEW SYNTHESIS AMONGST TAXONOMIC, ECOLOGICAL AND BIOGEOGRAPHICAL APPROACHES IN CARABIDOLOGY
Proceedings of the XIII European Carabidologists Meeting, Blagoevgrad, August 20-24, 2007, pp. 241-254.

Sofia–Moscow

A history of ground-beetle faunas of West Siberia and the Urals during the Late Pleistocene to Holocene

Evgeniy Zinovyev

Institute of Plant and Animals Ecology, Urals Branch of the Russia Academy of Sciences, Ekaterinburg. E-mail: zin@etel.ru, zinoviev@mail66.ru

SUMMARY

Changes in ground-beetle faunas of the Urals and West Siberia during the past 130,000 years were elucidated based on fossil material obtained from more than 100 sites scattered over this region. During the warm phase of the Eremian Interglacial, the areas lying between 61 and 62° N latitude supported boreal insect faunas. Insect faunas from deposits dated to the Early Weichselian interval (100,000-55,000 yr BP), in the central part of the study region were different from present-day communities, being defined as "mixed", or tundra-steppe. A similar situation is described for the end of the Mid-Weichselian interstadial interval (from >33,000 to 23,000 yr BP). In the faunas dated from the Last Glacial Maximum (LGM) (Sartan Glaciation), only arctic and subarctic species were found. Sub-boreal and boreal species were missing from these faunas. The end of the LGM was characterised by climatic instability, causing discrepancies between palaeoentomological and palaeobotanical evidence. The subsequent processes of afforestation and water-logging which occurred after 8,000 yr BP in West Siberia and adjacent lands caused reductions in the ranges of arctic and arcto-boreal beetle species and the regional disappearance of some sub-boreal forms (*Poecilus ravus, Cymindis mannerheimi*) which have since shifted their distributions eastwards to the East Siberian steppes or the subalpine belt of the Pamirs and western Altai Mountains.

Keywords: Late Pleistocene, Holocene, sub-fossil insects, fauna change, West Siberia, Urals

INTRODUCTION

The study of the history of insect faunas is a highly important component of entomological research. One of the sources of information is palaeontological material coming from deposits of the recent geological past, i.e. the Pleistocene and Holocene.

Ground-beetles (family Carabidae) are one of the most important groups for such studies. Sub-fossil remains of these beetles are common in many Quaternary deposits, being mainly represented by isolated exoskeletal sclerites, such as elytra, pronota, and head capsules. Modern Carabidae exhibit profound diversity at the genus and species level, and many species have clear-cut habitat preferences. The same species are found, well-preserved, in Pleistocene fossil assemblages. Therefore these sub-fossils are largely considered as reliable environmental indicators for use in both palaeoenvironmental and palaeofaunistic reconstructions. Such studies have been performed in North America, Western Europe, Russia and some other parts of the globe (Coope, 1970, 1986; Kiselev, 1974, 1987; Schwert & Ashworth, 1988; Morgan et al., 1986; Buckland & Coope, 1991; Bos et al., 2004; Whitehouse, 2006).

The objective of the present study is the delineation of the history of the ground-beetle faunas of the Urals and West Siberia, Russia during the Late Pleistocene and Holocene (the last 130,000 yr BP). This phase of the Quaternary is highly important for our understanding of the processes of the development of modern insect faunas. Deposits dated from this interval are widely distributed over this region, most of them being terrestrial in origin. Late Pleistocene deposits have mainly been found in the valleys of West Siberia's larger rivers (Ob and Irtysh) and their tributaries (Agan, Tavda, Loz'va, Kul'egan, etc.). These organic-rich deposits often contain both plant macrofossils and insect remains. Radiocarbon analyses allows for a fairly accurate dating of samples from the last 40,000 yr BP.

The Late Pleistocene is subdivided into the Eemian (Kazantzevo) Interglacial (MIS 5), the Weichselian (Valday in European Russia, Zyryanka in West Siberia) Glaciation (MIS 4-2) and the Holocene (MIS 1). In turn, the Weichselian interval is subdivided into two cold periods (MIS 4 and MIS 2) (Table 1) separated by a middle phase with an unstable and cooler-than-present climate with various similar sharp cooling events (MIS 3) (Karginian Mega Interstadial in Siberia or Weichselian Middle Pleniglacial in Europe).

MATERIAL AND METHODS

The work is based on sub-fossil material obtained from more than 100 sites scattered over the large territories of the Urals, West Siberia and the adjacent Pechora River Valley (Fig. 1). Most of these localities are dated to the Late Pleistocene or Holocene. Sub-fossil insect remains were found in deposits exposed in quarries, along river banks

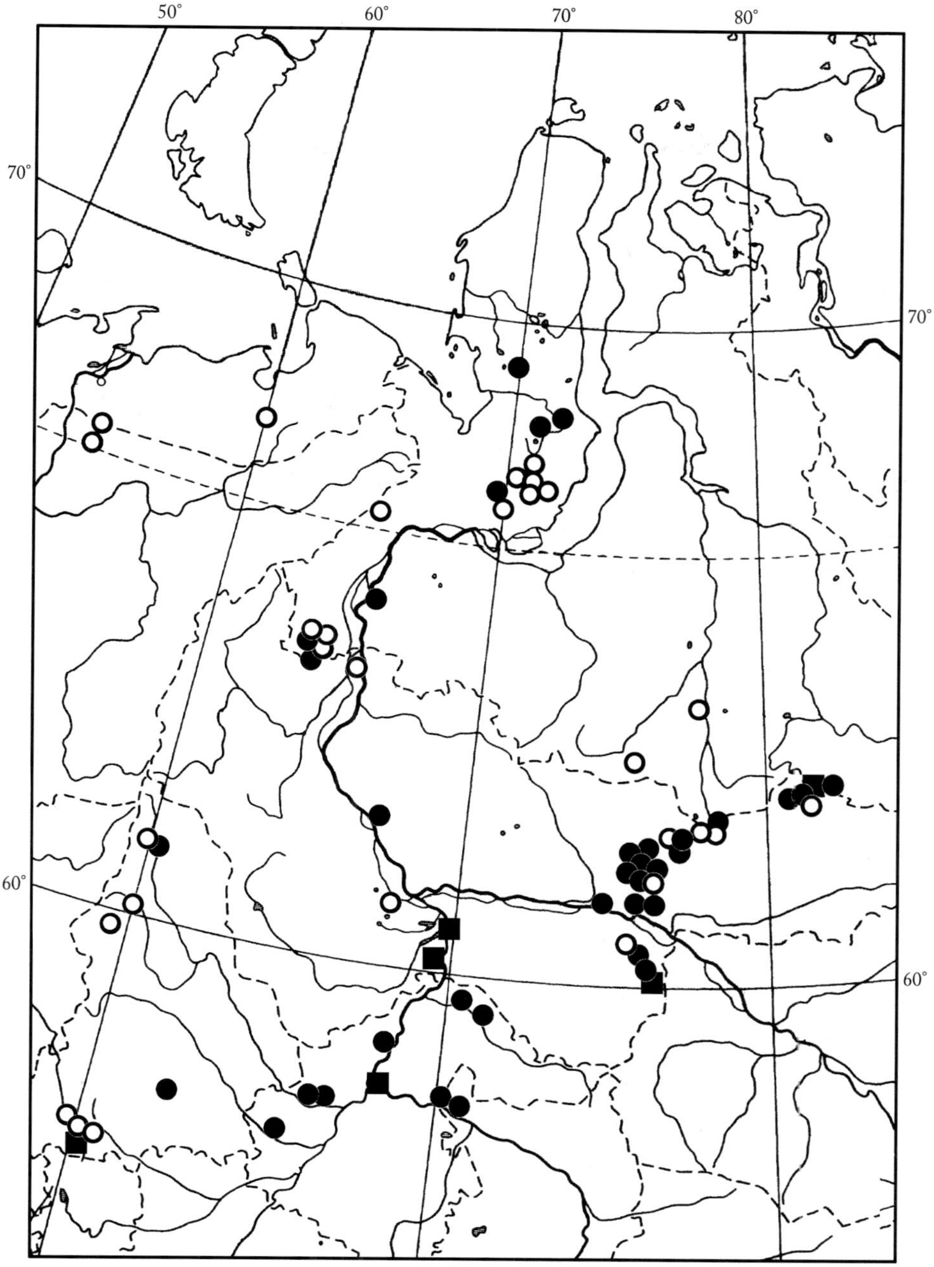

The localities of sub-fossil insects, dated by:
■ – Early and Middle Neopleistocene;
● – Late Neopleistocene;
O – Holocene

Fig. 1. Geographical location of the study sites in the Urals, Western Siberia and adjacent regions.

and in peat bogs. Field sampling was made using the standard techniques described by Kiselev (1987). Geological descriptions of the sites and their provisional dating were provided by geologists; some samples were radiocarbon-dated. The Holocene deposits associated with sub-fossil wood remains were dated dendrochronologically. One site (Karymkary) was dated by thermoluminescence. Both the laboratory treatment and subsequent determination of fossil specimens were performed at the Institute of Plant and Animal Ecology in Ekaterinburg. The classification of the sub-fossil insect faunas used is that proposed by the author (Zinovyev, 2006).

Table 1. Chronology of the Late Neopleistocene (after Van Andel & Tsedakis, 1996; Velichko et al., 2005; Volkova et al., 2005).

<table>
<tr><th rowspan="2"></th><th rowspan="2"></th><th rowspan="2">Time interval, yr BP</th><th rowspan="2">Marine Isotope Stage</th><th colspan="4">Period</th><th rowspan="2">Climate</th></tr>
<tr><th colspan="2">Europe</th><th>European Russia</th><th>West Siberia</th></tr>
<tr><td rowspan="6">NEOPLEISTOCENE</td><td rowspan="6">UPPER (LATE)</td><td></td><td>Stage 1</td><td colspan="4">HOLOCENE</td><td>Warm</td></tr>
<tr><td>10,000
22,000-
23,000</td><td>Stage 2</td><td rowspan="3">Weichselian (Valday, Zyryanka) Glaciation</td><td>Last Glacial Maximum</td><td>Late Valday</td><td>Sartan (Late Zyryanka)</td><td>Very cold</td></tr>
<tr><td>50,000-
55,000</td><td>Stage 3</td><td>Middle Weichselian Pleniglacial</td><td>Middle Valday</td><td>Karginian</td><td>Generally instable and cooler-than-present with various other sharp cooling events</td></tr>
<tr><td>115,000</td><td>Stage 4</td><td>Lower Weichselian Pleniglacial</td><td>Early Valday</td><td>Ermakovo (Early Zyryanka)</td><td>Cold</td></tr>
<tr><td></td><td>Stage 5a
Stage 5b
Stage 5c
Stage 5d</td><td rowspan="2">Eemian Intergl.</td><td>Eemian cold phase</td><td>Tchermenino</td><td rowspan="2">Kasantzevo</td><td>Cold</td></tr>
<tr><td>130,000</td><td>Stage 5e</td><td>Eemian Interglacial</td><td>Mikulino</td><td>Warm</td></tr>
</table>

RESULTS

The fossil insect data reported here represent the most important periods of the Late Pleistocene and Holocene, although their chronological distribution is not uniform. Thus, palaeoentomological material from West Siberia dating from MIS 5e (Table 1) is very poor. We have only two sites associated with that time. One of these (Karymkary, 62°03'N, 67°22'E) has a thermolumenescence age of 130,000±31,000 yr BP (Arkhipov & Volkova, 1994). Insect remains from that fossil peat bog were scarce; some were determined as *Trechus secalis*, a species not found in the study region during other parts of the Pleistocene. Neither arctic nor arcto-boreal carabids (*Pterostichus (Cryobius)* spp.) were recovered from this site, but they were found in horizons immediately below the Eemian peat, dated to the end of the Dnepr Glaciation (MIS 6) (Arkhipov & Volkova, 1994).

The other site dated to the Eemian Interglacial, Loz'va-2 (61°04'N, 60°33'E), is located in the upper reaches of Loz'va River, in the northern Urals. A similar fossil insect fauna, referred to the boreal type, was found there. This fauna included such thermophilous carabid species as *Trechus secalis* and *Oxypselaphus obscurum*.

A remarkable trait of these faunas is the absence of the arctic and arcto-boreal species that were abundant in faunas dating to the cold phases of the Late Pleistocene, such as *Pterostichus (Cryobius)* spp., *P. costatus,* and *Curtonotus alpinus*. Despite the scant information obtained from the Eemian insect faunas we can suggest that the carabid faunas of the MIS 5e were similar to modern insect communities inhabiting these territories. These data agree well with the conclusions concerning the composition of natural communities at that time. Thus, according to the palaeontological evidence (Arkhipov & Volkova, 1994; Van Andel & Tsedakis, 1996; Adams et al., 1999), the initial phase MIS 5 had a warmer-than-present climate. At that time both thermophilous vegetation and animals (including ground-beetles) shifted northwards at least in Europe (Coope, 1970; Nazarov, 1989; Van Andel & Tsedakis, 1996) and West Siberia (Arkhipov & Volkova, 1994); at the same time a "boreal" transgression occurred (Arkhipov & Volkova, 1994).

Some of the fossil assemblages date to a cold phase of MIS 5 (110,000 – 105,000 yr BP, MIS 5d). All the faunas of this time, situated in the lower reaches of Ob River, belong to the arctic type, characterized by the dominance of such cryophilous carabids as *Curtonotus alpinus*, and *Pterostichus (Cryobius)* spp.

Information concerning the ground-beetles that inhabited the study area in the Early Valdai Glaciation (MIS 5a and 4, 105,000 – 50,000 yr BP) is likewise incomplete. We have only three faunal assemblages from this interval, found at sites situated in the Agan (the Agan-3093 site), Tavda (the Andryushino site) and Tura River (the Mal'kovo site) valleys.

According to the regional literature, this time was characterised by dry and cold climates. It was at this time that the Scandinavian ice sheet developed, but it did not yet reach the Norwegian coast (Van Andel & Tzedakis 1996). Cold climatic conditions and open-ground tundra landscapes prevailed both in Europe and West Siberia (Arkhipov & Volkova, 1994). Entomological data obtained from sites in Eastern Europe agree with this reconstruction. Thus, Nazarov (1989) described cryophilous insect faunas from Early Valday Glaciation sites in Byelorussia, where arctic and subarctic ground-beetles (*Pterostichus pinguedineus, P. tundrae, Bembidion dauricum, Curtonotus torridus*) were dominant.

Arctic insect faunal assemblages were described from the region of the middle reaches of Ob River. Thus, at the Agan-3093 site, situated in the Agan River Valley (61°44'N, 76°12'E), only arctic and subarctic beetle species were found. These faunas include the carabids *Diacheila polita, Pterostichus pinguedineus, P. costatus, P. (Cryobius)* spp. and *Curtonotus alpinus*, and the rove beetle *Tachinus arcticus*. This fauna is very similar to the modern insect communities found on the tundra of the Yamal Peninsula.

The most interesting faunas from this cold interval are those from the Mal'kovo (57°09'N, 66°01'E) and Andryushino (57°41'N, 66°08'E) localities situated in the Tura and Tavda river valleys, respectively (about 70,000-60,000 yr BP). The insect fauna of the Mal'kovo site, lying near the city of Tyumen', was described by Kiselev (1974), that of the Andryushino site was described by Zinovyev et al. (2007). The insect assemblages from these sites share some features in common. Firstly, the faunas of these sites are characterised by a mixture of arctic and arcto-boreal species (*Pterostichus (Cryobius)* spp., *Diacheila polita, Curtonotus alpinus* etc), sub-boreal steppe species (*Poecilus ravus* (=*Pterostichus motschulskyi*), *P. major, P. hanhaicus*) and sub-alpine species of carabids (*Cymindis mannerheimi*). In addition, numerous remains of the weevil *Otiorhynchus politus* and some halophilous beetles (the darkling beetle *Belopus procerus* at Mal'kovo and *Pogonus* spp. at Andryushino) were found there. One of the specific traits of these faunas is the abundance of weevil remains that are morphologically similar to *Otiorhynchus politus*. This species is currently not found in the East-Siberian tundra-steppe, but inhabits both the sub-alpine belt of the South Urals, East Kazakhstan and the plain territories of European Russia and West Siberia (Korotyaev, 1980). A second common feature is the lack of boreal species such as *Trechus secalis* and *Oxypselaphus obscurum,* found in the Eemian interglacial faunas.

Similar insect faunas were described for the end phase of the Middle Weichselian Interstadial interval (MIS 3; 55,000-22,000 yr BP, Table 1). This interval was characterised by an unstable, cooler-than-present climate that fluctuated greatly on time spans of a few thousand years (Arkhipov & Volkova, 1994; Van Andel & Tsedakis, 1996; Adams et al., 1999; Bos et al., 2004). West Siberian sites with insect faunas from this interval date from 33,000 to 23,000 ^{14}C yr BP. These sites are spread over the vast territories of the study area between 67° N and 57° N.

At the sites lying north of 61°N latitude, arctic and arcto-boreal species (including *Pterostichus costatus P. sublaevis*, the subgenus *Cryobius* of the genus *Pterostichus*) dominated during this interval, whereas sub-boreal steppe species were either absent or rare. Thus, only fragments of a single *Poecilus ravus* specimen were found in the insect faunas of Aganskyi Uval low-hills (Aganskyi uval-1290/2 site, 61°22'N, 76°45'E). In the lower reaches of the Ob River, no sub-boreal carabids were found in faunas from this interval, except *Carabus sibiricus* and two species of leaf beetles (*Chrysolina perforata* and *C. aeruginosa*).

Between 61° N and 58° N, the fossil beetle faunas from this interval are of the subarctic type, including a few sub-boreal steppe species (*Poecilus ravus*, and the carrion-beetle *Blithophaga sericea*). These thermophilous insects were found at the Kul'egan-2247 site (60°25'N, 75°50'E), dating to the final phase of the interstadial.

The MIS 3 interstadial faunas from sites situated south of 59°N latitude contain carabid species assemblages similar to those of the Early Valday (MIS 4) faunas from the Mal'kovo and Andryushino sites. These faunas are of a "mixed" type characterised by combinations of species not presently found together. The main feature of this faunal type is combination of arctic, boreal, arcto-boreal and sub-boreal insects. These faunal assemblages contained ground-beetles of the *Pterostichus (Cryobius)* group, *Curtonotus alpinus*, and the sub-boreal species *Poecilus (Derus) ravus*, *P. (Derus) major*, *P. (Derus) hanhaicus*. These faunas could be classified as indicative of tundra-steppe, but their species composition differs from known relict tundra-steppe communities that live today in Eastern Siberia because they contain the weevil *Otiorhynchus politus*, some halophilous species of the genus *Pogonus* and the sub-alpine carabid *Cymindis mannerheimi*. Another characteristic species in these faunas is the pill-beetle *Morychus viridis*. This beetle is abundant in both the modern relict steppe and Pleistocene insect faunas of Eastern Siberia (Kuzmina & Korotyaev, 1986; Berman, 1990).

In the Urals and West Siberia, insect faunas from the LGM interval (22,000 to just before 13,000 ^{14}C years ago) are likewise poorly known. We have a few sites from the middle reaches of Ob River, i.e. Agan-1082/1 (62°04'N, 77°34'E) and Kul'egan-2247 (60°25'N, 75°50'E). According to palaeobotanical evidence, the LGM was characterised by very severe climates: large ice sheets were present over much of northern Europe; forests and woodlands were almost non-existent, except for isolated pockets of woody vegetation in and around the mountain ranges of Southern Europe (Van Andel & Tsedakis, 1996; Adams et al., 1999). Instead, a sparse grassland or semi-desert covered most of Southern Europe, whilst a mixture of the dry, open 'steppe tundra' and polar desert covered the parts of northern Europe and West Siberia not covered by ice sheets (Arkhipov & Volkova, 1994; Adams et al., 1997).

The LGM insect faunas of this study region are all of the arctic type, dominated by arctic and subarctic beetles (*Pterostichus costatus*, *Pterostichus* cf. *pinguedineus*, *Curtonotus alpinus*, *Tachinus* cf. *arcticus*). There were no occurrences of sub-boreal insects even at Kul'egan-2247, in the stratum dated 21,815±225 yr BP. However, remains of *Poecilus ravus* were found in adjacent layers.

At a site situated in the upper reaches of Agan River (the Agan-1082/2 site, 62°04'N, 77°34'E), a subarctic insect assemblage dating to ca. 15,000 yr BP has been described. This fauna includes some boreal beetles, including the ground-beetle *Chlaenius costulatus* and the weevil *Hylobius albosparsus*. Both of these species are absent from LGM faunal assemblages. Based on stratigraphic position, a faunal assemblage from the Nadtzy locality (low reaches of the Irtysh River near the town of Tobol'sk, 58°37'N, 68°35'E) may date to the same period. This fauna included arctic (the ground-beetle *Curtonotus alpinus*, the leaf-beetle *Chrysolina subsulcata*), subarctic (*Pterostichus (Cryobius)* spp.) and boreal insects (*Chlaenius costulatus*).

The final phase of the Last glaciation and the beginning of the Holocene (from 13,000 to ca. 9,000 ^{14}C years ago) is one of the turning points in the history of the biota as a whole and of insect faunas in particular. This time is known to have experienced rapid and frequent climatic changes. Thus, a rapid warming and moistening of the climate occurred in Europe shortly before 13,000 ^{14}C yr BP (Atkinson et al., 1987; Adams et al., 1999). Between 13,000 and 12,000 ^{14}C yr BP in Europe there was a change in herbaceous communities from dry and cold-climate steppe-tundra to steppe, with a slower response from tree species. The cold and dry Younger Dryas period (about 10,800-10,000 ^{14}C yr BP.) caused a temporary disappearance of the woodland cover that had previously extended over much of Europe (both north and south), and its replacement by dry steppe and steppe-tundra (Velichko, 1993; Laval et al. 1991; Starkel, 1991). After 10,000 ^{14}C yr BP, the pollen records show a gradual warming began with the thermal maximum during 8,000-5,000 ^{14}C yr BP (Atlantic warm phase of the Holocene). Fossil beetle assemblages from Europe indicate that this amelioration was extremely rapid, with temperatures reaching modern levels within a century after the end of the Younger Dryas oscillation (Atkinson et al., 1987; Galliard & Lemdahl, 1994; Coope & Lemdahl, 1995; Ponel et al., 2001).

We have fossil data from seven sites situated in the central and northern parts of the study area and radiocarbon dated from 12,000 to 9,000 yr BP (Table 2). The rapid climatic changes that occurred at the Pleistocene–Holocene boundary are demonstrated in this study region by two phenomena:

1. Essential differences in the structure of insect assemblages of synchronous (or presumably synchronous) sites cannot be fully explained by their latitudinal position. For example, the sub-fossil faunas of the subarctic type were described at 62° N latitude while insect complexes of the more thermophilous boreal type was found north of 64°N (Table 2). In particular, this concerns a period of 10,000–9,000 ^{14}C yr BP (Table 2).
2. Differences between the sub-fossil insect and palaeobotanical records are noteworthy. Thus, in the samples from the Ngoyun site (Middle Yamal), dated to 11,226±172 and 10,688±240 yr BP, arctic insect faunas (including such species as *Pterostichus sublaevis*, *P. vermiculosus*, *P. costatus*, *Amara glacialis*) were found in sediments containing the remains of woody plants (Table 2). It may be explained by the taphonomic factors. Probably, the woody remains were transported by water to the sample site from the southern regions of Yamal penninsula, in the same way that wood from the taiga certainly floats down river to the Arctic in large quantities today. A different situation was described for Kul'egan-2241 (10,700±325 yr BP), where the remains of boreal beetles were found, but no remains of trees were found (Zinovyev, 2005). Elias (1982) described a similar situation from Middle Holocene fossil assemblages from Ennadai Lake, Northwest Territories, Canada, in which boreal insects were found, even though no macrofossils or pollen remains of tree species were found. In this case, Elias (1982) interpreted the environment as being sufficiently cold to stop the local conifers from pollinating on a regular basis. However, the presence of coniferous bark beetles demonstrated that the trees must have been very close by.

Table 2. Chronological position of the study sites dated by the end of the Late Pleistocene and the beginning of the Holocene.

Period	Sites where sub-fossil insect faunas were found	(^{14}C data)	Coordinates		Faunal type
			N	E	
12,000-11,000 yr BP (Allerød)	Ngoyun	11,226±172 (IPAE-176)	68°32'	72°06'	Arctic (with the presence of woody vegetation reconstructed on the basis of palaeobotanical data)
	Agan-4068/2	11,400±350 (IPAE-98)	62°06'	77°55'	Subarctic
11,000-10,000 (Younger Dryas)	Ngoyun	10,688±240 (IPAE-175)	68°32'	72°06'	Arctic (with the presence of woody vegetation reconstructed on the basis of palaeobotanical data)
	Kul'egan-2241	10,700±325 (IPAE-94)	60°30'	75°45'	Boreal (no remains of trees found)
10,000 - 8,000 (Preboreal period)	Nyulsaveyto	(ca 9,000 yr BP)	67°32'	70°10'	Subarctic
	Vansevat	(ca 9,000 yr BP)	64°10'	66°03'	Boreal
	Agan-4068/2	9,770±300 (IPAE-97)	62°06'	77°55'	Subarctic
	Lugovskoye	9,685±95 (SOAN-4941)	60°57'	68°32'	Subarctic with a single subboreal species (*Poecilus ravus*)
8,000 – 7,000 (the beginning of the Atlantic period)	Vansevat	8350±300 (B-7064)	64°10'	66°03'	Intrazonal
	Nyulsaveyto	8,179±231 (IPAE-72) 8,182±227 (IPAE-79)	67°32'	70°10'	Intrazonal with single boreal species

The ground-beetle faunal assemblages dated 9685±95 yr BP from the Lugovskoye site, near the town of Khanty-Mansiysk (Table 2) is particularly interesting. These insect remains were found in deposits containing numerous, well-preserved bones of mammoth and human hunters (Zenin et al., 2003; Leshchinskiy, 2006). This sub-arctic ground-beetle fauna includes *Poecilus ravus,* found today in the Dahuro-Mongolian region. Its presence in this fossil assemblage constitutes its most recent occurrence in Western Siberia. However, a well-preserved pronotum identified to *Poecilus* (*Derus*) group was found at Loz'va-1 (61°05′N, 60°33′E), a site located in the upper reaches of Loz'va

River, northern Urals. This specimen is associated with a radiocarbon date of 5770±60 yr BP (Zinovyev & Fadeyev, 2002).

The Atlantic warm phase of the Holocene (8,000-5,000 ^{14}C yr BP) can be characterised by warmer-than-present climates that allowed forest to spread further north. There are various sources of evidence for warmer summer and winter temperatures across northern Europe during the mid-Holocene (Vork & Thomsen, 1996). Insects from the middle Holocene (from 8,000 yr BP to present) were found in different parts of the Urals, West Siberia and the Pechora River Valley. Most of these faunal assemblages are similar to the modern communities from the central and southern parts of this territory. But in the northern part of the study area (Yamal Peninsula), the Middle Holocene faunas reflect warmer-than-present climatic conditions (such as conifer taiga forests). These insect faunas were associated with the remains of sub-fossil wood in the Portsayakha and Yada-Yakhodyakha river valleys (southern part of the Yamal Peninsula) and include such boreal components as bark-beetles (Ipidae), the weevil *Hylobius albosparsus* and the carabids *Amara brunnea* and *Pterostichus adstrictus* (Zinovyev et al., 2001).

Late Holocene (< 5,000 yr BP) insect assemblages are similar in composition to the modern insect communities from the study area. At the same time, some sub-boreal species which inhabited these territories during the Late Pleisticene (*Poecilus ravus*, *P. hanhaicus* and *Pseudotaphoxenus dauricus*) have subsequently shifted their distributions eastwards to the Eastern Siberian relict steppes and the subalpine belt of the Pamirs and western Altai Mountains (*Cymindis mannerheimi*) (Fig. 2). However, individual sub-boreal

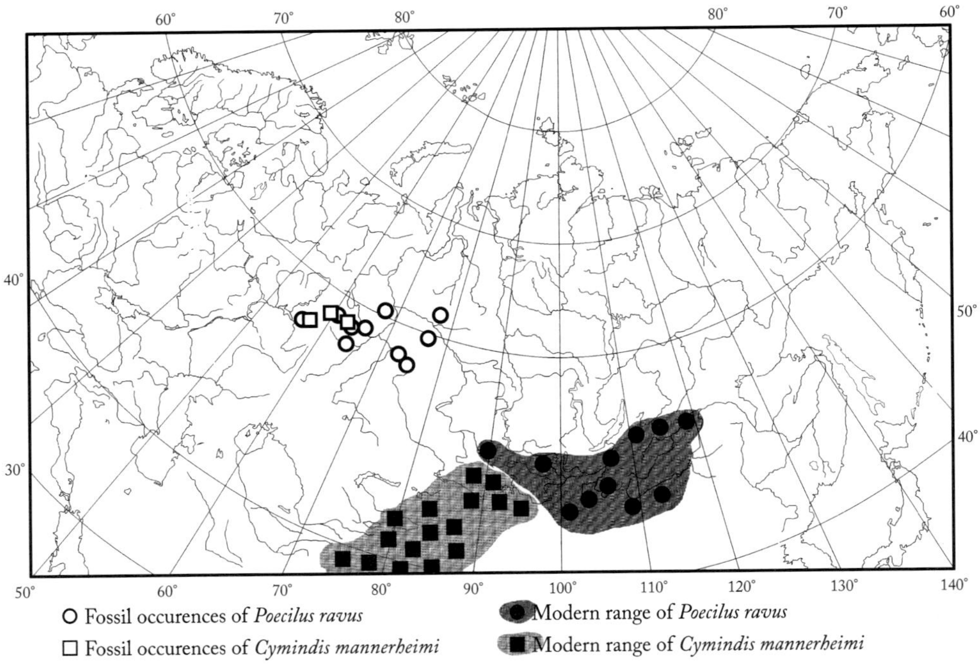

Fig. 2. Fossil occurrences of some sub-boreal species in comparison with their modern ranges

steppe species could have survived *in situ* in refuges in the Urals Mountains and in the lowland tundra of the Yamal Peninsula (*Carabus sibiricus*) or other regions of Western Siberia (*Polystichus connexus*) (Zinovyev, 2006).

DISCUSSION

The climatic fluctuations which occurred in the Late Pleistocene and Holocene brought marked changes in the insect faunas of the Urals and Western Siberia. Thus, during the warm phase of the Eemian Interglacial (MIS 5e), the territories lying between 61 and 62° N latitude supported boreal insect faunas, including such species as *Trechus secalis* and *Oxypselaphus obscurum*. These faunas are similar to the insect communities inhabiting this territory today. This agrees with palaeobotanical evidence which shows a wide distribution of woody vegetation all over the Urals and Western Siberia. The subsequent cold periods of MIS 5 can be characterised by the presence of arctic faunas in the central part of the study area and of subarctic elements in the central part of Western Siberia (Demyanka River valley).

As mentioned above, the Weichselian (Valday, Zyryanka) Glaciation is subdivided into two cold phases (MIS 4 and MIS 2, the LGM) and a long interstadial interval (MIS 3). However, similar beetle faunas have been found from both the MIS 4 and MIS 3 intervals, suggesting that some climatic stability seems to have existed over the study area during at least the early and middle parts of the Weichselian Glaciation.

Thus, in the deposits dated to MIS 4 (100,000-55,000 yr BP) we observe the presence, in the central part of the study area (upper reaches of Tavda and Tura rivers), of insect faunas which differ from any modern insect communities and can be defined as "mixed", or tundra-steppe faunas. These faunas comprise arctic, subarctic, sub-boreal steppe (including halophilous) species with the dominance of the weevil *Otiorhynchus politus* and some morphologically similar congeners. A similar situation is described for late MIS 3 faunas (>33,000 to 23,000 yr BP), in which the ranges of sub-boreal steppe insects (the ground-beetle *Poecilus ravus*, the carrion-beetle *Blitophaga sericea*) extended north to 60-61° N. Further north these insects were either absent or very rare (for example, *Carabus sibiricus*, which inhabits the south tundra belt of the Yamal Peninsula).

LGM beetle faunas contain only arctic and subarctic species (*Curtonotus alpinus, Pterostichus costatus, P. (Cryobius)* cf. *pinguedineus* etc.). This can be explained by the severe climatic conditions of that interval. No sub-boreal (*Poecilus ravus, P. hanhaicus* etc.) or boreal insects were found in the LGM deposits.

The period of the end of the Last glaciation can be characterised by climatic instability which affected the development of insect faunas. Individual boreal species, including xylophagous beetles, were found in deposits dated as early as 15,000 yr BP. A mixture of arctic and subarctic insects were found in faunas dating to the Pleistocene-Holocene transition (12,000-10,000 yr BP), but in the central part of the study area boreal insects were found in fossil assemblages containing no evidence of woody vegetation.

The subsequent processes of afforestation and water-logging which occurred after 8,000 yr BP in Western Siberia and adjacent lands caused reductions in the ranges of arctic and arcto-boreal species and the disappearance of some sub-boreal species. This latter group has since shifted eastwards to the relict steppes of Eastern Siberia or the subalpine belt of the Pamirs and western Altais. During the same period, there occurred expansions of polyzonal and boreal species over the territories of the Urals and Western Siberia (*Calathus micropterus*, *Pterostichus oblongopunctatus* etc.). At the same time, some 'boreal' carabid species (*Pterostichus mannerheimi*, *P. adstrictus*, *Trechus rivularis*) inhabited this territory during the cold phases of the Late Pleistocene and joined the modern insect complexes in the Holocene.

The insect faunas described here closely match the modern faunas found in biological communities in the Urals and West-Siberian Plain (Arkhipov & Volkova, 1994; Volkova et al., 2005 etc.). Faunas associated with warm periods of the Late Pleistocene and Holocene contained boreal and intrazonal insect species, whereas cryophilous (arctic, subarctic and "mixed", or tundra-steppe) insects were identified from the deposits dating to cold periods. During the Weichselian Glaciation in the Urals and Western Siberia, "exotic" beetles inhabited steppe and mountain steppe communities, whereas their modern distributions give no indications of their past distributions. These data also correspond to changes in insect complexes of Europe, where alternating warm- and cold-adapted entomofaunas likewise responded to climatic fluctuations (Nazarov, 1984, 1989; Coope, 1986, Buckland & Coope, 1991). Thus similar trends have developed in the insect faunas of Europe and Western Siberia, including the history of ground-beetle faunas.

ACKNOWLEDGEMENTS

We thank Dr A. Borodin, Dr V. Stefanovsky, Mr N. Erokhin (Institute of Plant and Animal Ecology Ekaterinburg) and Mr A. Yaskov (Khanty-Mansiysk) for their help in sampling the fossil material. Special thanks go to Dr B. Korotyaev, Dr I. Kabak (Zoological Institute of the Russian Academy of Sciences, St. Petersburg) for the assistance in determining the fossil remains, and to Dr P. Kosintsev and Dr R. Khantemirov (Institute of Plant and Animal Ecology Ekaterinburg) for the help in dating this material. Dr Sergei I. Golovatch (Moscow) kindly checked the English of an advanced draft. This work was supported by the Russian Foundation for Basic Research (project 06-04-49118) and the administration of the "Samarovskiy Chugas" National Park, Khanty-Mansiysk.

REFERENCES

Adams, J.M. & Faure, H. (1997). Paleovegetation maps of the world since the Last Glacial; an aid to archaeological understanding. – Journal of Archaeological Science 24: 623-647.

Adams, J., Maslin, M. & Thomas, E. (1999). Sudden climate transitions during the Quaternary. – Progress in Physical Geography 23 (1): 1-36.

Arkhipov, S.A. & Volkova, V.S. (1994). Geological History, Landscapes and Climates of the Pleistocene in West Siberia. Novosibirsk, Siberian Branch of the Russian Academy of Sciences. (In Russian).

Atkinson, T.C., Briffa, K.R. & Coope, G.R. (1987). Seasonal temperatures in Britain during the past 22,000 years, reconstructed using beetle remains. – Nature 325: 587-592.

Berman, D.I. (1990). Ecology of *Morychus viridis* (Coleoptera, Byrrhidae), a moss beetle from Pleistocene deposits in the northeastern USSR. – In: V. M. Kotlyakov & V. E. Sokolov (eds) Arctic Research: Advances and Prospects. Proceedings of the Conference of Arctic and Nordic Countries on Coordination of Research in the Arctic, 281-288. Nauka, Moscow. (In Russian).

Bos, J.A.A., Dickson, J.H., Coope, J.R. & Jardine, W.G. (2004). Flora, fauna and climate during the Weichselian Middle Pleninglacial – palynological, macrofossil and coleopteran investigations. – Palaeogeography, Palaeoclimatology, Palaeoecology 204: 65-100.

Buckland, P.C. & Coope, G.R. (1991). A Bibliography and Literature Review of Quaternary Entomology. J. Collis Publications, University of Sheffield.

Coope, G.R. & Lemdahl, G. (1995). Regional differences in the Lateglacial climate of Northern Europe based on coleopteran analysis. – Journal of Quaternary Science 10: 391-395.

Coope, G.R. (1970). Interpretations of Quaternary insect fossils. – Annual Review of Entomology 5: 97-120.

Coope, G.R. (1986). The invasion of Northern Europe during the Pleistocene by Mediterranean species of Coleoptera. – In: Biological Invasions in Europe and the Mediterranean Basin: Workshop Montpellier, 21-23 May, 1986. Dordrecht, 1990, p. 203-215.

Elias, S.A. (1982). Holocene insect fossils from two sites at Ennadai Lake, Keewatin, Northwest Territories, Canada. – Quaternary Research 17: 371-390.

Gaillard, M.J. & Lemdahl, G. (1994). Lateglacial insect assemblages from Grand-Marias, south-western Switzerland – climatic implications and comparison with pollen and plant macrofossil data. – Dissertationes Botanicae 234: 287-308.

Kiselev, S.V. (1974). Late Pleistocene Coleoptera of Transuralia. – Paleontological Journal 7: 507-510. (Translated from Russian, Paleontologicheski zhurnal, 4: 70-73 (1973)).

Kiselev, S.V. (1987). Field sampling for entomological analysis. – In: Complex Biostratigraphic Investidations: Manual. Moscow Univ. Press: 21-26. (In Russian).

Korotyaev, B.A. (1980). Materials on the weevil fauna (Coleoptera, Curculionidae) of the North-East of the USSR. – In: Entomological Investigations of the North-East of the USSR. Vladivostok, Far East Science Centre of the USSR Academy of Sciences: 23-50. (In Russian).

Kuzmina, S. & Korotyaev, B. (1987). New species of the pill beetle genus *Morychus* Er. (Coleoptera, Byrrhidae) from the Northwest of the USSR. – Entomological Review 66: 342-344. (In Russian).

Laval, H., Medus, J. & Roux, M. (1991). Palynological and sedimentological records of Holocene human impact from the Etang de Berre, southeastern France. – The Holocene 1: 269-272.

Leshchinskiy, S.V. (2006). Paleoecological studies, taphonomy and genesis of the Lugovskoye site. – Archaeology, Ethnography and Anthropology of Eurasia 25 (1): 33-40. (In Russian).

Morgan, A.V., Morgan, A., Nelson, R.E. & Pilny, J.J. (1986) Current status of knowledge on the past and present distribution of the genus *Blethisa* (Coleoptera: Carabidae) in North America. – Coleopterists' Bulletin 40: 105-115.

Nazarov, V.I. (1989). The climate of certain stages of the Byelorussian Pleistocene based on palaeoentomological data. Palaeoclimate and Glaciations in the Pleistocene. – In:

Proceedings of the All-Union Meeting "Climate and Glaciations in the Anthropogene", Nauka, Moscow: 70-75. (In Russian).

Ponel, Ph., Parchoux, F., Andrieu-Ponel, I. & de Beaulieu, J.-L. (2001). A Late-Glacial–Holocene insect succession from Vallée des Merveilles, French Alps, and its paleoecological implications. – Arctic, Antarctic and Alpine Research 33 (4): 481-484.

Schwert, D.P. & Ashworth, A.C. (1988). Late Quaternary history of the northern beetle fauna of North America: a synthesis of fossil and distributional evidence. – Memoirs of the Entomological Society of Canada 144: 93-107.

Starkel, L. (1991). Environmental changes at the Younger Dryas – Preboreal Transition and during the early Holocene: some distinctive aspects in Central Europe. – The Holocene 1: 234-242.

Van Andel, T.H. & Tsedakis, P.C. (1996). Palaeolithic landscapes of Europe and environs: 150,000-25,000 years ago: an overview. – Quaternary Science Reviews 15: 481-500.

Velichko, A.A. (1993). Evolution of Landscapes and Climates of Northern Eurasia. – Late Pleistocene–Holocene Elements of Prognosis. Vol.2. Moscow, Nauka Publ. (In Russian).

Velichko, A.A., Pisareva, V.V., Morozova, T.D., Faustova, M.A., Nechaev, V.P. & Gribchenko, Yu.N. (2005). Correlation of natural events of the Glacial and Periglacial Pleistocene in Eastern Europe: approaches to a solution. – In: Quarter-2005. Proceedings of the 4th All-Russian Meeting on the Study of the Quaternary. Syktyvkar: 64-66. (In Russian).

Volkova, V.S., Khazina, I.V. & Babushkin, A.E. (2005). Stratigraphy of the Pleistocene in West Siberia and a paleoclimatic scale. – In: Quarter-2005. Proceedings of 4th All-Russian Meeting on the Study of the Quaternary. Syktyvkar: 77-78. (In Russian).

Vork, K.A. & Thomsen, E. (1996). Lusitanean/Mediterranean ostracods in the Holocene of Denmark: implications for the interpretation of winter temperatures during the postglacial temperature maximum. – The Holocene 6: 423-432.

Whitehouse, N.J. (2006). The Holocene British and Irish ancient forest beetle fauna: implication for forest history, biodiversity and faunal colonization. – Quaternary Science Reviews 25: 1755-1789.

Zenin, V.N., Maschenko, E.N., Leshchinskiy, S.V., Pavlov, A.F., Grootes, P.M. & Nadeau, M.-J. (2003). The first direct evidence of mammoth hunting in Asia (Lugovskoye site, Western Siberia). – In: 3rd International Mammoth Conference: Program and Abstracts. – Yukon, 2003: 152-153 (Occasional Papers in Earth Sciences, No. 5).

Zinovyev, E.V. (2005). Early Holocene entomocomplexes from the middle reaches of the Ob River in West Siberia. – Euroasian Entomological Journal 4 (4): 283-292. (In Russian).

Zinovyev, E.V. (2006). Problems of ecological interpretation of Quaternary insect faunas from the central part of northern Eurasia. – Quaternary Science Reviews 25: 1821-1840.

Zinovyev E.V., Gilev, A.V. & Khantemirov, R.M. (2001). Changes in the entomofauna of the southern Yamal Peninsula in connection with shifts of the northern timberline in the Holocene. – Entomological Review 81 (9): 1146-1152. (Translated from: Entomologicheskoe Obozrenie 80: 843-851).

Zinovyev, E.V., Korona, O.M., Stefanovsky, V.V. (2007). Reconstruction of Late-Neopleistocene sediment deposits at the 'Andryushino' site on the basis of entomological and carpological data. – Urals Geological Journal 56 (2): 27-43. (In Russian).

Zinovyev, E.V. & Fadeyev, F.A. (2002). Reconstruction of Holocene sediment deposits at the Loz'va-1 site (North Urals) based on insect data. – In: Urals Fauna in the Pleistocene and Holocene Times, pp. 24-36. Scientific papers. Ekaterinburg University. (In Russian).

L. Penev, T. Erwin & T. Assmann (Eds) 2008
BACK TO THE ROOTS AND BACK TO THE FUTURE. TOWARDS A NEW SYNTHESIS AMONGST TAXONOMIC, ECOLOGICAL AND BIOGEOGRAPHICAL APPROACHES IN CARABIDOLOGY
Proceedings of the XIII European Carabidologists Meeting, Blagoevgrad, August 20-24, 2007, pp. 255-256.

Corrigenda to the book "The Genus Carabus in Europe. A Synthesis", edited by H. Turin L. Penev & A. Casale (Pensoft & EIS, 2003)

Achille Casale[1], Hans Turin[2] & Lyubomir Penev[3]

[1] Università di Sassari, Dipartimento di Zoologia e Genetica Evoluzionistica, Via Muroni 25, 07100 Sassari, Italy
[2] Esdoorndreef 29, 6871 LK, Renkum, The Netherlands. E-mail: h.turin@hccnet.nl
[3] Central Laboratory for General Ecology, Yuri Gagarin Street 2, 1113 Sofia, Bulgaria. E-mail: info@pensoft.net

The collective monograph *The Genus Carabus in Europe. A Synthesis* was published five years ago as a result of the many years efforts of a large team of researchers from the "true" European territory, from the Atlantic to the Urals. In this respect, the book was one of the first truly Pan-European research projects, initiated by Lyubomir Penev and Hans Turin as early as in 1987, soon after the changes in the former Soviet Union started. The book intended to summarize the huge knowledge accumulated on different aspects of taxonomy, faunistics, biogeography, biology, ecology and conservation of *Carabus* in Europe.

After the book was published, some colleagues and the editors themselves discovered a few errata and omissions that inevitably happened during the elaboration of the book. We apologise for these errors, for which we take the full responsibility as editors and/or authors.

The present *corrigendum* is addressed to all *Carabus* workers who continue to use the book. Naturally, it should not be accepted as a complete *addendum* to the monograph, a task hardly achievable due to growing number of papers on this fascinating genus published every year.

CORRIGENDA

p. 11, Fig. 1-1: owing to an error in re-drawing the original transect furnished by F. Weber, the altitude of the mountains is not precise.

p. 16, *clatratus* **c.** ssp. *auraniensis*: DELETE "SE-France" (this is referred to: **e.** ssp. *arelatensis*)
p. 132, 31 (B), "The larva of *C. planatus* Chaudoir is unknown", is wrong. Correct to: "The larva of *C. planatus* was described by Casale et al., 1982". For additional information, see Busato & Casale (2005).
p. 389, **T-25–28**: due to an error in re-drawing the original version provided by F. Weber, the figures do not contain information on the ecological conditions of the area.
p. 393, **T–35**: due to an error in re-drawing the original version provided by J. Pawlowski, three mistakes have to be corrected: 1, the signification of the highest alpine zone (Trifido-Supinetum), 1,650-1,725 m a.s.l., is lacking; 2, in the marginal text, "*C. fabricii*, in Poland only found here" is not correct, because this species is recorded also from the Polish side of the Tatra Mts (see p. 248); 3, final part of the text: "is probably endangered in refugia between 1,600-2,170 m", is referred to Tatra Mts, not to Babia Góra Mt.
p. 410, **T-68**, Greece, "Lanina lake": correct to "Ianina lake" (also Ioanina).
p. 425, **T-100**, Iberian Peninsula: owing to an error in re-drawing the original version provided by J. Serrano, "*montivagus*" must be DELETED.
p. 457, left column, "10. Division Ortocarabigenici": correct to "Oreocarabigenici".
p. 461, left column, Obydov, Note: "*Callisthenes* in the widest sense": correct to "*Calosoma* in the widest sense".
General note: Pawlowski (pers. comm.) rightly noted that in the Species Accounts chapter, the placement of the eastern Carpathian Mts within the Russian Plain is not correct. Indeed, it is true, however we adopted this division for a reason of convenience. The territory encompassed by the term "Russian Plain" covers the whole European territory of the former Soviet Union, including Ukraine (together with its Carpathian region), Moldova, Byelorussia, Latvia, Lithuania and Estonia.

ACKNOWLEDGEMENTS

We thank B. Lassalle (France), J. Pawlowski (Poland), J. Serrano (Spain), and F. Weber (Germany) for pointing out errors and omissions in the book as well as all other colleagues who commented it in various aspects.

REFERENCES

Busato, E. & Casale, A. (2005). Note sul ciclo biologico e sulla morfologia pre-immaginale di *Carabus* (*Macrothorax*) *planatus* Chaudoir, 1843, specie endemica dell'Appennino siculo (Coleoptera, Carabidae). Studi Trent. Sci. nat., Acta Biologica, 81 (2004): 177-187.
Casale, A., Sturani, M. & Vigna-Taglianti, A. (1982). Fauna d'Italia. XVIII. Coleoptera Carabidae I. Introduzione, Paussinae,Carabinae. Ed. Calderini (Bologna): 1-499.
Turin, H., Penev, L. & Casale, A. (Eds) (2003). The Genus *Carabus* in Europe. A Synthesis. Fauna Europaea Evertebrata No 2. Pensoft Publishers, Sofia-Moscow & European Invertebrate Survey, Leiden, xvi + 512 pp.

Biology and Conservation

L. Penev, T. Erwin & T. Assmann (Eds) 2008
Back to the Roots and Back to the Future. Towards a New Synthesis amongst Taxonomic, Ecological and Biogeographical Approaches in Carabidology
Proceedings of the XIII European Carabidologists Meeting, Blagoevgrad, August 20-24, 2007, pp. 259-284.

Evolution of biennial life cycles in ground beetles (Coleoptera, Carabidae) of the Western Palaearctic

Andrey V. Matalin

Zoology & Ecology Department, Moscow State Pedagogical University, Kibalchicha str. 6, build. 5, Moscow 129164, Russia. E-mail: a_matalin@tochka.ru

SUMMARY

Information about the life cycles of 400 Palaearctic carabid species was acquired on the basis of field study and analysis of literature. Five types of biennial development were defined: a biennial life cycle of spring-summer species; a biennial life cycle of summer-autumn species which developed as facultative (only part of the population develops within two years) or obligatory (biennial development occurs in the whole population); biennial life cycle of autumn species which may have appeared in two similar variants – facultative and obligatory. So, biennial life cycles have independently evolved in carabid beetles by at least three means: among spring breeders (type 1, according to Thiele, 1977) with gonad dormancy during hibernation and gonad maturation; males in SD, females after changing SD to LD; among autumn breeders (type 3) without gonad dormancy but with temperature controlled larval dormancy during hibernation; among autumn breeders (type 4) with gonad dormancy during aestivation and gonad maturation after changing LD to SD. The similarity and differences between latitudinal and altitudinal trends in biennial development of Carabidae are discussed. In similar, extreme conditions, the parallel variants of biennial development are observed. However even in favorable environments, some individuals may develop over two years. According to the assimilated data for this study, the proportion of species with biennial life cycles in different habitats within the temperate zone is rather high and biennial development is not unusual.

Keywords: Carabidae, biennial life cycles, Western Palaearctic

INTRODUCTION

The study of life cycles is important in both applied and theoretical biology. Investigations of the features of life cycles under specific conditions allows for detailed study of the structure and dynamics of populations of individual species, as well as of whole communities. Moreover, analysis of the patterns of development along natural gradients makes life cycles an important topic for the discussion of evolution among superspecific taxa (Thiele, 1977; Paarmann, 1979; Sharova, 1981; Hůrka, 1986; Makarov, 1989; den Boer & van Dijk, 1996; Matalin, 1998a, 2007; Sota & Ishikawa, 2004).

The problem of the evolution of multi-annual life cycles is one of the most interesting and challenging in modern biology. Among the Coleoptera, multi-annual development patterns are not rare. The longest development, of up to four years, is observed in the larvae of several Lucanidae, Scarabaeidae, Cerambycidae and Elateridae species (Bey-Bienko, 1980; Klausnitzer, 1981). In some Chrysomelidae, especially *Leptinotarsa decemlineata* L., such a phenomenon is known as "super-pause" (Bey-Bienko, 1980; Koval', 2005). In this case, the beetles of a new generation reproduce only two or even three years after pupation. During this period, they remain inert, within their pupation chamber. The development of some arctic Chrysomelidae, such as *Chrysolina subsulcata* Mann., typically lasts for three or four years (Chernov, 1974, 1978), and according to some estimates can extend for up to even five or six years (Khruleva, 1994). An extended developmental period is also known for the lycid, *Pyropterus nigroruber* DeGeer, which has been observed in captivity for five years before pupating (Bourgeois, 1882).

In the Carabidae, however, development does not seem to exceed two years. There is indirect evidence of a possibly triennial development in two species of *Amblystogenium*, which inhabits several sub-antarctic islands (Davies, 1987), *Amara quenseli* (Schönh.) in the Alps (De Zordo, 1979), and some *Carabus* in the barren tundra (Korobeinikov, 1991). Also according to Shelford (1908), the development in some species of Holarctic tiger-beetles lasts for up to three years. But such cases are rather exceptional, because the long development of tiger-beetle larvae is closely related to the features of their life. Stable habitats and food resources are of greater significance to them in comparison with other carabid genera.

Biennial life cycles have traditionally been considered as only occurring among carabid beetles under extreme conditions only, such as high latitudes (Houston, 1981; Refseth, 1980, 1984, 1988; Andersen, 1969; Korobeinikov, 1991; Sharova & Filippov, 2003; Filippov, 2006a-b), high mountains (De Zordo, 1979; Jakuczun, 1979; Brandmayr & Zetto Brandmayr, 1986; Butterfield, 1986, 1996; Chemini & Pizzolotto, 1992; Hemmer & Terlutter, 1991; Schatz, 1994; Sparks et al., 1995; Khobrakova & Sharova, 2005; Sharova & Khobrakova, 2005) and some arid landscapes (Shelford, 1908; Hamilton, 1925).

But how strongly is the biennial life cycle associated with such difficult environments? And how widely is the biennial pattern of development represented amongst the

Carabidae within the temperate belt? Before answering these questions a short review of the evolution of our knowledge concerning the life cycles in Carabidae, including biennial ones, will be provided.

The first classification of the reproductive rhythms of carabid beetles was elaborated in 1939 by Sven Larsson. It was based on a careful analysis of museum collections, primarily representing the Danish fauna. Larsson studied material of more than 270 carabid species caught by pitfall trapping. Using the three parameters of reproduction time, period of imaginal activity and developmental time, he established six groups among the ground beetles: 1) spring reproduction either with high F+, 2) moderate F(+) or 3) low autumn activity F(÷), 4) without autumn activity F÷, 5) with autumn reproduction H and 6) two-year long development 2F+.

The proportions of "spring" and "autumn breeders" were about three to one, while biennial development was established for two species only, *Nebria livida* (L.) and *Omophron limbatum* (F.). Larsson correctly observed that the "spring breeders" reproduce during the first half of the vegetation season, hibernating at the stage of an immature imago. In contrast, the "autumn breeders" propagate during the second half of the vegetation season, overwintering as larvae. Paradoxically, he formally grouped the species with a biennial development among "spring breeders", i.e. 2F+. Such an interpretation could be considered appropriate for *O. limbatum*, but is quite inappropriate for *N. livida*. This has become clear after study of the physiological basis of gonad maturation.

Much study on insect physiology, including that of reproduction, carried out during the latter half of the last century, has greatly extended our knowledge of the life cycles of ground beetles. Thus, the book of Hans-Ulrich Thiele, "Carabid beetles in their environments", which appeared in 1977, recognised six types of annual reproduction rhythms in Carabidae: 1) spring breeders without larval dormancy either with obligatory dormancy in the adult (parapause), mainly governed by photoperiod or 2) facultative dormancy in the adult, governed by photoperiod (photoperiodic quiescence) ; 3) autumn breeders with larval hibernation, or parapause, either without dormancy in the course of adult development or 4) with adult photoperiodic aestivation or parapause; 5) species with unstable conditions of hibernation and potentially lacking dormancy; or 6) species requiring two years to develop.

Thiele (1977) discussed a two-year long pattern of development using three species as examples. In 1972, van Dijk suggested that during the short period of reproduction observed at high latitudes, *Calathus melanocephalus* (L.) may need two years to mature. In 1973, Luff, based on his own observations and the unpublished data of Houston, demonstrated that the development of *Pterostichus madidus* (F.) in sub-arctic conditions is extended for two years. An experimentally proven two-year long development has only been documented for a single species, *Abax ovalis* Duft. In 1975, Lampe described its life cycle in detail, stating that the females of *A. ovalis* show gonad dormancy during hibernation. Maturation is observed only after the day changes from short to long. Contrastingly, males have no photoperiodic gonad dormancy, their maturation being controlled by temperature fluctuations during hibernation.

In 1979, Wilfried Paarmann established a further five types of annual reproduction rhythms among the Carabidae. Two of these are observed in the North African subtropics, and the other three in the Central African tropics. In addition, Paarmann discussed for the first time the possible pathways of evolution of annual reproduction rhythms under the seasonal climate of the temperate zone. In contrast to Thiele (1977), he distinguished two modifications of biennial development in Carabidae. Paarmann suggested that the biennial development discussed by van Dijk (1972) and Luff (1973) could have evolved from the reproductive cycle of "autumn breeders," with obligatory larval hibernation but without dormancy during adult development (Type 3, according to Thiele, 1977). On the other hand, he considered that the biennial cycle of *A. ovalis* (Lampe, 1975) might not have originated from the same type of annual cycle, because the control of gonad dormancy and adult maturation is drastically different in this case. Paarmann isolated this life cycle as a separate modification.

In 2004, Teiji Sota and Ryosuke Ishikawa reconstructed the phylogenetic and life-history evolution within the genus *Carabus* (sensu lato) on the basis of molecular data. They suggested that the biennial life cycles in Carabina could have evolved from "spring breeders". As a direct shift from spring to autumn breeders in warm temperate climates is unlikely, they surmised that larval overwintering might have evolved among spring breeders in cool temperate conditions with a short warm season,resulting in the evolution of species with summer breeding. After that, such species with larval hibernation might have either colonized sub-arctic habitats, where they develop for two years, or re-colonized the warmest habitats, where a reproduction summer aestivation parapause in adult is observed, i.e. this evolved adaptation could confer a fitness advantage under certain climatic conditions.

So, the evolutionary scenario of Sota and Ishikawa (2004) contradicts that proposed by Paarmann (1979). On the other hand, my own observations regarding the time of development of overwintering larvae in several species of ground beetles with spring reproduction rather support Paarmann's hypothesis.

MATERIALS AND METHODS

During the preparation of the current paper, numerous publications referring to the seasonal dynamics of activity, the demographic structure of populations, gonad maturation and larval development in Palaearctic Carabidae have been critically reviewed (Gilbert, 1956; Briggs, 1965; Inyaeva, 1965; Thiele, 1977; Murdoch, 1967; Vlijm & van Dijk, 1967; Andersen, 1969; Penney, 1969; Krehan, 1970; Kasandrova & Sharova, 1971; Potapova, 1972; van Dijk, 1972, 1979, 1994; Hůrka, 1973, 1975; Paarmann, 1974, 1976a-b, 1990, 1994; Bauer, 1974; Ferenz, 1975, 1977; Kůrka, 1975; Lampe, 1975; van Heerdt et al., 1976; Jørum, 1976, 1980, 1985; Jones, 1979; Sharova & Dushenkov, 1979; Houston, 1981; Zetto Brandmayr, 1983; Refseth, 1984, 1988; Sota, 1985, 1986, 1987, 1994, 1996; Loreau, 1985; Butterfield, 1986, 1996; Brandmayr & Zetto Brandmayr, 1986; van Schaick Zillesen et al.,

1986; Nekuliseanu, 1987, 1990, 1994; Nelemans, 1987; Wallin, 1987; Kruchkova & Panov, 1988; Makarov & Chernyakhovskaya, 1989, 1990; Dushenkov & Chernyakhovskaya, 1990; Chernyakhovskaya, 1990; Makarov, 1994; Chemini & Pizzolotto, 1992; Ernsting et al., 1992; Basedow, 1994; Cárdenas, 1994; Schatz, 1994; Cárdenas & Hidalgo, 1995, 1998, 2004; Sharova & Denisova, 1995, 1997; Chaabane et al., 1996; Ortuño & Marcos, 1997; Purvis, 1998; Turin, 2000; Sharova & Filippov, 2003; Turin et al., 2003; Khobrakova & Sharova, 2005; Sharova & Khobrakova, 2005; Filippov, 2006, 2006a). As a result, data on the life cycles of more than 250 carabid species have been obtained.

Moreover, my own long-term observations from 1982 to 2007 in 20 regions of the former USSR and Russia have been summarized as well (Karpova & Matalin, 1990; Matalin, 1994, 1997a-c, 1998a-b, 2003, 2006, 2007; Matalin & Budilov, 2003; Matalin & Makarov, 2006; Matalin et al., 2007). This has provided detailed material on the life cycles of an additional 150 species.

For discussion of developmental variations, the criteria annual and biennial development are defined. In the case of annual life cycles, the beetles of a new generation emerge every year. Therefore, the hibernation of parental and daughter generations is observed in the same ontogenetic phase (Figs 2A, 4A). In the case of biennial life cycles, the beetles of the daughter generation appear once in two years. Their first hibernation is observed in a different ontogenetic stage compared to the parental generation (Figs 2B, 4B-C).

RESULTS

At the present time, the following modifications of biennial development are recognised among carabid species in the temperate climates of the Western Palaearctic.

The first such species to be considered is *Harpalus affinis* Schrank (Matalin, 1998b). According to Larsson (1939), *H. affinis* is a "spring breeder" with adult hibernation. In the steppe zone, the seasonal dynamics of its activity are characterized by two peaks, one at the end of May to the middle of June, the second in the middle of August to early September (Fig. 1). The first peak represents overwintering beetles. Their reproduction commences in mid-May and ceases towards the beginning of August. The first instar larvae appear 12 days after egg-laying in late spring. Their development averages 14-15 days. At the end of June to early July, the first instar larvae molt to the second instar. The second instar development lasts c. 20 days. On average, the third instar larvae complete their development in 26 days. The first pupae are to be found at the beginning of August. The average duration of pupation is 16 days. So the development of *H. affinis* from adult to adult takes about 90 days. Similar developmental periods are observed in the first instar larvae taken by soil sampling at the end of May (Table 1). From the end of August, mass emergences of beetles of the new generation can be observed. This generation forms the second peak of locomotor activity. Later on, the young adults, as well as some beetles of the parental generation, hibernate. In this case the life cycle of *H. affinis* can be characterized as annual (Fig. 2A).

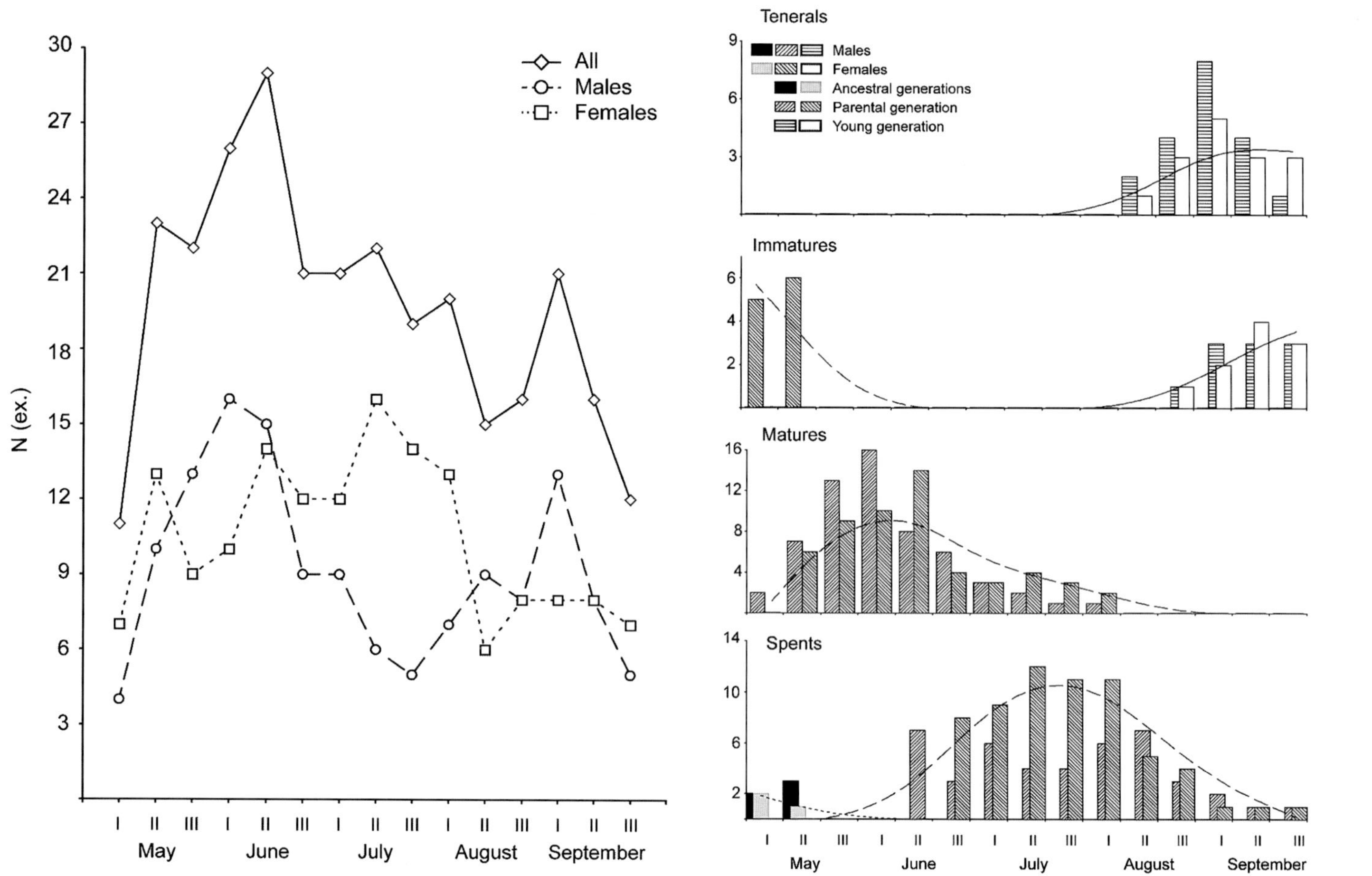

Fig. 1. Seasonal dynamics of activity, and sex and age structure of populations of *Harpalus affinis* in the steppe zone (south-western Moldavia, 1990).

Table 1. Developmental time of pre-imaginal stages of *Harpalus affinis* in south-western Moldavia (data from 1990, semi-natural conditions).

Developmental stages	Non-overwintering		Overwintering	
	Calendar times of development	Average duration of development ± SD	Calendar times of development	Average duration of development ± SD
Eggs	21.V -14.VI	14.5±8.7 (n = 4)		
L_1	10.VI -27.VI	16.7±6.4 (n = 4)		
L_2	21.VI -19.VII	21.8±2.4* (n = 7)	12.V -3.VII	49.5±0.5* (n = 2)
L_3	12.VII -12.VIII	NS25.5±0.6** (n = 7)	14.V -28.VII 30.VI -2.VIII (after L_2)	69.2±2.8** (n = 12) NS29.5±2.1 (after L_2) (n = 2)
Pupa	5.VIII -27.VIII	NS15.3±0.5*** (n = 7)	16.V -27.VIII 28.VII -26.VIII (after L_2) 20.VII -30.VIII (after L_3)	102±2.0*** (n = 2) NS20.5±6.4 (after L_2) (n = 2) NS28.9±7.5 (after L_3) (n = 12)

Notes: *, **, *** - significant differences ($p<0.05$, T-test); NS – not significant differences.

In *H. affinis,* also older instar larvae, as well as pupae, hibernate in addition to mature and immature adults. This is well supported both by personal soil sampling data taken in early May and literature sources (Briggs, 1965; Inyaeva, 1965; Budiolov, 1990; Dushenkov, Chernyakhovskaya, 1990). According to my own observations, the ratio of overwintering pre-imaginal stages of *H. affinis* in the steppe zone is: 1 pupa, 2 second instars larvae, 20 third instars larvae. It is noteworthy that the development of overwintering larvae

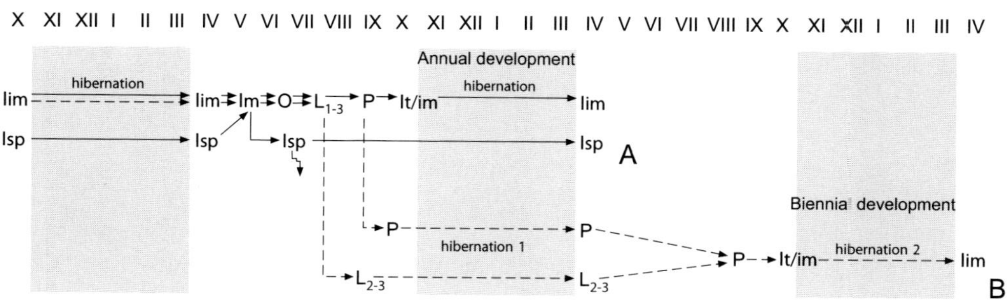

Fig. 2. A scheme of the facultatively biennial life cycle with spring-summer reproduction (an example of *Harpalus affinis*). I*t/im* - teneral; I*im* - immature; I*m* - mature; I*sp* - spent beetles; O - eggs; L_{1-3} - larvae of first-third instars, respectively; P - pupae.

and pupae is greatly prolonged. On average, the second instar larvae molt into the third instar 58 days after hibernation. After that, the development of the third instar larvae usually lasts 25 days, and the pupae, 23 days. So the time of appearance of young beetles, starting from the overwintering second instar larvae, averages 106 days. The development of the overwintering third instar larvae takes longer, and they can be encountered in the soil at the beginning of May, but on average, they pupate only in mid to late July, i.e. in 68 days. The emergence of young beetles is observed after another 36 days, in the mid to late August. So the average developmental time from the overwintering third instar larvae until the beetles of the new generation, takes 105 days. The maximum delay in development is observed in the hibernated pupae. In 1991, two pupae were collected on the 16th of May. The beetles (a male and a female) emerged only on the 23rd and 27th of August, that is, after 100 and 104 days, respectively (Table 1).

In spite of the delayed development of the overwintering pre-imaginal stages, the emergence of teneral beetles is observed at the same time as mentioned above, that is, from the end of August to early September. On the other hand, these beetles do not reproduce in the same season. So the total duration of their development period increases to two years (Fig. 2B).

Thus, in the populations of *H. affinis* in Moldova, two subpopulations with different seasonal rhythms are formed. Due to this, the life cycle of *H. affinis* is characterized as multivariant. In the case of adult hibernation, the development is completed in one year. But in the case of larval and pupal hibernation, the development is increased to two years. Since only a part of the whole population develops during two years, the life cycle of *H. affinis* is realized as facultatively biennial with spring-summer reproduction (Fig. 2). The extended period of reproduction seems to be the cause of *H. affinis* hibernation in the atypical ontogenetic phases. As a result, desynchronization of the life cycle is observed. Yet the completion of the development of both hibernated and non-hibernated larvae at the same time supports its life cycle synchronization. It is noteworthy that these two sub-populations are not isolated from each other. The daughter generation which hibernates in the different ontogenetic stages can develop in either of the two ways described above. Thus, the biological integrity of the species is maintained.

An analysis of the sex and age structures of the population showed that gonad maturation in *H. affinis* males is completed in short-day conditions. On the contrary, the gonad maturation in females is observed only after the daytime changes from short to long (Fig. 1). Hence the above pattern of a biennial life cycle could have evolved in response to the annual reproduction rhythms of "spring breeders" (Type 1, according to Thiele, 1977), with the same mechanism of gonad maturation. A similar life cycle is observed in *Harpalus distinguendus* (Duft.), *Ophonus azureus* (F.), *Chlaenius vestitus* (Payk.), *Omophron limbatum* (F.) and possibly also in *Abax ovalis* Duft.

As a second example, *Pterostichus melanarius* (Ill.), one of the most common and abundant Palaearctic ground beetle species (Matalin, 2006), has been chosen. According to Larsson (1939), *P. melanarius* is an "autumn breeder" with larval hibernation. The overwintering third instars occur in pitfall traps and soil samples at the beginning of the vegetation season. In mixed forests, the duration of larval development after hibernation averages 24 days, and the pupal stage only seven days. As a rule, the beetles of a new generation emerge at the end of June/early July. The reproduction period continues from the end of July until mid-September (Fig. 3). The first instar larvae appear at the beginning of August. The second instar larvae are to be found at the end of August while the third instars are to be seen in the middle of September. The larvae of the older instars hibernate. So, the life cycle of *P. melanarius* in such conditions is annual, because the emergence and reproduction of teneral beetles are observed during the same season and the hibernation of parental and daughter generations take place during the same ontogenetic phase (Fig. 4A).

However some reports (Jørum 1980; Wallin, 1987; Kryuchkova & Panov, 1988; Nekuliseanu, 1990; Sharova & Filippov, 2003) present direct evidence of a possible biennial development of *P. melanarius*. In addition, some indirect data concerning a two-year long development in at least part of *P. melanarius* populations are contained in virtually any publication dealing with the ecology of this species (Lindroth, 1945; Skuhravý, 1959; Greenslade, 1965; Vasil'eva, 1978; Jones, 1979; Desender et al., 1985; Kålås, 1985; Makarov & Chernyakhovskaya 1989; Sharova & Denisova 1997; Fadl & Purvis, 1998). Both larvae and adults are known to hibernate all over the distribution area of *P. melanarius*. The imagines are represented by mature individuals of the ancestral generations, as well as by non-copulated individuals of the new generation. A delayed maturation of the teneral beetles is due to their late emergence in mid to late August, which implies a less than minimum sum of effective temperatures. According to Krehan (1970), *P. melanarius* females and males attain sexual maturity in more or less three weeks at about 20°C. At lower temperatures, a thermic quiescence occurs.

The overwintered beetles form an early summer peak of locomotor activity. At the same time, specimens of ancestral generations start their reproduction at least ten days earlier than the immature ones (Fig. 3). In mixed forests, oviposition is observed from the end of June to early July. On average, eggs develop for 9.5 days, the first instar larvae for 17.5 days, and the second instars are found in mid to late July. During August, some of them molt into the third instar, whereupon the older instar larvae hibernate. In this case, emergence and reproduction of teneral beetles are observed over a year. Besides this, they pass the first hibernation as adults while the parental generation hibernates as larvae. Thus, the overwintering immature specimens of *P. melanarius* need two years to develop (Fig. 4B).

As with *H. affinis*, the populations of *P. melanarius* show two temporally, partly overlapping, sub-populations with different seasonal rhythms. Larval hibernation ensures an annual development, though the hibernation of immature adults maintains a two-year long development. Thus, over most of the area of natural habitats, ranging from the steppe to the middle taiga (= boreal forest) zones, the life cycle of *P. melanarius* is multivariant and could be defined as facultatively biennial with summer-autumn reproduction. The prolonged period of emergence of the new generation together with the heterogeneous maturation of the beetles, is the cause of desynchronisation of *P. melanarius*' life cycle, though here the larval hibernation maintains its overall synchronisation.

In adverse conditions, however, such as those observed near the northern range limits of *P. melanarius*, the life cycle is sharply different. In the northern taiga zone, its reproduction

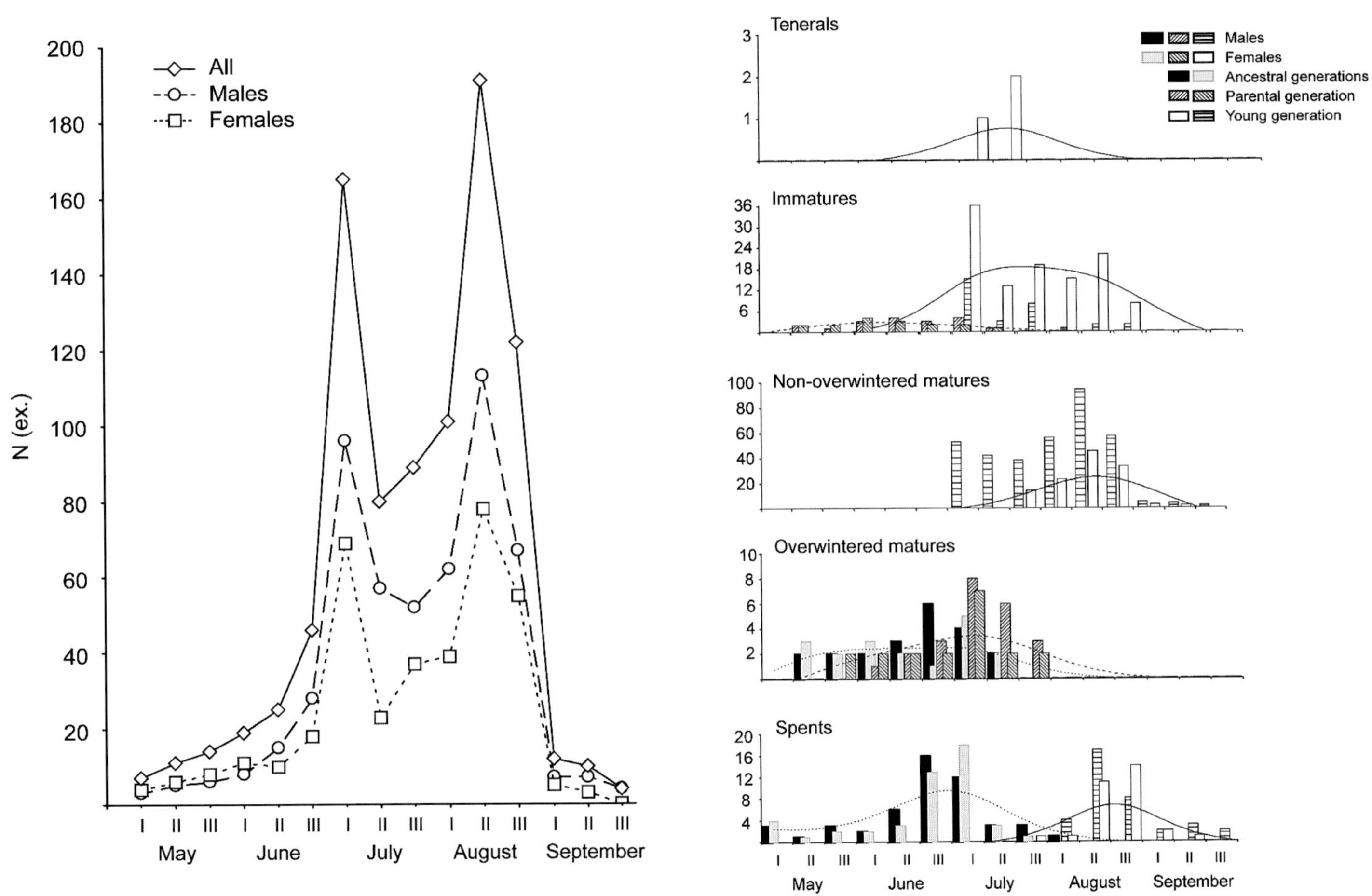

Fig. 3. Seasonal dynamics of activity, and sex and age structure of populations of *Pterostichus melanarius* in the steppe zone (south-western Moldavia, 1990).

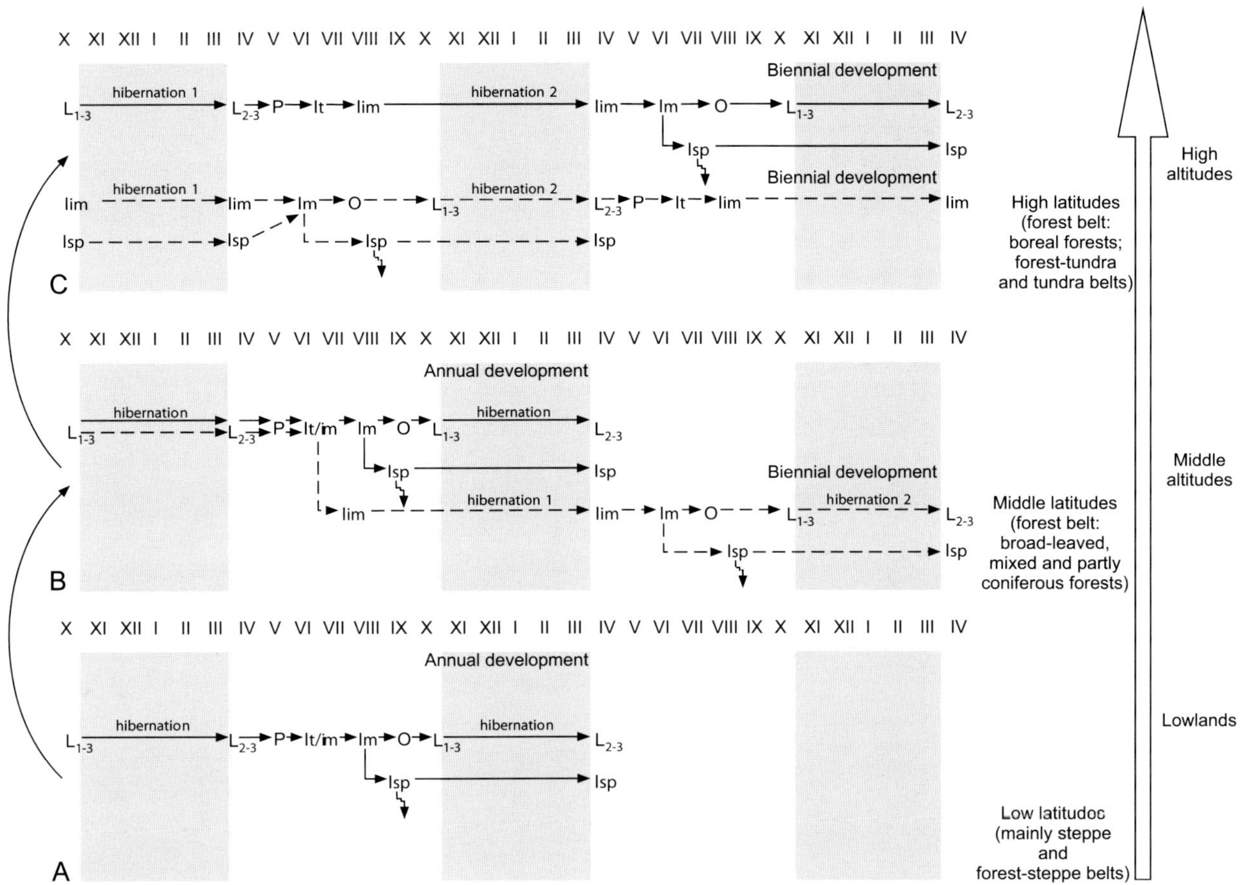

Fig. 4. A scheme of modifications of the *Pterostichus melanarius* life cycle across a climatic gradients. A - annual life cycle with summer-autumn reproduction; B - facultatively biennial life cycle with summer-autumn reproduction; C - obligate biennial life cycle early summer reproduction; other explanations as in Fig. 2.

lasts from mid-June until early August. Then only the overwintered beetles, both mature and immature, breed. The new generation cohort that emerges from the overwintering larvae during the current season remain sexually immature until the following year (Sharova & Filippov, 2003: fig. 3B). In this case, both adult and larval hibernation is the cause of a two-year long development. As a result, an almost complete separation of the sub-populations overwintering in different ontogenetic phases is observed. However, individuals from the ancestral generations represent a connecting link between the sub-populations, thus maintaining the biological integrity of the species. In such conditions, the life cycle of *P. melanarius* is altered from facultatively biennial with summer-autumn reproduction to obligate biennial with early-summer reproduction. A new generation in either sub-population appears only once in two years. But thanks to the emergence alternately every second year of teneral beetles in each subpopulation, a new generation in the population as a whole is produced every year. The development of beetles in each subpopulation is realized on the basis of the same seasonal rhythms shifted by a year (Fig. 4C). Earlier, such populations were termed as by-side (Matalin, 2006). In this way, in the northern taiga the multivariant pattern of the *P. melanarius* life cycle is replaced by a monovariant one.

Gonad maturation of both males and females of *P. melanarius* is well known to be independent from photoperiod (Krehan, 1970; Thiele, 1977). However, normal completion of metamorphosis is observed only in overwintering larvae (Hůrka, 1975; Thiele, 1977). Then the described biennial life cycle could have evolved on the basis of annual reproduction rhythms of "autumn breeders" (Type 3, according to Thiele, 1977), with a similar mechanism of gonad maturation.

The life cycle of *P. madidus* is just as in *P. melanarius*. According to Luff (1973), in the south of Northumberland (UK), the life cycle of this species is facultatively biennial, because some immature specimens hibernate and are then to be found in May. Still, according to unpublished data of Houston (quoted by Luff, 1973), in the Moor House National Reserve (UK), all populations of *P. madidus* develop in two years. In this case, its life cycle is obligate biennial. Similar results have been obtained by Butterfield (1996) after a study of the *P. madidus* life cycle across an altitudinal gradient in Great Dun Fell (UK). In addition, similar life cycles are observed in numerous carabid beetles, such as *Carabus glabratus* Payk., *C. violaceus* L., *C. aurolimbatus* Dej., *Broscus cephalotes* L., *Pterostichus niger* (Shall.), *Harpalus rufipes* Degeer and so on.

As an example of the third type of biennial development among carabid beetles, *Carabus problematicus* Herbst has been selected. Larsson (1939) and later Lindroth (1945), considered *C. problematicus* to be a "spring breeder." However this is not entirely true. In fact this species is an "autumn breeder" with larval hibernation. The dynamics of its locomotor activity are characterized by two peaks. The first is observed in late spring to early summer, while the second is in late summer to mid-autumn (van der Drift, 1951, 1958; Greenslade, 1965; Thielle, 1977; Houston, 1981; Jørum, 1985; Loreau, 1985; Butterfield, 1986, 1996; Sparks et al., 1995). In the temperate zone, the first peak is due to the activity of both the overwintering parental generation and the freshly emerged teneral beetles. The subsequent activity decline is related to a short aestivation period, only then followed by reproduction. The larvae appearing during the autumn hibernate. In this case, the life cycle of *C. problematicus* is annual.

But in sub-arctic conditions, its life cycle is sharply altered. According to Houston (1981: fig. 3), in the Moor House National Reserve (UK), the first activity peak in the population of *C. problematicus* is observed in June. At this time, the overwintering beetles both from the parental and daughter generations are active. However, there is no reproduction until the end of July. As a result, the life cycle of *C. problematicus* is strongly synchronized, with oviposition observed during the same period. The peak of reproduction activity is restricted to the middle to the end of August and this coincides with the emergence of the new generation. The locomotor activity finishes at the end of September, whereafter the larvae, some of the adults of the parental generation as well as all beetles of the new generation hibernate. The development of the overwintering larvae and adults lasts two years. In these conditions the life cycle of *C. problematicus* is obligate biennial with late summer reproduction.

Presumably there must be a gradual transition between these extremely different life cycle patterns. Under certain conditions, cohorts of both the annual and biennial development cycles can coexist in the same population. In this case, the life cycle is facultatively biennial with autumn reproduction. This is well-supported by studies on the *C. problematicus* population structure across an altitudinal gradient in northern England (Butterfield, 1986, 1996) and northern Wales (Sparks et al., 1995). In the lowlands, the life cycle of *C. problematicus* is either annual or a small proportion of specimens develops over two years. At intermediate altitudes (100-300 m), the proportion of individuals with biennial development is greater. At high altitudes (500-850 m), the life cycle is purely biennial. It is noteworthy that, as in *H. affinis* and *P. melanarius*, the facultative biennial life cycle of *C. problematicus* is multivariant, because both subpopulations show different seasonal rhythms.

According to the dominant opinion, *C. problematicus* exhibits a larval hibernation parapause, as well as an adult photoperiodic aestivation parapause (van der Drift, 1951, 1958; Thielle, 1977; Houston, 1981; Jørum, 1985; Loreau, 1985; Butterfield, 1986). Gonad maturation in this species is observed only after the transition from long-day to short-day conditions. Thus, this kind of biennial development could have evolved on the basis of annual reproduction rhythms of "autumn breeders" (Type 4, according to Thiele, 1977). The same type of life cycle is found in *Carabus hortensis* L., *C. exaratus* Quens., *Calathus erratus* (Sahlb.), *C. fuscipes* (Goeze), *Curtonotus alpinus* (Payk) etc.

DISCUSSION

An analysis of the geographic variability of the sex and age population structures in Carabidae shows that a development over two years is quite often observed among species with phenologies differing in various natural zones, ranging from the tundra to the steppe (Jørum, 1980, 1985; Houston, 1981; Butterfield, 1986, 1996; Sharova & Denisova, 1997; Matalin, 1997c, 1998b, 2006; Matalin & Budilov, 2003; Sharova & Filippov, 2003; Filippov, 2006a-b). Numerous carabid species have quite broad distributions. So they have to cope with a great variation in weather and climatic conditions, as well as with different landscape-habitat and vegetation patterns. As a result, across such considerable gradients, regular changes can be observed in the duration of locomotor activity, breeding time and development of pre-imaginal stages, as well as adult maturation.

With increasing latitude or altitude, the reproduction period in "spring breeders" is not only shortened but also shifts to a later time. Spring reproduction is replaced by a spring-summer or early-summer one. Amongst "autumn breeders," the period of reproduction is shifted to an earlier date, being observed in late, mid or even early summer (Briggs, 1965; Jørum, 1980, 1985; Sota, 1985, 1994, 1996; Butterfield, 1986, 1996; Refseth, 1988; Sparks et al., 1995; Matalin, 1997c, 1998b, 2006; Sharova & Filippov, 2003; Khobrakova & Sharova, 2005; Filippov, 2006a-b). Because of the shorter vegetation season, the oviposition time is greatly reduced, while the duration of larval development and adult maturation is increased. For the reasons given above, hibernation may be observed in atypical ontogenetic phases, such as larva and pupa for "spring breeders" and as immature beetles for "autumn breeders". Therefore, in numerous carabid beetle species some individuals of the population develop in two years.

In southern regions or over the plains, most individuals show annual life cycles and the proportion of beetles with a biennial development is small. When moving to higher altitudes or latitudes, the proportion of species with biennial development increases. At the northern range peripheries, and at high altitudes, entire populations demonstrate biennial life cycles. So not only facultatively biennial life cycles might have evolved across gradients of climate conditions, but also their gradual transformations into obligate biennial ones (Fig. 4).

At the same time, in various habitats of the same climatic zone, even with relatively favourable conditions, different types of biennial life cycle can be realized. According to Sharova and Denisova (1997), hibernated specimens of *P. melanarius*, both mature and immature, reproduce in oak woods of the Tambov area of Russia at the end of spring to early summer. On the other hand, from mid-summer until early autumn, the teneral beetles of the new generation breed as well (Sharova & Denisova, 1997: fig 4b). In this case, the life cycle is facultatively biennial and multivariant. In contrast, only hibernated beetles reproduced in pine forests. Specimens of the new generation, which emerged from the overwintering larvae, hibernated as immature adults (Sharova & Denisova, 1997: fig 4a). Thus, the life cycle in this case is obligate biennial and monovariant. Besides this, a facultatively biennial life cycle adapting to an obligate biennial one has been observed under anomalous weather conditions. Thus, according to Jørum (1980), in Denmark in years with a long and cold winter, and a short and cool summer, the life cycles of *P. melanarius* and *P. niger* are solely obligate biennial.

According to this evidence, one can derive a very important conclusion. In the temperate zone of the Western Palaearctic, biennial life cycles among carabid beetles have evolved independently at least three times (Fig. 5). Under these conditions, the following versions of biennial development may have developed:

- Facultatively biennial with spring-summer reproduction

(on the basis of the annual rhythm of "spring-breeders" – Type 1, according to Thiele, 1977);

- Facultatively biennial with summer-autumn reproduction

and then

- Obligate biennial with early summer reproduction

(on the basis of the annual rhythm of "autumn breeders" – Type 3, according to Thiele, 1977);

- Facultatively biennial with autumn reproduction

and then

- Obligate biennial with late summer reproduction

(on the basis of the annual rhythm of "autumn breeders" – Type 4, according to Thiele, 1977).

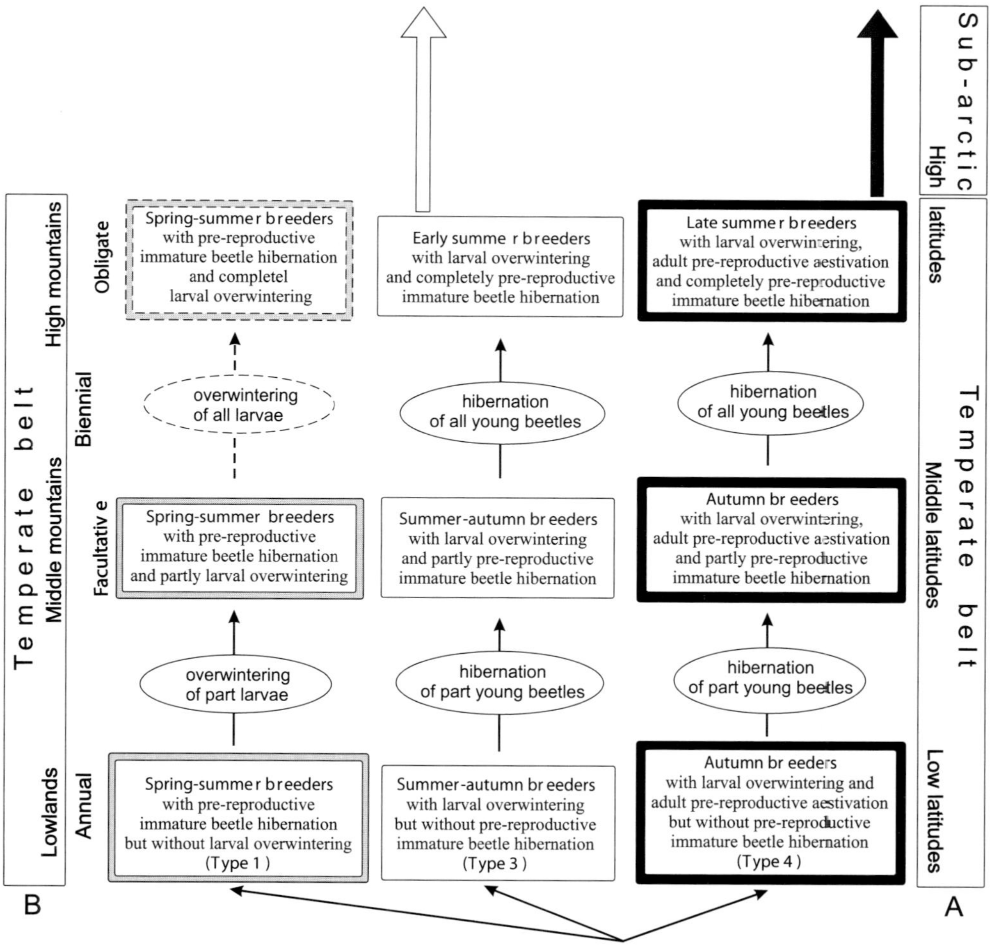

Fig. 5. A scheme of possible evolution of biennial life cycles among Carabid beetles of Western Palaearctic. A - latitudinal trend; B - altitudinal trend; large arrows indicate immigration of "autumn breeders," with different biennial life cycles in sub-arctic regions; dotted lines indicate a possible pathway of biennial life cycle among "spring breeders."

The existence of an obligate biennial life cycle among "spring breeders" remains open to question. At present there is no firmly established evidence that would support such a type of biennial development. Colonization of northern habitats by numerous carabid species with spring or spring-summer reproduction e.g. all *Harpalus*, most *Amara* and *Pterostichus* spp. is limited first of all by the lack of suitable habitats, coupled with both poor food resources and the developmental features of overwintering larval stages. Unlike species with autumn reproduction and hibernating larvae, "spring breeders" have no pre-adaptations for larval overwintering. The hibernation of pre-imaginal stages amongst "spring breeders" can only be successful in relatively mild conditions, such as those observed in the southern regions, partly also in the middle part of the temperate zone, as well as in regions with a mild coastal climate.

The development of overwintering pre-imaginal stages amongst "spring breeders" seems to be controlled by photoperiod. Indeed, successful completion of metamorphosis in this case is only observed under daylight changing from long to short, or even under short-day conditions alone. This is suggested by the development of overwintering *H. affinis* larvae and pupae in the steppe zone as described above (Table 1). Apparently, under constant illumination conditions at high latitudes, the development of overwintering larvae and pupae of "spring breeders" is suppressed.

In any case, in the northern taiga, sub-arctic and barren tundra, all carabid beetles with initial spring reproduction fully complete their annual development even during the short vegetation season (Andersen, 1969, 1984; Butterfield, 1996; Sharova & Filippov, 2003; Filippov, 2006a, 2007). So, at high latitudes, among ground beetles with biennial development, only "autumn breeders" are found (Lindroth, 1945; Refseth, 1988; Butterfield, 1996; Filippov, 2007).

However, by considering the characteristic sex and age structure of a population, I am able to postulate a mechanism by means of which a certain type of biennial life cycle might have evolved. A study by Houston (1981) is a good example of this. He studied the phenology and life cycles of *C. glabratus* and *C. problematicus* under sub-arctic conditions.

Reproduction in hibernated specimens of *C. glabratus* has been observed in June to July, while the emergence of teneral beetles - in August to September (Houston, 1981: fig. 2). Under sub-artic conditions, its life cycle is obligate biennial with early-summer reproduction. In southern and central Europe, this species shows a summer-autumn reproduction pattern without gonad dormancy (Stiprais, 1970; Hůrka, 1973; Grüm, 1975; Feoktistov & Dushenkov, 1982; Dülge, 1994; Günther & Assmann, 2000; Turin et al., 2003). Thus, the biennial life cycle in *C. glabratus* could have evolved on the basis of the annual, Type 3 reproduction rhythm of Thiele, (1977). Similar patterns are observed in *Carabus violaceus* (Refseth, 1984), *C. aurolimbatus* (Sharova & Filippov, 2003), *Pterostichus melanarius* and *P. niger* (Jørum, 1980; Sharova & Filippov, 2003), as well as in *P. madidus* (Luff, 1973; Butterfield, 1996).

Contrastingly, in *Carabus problematicus* the reproduction of overwintering beetles and the emergence of a new generation are observed concurrently, from late July until mid-September (Houston, 1981: fig. 3). Thus, in the sub-arctic, the life cycle of *C. problematicus* is obligate biennial with late summer reproduction. However, in middle Europe this species is characterized by autumn reproduction with gonad dormancy during aestivation and by gonad maturation after daylength changing from long to short (van der Drift, 1951, 1958; Thiele, 1977; Jørum, 1985; Loreau, 1985; Turin et al., 2003). The biennial life cycle of *C. problematicus* could have evolved on the basis of the Type 4 annual reproduction rhythm, as described by Thiele (1977). At high latitudes, the life cycles of *Carabus hortensis* (Refseth, 1988), *Calathus erratus* (Kålås, 1985; Refseth, 1988) and *C. fuscipes* (Kålås, 1985; Butterfield, 1996) are similar.

Because the vegetation season at high latitudes and altitudes is short, whilst the reproduction time in all carabid species is markedly shortened, reproductive monotony and a restricted spectrum of life cycles in different habitats is observed. As a result, the contradiction arises that phenologically, the species are biennials with "early-summer" or "late-summer" reproduction, whereas their life cycles could have evolved starting from either of the two annual autumn breeder rhythms. Perhaps this paradox led to the mistake in Sota and Ishikawa's (2004) hypothesis.

My studies allow me to conclude that biennial life cycles among "spring breeders" (Type 1, according to Thiele, 1977) and "autumn breeders" (Type 4, according to Thiele, 1977), could have originated independently from one another (Fig. 5). The most important feature in both cases is the hibernation of some individuals in atypical ontogenetic stages. The developmental time of such individuals is then increased to two years and the life cycle is thus transformed from annual to facultatively biennial. As one can easily see, the reproduction time both in spring and autumn breeders remains unchanged. Under cooler environmental conditions, further transformations of a biennial life cycle in autumn breeders can be observed. This results from the replacement of a facultative biennial development with an obligate biennial one. In this case, the period of reproduction is shifted from autumn to late summer. A similar transformation is observed in a different group of "autumn breeders" (Type 3 of Thiele, 1977).

In theory, an obligate biennial life cycle among "spring breeders" is possible in mountains, because the photoperiodic conditions there do not change so sharply. Here lies the principal difference between two general trends in biennial development in carabid beetles: latitudinal and altitudinal.

At high latitudes, obligate biennial life cycles seem to have eveolved on the basis of annual reproduction rhythms of "autumn breeders," Types 3 and 4 only (Fig. 5A). In contrast, at high altitudes, in addition to the types mentioned above, an obligate biennial life cycle in "spring breeders," Type 1, is theoretically possible (Fig. 5B). However, as is the case with high latitudes, no reliable data exist for mountains which would support obligate biennial carabids with a spring or spring-summer reproduction pattern (De Zordo, 1979; Jakuczun, 1979; Sota, 1985, 1994, 1996; Brandmayr & Zetto Brandmayr, 1986; Butterfield, 1986, 1996; Chemini & Pizzolotto, 1992; Hemmer & Terlutter, 1991; Schatz, 1994; Sparks et al., 1995; Khobrakova & Sharova, 2005; Sharova & Khobrakova, 2005).

At present, we have sufficient knowledge concerning both gonad maturation and the exogeneous regulation of annual reproduction rhythms in carabid beetles. However, there is an obvious knowledge gap regarding the study of the main features of larval development regulation. Further collection of such data would expand our understanding biennial life cycles. The biennial development patterns as observed in some individuals in the populations of numerous carabids in different habitats within the temperate zone are clearly of adaptive importance. The long reproduction period and asynchronous appearance of the subsequent generations causes age heterogeneity. These factors stabilize the population structure and decrease the risk of extinction of local populations (Weber & Klenner, 1987; Matalin, 1998b) and it is probably not coincidental that the majority of brood-watching Pterostichines, many of which inhabit mountainous habitats, have a biennial life cycle (Brandmayr, 1977). Individuals with a prolonged development form a strategic reserve in the whole population in force-majeure situations. Maybe thanks to such features as their life cycles, these species show vast distributions and occur in high numbers, not only in natural but also in man-made habitats, including agrocoenoses. According to my estimates, the proportions of such species in different communities within the temperate zone is very high, well over one-third of the fauna of dominants. So a biennial development in Carabidae is rather the rule than the exception.

ACKNOWLEDGEMENTS

I would like to thank all of my carabidologist colleagues who have helped with this study and who have contributed material for analysis and discussion. In particular, I would like to express my gratitude to my teacher, Professor Inessa Khristianovna Sharova, as well as to my colleague and friend, Professor Kirill Vladimirovich Makarov (both from the Moscow State Pedagogical University, Russia), for the important critiques and fruitful discussions of the presented contribution. I would also like to address special thanks to Professor Sergei Il'ich Golovach (Institute of Problems of Ecology and Evolution, Moscow, Russia) and Dr. Stephen Venn (University of Helsinki, Finland) for a critical review of the text.

This study was financially supported by the Russian Foundation for Basic Research (projects No. 06-04-49456, 07-04-08381) as well as the Presidential Support Program for Leading Academic Schools (project No. НШ-2154.2003.4).

REFERENCES

Andersen, J.M. (1969). Habitat choice and life history of Bembidiini (Col., Carabidae) on river banks in central and northern Norway. – Norsk ent. Tidsskr. 17: 17-65.

Andersen, J.M. (1984). A re-analysis of the relationships between life cycle patterns and the geographical distribution of Fennoscandian carabid beetles. – J. Biogeography. 11: 479-489.

Basedov, Th. (1994). Phenology and egg production in *Agonum dorsale* and *Pterostichus melanarius* (Col., Carabidae) in winter wheat fields of different growing intensity in

Northern Germany. – In: Carabid beetles: ecology and evolution (Desender, K., Dufrêne, M., Loreau, M., Luff, M.L. & Maelfait, J-P., eds). Kluwer Acad. Publ., Dordrecht-Boston-London, p. 101-108.

Bauer, T. (1974). Ethologische, autökologische und ökophysiologische Untersuchungen an *Elaphrus cupreus* Dft. und *Elaphrus riparius* L. (Coleoptera, Carabidae). – Oecologia. 14: 139-196.

Bey-Bienko, G.Ya. (1980). General entomology (3-d edition). – High School Press, Moscow, 416 pp. [in Russian].

Brandmayr, P. (1977). Ricerche etologiche e morfofunzionali sulle cure parentali in Carabidi Pterostichini (Coleoptera, Carabidae:Pterostichini). – Redia, 60:275-316.

Brandmayr, P. & Zetto Brandmayr, T. (1986). Phenology of ground beetles and its ecological significance in some of the main habitat types of Southern Europe. – In: Carabids beetles: their adaptations and dynamics (den Boer, P.J., Mossakowski, D., Luff, M.L. & Weber, F., eds). Gustav Fisher Verlag, Stuttgart-New York, p. 195-220.

Briggs, J.B. (1965). Biology of some ground beetles (Col. Carabidae) injurious to strawberries. – Bull. Entomol. Res. 56: 79-93.

Budilov, V.V. (1990). Autumnal distribution of ground beetles in agrocenosis of Mordovinian ASSR. – In: Structure and dynamics of populations of edaphic and herpetobiotic invertebrates (Sharova, I.Kh., ed). Moscow State Pedagogical Institute Press, Moscow, 1: 51-57 [in Russian].

Bourgeois, J. (1882). Monographie des Lycides de l'Ancirnt-Monde. – L'Abeille. 20: 1-120.

Butterfield, J. (1986). Changes in life-cycle strategies of *Carabus problematicus* over a range of altitudes in Northern England. – Ecol. Entomol. 11: 17-26.

Butterfield, J. (1996). Carabid life cycle strategies and climate change: a study on an altitudinal transect. – Ecol. Entomol. 21: 9-16.

Cárdenas, A.M. (1994). On the life history of *Calathus granatensis* (Coleoptera, Carabidae) in southwest Iberian Peninsula. – In: Carabid beetles: ecology and evolution (Desender, K., Dufrêne, M., Loreau, M., Luff, M.L. & Maelfait, J-P., eds). Kluwer Acad. Publ., Dordrecht-Boston-London, p. 109-116.

Cárdenas, A.M. & Hidalgo, J.M. (1995). Datos sobre la biología de reproductión dasarrollo larvario de *Carabus* (*Hadrocarabus*) *lusitanicus* (Fabricius, 1801) (Coleoptera: Carabidae). – Elytron. 9: 139-145.

Cárdenas, A.M. & Hidalgo, J.M. (1998). Data on the biological cycle of *Steropus globosus* (Coleoptera Carabidae) in the south west of Iberian Peninsula. – Vie Milieu. 48: 35-39.

Cárdenas, A.M. & Hidalgo, J.M. (2004). Life cycles of riparian ground beetles (Coleoptera, Carabidae): strategies for reproduction in the bank of a temporary stream in the southern Iberian peninsula. – Vie Milieu. 54: 47-56.

Chaabane, K., Loreau, M. & Josens, G. (1996). Individual and population energy budgets of *Abax ater* (Coleoptera, Carabidae). – Ann. Zool. Fenn. 33: 97-108.

Chemini, C. & Pizzolotto, R. (1992). Comunità di carabidi in siti forestali dei Monti Lessini (Trentino) (Coleoptera: Carabidae). – Studi Trentini Sci. Nat. Acta Biol. 67: 197-227.

Chernov, Yu.I. (1974). Some regularities of the adaptation of terrestrial animals to landscape-zonal conditions. – Zh. Obshch. Biol. 35: 846-857 [in Russian].

Chernov, Yu.I. (1978). Adaptive features of the life cycles of the tundra zone insects. – Zh. Obshch. Biol. 39: 394-402 [in Russian].

Chernyakhovskaya, T.A. (1990). Seasonal dynamics of activity and structure of populations of carabid species *Pterostichus niger* Schaller in different biotopes. – In: Structure and dynamics of populations of edaphic and herpetobiotic invertebrates (Sharova, I.Kh., ed). Moscow State Pedagogical Institute Press, Moscow, 1: 44-49 [in Russian].

Davies, L. (1987). Long adult life, low reproduction and competition in two sub-Antarctic carabid beetles. – Ecol. Entomol. 12: 149-162.

den Boer, P.J. & van Dijk, Th.S. (1996). Life history patterns among carabid species. – Tijdschr. Ent. 139: 1-16.

Desender, K., van den Broeck, D. & Maelfait, J.-P. (1985). Population biology and reproduction in *Pterostichus melanarius* Ill. (Coleoptera, Carabidae) from a heavily grazed pasture ecosystem. – Med. Fac. Landbouww. Rijksuniv. 50: 567-575.

De Zordo, I. (1979). Phänologie von Carabiden im Hochgebirge Tirols (Obergurgl, Österreich) (Insecta: Coleoptera). – Ber. Nat. -med. Ver. Innsbruck. 66: 73-83.

Dülge, R. (1994). Seasonal activity of carabid beetles in wooded habitats in northwest Germany (Coleoptera, Carabidae). – In: Carabid beetles: ecology and evolution (Desender, K., Dufrêne, M., Loreau, M., Luff, M.L. & Maelfait, J-P., eds). Kluwer Acad. Publ., Dordrecht-Boston-London, p. 125-131.

Dushenkov. V.M. & Chernyakhovskaya, T.A. (1990). Carabid larvae in agricultural areas of the Moscow region. – In: Structure and dynamics of populations of edaphic and herpetobiotic invertebrates (Sharova, I.Kh., ed). Moscow State Pedagogical Institute Press, Moscow, 1: 84-99 [in Russian].

Ernsting, G., Isaaks, J.A. & Berg, M.P. (1992). Life cycle and food availability indices in *Notiophilus biguttatus* (Coleoptera, Carabidae). – Ecol. Entomol. 17: 33-42.

Fadl, A. & Purvis, G. (1998). Field observations on the lifecycles and seasonal activity patterns of temperate carabid beetles (Coleoptera: Carabidae) inhabiting arable land. – Pedobiologia. 42: 171-183.

Feoktistov, V.F. & Dushenkov, V.M. (1982). Seasonal dynamics of activity of the ground beetles (Coleoptera, Carabidae) in forests of different types at the southern taiga boundary. – Zool. Zh. 61: 227-232 [in Russian].

Ferenz, H.-J. (1975). Photoperiodic and hormone control of reproduction in male beetles, *Pterostichus nigrita*. – J. Insect Physiol. 21: 331-341.

Ferenz, H.-J. (1977). Two-step photoperiodic and hormonal control of reproduction in the female beetle *Pterostichus nigrita*. – J. Insect Physiol. 23: 671-676.

Filippov, B.Yu. (2006a). Seasonal aspects of life cycles in the ground beetles *Carabus granulatus* and *C. glabratus* (Coleoptera, Carabidae) in the northern taiga. – Zool. Zh. 85: 1076-1084 [in Russian].

Filippov, B.Yu. (2006b). Seasonal aspects of life cycles in the ground beetles *Calathus melanocephalus* and *C. micropterus* (Coleoptera, Carabidae) in the northern taiga. – Zool. Zh. 85: 1196-1204 [in Russian].

Filippov, B.Yu. (2007). Seasonal life cycles in some species of ground beetles in southern tundra. – Russ. Entomol. J. 16: [in press].

Gilbert, O. (1956). The natural histories of four species of *Calathus* (Coleoptera, Carabidae) living on sand dunes in Anglesey, North Wales. – Oikos. 7: 22-47.

Greenslade, P.J.M. (1965). On the ecology of some British carabid beetles with special reference to life histories. – Trans. Soc. Brit. Ent. 16: 149-179.

Grüm, L. (1975). Growth rate of the *Carabus* L. larvae (Col. Carabidae). – Ekol. Polska, Ser. A. 23: 667-671.

Günther, J.M. & Assmann, Th. (2000). Competition in the woodland? Phenology, body mass and body length of coexisting *Carabus* species – preliminary results (Coleoptera, Carabidae). – In: Natural history and applied ecology of Carabid beetles (Brandmayr, P., Lövei, G.L., Zetto-Brandmayr, T., Casale, A. & Vigna Taglianti, A, eds). Pensoft, Sofia-Moscow, p. 185-195.

Hamilton, C.C. (1925). Studies on the morphology, taxonomy and ecology of the larvae of Holarctic tiger beetles (family Cicindelidae). – Proc. U.S. Nat. Mus. 65: 1-87.

Hemmer, J & Terlutter, H. (1991). Untersuchungen zur Höhenverbreitung von Carabiden am Alpen-Nordrand (Coleoptera, Carabidae). – NachrBl. Bayer. Ent. 40: 72-79.

Houston, W.W.K. (1981). The life cycles and age of *Carabus glabratus* Paykull and *C. problematicus* Herbst. (Col.: Carabidae) on moorland in northern England. – Ecol. Entomol. 6: 263-271.

Hůrka, K. (1973). Fortpflanzung und Entwicklung der mitteleuropäischen *Carabus* und *Procerus*-Arten. – Studie ČSAV. 9: 1-78.

Hůrka, K. (1975). Laboratory studies on the life cycle of *Pterostichus melenarius* (Ill.) (Coleoptera, Carabidae). – Věst. Čs. Společ. Zool. 39: 265-274.

Hůrka, K. (1986). The developmental type of Carabidae in the temperate zones as a taxonomic character. – In: Carabids beetles: their adaptations and dynamics (den Boer, P.J., Mossakowski, D., Luff, M.L. & Weber, F., eds). Gustav Fisher Verlag, Stuttgart-New York, p. 187-193.

Inyaeva, Z.I. (1965). Ground beetles on the field crops. – Thesis of diss. PhD. Moscow State University Press, Moscow, p. 1-20 [in Russian].

Jakuczun, L. (1978). Fenologia rozdoru czterech wysokogórskich gatunkow chrząsczy biegaczowatych (Coleoptera, Carabidae) w Tatrach. – Polskie pismo Entomol. 48: 573-581.

Jones, M.G. (1979). The abundance and reproductive activity of common carabidae in a winter wheat crop. – Ecol. Entomol. 4: 31-43.

Jørum, P. (1976). Life cycle and population density of *Nebria brevicollis* F. (Coleoptera: Carabidae) in a Danish beech forest. – Vidensk. Medd. Dansk. nat. Foren. 139: 245-261.

Jørum, P. (1980). Life cycles and annual activity patterns of *Pterostichus melanarius* (Illig.) and *P. niger* (Schall.) (Coleoptera: Carabidae) in a Danish beech wood. – Ent. Medd. 48: 19-25.

Jørum, P. (1985). Life cycle and phenology of *Carabus problematicus* Herbst, 1786 in Denmark (Coleoptera: Carabidae). – Ent. Medd. 53: 27-30.

Kålås, J.A. (1985). Species composition and seasonal activity patterns of Carabidae (Col.) in a small deciduous forest in western Norway. – Fauna Norvegica. Ser. B. 32: 28-32.

Karpova, V.E. & Matalin, A.V. (1990). Functional age structure of *Anisodactylus signatus* Panz. populations (Coleoptera, Carabidae) in the steppe zone. – In: Advances of entomology in USSR: Coleoptera (Medvedev, G.S. & Korotyaev, B.A., eds). Zoological Institute of USSR Acad. Sci. Press, Leningrad, p. 62-63 [in Russian].

Kasandrova, V.I. & Sharova, I.Kh. (1971). Development of *Amara ingenua*, *Anisodactylus signatus* and *Harpalus distinguendus* (Coleoptera, Carabidae). – Zool. Zh. 50: 215-220 [in Russian].

Khobrakova, L.Ts. & Sharova, I.Kh. (2005). Life cycles of ground beetles (Coleoptera, Carabidae) from the mountain taiga and mountain forest-steppe in the Eastern Sayan. – Izv. Ross. Acad. Nauk, Ser. Biol. 6: 688-693 [in Russian].

Khruleva. O.A. (1994). Life cycle of the leaf beetle *Chrysolina subsulcata* (Coleoptera, Chrysomelidae) on Wrangel Island. – Zool. Zh. 73: 29-38 [in Russian].

Klausnitzer, B. (1991). Wunderwelt der Käfer. – Edition Leipzig, 212 ss.

Koval', A.G. (2005) Ground beetles (Coleoptera, Carabidae) of fields of vegetable nightshade (species composition, ecology, biology, entomophages of Colorado beetle). – Thesis of diss. PhD. All-Russian Institute of Plant Protection Press, St.-Petersburg, p. 1-22 [in Russian].

Korobeinokov, Yu.I. (1991). About some features of seasonal dynamics activity and reproduction of ground beetles in sub-Arctic conditions. – In: The problems of soil zoology (Mordkovich, V.G., Stebaeva, S.K. & Grishina, L.G., eds). Eastern-Siberian Forest Factory Press, Novosibirsk, p. 127-128 [in Russian].

Krehan, I. (1970). Die Steuerung von Jahresrhythmik und Diapause bei Larval- und Imago-überwintern der Gattung *Pterostichus* (Col., Carab.). – Oecologia. 6: 58-105.

Kůrka, A. (1975). The life cycle of *Bembidion tibiale* (Coleoptera, Carabidae). – Acta Entomol. Bohemoslov. 72: 374-382.

Kryutchkova, E.E. & Panov, A.A. (1988). Activity alterations in cerebral neurosecretory cells in three species of ground-beetles (Coleoptera, Carabidae) related to the annual reproduction cycle. – Zool. Zh. 67: 1661-1669 [in Russian].

Lampe, K.H. (1975). Die Fortpflanzungsbiologie und Ökologie des Carabiden *Abax ovalis* Dft. und der Einfluss der Umweltfaktoren Bodentemperatur, Bodenfeuchtigke und Photoperiode auf die Entwicklung in Anpassung an die Jahreszeit. – Zool. Jb. Syst. 102: 128-170.

Larsson, S.G. (1939). Entwicklungstypen und Entwicklungszeiten der dänischen Carabiden. Kommission Hos P. Haase & Søns Forlag, København. – Entomologiske Meddelelser. 20: 277-562.

Lindroth, C.H. (1945). Die fennoskandischen Carabidae, Eine tiergeographische Studie. I. Spezieller Teil. – Bröderna Lager-ström Boktryckare, Göteborgs, 709 p.

Loreau, M. (1985). Annual activity and life cycles of carabid beetles in two forest communities. – Holarctic Ecology. 8: 228-235.

Luff, M.L. (1973). The annual activity pattern and life cycle of *Pterostichus madidus* F. (Col. Carabidae). – Ent. Scand. 4: 259-273.

Makarov, K.V. (1989). Morpho-ecological characteristics of life cycles in the ground beetles of genus *Carabus* (L.) Thoms. – In: Life forms and strategy of organisms and their use for bioindication of environment conditions (Bogach, Ya.Ya., ed). Institute of Landscape Ecology of Acad. Sci. of Czech. SSR, České Budějovice, p. 22-23 [in Russian].

Makarov, K.V. (1994). Annual reproduction rhythm of ground beetles: a new approach to the old problem. – In: Carabid beetles: ecology and evolution (Desender, K., Dufrêne, M., Loreau, M., Luff, M.L. & Maelfait, J-P., eds). Kluwer Acad. Publ., Dordrecht-Boston-London, p. 177-182.

Makarov, K.V. & Chernyakhovskaya, T.A. (1989). Variability of the seasonal dynamics activity of the ground beetle *Pterostichus melanarius* Ill. (Coleoptera, Carabidae) in different types of forest. – In: Ecological aspects of adequate nature management (Tyuryapin, I, Ventin'sh, Ya., Il'in, V., Tarvid, I., Utrans, A. & Sheshukova, T., eds). Institute of Biology of Acad. Sci. of Lithuanian SSR Press, Riga, p. 55-56 [in Russian].

Makarov, K.V. & Chernyakhovskaya, T.A. (1990). Phenology of development and population structure of *Loricera pilicornis* (F.) (Coleoptera, Carabidae) in an agrocenosis. – In: Structure and dynamics of populations of edaphic and herpetobiotic invertebrates (Sharova, I.Kh., ed). Moscow State Pedagogical Institute Press, Moscow, 1: 21-32 [in Russian].

Matalin, A.V. (1994). The strategy of dispersal behaviour in some Carabidae species of southeastern Europe. – In: Carabid beetles: ecology and evolution (Desender, K., Dufrêne, M., Loreau, M., Luff, M.L. & Maelfait, J-P., eds). Kluwer Acad. Publ., Dordrecht-Boston-London, p. 183-188.

Matalin, A.V. (1997a). Peculiarities of spatial and temporal differentiation of carabids (Coleoptera, Carabidae) in the steppe zone. – Entomol. Rev. 77: 1155-1166.

Matalin, A.V. (1997b). Life cycles of carabids of the genus *Stenolophus* (Coleoptera, Carabidae) in the steppe zone of Europe. – Entomol. Rev. 77: 1181-1190.

Matalin, A.V. (1997c). Specific features of life cycle of *Pseudoophonus* (s. str.) *rufipes* Deg. (Coleoptera, Carabidae) in southwestern Moldova. – Biol. Bull. 24: 371-381.

Matalin, A.V. (1998a). Life cycles of ground beetles of the tribe Harpalini (Coleoptera, Carabidae) on the European subcontinent: the possible ways of their origin and transformation. – In: The problems of entomology in Russia (Medvedev, G.S., ed). Zoological Institute of Russ. Acad. Sci. Press, St.-Petersburg, 2: 25-26 [in Russian].

Matalin, A.V. (1998b). Polyvariance of *Harpalus affinis* Schrank (Coleoptera, Carabidae) and its adaptive significance. – Biol. Bull. 25: 402-410.

Matalin A.V., 2003. Variations in flight ability with sex and age in ground beetles (Coleoptera, Carabidae) of south-western Moldova. – Pedobiologia. 47: 311-319.

Matalin, A.V. (2006). Geographic variability of the life cycle in *Pterostichus melanarius* (Coleoptera, Carabidae). – Entomol. Rev. 86: 409-422.

Matalin, A.V. (2007). Typology of life cycles of ground beetles (Coleoptera, Carabidae) from Western Palaearctic. – Zool. Zh. 86: 1196-1220 [in Rusian].

Matalin, A.V. & Budilov, P.V. (2003). Geographical variability of sexual and age structure of populations and the life cycle in *Broscus cephalotes* (Coleoptera, Carabidae). – Entomol. Rev. 83: 1037-1045.

Matalin, A.V. & Makarov, K.V. (2006). Life cycle of the halophilic carabid beetle *Pogonus* (s. str.) *transfuga* Chaudoir, 1871 (Coleoptera: Carabidae) in the Elton region. – In: Biodiversity and problems of nature management in Elton region (Kursakova, N.A., Nekrutkina, Yu.A., Sokhina, E.N. & Chernobai, V.F. eds). PrinTerra, Volgograd, p. 40-46.

Matalin, A.V., Makarov, K.V. & Bokhovko, E.E. (2007). Estimation of abundance and spatial distribution of the ground beetles (Coleoptera, Carabidae) in natural and anthropogenic conditions. – In: Problems and perspectives of general entomology (Zamotailov, A.S. ed). KubGAU Press, Krasnodar, p. 224-225 [in Russian].

Murdoch, W.W. (1967). Life history patterns of some British Carabidae (Coleoptera) and their ecological significance. – Oikos. 19: 25-32.

Nekuliseanu, Z.Z. (1987). Biology of *Pogonus litoralis* Duft. (Coleoptera, Carabidae). – In: The problems of soil zoology (Kurashvili, B.E., ed). Metscniereba, Tbilisi, p. 199 [in Russian].

Nekuliseanu, Z.Z. (1990). Specific biological features of some ground beetles of genus *Pterostichus* Bon. (Coleoptera, Carabidae) in Moldova. – In: Advances of entomology in USSR: Coleoptera (Medvedev, G.S. & Korotyaev, B.A., eds). Zoological Institute of USSR Acad. Sci. Press, Leningrad, p. 96-98 [in Russian].

Nekuliseanu, Z.Z. (1994). Biologia unor Carabidae ale genului *Amara* Bon. (Coleoptera, Carabidae) din Moldova. – Bul. Acad. Ştiin. Rep. Moldova. Ştiin. boil. chim. 1: 37-41 [in Moldavian].

Nelemans, M.N.E. (1987). On the life history of the carabid beetle *Nebria brevicollis* (F.). Egg production and larval growth under experimental conditions. – Neth. J. Zool. 37: 26-42.
Ortuño, V.M. & Marcos, J.M. (1997). Carabidofauna (Insecta, Coleoptera) del área natural de Salburua (Vitoria, Álava). – Ecologia. 11 457-482.
Paarmann, W. (1974). Der Einfluss von Temperatur und Lichtwechsel auf die Gonadenreifung des Laufkäfers *Broscus laevigatus* Dej. (Coleoptera, Carabidae) aus Nordafrika. – Oecologia. 15: 87-92.
Paarmann, W. (1976a). The annual periodicity of the polyvoltine ground-beetle *Pogonus chalceus* Marsh. (Col. Carabidae) and its control by environmental factors. – Zool. Anz. 196: 150-160.
Paarmann, W. (1976b). Die Bedeutung exogener Faktoren für die Gonadenreifung von *Orthomus barbarus atlanticus* (Coleoptera, Carabidae) aus Nordafrika. – Ent. Exp. et Appl. 19: 23-26.
Paarmann, W. (1979). Ideas about the evolution of the various annual reproduction rhythms in carabid beetles of the different climatic zones. – In: On the evolution of behaviour in Carabid beetles (den Boer, P.J., Thiele, H.U. & Weber, F., eds). H. Veeneman & Zonen B.V. Publ., Wageningen. 18: 119-132.
Paarmann, W. (1990). *Poecilus lepidus* Leske (Carabidae, Coleoptera), a species with the ability to be a spring and autumn breeder. – In: The Role of Ground Beetles in ecological and environmental studies (Stork, N.E., ed). Intercept Publishers, Andover-Hampshire, p. 259-267.
Paarmann, W. (1994). Temperature and photoperiodic influence on development stages and adults of the subarctic carabid beetles *Pterostichus adstrictus* (Coleoptera). – In: Carabid beetles: ecology and evolution (Desender, K., Dufrêne, M., Loreau, M., Luff, M.L. & Maelfait, J-P., eds). Kluwer Acad. Publ., Dordrecht-Boston-London, p. 201-205.
Penney, M.M. (1969). Diapause and reproduction in *Nebria brevicollis* (F.) (Coleoptera, Carabidae). – J. Anim. Ecol. 38: 219-233.
Potapova, N.A. (1972). Seasonal dynamics of activity of Carabidae (Coleoptera) in semidesert of the north-west Kazakhstan. – Zool. Zh. 60: 1651-1658 [in Russian].
Refseth, D. (1980). Differences in seasonal activity pattern and breeding time of *Patrobus atrorufus* (Carabiade) in central Norway. – Holarctic Ecology. 3: 87-90.
Refseth, D. (1984). The life cycle and growth of *Carabus glabratus* and *C. violaceus* in Budalen, central Norway. – Ecol. Entomol. 9: 449-455.
Refseth, D. (1988). Annual patterns of activity, reproduction and development in some Norwegian Carabidae (Col.). – Fauna Norvegica. 35: 21-30.
Schatz, I. (1994). Life strategy of an alpine Carabid: *Pterostichus jurinei* (coleoptera, Carabidae). – In: Carabid beetles: ecology and evolution (Desender, K., Dufrêne, M., Loreau, M., Luff, M.L. & Maelfait, J-P., eds). Kluwer Acad. Publ., Dordrecht-Boston-London, p. 213-217.
Sharova, I.Kh. (1981). Morpho-ecological typification of life cycles of ground beetles (Coleoptera, Carabidae). – Proc. All-Union Entomol. Soc. 63: 59-61 [in Russian].
Sharova, I. Kh. & Denisova, M.I. (1995). Life cycles and reproduction rhythms of *Calathus erratus* C.R. Sahlberg and *C. micropterus* Duftschmidt (Coleoptera, Carabidae) in the forest of Russian forest-steppe zone. – Russ. Ent. J. 4: 159-165.
Sharova, I.Kh. & Denisova. M.I. (1997). Seasonal dynamics of carabid populations from the genus *Pterostichus* (Coleoptera, Carabidae). – Zool. Zh. 76: 418-427 [in Rissian].

Sharova, I.Kh. & Dushenkov, V.M. (1979). Types of development and types of seasonal activity in ground beetles. – In: Fauna and ecology of invertebrate animals (Chernova, N.M., ed), Moscow State Pedagogical Institute Press, Moscow, p. 5-25 [in Russian].

Sharova, I.Kh. & Filippov, B.Yu. (2003). Peculiarities of ground beetles (Coleoptera, Carabidae) life cycles in northern taiga. – Zool. Zh. 82: 229-238 [in Russian].

Sharova, I.Kh. & Khobrakova, L.Ts. (2005). Features of the life cycles of *Pterostichus montanus* (Motschulsky, 1844) and *Carabus loschnikovi* (Fischer-Waldheim, 1822) in conditions of the mountain taiga zone in the Eastern Sayan. – Izv. Ross. Acad. Nauk, Ser. Biol. 1: 36-46 [in Russian].

Shelford, V.E. (1908). Life-histories and larval habits of the tiger beetles (Cicindelidae). – J. Linn. Soc. London. 30: 157-184.

Skuhravý, V. (1959). Příspěvek k bionomii polních střelíkovitých (Col. Carabidae). – Rozpr. Českosl. Akad. Věd. Řada matemat. a přírodn. věd. 69: 3-64.

Sota, T. (1985). Life history patterns of carabid beetles belonging to the subtribe Carabina (Coleoptera, Carabidae) in the Kinki district, W. Japan. – Kontyû. 53: 370-378.

Sota, T. (1986). Effects of temperature and photoperiod on the larval development and gonad maturation of a carabid beetle, *Carabus yaconinus* (Coleoptera: Carabidae). – Appl. Ent. Zool. 21:89-94.

Sota, T. (1987). Effects of temperature and photoperiod on the larval hibernation and adult aestivation of *Leptocarabus kumagaii* (Coleoptera: Carabidae). – Appl. Ent. Zool. 22: 617-623.

Sota, T. (1994). Variation of carabid life cycles along climatic gradients: an adaptive perspective for life-history evolution under adverse conditions. – Series Entomologica. 52: 91-112.

Sota, T. (1996). Altitudinal variation in life cycles of carabid beetles: life-cycle strategy and colonization in Alpine zones. – Arctic Alpine Res. 28: 441-447.

Sota, T. & Ishikawa, R. (2004). Phylogeny and life-history evolution in *Carabus* (subtribe Carabina: Coleoptera, Carabidae) based on sequences of two nuclear genes. – Biol. J. Linn. Soc. 81: 135-149.

Sparks, T.H., Buse, A. & Gadsden, R.J. (1995). Life strategy of *Carabus problematicus* (Coleoptera, Carabidae) at different altitudes on Snowdon, north Wales. – J. Zool. 236: 1-10.

Stiprais, M.A. (1970). Growing of *Carabus glabratus* Payk. – LVU Zool. Muz. Raksiti. 4: 103-106 [in Russian].

Thiele, H.-U. (1977). Carabid beetles in their environments. A study on habitat selection by adaptations in physiology and behaviour. Springer-Verlag, Berlin-Heidelberg-New York, XVII+369 p.

Turin, H. (2000). De Nederlandse loopkevers: verspreiding en oecologie (Coleoptera: Carabidae). – Nat. Natuurhistorisch Mus. Natur., Tekeningen, 666 p.

Turin, H., Penev, L. & Casale A. (Eds) (2003). The genus Carabus in Europe. A synthesis. – Pensoft, Sofia-Moscow, 511 p.

van der Drift, J. (1951). Analysis of the animal community in a beech forest floor. – Tijdschr. Ent. 94: 1-168.

van der Drift, J. (1958). Life cycle and phenology of *Carabus problematicus* Herbst. – In: Proceeding of X International Congress of Entomology. 2: 661.

van Dijk, Th.S. (1972). The significance of the diversity in age composition of *Calathus melanocephalus* L. (Coleoptera, Carabidae) in space and time at Schiermonnikoog. – Oecologia. 10: 111-136.

van Dijk, Th.S. (1979). Reproduction of young and old females in two carabid beetles and the relationship between the number of eggs in the ovaries and number of eggs laid. – In: On the evolution of behaviour in Carabid beetles (den Boer, P.J., Thiele, H.U. & Weber, F., eds). H. Veeneman & Zonen B.V. Publ., Wageningen. 18: 167-183.

van Dijk, Th.S. (1994). On the relationship between food, reproduction and survival of two carabid beetles: *Calathus melanocephalus* and *Pterostichus versicolor*. – Ecol. Entomol. 19: 263-270.

van Heerdt, P.F., Blokhuis, P. & van Haaften, C. (1976). The reproductive cycle and age composition of a population of *Pteropstichus oblongopunctatus* F. in the Netherlands (Coleoptera: Carabidae). – Tijdschr. Ent. 119: 1-13.

van Schaick-Zillesen, P.G., Brunsting, A.M.H. & Siepel, H. (1986). The role of larvae in the synchronization of the life cycle of *Pterostichus oblongopunctatus*. – In: Carabids beetles: their adaptations and dynamics (den Boer, P.J., Mossakowski, D., Luff, M.L. & Weber, F., eds). Gustav Fisher Verlag, Stuttgart-New York, p. 181-185.

Vasil'eva, R.M. (1978). Specific features of development of some species of ground beetles in the Bryansk province. – In: Fauna and ecology of invertebrates (Naumov, S.P., ed), Moscow State Pedagogical Institute Press, Moscow, p. 40-52 [in Russian].

Vlijm, L. & van Dijk, Th.S. (1967). Ecological studies on carabid beetles II. General pattern of population structure in *Calathus melanocephalus* L. at Sciermonnikoog. – Zeitschr. Morph. Ökol. Tiere 58: 396-404.

Wallin, H. (1987). Distribution, movements and reproduction of Carabid beetles (Coleoptera, Carabidae) inhabiting cereal fields. Plant Protection Reports and Dissertations of the Swedish Univ. for Agric. Sci., Uppsala, 25 + 109 p.

Weber, F. & Klenner, M. (1987). Life history phenomena and risk of extinction in a subpopulation of *Carabus auronitens*. – Acta Phytopath. Entom. Hung. 22: 321-328.

Zetto Brandmayr, T. (1983). Life cycle, control of propagation rhythm and fecundity of *Ophonus rotundicollis* Fairm.et. Lab. (Coleoptera, Carabidae, Harpalini) as an adaption to the main feeding plant *Daucus carota* L. (Umbelliferae). – In: The synthesis of field studies and laboratory experiment (Brandmayr, P., den Boer, P.J. & Weber, F., eds). Centre for Agricultural Publishing and Documentation (PUDOC), Wageningen, p. 93-103.

L. Penev, T. Erwin & T. Assmann (Eds) 2008
BACK TO THE ROOTS AND BACK TO THE FUTURE. TOWARDS A NEW SYNTHESIS AMONGST
TAXONOMIC, ECOLOGICAL AND BIOGEOGRAPHICAL APPROACHES IN CARABIDOLOGY
Proceedings of the XIII European Carabidologists Meeting, Blagoevgrad, August 20-24, 2007, pp. 285-304.

Sofia–Moscow

Adaptive radiation of carabid larvae (Coleoptera, Carabidae)

Inessa Kh. Sharova

Department of Ecology & Zoology, Moscow State Pedagogical University,
Kibalchicha str. 6, Build. 5, Moscow 129164, Russia

SUMMARY

Synthesis of numerous literature sources on carabid larvae morphology and adaptations made possible to develop a new version of life forms system, based on previously stated hierarchical foundations. It reflects a large variety of carabid larvae adaptations, connected with type of feeding, capturing prey, life in different levels of biogeocenosis etc. On the basis of this updated system, a hypothesis of main trends of adaptive radiation of carabid larvae is set up. The general trend is the diversification of zoophags, which reflects the expansion in all possible habitat's layers: phytobionts, epigeobionts, stratobionts, geobionts and ambush-predators. The second trend is the transformation from zoophagy through mixophyto-, to phytophagy. During the comparison of the adaptive radiation of Carabidae adults and larvae, there were revealed the trends of conjugated evolution, as well as distinct amphygenesis.

Two systems of larval Carabidae life forms are present now (Sharova, 1981; Zetto Brandmayr et al., 1998). Synthesis of the extensive literature on their morphology and adaptations made it possible to develop a new system, based on previously stated hierarchical foundations (Sharova, 1981). It reflects a large variety of adaptations of larval Carabidae, connected with the type of feeding, capturing prey, life in different levels of ecosystem, etc. On the basis of this updated system, a hypothesis of the main trends of adaptive radiation of carabid larvae is set up below. This renewed system of life forms of larval Carabidae was constructed on the basis of authors and their foreign colleagues' research. It is formed by principles of hierarchy and adaptive convergence (Sharova, 1981, 2002).

Keywords: Carabidae, larvae, adaptive radiation, life forms, classifications

CONSPECT OF THE SYSTEM OF LIFE FORMS OF CARABIDAE LARVAE

I class: ZOOPHAGES
- 1 subclass Phytobionts (runners or climbers)
  - A series Epiphytobionts with elongated legs and sensitive urogomphi (type – *Agra*)
  - B series Intraphytobionts with shortened legs and reduced urogomphi (type – *Dromius*)
- 2 subclass Epigeobionts-runners, hemicryptobionts
  - A group Campodeiform with elongated urogomphi (type – *Drypta*)
  - B group Cychrus-like forms with shortened urogomphi, flattened body and broad tergites (type – *Cychrus*)
- 3 subclass Stratobionts- runners, campodeiform.
  - A series Hemicryptobionts (type – *Nebria*)
  - B series Cryptobionts (type – *Synuchus*)
- 4 subclass Geobionts (runners-burrowers, burrowers only)
  - A series Soil cave dwelling-forms (type – *Dyschirius*, *Clivina*)
  - B series Burying-forms (type – *Carabus*)
  - C series Burrowing-forms (type – *Broscus*)
- 5 subclass Ambush predators (burrowers, or gnawing-forms)
  - A series S-like (type – *Cicindela*)
  - B series Sacciform (type – *Orthogonius*)

II class MIXOPHYTOPHAGES (running-burying forms, burrowers)
- 1 subclass Mixophytophages stratobionts, running-burying forms (type – *Stenolophus*)
- 2 subclass Mixophytophages stratogeobionts, running-burrowing forms (type – *Curtonotus*)
- 3 subclass Mixophytophages geobionts, burrowers, cryptobionts (type – *Harpalus*)

III class PHYTOPHAGES (running-burrowing forms, burrowers)
- 1 subclass Epigeobionts, running-burrowing forms (type – *Zabrus*)
- 2 subclass Stratogeobionts, running-burrowing forms (type – *Ophonus*)
- 3 subclass Geobionts, burrowers, cryptobionts (type – *Ditomus*)

IV class MYCETOPHAGES (gnawing forms (type – *Mormolyce*))

V class SYMPHYLOUS (slow-moving cryptobionts)
- 1 subclass Myrmecophilous (mainly symbionts) (type – *Paussus*)
- 2 subclass Termitophilous (mainly inquilines) (type – Ozaenini)

VI class ECTOPARASITOIDS (slow-moving cryptobionts (type – *Brachinus*))

VII class APHAGES (troglobionts (type – *Aphaenops*))

REVIEW OF SYSTEM OF CARABIDAE LARVAE LIFE FORMS

I class. ZOOPHAGES

Zoophages – predatory larvae, comprising the majority of Carabidae. Usually they have sharp elongated mandibles with large retinaculum. The body shape is mostly campodeiform. Legs elongated, can be adapted for running, borrowing, climbing or rarely, prehensile.

Zoophagous larvae have various adaptive habits depending on trophic specialization, layer of habitat, type of locomotion and feeding, and the degree of cryptic life-style. The complex of adaptations to layer of habitat and type of locomotion are the main criteria for the establishment of subclasses within this large larval class (Sharova, 2002).

1 subclass. Phytobionts (runners and climbers, or runners only).

Predatory larvae inhabit plant surfaces. Most of them have running-climbing legs with notched tarsungulus, often with pulvilla. Only minute subcortical forms have running legs.

Among phytobionts, two groups can be distinguished (Figs 1-2) by adaptations to movement on plant surfaces – epiphytobionts, with sclerotized body, well developed sensory organs and climbing legs; and flattened intraphytobionts, which inhabit subcortical surfaces and have adaptations to movement in narrow spaces. They are characterized by hemi cryptic life-style.

Epiphytobionts – habitats on different plants of tropical and subtropical forest regions. Usually they are brightly colored, with hyperchaetosis, elongated legs and urogomphi (*Parena*, *Agra*, *Catascopus*), sensory organs well developed (van Emden, 1942; Habu & Sadanaga, 1967; Kobayashi, Kudagamage & Nugaliyadde, 1995; Arndt, Kirmse & Erwin, 2001).

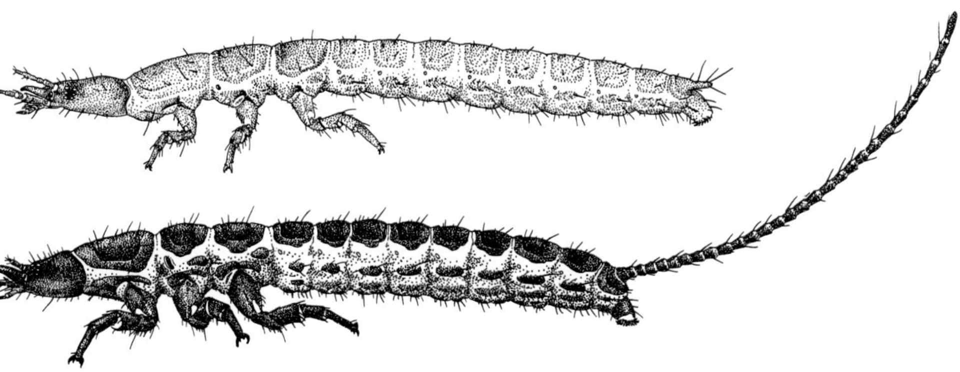
1
2

Figs 1-2. Habitus of phytobiont carabid larvae: 1 – epiphytobiont (*Agra* sp., reconstruction after Arndt et al., 2001); 2 – intraphytobiont (*Dromius* sp.)

Intraphytobionts composed of larvae with narrowed body and reduced urogomphi (ex.- *Dromius*) –(Gardner, 1936; van Emden, 1942; Mahar, Stehr & Simmons, 1983; Arndt, 1989; Casale, Giachino & Pantaleoni, 1996) and flattened larvae of Physocrotaphini (Moore, 1998).

2 subclass. Epigeobionts-runners (hemicryptobionts)

These larvae mainly inhabit the soil surface. They have long, running legs, sclerotized body, well developed eyes and antennae (Figs 3-4). This subclass can be subdivided into groups of larvae with different habitus and life-style: campodeiform (*Drypta*-like) and flattened, *Cychrus*-like (van Emden, 1942; Sharova, 1981; Casale & Vigna Taglianti, 1991; Arndt, & Drechsel, 1998)

3 subclass. Stratobionts-runners

They inhabit litter, upper layers of soil, holes and caves. Legs and urogomphi usually of average length or shortened. They are active runners, and catch their prey on soil surface, litter, or in mammal holes.

Among stratobionts two series are distinguished: hemicryptobionts and cryptobionts (Figs 5-6).

A series: Hemicryptobionts (type – *Nebria*) hunt on soil surface (Bauer, 1982), litter and in cavities of upper layers of soil. They have average to elongated, sclerotized body, well developed eyes and elongated urogomphi (Andersen, 1970; Luff, 1972; Spence

3

4

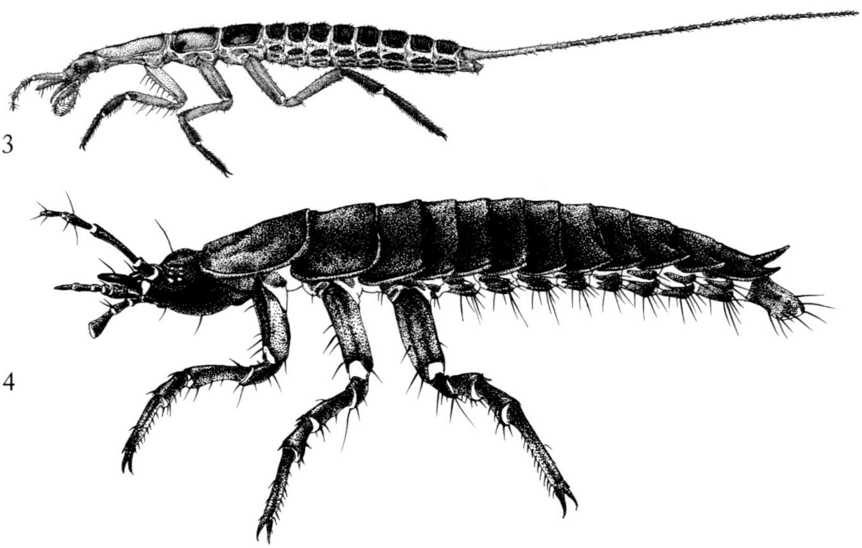

Figs 3-4. Habitus of epigeobiont carabid larvae: 3 – campodeiform (*Galerita* sp.); 4 – *Cychrus*-like (*Cychropsis* sp.).

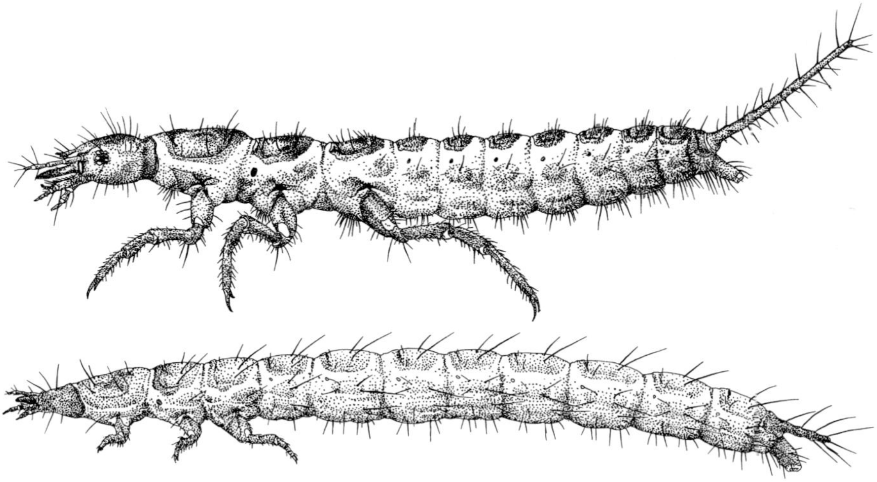

Figs 5-6. Habitus of stratobiont carabid larvae: 5 – hemicryptobiont (*Nebria livida* (Linnaeus, 1758)); 6 – cryptobiont (*Synuchus nivalis*).

& Sutcliffe, 1982; Liebherr, 1983; Bousquet, 1987; Bousquet & Smetana, 1991). The Hemicryptobiont series is composed of three groups, which differ in adaptations for living in different soil layers: surface or litter, litter only and litter or soil stratobionts (Davies, 1963; Hůrka, 1966, 1998; Grebennikov & Bousquet, 1999).

B series: Cryptobionts (type – *Synuchus*). Larvae with cryptic life-style, which inhabit litter and soil. They have depigmented body, only head and pronotum sclerotized, the remaining tergites soft and pale. Eyes small, usually partly reduced. Among cryptobionts three groups are distinguished: litter-soil inhabitants, bothrobionts and troglobionts. (Jeannel, 1926, 1941; Boldori, 1934, 1935, 1958; Lindroth, 1956; Sharova, 1958, 1981; May, 1963; Bourne, 1975; Thompson, 1979; Avon, 1995; Grebennikov, 1996, 2002; Zamotajlov, 2001; Makarov & Koval, 2003)

4 subclass. Geobionts (runners-burrowers, burrowers only)

These larvae build galleries in soil. They have various body shapes and styles of burrowing. The characteristic feature of all geobionts is the presence of stout running-burrowing or burrowing type of legs with rigid setae. The head is usually of the burrowing type, often stout urogomphi are present. Among geobionts there are hemicryptobionts with pigmented bodies, hunting on the surface, as well as cryptobionts– depigmented, always living and hunting in soil. The Geobiont subclass can be divided into three series (Figs 7-9):

A series: soil diggers (Luff, 1978; Vanek, 1984; Bousquet, 1988) without stout urogomphi and digging legs, but with burrowing type of head (soil hemicryptobionts (type – *Dyschirius*) and cryptobionts (type – *Clivina*).

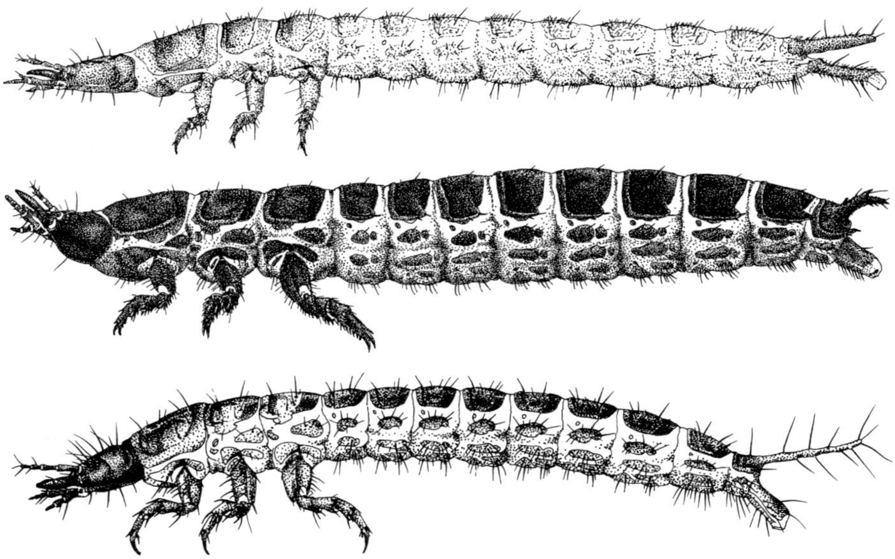

Figs 7-9. Habitus of geobiont carabid larvae: 7 – dweller of soil passages (*Clivina fossor* (Linnaeus, 1758)); 8 – burying form (*Calosoma denticolle* Gebler, 1833); 9 – burrowing form (*Broscus cephalotes* (Linnaeus, 1758)).

B series: soil hemicryptobionts with massive urogomphi, running-burrowing type of legs (van Emden, 1942; Lindroth, 1954; Sharova, 1957, 1981; Sturani, 1962; Luff, 1969; Hůrka, 1971; Mikhailov, 1978; Paarmann, 1979; Goulet, 1983; Rougemont, 1983; Makarov, 1987, 1992; Shilenkov & Berlov, 1987), include groups of surface-litter-soil (type – *Carabus*) or litter-soil dwellers (type – *Elaphrus*), and psammophilous sand burrowers (type – *Anthia*).

C series: soil dwellers with burrowing legs and elongated urogomphi (Sharova, 1958; 1981; Andersen, 1968; Harris, 1978; Luff, 1978; Sharova, Makarov, 1984; Makarov, 2005), include the following groups: surface-soil campodeiform hemicryptobionts (type – *Broscus*), psammophilous littoral burrowers, cryptobionts (type – *Omophron*), soil-interstitial hemicryptobionts (type – *Scarites terricola*) and psammophilous sand cryptobionts (type – *S. bucida*).

5 subclass. Hole-ambuscaders

Highly-specialized Carabidae larvae, which inhabit self-created holes for catching mobile insects. Head, prothorax and legs sclerotized, other parts of body, which are permanently situated in a hole – poorly sclerotized, depigmented, with supporting spines, urogomphi absent. Body S-shaped or sacciform.

A series: S-shaped hole-ambuscaders (Fig. 10). Body S-shaped, with supporting hooks or spines on middle abdominal tergites, legs of supporting-type (van Emden,

10

11

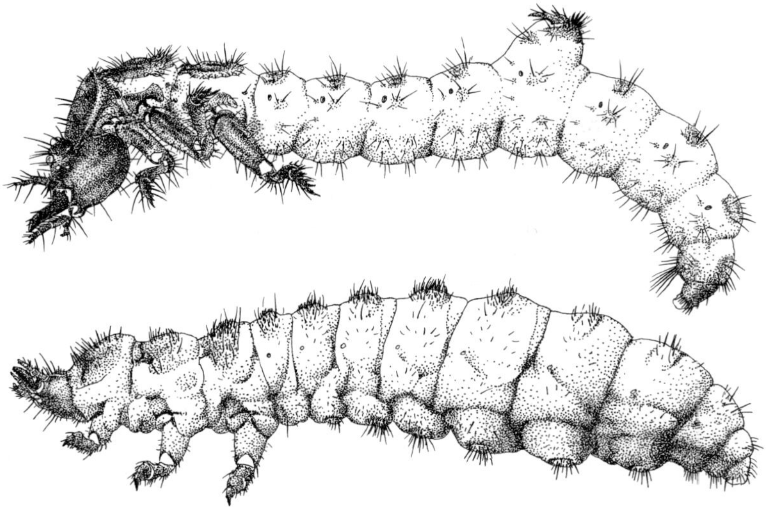

Figs 10-11. Habitus of ambush predators: 10 – S-like (*Cicindela sahlbergi* Fischer-Waldheim, 1824); 11 – sacciform (*Orthogonius* sp.).

1935; Ghilarov & Sharova, 1954; Moore, 1974; Palmer, 1976; Puchkov, 1993; Zetto Brandmayr, Marano & Paarmann, 1993; Arndt & Cassola, 2000; Putchkov & Dolin, 2005). During locomotion in a hole the larva leans on one side with the help of dorsal spines, and on the other – with legs and terminal part of the body. They can inhabit soil – soil geobionts (type – *Cicindela*), wood – gnawing dendrobionts (type – *Collyris*), ant hills – inquilines-myrmecophages (type – *Graphipterus*).

B series: Sacciform hole-ambuscaders, myrmecophilous or termitophilous (Fig. 11). They have sclerotized head, thorax and legs, sacciform abdomen, with soft and depigmented integument (Gardner, 1936; Moore, 1974; Erwin, 1981; Baehr, 1997; Makarov, 1998). Two groups can be distinguished: termitophagous inquilines (type – *Orthogonius*) living in horizontal holes and myrmecophagous inquilines (type – *Pseudomorpha*) inhabiting vertical holes.

II class. MIXOPHYTOPHAGES (running-burying forms, burrowers)

These larvae have mixed feeding including phytophagy, sporophagy and partly – zoophagy. To this group belongs the majority of Zabrini and Harpalini larvae. They have short and stout mandibles with broad molar parts, sometimes with additional denticles distally. A masticatory lobe of the maxilla is well developed. In Harpalini larvae, the inner part of the cardo usually forms a conspicuous, granulose or spinose knob for grinding plant fibers (van Emden, 1942). Head broad, without cervical part. Body campodeiform, fleshy, legs usually of running-digging type with stout spines. Urogomphi mostly short, barely longer than pygopodium. Among the mixophytophages, several subclasses are recognized, based on adaptations to living in different soil layers.

1 subclass. Running-burying stratobionts

Larvae of minute mixophytophages (Fig. 12), inhabit litter and upper layers of soil. They are less specialized to phytophagy and burrowing. Legs of running type, but with stout tarsungulus and spines; urogomphi elongated, mandibles short, not massive (Hůrka, 1975; 1997; Arndt, 1990, 1991; Matalin, 1995; 1998; 2001). This subclass consists of litter-soil hemicryptobionts (type – *Bradycellus*) and litter-soil cryptobionts (type – *Stenolophus*).

2 subclass. Running-burrowing stratogeobionts

Larvae of this subclass are surface-runners (Fig. 13), and easily burrow in loose soil (Forsskåhl, 1966; Bílý, 1975; Hůrka & Ducháč, 1980; Zetto Brandmayr, Marano & Pizzolotto, 1994). Consists of only surface-litter-soil hemicryptobionts (type – *Curtonotus*).

3 subclass. Geobionts, burrowers, cryptobionts

Soil dwellers with cryptic life-style (Fig. 14), specialized for living in soil and tunneling (Sharova, 1967; Habu & Sadanaga, 1970a, b; Bousquet & Tchang, 1992; Makarov &

12

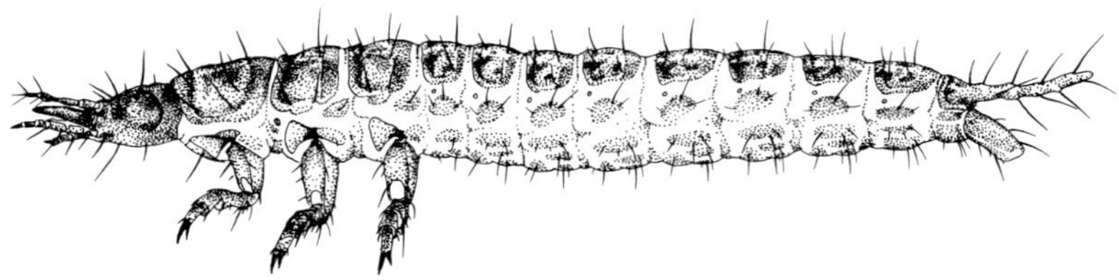

13

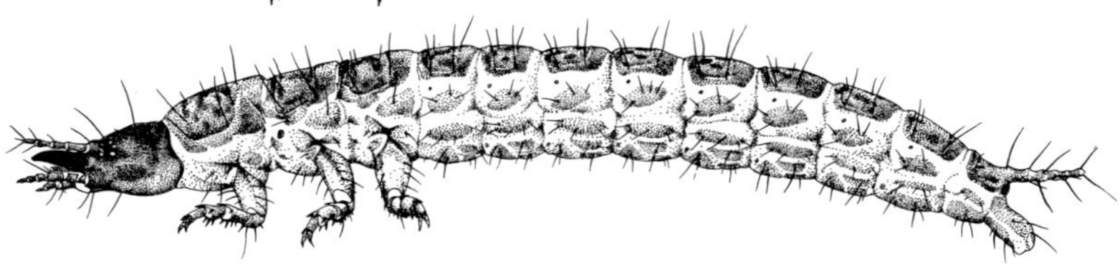

14

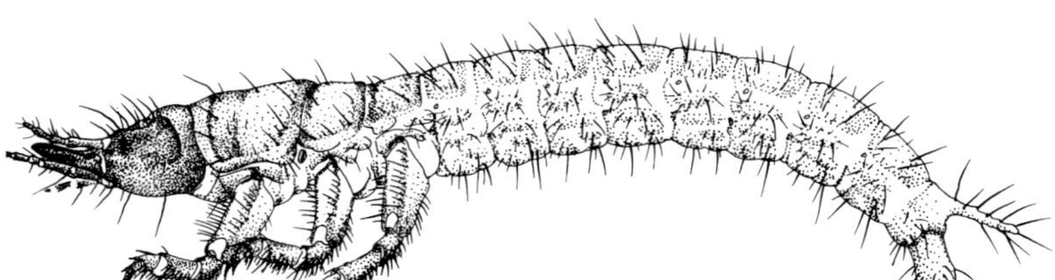

Figs 12-14. Habitus of mixophytophagous carabid larvae: 12 – stratobiont (*Dicheirotrichus ustulatus* (L. Redtenbacher, 1858)); 13 – stratogeobiont (*Curtonotus propinguus* (Ménétrié, 1832)); 14 – geobionts (*Harpalus* (*Haploharpalus*) sp.).

Brinev, 2001; Makarov, Shilenkov, 2001). Includes the following groups: soil harpaloids (type – *Harpalus*) and psammophilous (type – *Harpalodema*).

III class. PHYTOPHAGES (running-burrowing forms, burrowers)

This group consists of Zabrini and Harpalini larvae, specialized to feeding on plant tissues. Some of them still have the ability to be mixophytophagous, which might be evidence of evolutionary change from mixophagy to phytophagy. Phytophages have more specialized mouthparts: massive mandibles with apex often bifurcated; maxilla with rigid setae and large lacinia. Legs of running-burrowing or burrowing type. This class consists of three subclasses.

1 subclass. Running-burrowing epigeobionts

This subclass composed of a single group – zabroid hemicryptobionts (type – *Zabrus*). These larvae (Fig. 15) have adaptations to locomotion on surface and in soil. Body massive, well sclerotized (especially head, prothorax and legs), urogomphi short. Legs are of running-burrowing type with strong tarsungulus and supporting spines. Eyes well developed. They are typical phaeic hemicryptobionts, feeding on young cereals; most of them are wheat pests (Znojko, 1929; Arabadzhiev, Balevski, Drenski, Zakharieva & Radev 1953; Sharova, 1981; Makarov, Gurgenidze & Rekk, 1991).

2 subclass. Running-burrowing stratogeobionts

Consists of ophonoid hemicryptobionts (type – *Ophonus*). This group includes *Ophonus*, some *Harpalus* and *Amara* larvae, feeding mostly on plant seeds (Zetto Brandmayr & Brandmayr, 1975; Brandmayr & Zetto Brandmayr, 1981; Saska & Jarošik, 2001; Saska, 2004 a,b). Larvae are campodeiform with running-borrowing type of legs and elongated urogomhi (Fig. 16). Mandibles often with retinaculum shifted apically, and with bifurcated apex. For several species parental care is described.

3 subclass. Slow-moving geobionts, burrowers

Consists of ditomoid cryptobionts (type – *Ditomus*). These larvae are obligate seed-feeding phytophages, developing in holes with seeds, stored by adults (Brandmayr & Zetto Brandmayr, 1974; Sharova & Makarov, 1983). Body eruciform, S-shaped (*Ditomus*, *Chilotomus*), or cylindrical (*Machozetus*). Only the head, prothorax and legs are sclerotized (Fig. 17). Urogomphi absent, eyes and sensorial appendage of the third antennomere reduced.

IV class. MYCETOPHAGES (gnawing cryptobionts)

Larvae of Mormolycini develop in polyporous fungi of tropical forests of Indo-Malayan region (van Emden, 1942; Lieftinck & Wiebes, 1968). It seems that they originated from mixophytophages: head and prothorax sclerotized, the rest part of the body soft, legs poorly developed, urogomphi short (Fig. 18).

V class. SYMPHYLES (slow-moving cryptobionts)

Includes larvae of Paussini, Ozaenini – symbionts of ants and termites. Originated from zoophages. Sensory organs distinctly reduced, abdominal terminus with well developed

15

16

17

18

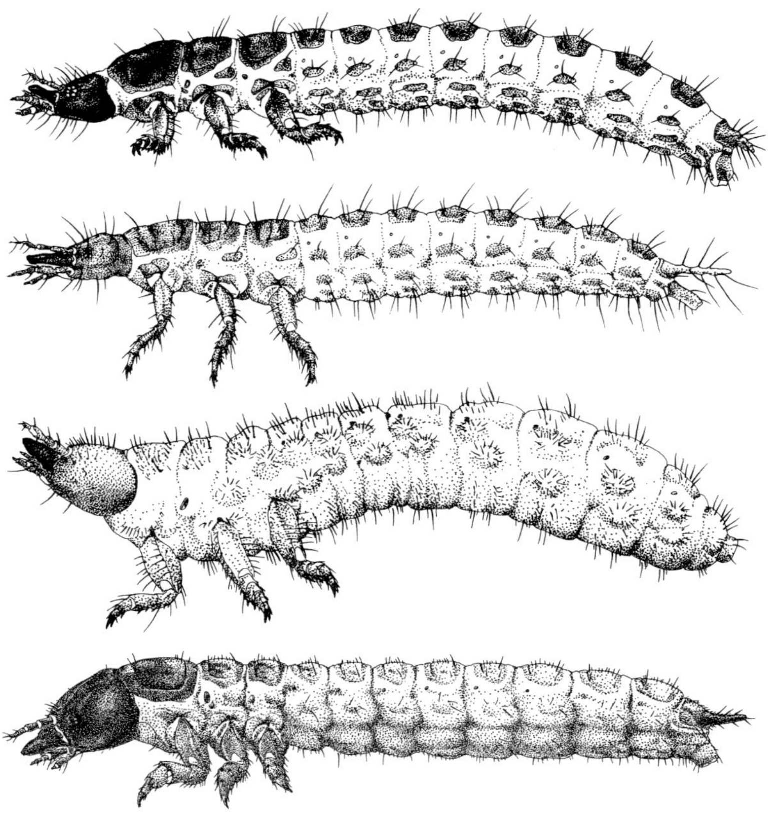

Figs 15-18. Habitus of mixophytophagous (15-17) and mycetophagous carabid larvae: 15 – epigeobiont (*Zabrus* (s.str.) *tenebrioides* (Goeze, 1777)); 16 – stratogeobiont (*Ophonus* sp.); 17 – geobiont (*Ditomus calydonius* (P. Rossi, 1790), after Brandmayr, 1975); 18 – mycetophagous (*Mormolyce phylloides* author, reconstruction after van Emden, 1942; Lieftnick & Wiebes, 1968).

glands (Figs 19-20), from which discharges substances that attract their hosts. Hosts take care of their symbionts. Two monotypical subclasses are distinguished: myrmecophylous (mainly symbionts) – Paussini larvae, which inhabit ant hills and receive food from their hosts (Bøving, 1907; van Emden, 1922; Paulian, 1947; Di Giulio & Moore 2004); and termitophilous inquilines – Ozaenini larvae (*Goniotropis* etc.) with glandular appendages on caudal disc. They attract insects with glandular discharges and feed on them (Vigna Taglianti et el., 1998; Moore & Di Giulio, 2006).

VI class. ECTOPARASITOIDS (slow-moving cryptobionts)

Development of these larvae (*Lebia*, *Brachinus*) is characterized by hypermetamorphosis and change of life forms. First-instar larvae – campodeiform "triungulins", search for prey (beetle pupae). Late instar larvae (Fig. 21) – ectoparasitoids, feeding on prey tissues

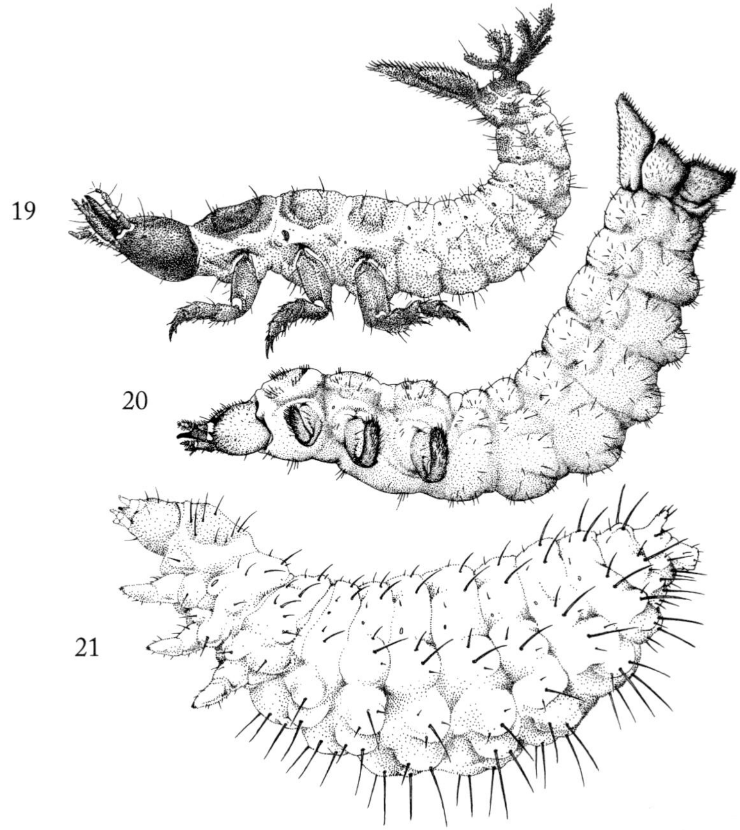

Figs 19-21. Habitus of symphilous and ectoparasitoid carabid larvae: 19 – termitophilous larva (*Pachyteles* sp., reconstruction after Vigna Taglianti et al., 1998; Di Giulio & Vigna Taglianati, 2001); 20 – myrmecophilous larva (*Paussus* sp.); 21 – ectoparasitoid larva (*Brachinus elegans* (Chaudoir, 1842)).

(Silvestri, 1904; van Emden, 1919; Lindroth, 1954; Erwin, 1967; Habu & Sadanaga, 1971; Capogreco, 1989; Saska & Honěk, 2004; Makarov, Bokhovko, 2005). They differ strikingly from triungulinine-like first instars by numerous reductions and adaptive features.

VII class. APHAGES (troglobionts)

The highly specialized troglobiontic Trechini (*Aphaenops*, etc.) are characterized by viviparity of a single larva (developed in the female body), which is aphagous and soon pupates. It has a white, fleshy body, short legs, and reduced mouthparts and urogomphi (Deleurance & Deleurance, 1964).

–

The system of carabid larval life forms, outlined above, exemplifies the major trends of adaptive radiation in Carabidae larvae. From primitive plesiomorphic zoophagous larvae with hemicryptic life-style in soil, the following trends of their adaptive evolution can be traced (Fig. 22).

The general trend is the diversification of zoophages, which reflects the expansion of possible habitat layers. Among zoophages, there are phytobionts with running and climbing-types of legs, epigeobionts – the dwellers of the ground surface with legs adapted for walking and running, and well developed sensory organs; stratobionts, occurring in litter, top layers of soil, burrows and caves; geobionts with digging legs – the dwellers of soil strata with hemi-cryptic and cryptic way of life. Especially notable among zoophages are hole-ambuscaders with S-shaped or sacciform body, which construct and/or inhabit holes in the soil, wood and ant hills.

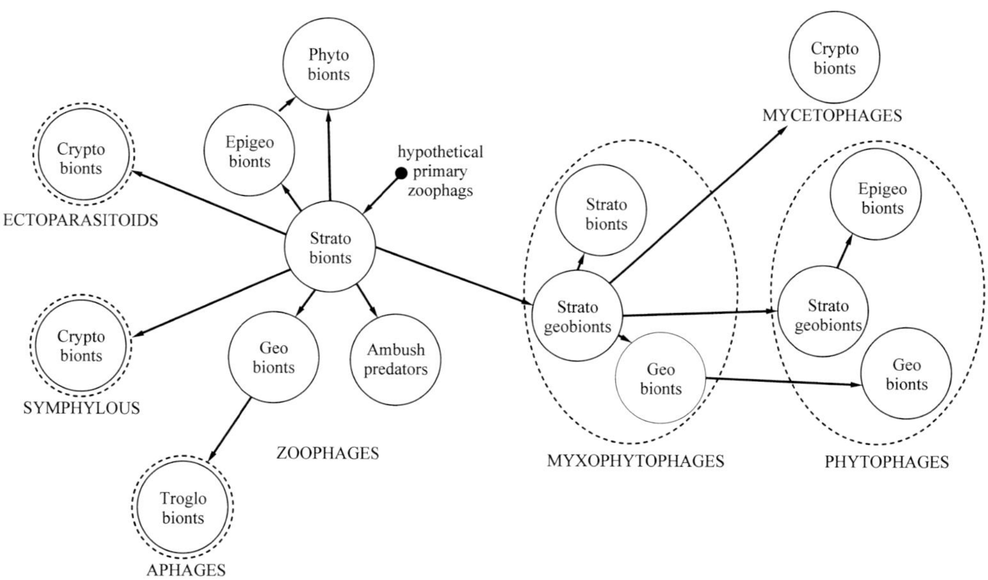

Fig. 22. Main trends of adaptive radiation of carabid larvae.

The second trend is the transformation from zoophagy through mixophyto-, to phytophagy. This is characteristic of Zabrini and Harpalini larvae. Among mixophytophages, the dwellers of various layers of soil can be distinguished: stratobionts, stratogeobionts and geobionts, which are characterized by obvious adaptations for burrowing and a cryptic life style.

Phytophages have adaptations for feeding both on the vegetative parts of plants (zabroid) and on seeds (ophonoid, ditomoid). Within the latter, a progressive development of parental care exists.

Other trends, leading to origin of the specific life forms, are derivative. Thus, regressive aphagous larvae (some troglobiontic Trechini with viviparity and embryonisation of development), symbionts and inquilines of ants and termites (Paussinae), ectoparasites with hypermetamorphosis (Brachinini and part of Lebiini) are evolved on the basis of various stratobionts-zoophages. As for mycetophages (Mormolycini) – they seem to be specialized mixophages. Comparison of the adaptive radiation of Carabidae adults and larvae has revealed examples of congruent evolution, but also trends differ between the larvae and adults.

Besides a purely theoretical use, this system of Carabidae larvae life forms allows a quantitative characterization of carabid complexes in different landscapes and regions. Spectra of Carabidae larvae life forms serve as bioindicators of soil, plant and climatic conditions.

ACKNOWLEDGMENTS

The author is grateful to all colleagues who took part in the discussion of the article matter. The English translation was accomplished by A. Zaitsev, and the great work of language editing was carried out by Dr. D. Pollock. The illustrations were made by K. Makarov. All this valuable help is highly appreciated.

REFERENCES

Andersen, J. (1968). The larva of *Miscodera arctica* Payk. (Col., Carabidae). – Norsk Entomologisk Tidsskrift. 15: 71-74.

Andersen, J.M. (1970). The larva of *Pelophila borealis* Payk., *Nebria gyllenhali* Schnh. and *N. nivalis* Payk (Col. Carabidae). – Astarte 3: 87-95.

Arabadzhiev, D., Balevski, A., Drenski, P., Zakharieva, B. & Radev, R. (1953). Вредните житни бегачи от рода *Zabrus* в България и борбата с тях. – Труда на Института по зоология. (2): 1-109. [in Bulgarian].

Arndt, E. (1989). Beschreibung der Larve von *Lionychus quadrillum* (Duft.) und Bemerkungen zur Larvalsystematik der mitteleuropäischen Gruppen des Subtribus Dromiina (Insecta, Coleoptera, Carabidae: Lebiini) Reichenbachia. Staatliche Museum Tierkunde Dresden. 27: 47-56.

Arndt, E. (1990). Die Larve von *Parophonus maculicornis* (Duft.): Beschreibung und Discussion ihrer Merkmale unter phylogenetischen Aspekt (Insecta, Coleoptera, Carabidae,

Harpalini). – Entomologische Abhandlungen Staatluche Museum Tierkünde (Dresden). 53: 143-150.

Arndt, E. (1991). Beschreibung der Larven mitteleuropäischer Stenolophina-Arten (Coleoptera, Carabidae, Harpalini). – Entomologische Nachrichten und Berichte. 35: 45-52.

Arndt, E. & Cassola, F. (2000). Description of Prothymine tiger beetle larvae from South East Asia (Coleoptera Cicindelidae). – Bolletino della Societa Entomologica Italiana. 132: 105-115.

Arndt, E. & Drechsel, U. (1998). Description of the larva of *Trichognathus marginipennis* Latreille, 1829 (Coleoptera, Carabidae). – Mitteilungen Münchener Entomologischen Gesellschaft. 88: 19-24.

Avon, C. (1995). La perte de l'instinct de chasse chez les larves anophtalmes de coléoptères hypogés (notamment chez le genre *Duvalius* Delarouzée) et études volumetriques comparatives des vésicules collectrices des glandes anales. – L'Entomologiste 51: 193-204.

Baehr, M. (1997). Revision of the Pseudomorphinae of the Australian region. 1. The genera *Pseudomorpha* Kirby, *Adelotopus* Hope, *Cainogenion* Notman, *Paussotropus* Waterhouse, and *Cryptocephalomorpha* Ritsema. Taxonomy, phylogeny, zoogeography (Insecta, Coleoptera, Carabidae). – Spixiana. Supplement 23: 1-508.

Bauer, T. (1982). Prey-capture in a ground-beetle larva. – Anim. Behav. 30: 203-208.

Bílý, S. (1975). Larvae of the genus *Amara* (subgenus *Celia* Zimm.) from Central Europe (Coleoptera, Carabidae). – Academia, Nakladatelství Èeskoslovenské akademie vìd, Praha.

Boldori, L. (1934). Appunti sulle larve degli Sphodrini. I. La larva dello *Sphodrus leucophthalmus* L. – Bolletino della Societa Entomologica Italiana. 66: 102-111.

Boldori, L. (1935). Appunti sulle larve degli Sphodrini. II. La larva di *Antisphodrus mairei* Peyer. – Bulletin de la Société Entomologique de France. 40: 150-157.

Boldori, L. (1958). Larve di Trechini. X. – Memorie della Societa Entomologica Italiana. 37: 149-161.

Bourne, J.D. (1975). Morphologie de la larve de *Trichophaenops cerdonicus* Ab. (Coleoptera Carabidae: Trechinae). – Revue Suisse de Zoologie. 82: 163-167.

Bousquet, Y. (1987). Description of the larva of *Helluomorphoides praeustus bicolor* Harris with comments on the relationships of the Helluonini (Coleoptera: Carabidae). – Canadian Entomologist. 119: 921-930.

Bousquet, Y. (1988). *Dyschirius* of America north of Mexico: descriptions of new species with keys to species groups and species (Coleoptera, Carabidae). – Canadian Entomologist. 120: 361-387.

Bousquet, Y. & Smetana, A. (1991). The tribe Opisthiini (Coleoptera: Carabidae): description of the larvae, note on habitat, and brief discussion on its relationships. – Journal of the New York Entomological Society. 99: 104-114.

Bousquet, Y. & Tchang, J.-P. (1992). Anisodactylinae larvae (Coleoptera: Carabidae: Harpalini); descriptions of genus–group taxa of eastern Canada and phylogenetic remarks. – Canadian Entomologist. 124: 751-783.

Bøving, A.G. (1907). Om Paussiderne og Larven till *Paussus Kannegieteri* Wasm. – Videnskabelige Meddelelser fra Dansk Naturhistorisk Foreninig i Kjøbenhavn. 9: 109-136.

Brandmayr, P. & Zetto Brandmayr, T. (1974). Sulle cure parentali e su altri aspetti della biologia di *Carterus* (*Sabienus*) *calydonius* Rossi, cun alcune considerazioni sui fenonemi di cura della prole sinoad oggi riscontrati in carabidi. – Redia. 55: 143-175.

Brandmayr, P. & Zetto Brandmayr, T. (1981). Identificazione di larve del genere *Ophonus* Dejean, 1821 (sensu novo) e note bionomiche. – Memorie della Societa Entomologica Italiana. 60: 67-103.

Capogreco, J.V. (1989). Immature *Lebia viridis* Say (Coleoptera: Carabidae): bionomics, descriptions, and comparisons to other *Lebia* species. – Coleopterist Bulletin. 43: 183-194.

Casale, A., Giachino, P.M. & Pantaleoni, R. (1996). Life history and pre-imaginal stages of *Dromius meridionalis* (Coleoptera: Carabidae: Dromiini) in Sardinia. – Acta Societatis Zoologicae Bohemicae. 60: 363-371.

Casale, A. & Vigna Taglianti, A. (1991). Morphologie larvaire de *Cychrus cylindricollis* Pini (Coleoptera: Carabidae) avec de notes sur la biologie et l'écologie de l'espèce. – Elytron. Suppl. 5: 95-106.

Davies, M.J. (1963). The larvae of some British *Notiophilus* species (Col., Carabidae). – The Entomologst's Monthly Magazine. 99: 206-209.

Deleurance, S. & Deleurance, E.P. (1964). Reproduction et cycle evolutif larvaire des *Aphaenops* (*A. cerberus* Dieck, *A. crypticola* Lindner), insectes Coléoptères cavernicoles. – Comptes Rendus Hebdomadaires des Seances de l'Academie des Sciences Paris. Serie D. Sciences Naturelles. 258: 4369-4370.

Di Giulio, A. & Moore, W. (2004). The first-instar larva of the genus *Arthropterus* (Coleoptera: Carabidae: Paussinae): implications for evolution of myrmecophily and phylogenetic relationships within the subfamily. – Invertebrate Systematics 18: 101-115.

Emden, F.I. van (1919). Beschreibung der Larve von *Pheropsophus hispanicus* Dej. – Suppl. Entomologische Mitteilungen. Berlin. 8: 34-38.

Emden, F.I. van (1922). Über die Larven der Paussiden und Beschreibung der Larve der *Paussus granulatus* Westwood (Col.). – Entomologische Blätter für Biologie und Systematik der Käfer. 18: 37-47.

Emden, F.I. van (1935). Die Larven der Cicindelinae. 1. Einleitendes und Alacosternale Phyle. – Tijdschrift voor entomologie. 78: 134-183.

Emden, F.I. van (1942). A key to the genera of larval Carabidae. – Transaction of the Royal Entomological Society of London. 92: 1-99.

Erwin, T.L. (1967). Bombardier beetles (Col. Car.) of North America Part II. Biology and behaviour of *Brachinus pallidus* Erwin in California. – Coleopterist Bulletin. 21: 41-55.

Erwin, T.L. (1981). A synopsis of the immature stages of Pseudomorphini (Coleoptera: Carabidae) with notes on tribal affinities and behavior in relation to life with ants. – Coleopterist Bulletin. 35 (1): 53-68.

Forsskåhl, B. (1966). The larve of *Amara alpina* Fbr. (Col. Carabidae). – Notulae Entomologicae. 46: 29-31.

Gardner, J.C.M. (1936). Immature stages of Indian Coleoptera (20. Carabidae). – Indian Forest Records N.S., Entomology. 2: 181-201.

Ghilarov, M.S. (1949). Osobennosti pochvy kak sredy obitaniya i ee znachenie v evolitsii nasekomykh. – Nauka, Moscow. [in Russian].

Ghilarov, M.S. & Sharova, I.Kh. (1954). Lichinki zhukov-skakunov (Cicindelidae). – Zoological Journal. 33: 598-615. [in Russian].

Goulet, H. (1983). The genera of Holarctic Elaphrini and species of *Elaphrus* Fabricius (Coleoptera: Carabidae): classification, phylogeny and zoogeography. – Quaestiones Entomologicae. 19: 219-482.

Grebennikov, V.V. (1996). Larvae of Bembidiini: subgenera *Synechostictus* and *Pseudolimnaeum* of the genus *Bembidion* and their taxonomic position (Coleoptera, Carabidae). – Zoosystematica Rossica. 5: 263-272.
Grebennikov, V.V. (2002). Description of the first-instar larva of *Geocharidius* (Coleoptera: Carabidae: Trechitae) with a discussion of the phylogeny of the subtribe Anillina. – European Journal of Entomology . 99: 523-527.
Grebennikov, V.V. & Bousquet, Y. (1999). Larvae of Pogonini (Coleoptera: Carabidae): genera *Pogonus*, *Pogonistes*, *Cardiaderus*, and *Thalassotrechus*. – Acta Societatis Zoologicae Bohemicae. 63: 427-441.
Habu A. & Sadanaga K. (1970b). Description of some larvae of the Carabidae found in cultivated fields and paddy fields (II). – Kontyû. 38: 24-41.
Habu, A. & Sadanaga, K. (1970a). Description of some larvae of the Carabidae found in cultivated fields and paddy fields (I). – Kontyû. 38: 9-23.
Habu, A. & Sadanaga, K. (1971). Description of some larvae of the Carabidae found in cultivated fields and paddy fields (III). – Kontyû. 39: 159-166.
Habu, A. & Sadanaga, K. 1967. Description of the larva of *Parena* (*Parena*) *perforata* (Bates) with some notes on its life history (Carabidae). – Kontyû. 35: 391-397. .
Harris, A.C. (1978). The larva of *Brullea antarctica* (Coleoptera: Carabidae: Broscinae). – New Zealand Entomologist. 6: 401-405.
Hůrka, K. (1966). Zur Kenntnis der Larven der mitteleuropäischen *Chlaenius*-Arten. – Acta Entomologica Bohemoslovaca. 63: 203-212.
Hůrka, K. (1971). Die Larven der mitteleuropäischen *Carabus*- und *Procerus*- Arten. – Rozpravy Československé akademie věd. Řada matematických a přirodních věd. 80: 1-136.
Hůrka, K. (1975). Larval diagnosis of the tribe Stenolophini and notes on the classification of the subfamily Harpalinae (Coleoptera, Carabidae). – Acta Entomologica Bohemoslovaca. 72 (7): 247-256.
Hůrka, K. (1986). Larval taxonomy and breeding type of Palearctic *Cymindis* (Coleoptera, Carabidae). – Acta Entomologica Bohemoslovaca. 83: 30-61.
Hůrka, K. (1997). The status of *Tetraplatypus* (Coleoptera: Carabidae: Stenolophina) and larval description of *Bradycellus ruficollis* and *B. verbasci*. – Acta Societatis Zoologicae Bohemicae. 61 (3): 191-197.
Hůrka, K. & Ducháč, V. (1980). Larval descriptions and the breeding type of the central European species of *Amara* (Curtonotus) (Coleoptera: Carabidae). – Acta Entomologica Bohemoslovaca. 77: 258-270.
Jeannel, R. (1926). Monographie des Trechinae. Les Larves des Trechinae. – L'Abeille. 32: 375-392.
Jeannel, R. (1941-42). Coléoptères Carabiques. I. – Faune de France. 39-40. Lechevalier, Paris.
Kobayashi, M., Kudagamage, C. & Nugaliyadde, L. (1995). Distribution of larvae of *Ophionea indica* Thunberg (Carabidae), a predator of the rice gall midge, *Orseolia oryzae* (Wood-Mason) in paddy fields of Sri Lanka. – Japan Agricultural Research Quarterly. 29: 89-93.
Liebherr, J.K. (1983). Larval description of *Calybe* (*Ega*) *sallei* (Chevrolat) with a preliminary assessment of lachnophorine affinities (Carabidae: Lachnophorini). – Coleopterist Bulletin. 37: 245-260.
Lieftinck, M.A. & Wiebes, J.T. (1968). Notes on the genus *Mormolyce* Hegenbach (Coleoptera, Carabidae). – Bijdragen tot de Dierkunde. 38: 59-68.

Lindroth, C.H. (1954). A revision of *Diachila* Motsch. and *Blethisa* Bon. – Lunds Universitets Årsskrift. 50: 1-28.
Lindroth, C.H. (1954). Die Larve von *Lebia chlorocephala* Hoffm. (Col. Carabidae). – Opuscula Entomologica. 19: 29-33.
Lindroth, C.H. (1956). A revision of the genus *Synuchus* Gyllenhal (Coleoptera: Carabidae) in the widest sence, with notes on *Pristosia* Motschulsky (*Eucalathus* Bates) and Calathus Bonelli. – The Transaction of the Royal Entomological Society of London. 108: 485-576.
Luff, M.L. (1969). The larvae of the British Carabidae (Coleoptera), I. Carabini and Cychrini. – Entomologist 102: 245-263.
Luff, M.L. (1972). The larvae of the British Carabidae (Coleoptera), II. Nebriini. – Entomologist 105: 161-179.
Luff, M.L. (1978). The larvae of the British Carabidae (Coleoptera) V. Omophronini, Loricerini, Scaritini and Broscini. – Entomologist's Gazette. 29: 265-287.
Mahar, J.M., Stehr, F.W. & Simmons, G.A. (1983). Descriptions of larvae and notes about the life habits of *Dromius piceus* Dejean (Coleoptera: Carabidae: Lebiini). – Coleopterist Bulletin. 37: 23-26.
Makarov, K.V. (1987). Morpho-ekologicheskie tipy lichinok zhukov-zhuzhelits roda *Carabus* L. – In: Problemy pochvennoi zoologii Tezisy dokladov IX Vsesojuznogo Sovestshchanija. Metsniereba, Tbilisi, p. 177-178. [in Russian].
Makarov, K.V. (1992). Lichinki zhukov-zhuzhelits roda *Carabus* (Coleoptera, Carabidae) fauny Rossii i sopredel'nykh stran. 1. – Entomologicheskoe obozrenie. 71: 752-774. [in Russian].
Makarov, K.V. (1998). Morphologiya termitophilnykh lichinok *Orthogonius* i *Helluodes* i polozhenie trib Orthogoniini i Helluodini v semeistve zhuzhelits. – Zoological Journal. 77: 1285-1296. [in Russian].
Makarov, K.V. (2005). Lichinki triby Scaritini (Coleoptera: Carabidae) fauny Rossii i sopredel'nykh stran. – Russian Entomological Journal. 14: 169-201. [in Russian].
Makarov, K.V. & Bokhovko, E.E. (2005). Preemstvennost struktur khetoma u razvivayushchikhsya s hypermetamorphosom lichinok *Brachinus* Weber (Coleoptera: Carabidae). – Russian Entomological Journal. 14: 263-274. [in Russian].
Makarov, K.V. & Brinev, A.E. (2001). Lichinki roda *Harpalodema* Reitter i ego polozhenie v sisteme triby Zabrini. – Russian Entomological Journal. 10: 273-288. [in Russian].
Makarov, K.V., Gurgenidze, L.N. & Rekk, N.G. (1991). Description of the larvae of the endemic caucasian species *Zabrus* (*Pelor*) *trinii* F.-W. (Coleoptera, Carabidae) and diagnosis of the subgenus *Pelor* by larval features. – Izvestija Academii Nauk GruzSSR 17: 103-109. [in Russian].
Makarov, K.V. & Koval, A.G. (2003). K biologii peshchernoi zhuzhelitsy *Jeannelius birsteini* Ljovuschkin , 1963. – Entomologicheskoe obozrenie. 82: 836-845. [in Russian].
Makarov, K.V. & Shilenkov, V.G. (2001). Lichinochnye stadii zhuzhelitsy *Daptus vittatus* Fischer (Coleoptera, Carabidae). – Russian Entomological Journal 10: 289-296. [in Russian].
Matalin, A.V. (1995). The larvae of the ground beetle *Bradycellus* (*Tachycellus*) *glabratus* (Coleoptera: Carabidae: Harpalini). – Zoosystematica Rossica 4: 279-291.
Matalin, A.V. (1998). Lichinki zhuzhelits roda *Dicheirotrichus* Jacq. fauny Rossii i sopredel'nykh stran. I. – Entomologicheskoe obozrenie. 77: 134-150. [in Russian].
Matalin, A.V. (2001). Lichinki zhuzhelits poda *Dicheirotrichus* Jacquelin du Val, 1857 (Coleoptera, Carabidae: Harpalini) fauny Rossii i sopredel'nykh stran. II. Lichinki

podroda *Trichocellus* Ganglbauer, 1892. – Russian Entomological Journal. 10: 309-322 [in Russian].

May, B.M. (1963). New Zealand cave fauna. – I. The larva of *Duvaliomimus mayae* Britton 1958 (Coleoptera: Carabidae: Trechinae). – Transaction Royal Society of New Zealand 3: 145-150.

Mikhailov, V.A. (1978). Lichinki zhuzhelits roda *Carabus* iz Srednej Asii. – Zoological Journal. 57: 851-856. [in Russian].

Moore, B.P. (1974). Two larval habits of two species of *Spallomorpha* Westwood (Coleoptera: Carabidae, Pseudomorphinae). – Journal of the Australian Entomological Society. 13: 179-183.

Moore, B.P. (1998). The enigmatic tribe Physocrotaphini Chaudoir (Helluodini auct.) (Coleoptera: Carabidae– In: Phylogeny and Classification of Caraboidea. XX I.C.E. (1996, Firenze, Italy) (Ball, G.E., Casale, A., & Vigna Taglianti, A., eds.). Museo Regionale di Scienze Naturali, Torino, p. 369-380.

Moore, W. & Di Giulio, A. (2006). Description and behaviour of *Goniotropis kuntzeni* larvae and key to genera of Paussinae larvae. – Zootaxa. 1111: 1-19.

Paarmann, W. 1979. A reduced number of larval instars as an adaption of the desert carabid beetle *Thermophilum* (*Anthia*) *sexmaculatum* to its arid environment. – In: On the evolution of behaviour in Carabid beetles. (Boer, P.J. den, Thiele, H.-U., Weber, F., eds). Miscellaneous papers Agricultural University Wageningen, H. Veeneman & Zonen B.V., Wageningen, p. 113-117.

Palmer, M. (1976). Natural history and behavior of *Pseudoxychila tarsalis* Bates. – Cicindela 8: 61-92.

Paulian, R. (1947). La larve de *Pseudozaena* (*Afrozaena*) *luteus* Hope (Col. Carab. Ozaenidae). – Bulletin du Museum National d'Histoire Naturelle. Section A. Zoologie Biologie et Ecologei Animales. 2: 335-339.

Putchkov, A.V. (1993). Lichinki zhukov-skakunov (Carabidae, Cicindelinae) podroda *Cicindela* s.str. Russkoi Ravniny i Kavkaza. – Zoological Journal. 72: 52-62. [in Russian].

Putchkov, A.V. & Dolin, V.G. (2005). Description of the larva of the tiger beetle *Pogonostoma majunganum* (Coleoptera, Cicindelidae). – Vestnik Zoologii. 39: 35-38.

Rougemont, G.M. de (1983). Contributions à la connaissance de la faune de l'Éthiopie 3. Les larves de Calosomes récoltées par Clarke et de Rougemont (Coleoptera, Carabidae). – Revue de Zoologie Africaine. 97: 641-646.

Saska, P. (2004). Carabid larvae as predators of weed seeds: granivory in larvae of *Amara eurynota* (Coleoptera: Carabidae). – Communications in Agricultural and Applied Biological Sciences. 69: 27-33.

Saska, P. (2004). Larvae of the nominotypical subgenus *Amara* (Coleoptera: Carabidae). – Acta Societatis Zoologicae Bohemicae. 68: 191-234.

Saska P. & Jarošik V. (2001). Laboratory study of larval food requirements in nine species of *Amara* (Coleoptera, Carabidae). – Plant Protect Science. 37: 103-110 .

Saska, P. & Honěk, A. (2004). Development of the beetle parasitoids, *Brachinus explodens* and *B. crepitans* (Coleoptera: Carabidae). – Journal of Zoology. 262: 29-36.

Sharova, I.Kh. (1957). Lichinki zhukov-krasotelov (*Calosoma*, Carabidae). – Zoological Journal. 36: 878-884 [in Russian].

Sharova, I.Kh. (1958). Lichinki zhukov-zhuzhelits, poleznykh i vrednykh v sel'skom khozyaistve. – Uchenye zapiski Moskovskogo Gosudarstvennogo Pedagogicheskogo Instituta im. V.I.Lenina. 12: 4-164. [in Russian].

Sharova, I.Kh. (1964). Semejstvo Carabidae – zhuzhelitsy. – In: Opredelitel' obitayushchikh v pochve lichinok nasekomykh (Ghilarov, M.S., ed.). Nauka, Moscow, p. 112-185. [in Russian].

Sharova, I.Kh. (1967). Lichinki rodov *Acinopus* Latr. i *Osimus* Motsch. (Carabidae, Coleoptera) i ikh sistematicheskoe polozhenie. – Zoological Journal. 46: 891-900. [in Russian].

Sharova, I.Kh. (1976). Zhiznennye formy lichinok zhuzhelits i ikh evolutsionnye vzaimootnosheniya. – In: Evolutionnaya morphologiya lichinok nasekomykh. Nauka, Moscow, p. 56-80. [in Russian].

Sharova, I.Kh. (1981). Zhiznennye formy zhuzhelits (Carabidae). – Nauka, Moscov. [in Russian].

Sharova, I.Kh. (2002). Ekologicheskaya radiatsiya pochvoobitayushchikh lichinok nasekomykh. – Problemy pochvennoj zoologii. Materialy III (XIII) Vserossijskogo Soveshchania po Pochvennoi Zoologii (Striganova, B.R., ed.). KMK Publ., Moscow, p. 197-198. [in Russian].

Sharova, I.Kh. & Makarov, K.V. (1983). Lichinka zhuzhelitsy *Machozetus lehmanni* Mén. (Coleoptera, Carabidae). – Entomologicheskoe obozrenie 62: 524-528. [in Russian].

Sharova, I.Kh. & Makarov, K.V. (1984). O lichinkakh zhuzhelits roda *Broscus* Panz. (Coleoptera, Carabidae). – Entomologicheskoe obozrenie. 63: 742-750. [in Russian].

Shilenkov, V.G. & Berlov, E.Ya. (1987). Opisanie lichinok krasotela *Callistenes elegans* Kirsch. (Coleoptera, Carabidae) iz Kazakhstana. – Entomologicheskoe obozrenie. 66: 584-590. [in Russian].

Silvestri, F. (1904). Contribuzione alla conoscenza della metamorfosi e dei costume della *Lebia scapularis* con descrizione dell'apparato sericiparo della larva. – Redia. 2: 68-84.

Spence, J.R. & Sutclife ,J.F. (1982). Structure and feeding in larvae of *Nebria* (Coleoptera, Carabidae). – Canadian Journal of Zoology. 60: 2382-2394.

Sturani, M. (1962). Osservazioni e ricerche biologiche sul genere *Carabus* Linnaeus (*sensu lato*) (Col. Car.). – Memorie della Societa Entomologica Italiana. 41: 85-202.

Thompson, R.G. (1979). A systematic study of the larvae in the tribes Pterostichini, Morionini, and Amarini (Coleoptera: Carabidae). – Bulletin Agricultural Station, University of Arkansas. 837: 1-105.

Vanek, S. (1984). Larvae of the Palaearctic species *Clivina collaris* and *Clivina fossor* (Coleoptera, Carabidae, Scaritini). – Acta Entomologica Bohemoslovaca. 81: 99-112.

Vigna Taglianti, A., Santarelli, F., Di Giulio, A. & Oliverio, V. (1998). Phylogenetic implications of larval morphology in the tribe Ozaenini (Coleoptera: Carabidae). – In: Phylogeny and Classification of Caraboidea. XX I.C.E. (1996, Firenze, Italy) (Ball, G.E., Casale, A., & Vigna Taglianti, A., eds.). Museo Regionale di Scienze Naturali, Torino, p. 273-296.

Zamotajlov, A.S. (2001). Contribution to the knowledge of the carabid genus *Deltomerus* Motschulsky, 1850 (Coleoptera, Carabidae) from the Middle East. – Russian Entomological Journal 10: 327-342.

Zetto Brandmayr, T. & Brandmayr, P. 1975. Biologia di *Ophonus puncticeps* Steph. (Cjleoptera, Carabidae): cenni sulla fitofagia delle larve e loro etologia. – Annali della Facoltà di Scienze Agrarie della Università degli Studi di Torino. 9: 421-430.

Zetto Brandmayr, T., Giglio, A., Marano, I. & Brandmayr, P. (1998). Morphofunctional and ecological features in carabis [should this be "carabid"?] (Coleoptera) larvae. – In: Phylogeny and Classification of Caraboidea. XX I.C.E. (1996, Firenze, Italy) (Ball, G.E., Casale, A., & Vigna Taglianti, A., eds.). Museo Regionale di Scienze Naturali, Torino,. p. 449-4990.

Zetto Brandmayr, T., Marano, I. & Paarmann, W. (1993). Hypotheses concerning the preying behaviour of the larvae of *Graphipterus serrator* Förskahl, a myrmecophilous carabid beetle. – Ethology, Ecology & Evolution. Special Issue. 3: 119-124.

Zetto Brandmayr, T., Marano, I. & Pizzolotto, R. (1994). Larval morphology and bionomy of *Amara (Leirides) alpestris* Villa (Coleoptera, Carabidae). – Gortania. Atti Museo Friulano di Storia Naturale. 16: 187-202.

Znojko D.V. 1929. Opyt kratkogo opredelitelya lichinok rodov zhuzhelits, vstrechayushchikhsya v SSSR, I opisaniya lichinok *Zabrus tenebrioides* Gz., *Harpalus pubescens*. – Zashchita Rastenij ot Vreditelej I Boleznej 6: 335-360. [in Russian].

L. Penev, T. Erwin & T. Assmann (Eds) 2008
Back to the Roots and Back to the Future. Towards a New Synthesis amongst Taxonomic, Ecological and Biogeographical Approaches in Carabidology
Proceedings of the XIII European Carabidologists Meeting, Blagoevgrad, August 20-24, 2007, pp. 305-323.

Life cycles in the ground-beetle tribe Pogonini (Coleoptera, Carabidae) from the Lake Elton region, Russia

Andrey V. Matalin & Kirill V. Makarov

Zoology & Ecology Department, Moscow State Pedagogical University, Kibalchicha Str. 6, Bld. 5, Moscow 129164, Russia. E-mail: a_matalin@tochka.ru, kvmac@inbox.ru

ABSTRACT

In 2006-2007, the main features of life cycles in seven species of the tirbe Pogonini in the semi-desert Lake Elton region, Russia were studied. Six species, *Cardiaderus chloroticus*, *Pogonus transfuga*, *P. meridianalis*, *Pogonistes rufoaeneus*, *P. angustus* and *P. convexicollis*, are spring breeders. Among them, *P. transfuga*, *P. meridionalis* and *P. rufoaeneus* are iteroparous while *C. chloroticus*, *P. angustus* and *P. convexicollis* are semelparous. The life cycles of all these species are monovariant monovoltine with spring or spring-summer reproduction and adult (immature or both immature and spent) hibernation. However, the life cycle in *C. chloroticus* is possibly bivoltine. In contrast, *P. cumanus* is an autumn-breeding iteroparous species, its life cycle being monovariant monovoltine with obligate larval hibernation and obligate adult aestivation, parapause. This is a surprising finding, because previously all Pogonini were considered to be "spring breeders" or "imaginal hibernators". Among the species studied, *P. transfuga* and *P. meridionalis* can be characterized as halophiles, because they live in habitats with moderate salinity. All the other species are pronounced halobionts because, with rare exceptions only, they are restricted to high-salinity habitats. Parasitic mites of the family Podapolipidae were found on *P. transfuga*, *P. meridionalis*, *P. cumanus* and *P. rufoaeneus* adults. Survival over winter and reproduction during a second season, together with the new generation specimens, is an important precondition for dispersal of Podapolipidae. On the other hand, this greatly restricts the range of their possible hosts.

Keywords: Carabidae, Pogonini, Podapolipidae, life cycle, population, spatial and temporal distribution, semi-desert zone, desert steppe, Western Palaearctic

INTRODUCTION

The carabid beetle tribe Pogonini is widespread and, according to different estimates, contains between 77 and 100 species worldwide (Kryzhanovsky, 1982; Lorenz, 1998). Of these, 56 species live in the Palaearctic region (Bousquet, 2003). Salinity tolerance is a major physiological characteristic of all Pogonini species (Heydemann, 1962) which allows them to live in habitats of variable salinity, ranging from the banks of brackish-water ponds, streams and rivers to marine coasts and even virtually lifeless salinas. The biology of most of the Pogonini species is still not clearly understood because, as a rule, only a few, irregular and short-term observations have been reported for these beetles. Furthermore, the species of some genera, e.g. *Cardiaderus* Dejean, 1828, *Diodercarus* Lutshnik, 1931, *Olegius* Komarov, 1996, *Sirdenus* Chaudoir, 1971, are cryptobionts, which are never abundant, and so their records are a matter of luck, as a rule.

Larvae have only been described for ten species from three genera of Palaearctic Pogonini (Jeannel, 1941; van Emden, 1942; Sharova, 1958, 1964; Larsson, 1968; Raynaud, 1976; Luff, 1985, 1993; Arndt, 1991; Grebennikov & Bousquet, 1999), and their population dynamics are effectively terra incognita. Detailed information concerning seasonal activity and habitat requirements are only available for four species of *Pogonus* Dejean, 1821: *P. chalceus* (Marsham, 1802), *P. transfuga* Chaudoir, 1870, *P. litoralis* (Duftschmid, 1812) and *P. luridipennis* (Germar, 1823) (Larsson, 1939; Paarmann, 1976; Desender, 2000; Turin, 2000; Matalin & Makarov, 2006). The reproductive cycle, geographic distribution and habitat requirements have been thoroughly studied in the first two species (Paarmann, 1976, 1979; Desender, 1985, 2000; Matalin & Makarov, 2006), and the development under laboratory conditions has been observed in the third species (Nekuliseanu, 1990).

MATERIAL AND METHODS

Vegetation

Carabid beetle communities on the banks of Lake Elton, Volgograd Region, Russia, were studied from 10 May until 31 October 2006, and from 1 April until 10 May 2007. The area is located near the Russia-Kazakhstan border (49°12.47'N, 46°39.75'E).

Lake Elton is situated inside the blind drainage Botkul-Bulukhta desert depression, which has a strongly pronounced salt-dome structure. Desert steppes are typical plant associations in most of the habitats there (Safronova, 2006). The most abundant plants in this landscape are *Artemisia lerchiana*, *A. pauciflora*, *A. austriaca*, *Kochia prostrata*, *Agropyron desertorum* and *Festuca valesiaca*. On salinas in floodplain terraces and in lakeside salt-marshes, hyper-halophilic communities are formed, where *Halocnemum strobilaceum*, *Atriplex cana*, *Anabasis salsa*, *Salicornia europaea*, *Salsola collina*, *S. tragus*, as well as *Artemisia santonicum*, *A. pauciflora*, *Suaeda physophora*, *Limonium suffruticosum*, *L. caspium* and *L. gmelinii*, are dominants. Dense reedbeds grow in river valleys.

Trapping

During both years, beetles were collected in three habitats: a lakeside salt-marsh, a salina on a floodplain terrace of the Khara River, and reedbeds along the right bank of the Khara River (Fig. 1).

Fig. 1. Locations of study habitats in the Lake Elton region.
A-B – lakeside salt-marsh; C-D – salina on a floodplain terrace of the Khara River; E-F – reedbeds along the Khara River; A, C, E – spring (late April – early May); B, D, F – summer (late July – early August).

Plastic pitfall traps of 0.5 l capacity and 72 mm upper diameter with 4% formalin as a fixative were used for collecting. In each habitat, traps were arranged along a transect at 10 m intervals. The traps were checked at 10 day intervals on the 10th, 20th and 30th (31st) of each month. The traps were maintained from November 1st, 2006 until of March 31st.

Material

During the trapping period, 12 species of Pogonini were collected, of which seven were abundant: *Cardiaderus chloroticus* (Fischer-Waldheim, 1823) – 73 ex., *Pogonus transfuga* Chaudoir, 1870 – 2,175 ex., *P. meridianalis* Dejan, 1828 – 142 ex., *P. cumanus* Lutshnik, 1916 – 77 ex., *Pogonistes rufoaeneus* (Dejean, 1828) – 107 ex., *P. angustus* (Gebler, 1830) – 29 ex. and *P. convexicollis* Chaudoir, 1871 – 47 ex. (Table 1).

Data analysis

Each specimen was dissected and the sex and age determined using a modified version of the method of Wallin (1987). Six physiological states in the adults of both sexes were distinguished, based on gonad conditions: teneral, immature, mature of either the first or second year and spent of either the first or second year of life. As additional criteria, the condition of the mandibles (Butterfield, 1986), as well as the surface of the pronotum and elytra, were evaluated. In females, the number of ripe eggs in the ovaries was estimated. Moreover, in each beetle the conditions of the hind-wings (den Boer, 1977) and wing muscles (i.e. dorso-ventral mesothoracal) (Tietze, 1963; Matalin, 1997b) were recorded, followed by the calculation of the index of potential migrants, *Ipm* (Matalin, 2003). Finally, the rate of infestation by parasites of each beetle was determined. Because data on the duration of development of pre-imaginal stages and number of eggs were not normally distributed, the differences between the median values were tested using Mann-Whitney U-test for independent n (Borovikov, 2001).

RESULTS

Activity and life cycles

The seasonal activities of all studied species were characterized by high fluctuations. In habitats with periodic floods (salina on a floodplain terrace and reedbeds) the pattern of activity in *P. transfuga* and *P. meridionalis* was more smooth, albeit sometimes sizeable amplitudes of population curves. In habitats with casual floods (lakeside salt-marsh), the abundances of all *Pogonistes* spp. and, especially, *C. chloroticus,* changed rapidly every

Table 1. Seasonal changes in demographic population structure of seven Pogonini species in three study habitats in the Lake Elton region.

Habitats	Species	Years	2006																	2007			
		Months	**May**		**June**			**July**			**August**			**September**			**October**			**April**			**May**
		Days	**20**	**31**	**10**	**20**	**30**	**10**	**20**	**31**	**10**	**20**	**31**	**10**	**20**	**30**	**10**	**20**	**31**	**10**	**20**	**30**	**10**
Lakeside salt-marsh	*Cardiaderus chloroticus*	teneral																					
		immature	/2	/1			1/		1/	1/2		4/3							/1	6/	/1		
		mature	1/2	2/1	1/1				1/	2/		2/	1/	1/							25/4		1/
		spent			/1							1/				1/2							
		Total	**1/4**	**2/2**	**1/2**		**1/**		**2/**	**3/2**		**7/3**	**1/**	**1/**		**1/2**			**/1**	**6/**	**25/5**		**1/**
	Pogonistes angustus	teneral								/2													
		immature					/1	3/7		3/		/2	/2	1/3	/1	1/	/2						
		mature																					
		spent												1/									
		Total					**/1**	**3/7**		**3/2**		**/2**	**/2**	**2/3**	**/1**	**1/**	**/2**						
	Pogonistes rufoaeneus	teneral							/2	1/													
		immature	/2	/ 4				/1	/1	/3	1/	/1	1/2	/1	1/	1/	/1	/1		1/5	/2	/2	/2
		mature	9/5	3/1	3/2	4/			/2	1/			2/								3/1		8/4
		spent P					/2		/3			2/1	4/2	3/	1/1	1/1							
		spent A		/1	1/																		
		Total	**9/7**	**3/6**	**4/2**	**4/**	**/2**	**/1**	**/8**	**2/3**	**1/**	**2/2**	**7/4**	**3/1**	**2/1**	**2/1**	**/1**	**/1**		**1/5**	**3/3**	**/2**	**8/6**
	Pogonistes convexicollis	Teneral				1/			2/				/1	/1									
		immature										/2	1/		1/			1/1		/4	/1		
		mature	5/5	8/	3/1	1/1															2/	/1	1/
		spent							/1				1/1										
		Total	**5/5**	**8/**	**3/1**	**2/1**			**2/1**			**/2**	**2/2**	**/1**	**1/**			**1/1**		**/4**	**2/1**	**/1**	**1/**
	Pogonus cumanus	teneral		/1					/1														
		immature			1/1					1/2	/1	1/1	/9	1/1	/1								
		mature P									3/4		17/11	17/9	5/1			1/					
		mature A	1/	1/	1/																		
		spent P														/2	2/1	1/					
		spent A	/1	/1							/1			/1							/1		
		Total	**1/1**	**1/2**	**2/1**				**/1**	**1/2**	**3/6**	**1/1**	**17/20**	**18/11**	**5/2**	**/2**	**2/1**	**2/**			**/1**		

Habitats	Species	Years	2006																	2007			
		Months	**May**		**June**			**July**			**August**			**September**			**October**			**April**			**May**
		Days	**20**	**31**	**10**	**20**	**30**	**10**	**20**	**31**	**10**	**20**	**31**	**10**	**20**	**30**	**10**	**20**	**31**	**10**	**20**	**30**	**10**
Salina on a floodplain terace	*Pogonus meridionalis*	teneral											2/										
		immature					1/	/1					1/1	1/									/1
		mature P	15/7	23/8	14/9	2/4	3/	1/2					1/							4/2	9/7		5/5
		mature A																			3/		½
		spent					1/2	/2					1/							1/			
		Total	**15/7**	**23/8**	**14/9**	**2/4**	**5/2**	**1/5**					**5/1**	**1/**						**5/2**	**12/7**		**6/8**
	Pogonus transfuga	teneral					/4	8/10	/7	3/7	1/3												
		immature		/4	/1	/1	2/	10/12	10/6	7/2	4/	1/3	2/1	3/		/1	/1			3/4			/1
		mature P	147/99	104/98	25/29	18/29	7/6	2/	2/	3/	4/									108/64	86/33	12/6	48/18
		mature A	/1	/1																128/91	64/63	11/12	24/23
		spent P		/3	/5	6/17	/7	5/12	7/9	2/2	2/3	3/3	1/2	9/6	/1	2/1	2/	6/6	5/3				
		spent A	/1	/1				/1	/2											6/12	/2	/1	/2
		Total	**147/101**	**104/107**	**25/35**	**24/47**	**9/17**	**25/35**	**19/24**	**15/11**	**11/6**	**4/6**	**3/4**	**12/6**	**/1**	**2/2**	**2/1**	**6/6**	**5/3**	**245/171**	**150/101**	**23/19**	**72/44**
Reedbeds	*Pogonus transfuga*	teneral						2/4	2/3	4/5	1/3												
		immature	/1				1/1	3/9	/5	2/3	/1				1/	1/	2/1	4/7	2/4	9/18		/1	/4
		mature P	22/30	16/29	19/16	5/1														40/24	22/18	20/19	48/37
		mature A																2/	2/	15/3	6/3	7/6	15/8
		spent P					1/	1/	/2	/1	1/							2/1	/1				
		spent A																		/4	/1		
		Total	**22/31**	**16/29**	**10/16**	**5/1**	**2/1**	**6/13**	**2/10**	**6/9**	**2/4**				**1/**	**1/**	**2/1**	**8/8**	**4/5**	**64/49**	**28/22**	**27/26**	**63/49**

Notes. Numerator - number of males, Denominator - number of females; P - parental generation; A - ancestral generations

30-40 days. In *C. chloroticus*, *Pogonus transfuga*, *P. meridionalis*, *Pogonistes rufoaeneus*, *P. angustus* and *P. convexicollis*, the maximum spring activity was pronounced. During this maximum, at least 50% of total numbers of each species were observed (Table 1). The period of maximum activity varied strongly from species to species. In *P. transfuga* and *P. meridionalis*, it lasted about 60 days from early April until the end of May; in *P. rufoaeneus* and *P. convexicollis*, it lasted for a period of less than 30 days in May; in *C. chloroticus* it lasted for only 10 days during mid April.

During the period of maximum locomotor activity, mature beetles of all the above mentioned species were more abundant than immature ones. *Pogonus transfuga*, *P. meridionalis* and *Pogonistes rufoaeneus* were found to be iteroparous, because some specimens belonged to the ancestral generations, i.e. had already oviposited during the preceding years. In *P. meridionalis* and *P. rufoaeneus*, the proportion of spent beetles in each of the ten-day catches did not exceed 12%; in *P. transfuga* the proportion amounted to 25-33% in the reedbeds and to 52-58% in the salinas. In contrast, *C. chloroticus*, *P. angustus* and *P. convexicollis* were semelparous species, because their populations included only beetles of the parental and daughter generations (Table 1).

The maximum number of ripe eggs in the females of most of the studied species did not exceed 9; only in *P. transfuga* were 20 eggs found. The number of eggs per female averaged between 2.1-2.8, except for both *P. meridionalis* and *P. transfuga* 4.7 and 5.3 eggs, respectively, were recorded (Fig. 2). The dynamics of oviposition correlated well

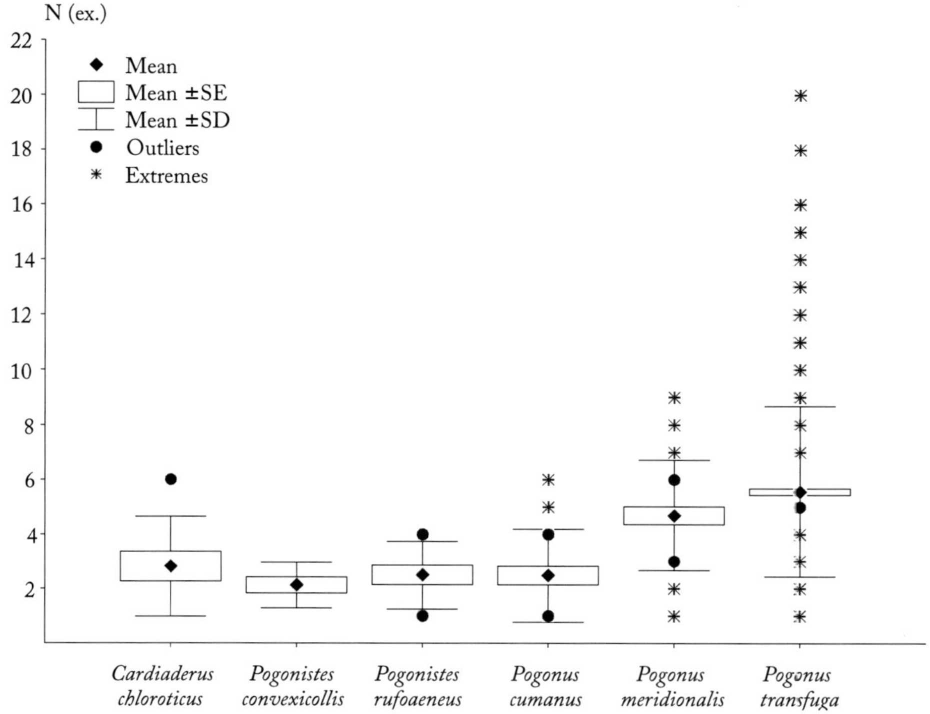

Fig. 2. Mean number of eggs in six species of Pogonini in the Elton region.

with seasonal activity in the habitats. In reedbeds, oviposition of *P. transfuga* occurred from mid May to early June in 2006 and from early to mid April in 2007, whereas in the salinas it occurred from mid May to mid June in 2006 and early April in 2007 (Fig. 3A-B). The average number of eggs per female was higher in the parental generation than in the ancestral generations. In the reedbeds, the differences were significant in some cases (Fig. 4A), whereas they were not significant in the salinas on the floodplain terrace (Fig. 4B). In reedbeds, sometimes the average number of eggs per female was significantly lower in the ancestral generations than in the parental generation (Fig. 4A). In contrast, in salina on a floodplain terrace, the average number of eggs per female of ancestral and parental generations was non-significant in all cases (Fig. 4B).

In *P. rufoaeneus* and *P. convexicollis*, the maxima of oviposition were recorded from early to mid May, in *P. meridionalis* from mid May to mid June in 2006 (Fig. 3C) and from mid April to early May in 2007.

The larval development in all these species was completed during one growing season. However, the duration of development strongly depended on the environment. The first larval instar of *P. transfuga* were observed from the beginning of May to early June. By mid May, some of them reached the third instar stage while the first pupae oc-

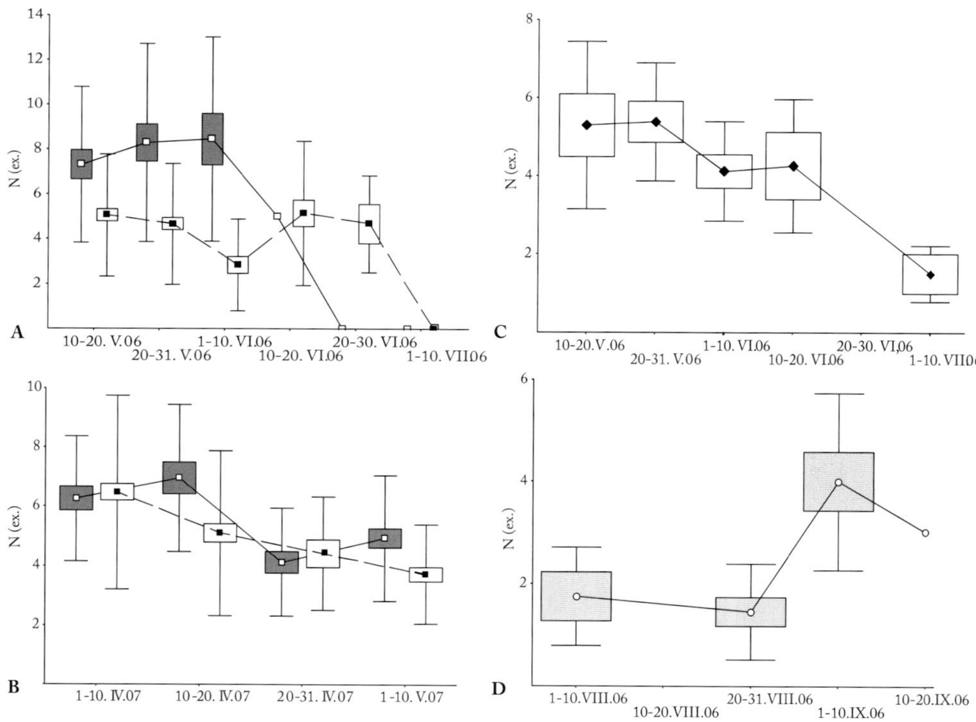

Fig. 3. Dynamics of oviposition (for means) in three species of *Pogonus* in the Elton region. A-B – *P. transfuga*; C – *P. meridionalis*; D – *P. cumanus*; boxes – means ± SE; whisker – means ± SD; dark grey boxes – reedbeds along the Khara River; open boxes – salinas on a floodplain terrace of the Khara River; light grey boxes – lakeside salt-marsh.

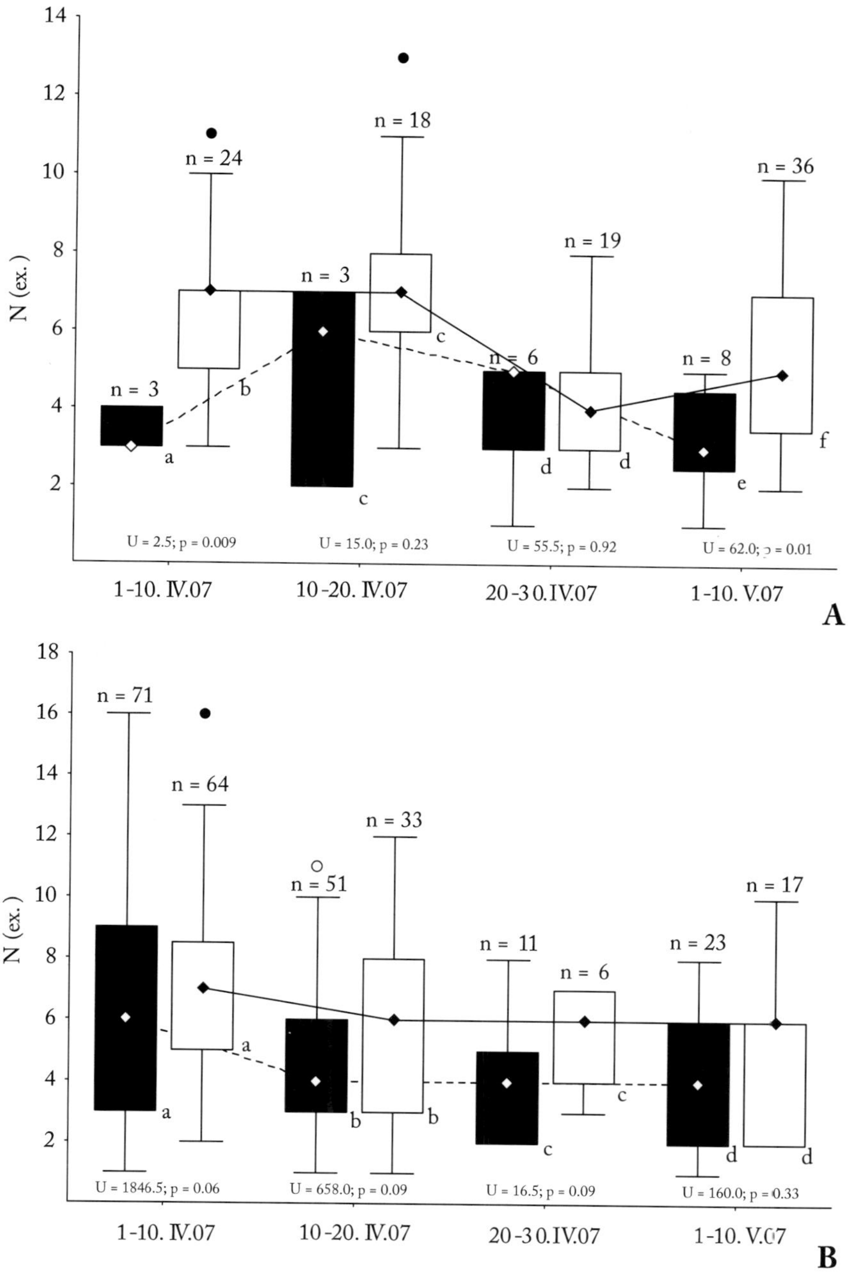

Fig. 4. Dynamics of oviposition (for medians) in females of parental and ancestral generations of *Pogonus transfuga* from the Elton region (data of 2007).
A – reedbeds along the Khara River; B – salina on a floodplain terrace of the Khara River; boxes – 25%-75%; whisker – non-outlier ranges; circles – outliers; open boxes – parental generation; filled boxes – ancestral generations; same letters indicating statistically non-significant differences; different letters indicating statistically significant differences ($p<0.05$; U-test).

curred from the end of May (Table 2). At the same time, in most of the species, females with ripe eggs could be found over a long period: in *P. convexicollis* up to mid June, in *P. transfuga*; to the end of June, in *P. meridionalis* until early July, and in *P. rufoaeneus* and *C. chloroticus* to the end of August. As a result, depending on the species, all instar larvae were encountered until mid August, or even up until early September. The duration of their development was considerably longer compared to the larvae which developed at the beginning of the growing season (Table 2).

The end of the reproduction period was characterized by an activity decline from the middle to the end of June. During this period, the proportion of spent beetles increased in the population while the first individuals from the new generation developed. A 1-1.5 month interval was found between the maximum reproduction and the appearance of the new generation. The new generation accounted for ¾ of the activity in the peaks of July-August. The peak of activity in September-November was more even. Immature adults from the daughter generation, as well as some beetles from the parental and ancestral generations, were mainly represented in this peak.

Variation in the life cycles was insignificant between the different habitats. For example, in *P. transfuga*, the activity of breeding beetles in the salina decreased more quickly than in the reedbeds (Table 1). In the salinas, the average number of eggs per female varied from 4.7 to 5.1 (2006) and 4.4 to 6.5 (2007). In the reedbeds it varied from 7.3-8.3 to 5.0 (2006) and 6.3-6.9 to 4.9 (2007)(Fig. 3A-B).

Hence, the life cycles of all these species are monovariant monovoltine with spring or spring-summer reproduction and adult (immature or both immature and spent) hibernation. However, the interpretation of the *C. chloroticus* life cycle is equivocal. This species is a spring-breeder. Under laboratory conditions, however, larvae reared from beetles that were transferred from the field to the laboratory in the beginning of June,

Table 2. Duration of pre-imaginal stages (in days) in several species of Pogonini in the Lake Elton region (data for 2006-2007, laboratory conditions).

Species		**Time of development**			
		Dates	**Min-Max**	**Average duration of development (mean ± SD)**	**Median**
Pogonus transfuga	L_1 (n = 6)	1-9.V.2007	6 - 9	7.0 ± 1.1	7.0*
	L_2 (n = 6)	7-17.V.2007	6 - 10	7.7 ± 1.5	8.0**
	L_3 (n = 5)	14-24.V.2007	7 - 11	8.6 ± 1.5	8.0***
Pogonistes rufoaeneus	L_1 (n = 2)	3-14.IX.2006		12.0 ± 0	12.0*
	L_3 (n = 1)	29.VII-14.VIII.2006		17.0	
Cardiaderus chloroticus	L_1 (n = 1)	28.VIII-3.IX.2006		7.0	
	L_2 (n = 4)	15.VIII-3.IX.2006	11 - 20	17.8 ± 4.5	20.0**
	L_3 (n = 3)	15.VIII-5.IX.2006	16 - 22	19.0 ± 3.0	19.0***

Notes: The medians with the same left symbols are compared with each other ($p<0.05$, U-test); * - U = 0; Z = -2.0; p = 0.045; ** - U = 0; Z = -2.56; p = 0.01; *** - U = 0; Z = -2.34; p = 0.025.

emerged in August. Under natural conditions, the last larvae of the third instar were found in October. Possibly, the life cycle of *C. chloroticus* is bivoltine, hence multivariant. The high activity of mature males from the end of July to early August supports this suggestion. The absence of females with ripe eggs from the catches was probably related to their low locomotor activity during the hot and dry second half of the summer.

In contrast, the maximum activity in *P. cumanus* was from the beginning of August to the end of September. Freshly emerged beetles from the new generation, as well as adults from the ancestral generations, were observed from mid May to early June. The proportion of the latter in the ten-day catches ranged from 100% to 67%. From mid-June to mid-July, no activity was recorded. The second peak of activity in non-breeding beetles from the new and ancestral generations was from the middle of July to mid of August. The share of the latter in the ten-day catches was reduced to 11%. The maximum reproduction occurred from the end of August to early September (Table 1).

The maximum number of ripe eggs amounted to 6, and the average per female was 2.5 (Fig. 2). The peak of oviposition was at the beginning of September (Fig. 3D).

Spent beetles and larvae from the *P. cumanus* population were found to enter hibernate from the end of September to the end of October.

Thus, *P. cumanus* is an iteroparous species with an autumnal breeding period. Its life cycle is monovariant monovoltine with obligatory larval hibernation and obligatory adult aestivation parapause.

Spatial and temporal pattern and dispersal potential

The spatial and temporal pattern of activity density was analyzed for seven Pogonini species with respect to their habitat. *P. transfuga* and *P. meridionalis* were more abundant in moderately to poorly salinated habitats. While *P. meridionalis* lived only on salinas on floodplain terraces, *P. transfuga* also inhabited reedbed thickets along river banks (Fig. 5). In the latter species, the peak of locomotor activity was observed 20 days earlier than that of the former species (Table 1).

All the other species were more abundant in habitats with high salinity (Table 1). *C. chloroticus*, *P. cumanus*, *P. rufoaeneus*, *P. angustus* and *P. convexicollis* inhabited only lakeside salt-marshes. *P. rufoaeneus* additionally inhabited floodplain salinas, as well using river banks as corridors (Fig. 5). Only *P. cumanus* was an "autumn" breeder (according to Larsson, 1939), while the others were "spring" breeders (Table 1). At least *C. chloroticus* preferred humid sites of salt-marshes with deep soil cavities. *Pogonistes* species inhabited different microhabitats, e.g. inside debris, under the salt crust, and in small soil crevices.

All studied specimens of the Pogonini were macropterous. However, their dispersal potential varied. More than 80% individuals of *C. chloroticus* and *P. rufoaeneus*, and more than 90 % specimens of *P. angustus* and *P. convexicollis*, had wing muscles. A decrease in *Ipm* value was observed during the period of maximum reproduction or at the end of the growing season. This decrease resulted from the activity of mature and spent beetles

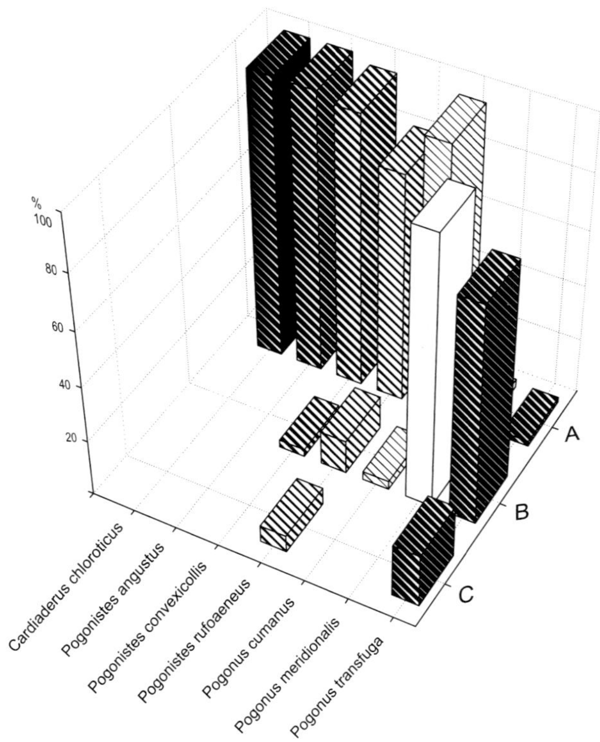

Fig. 5. Spatial distribution of seven species of Pogonini in the Elton region.
A – lakeside salt-marsh; B – salina on a floodplain terrace of the Khara River; C – reedbeds along the Khara River.

of the parental generation. In contrast, adults with fully developed wing muscles in the three *Pogonus* species were less abundant. The value of *Ipm* amounted to 0.5, 0.6 and between 0.32 and 0.86 in *P. cumanus, P. meridionalis* and in *P. transfuga,* respectively. In the latter two species, the maximum value of *Ipm* was recorded in May or in June-July, during the emergence of young beetles (Table 3).

Infestation by parasites

During our study, parasitic mites of the family Podapolipidae were reported for the first time on *P. transfuga*, *P. meridionalis*, *P. cumanus* and *P. rufoaeneus*. In some species, for example *P. transfuga*, 16% of the specimens were infested. In April and May, only beetles of the ancestral generations were infested, whereas in July-October specimens of the parental generation were infested (Table 4). Mites were invariably located ventrally at the base of the wings which is considered to be their typical location on their hosts (Regenfuss, 1972). Each wing carried one female and groups of eggs and nymphs. Asymmetric arrangements of mites were recorded in only a few cases.

Table 3. Seasonal changes in *Ipm* values in Pogonini species from different habitats in the Lake Elton region (pooled data for 2006-2007).

Habitats	Species	Sex	Months						
			IV	**V**	**VI**	**VII**	**VIII**	**IX**	**X**
Lakeside salt-marsh	*Cardiaderus chloroticus*	♂	0.62	1.0	1.0	1.0	1.0	1.0	1.0
		♀	0.71	1.0	1.0	1.0	1.0	1.0	1.0
	Pogonistes angustus	♂			1.0	1.0		1.0	
		♀			1.0	1.0	0.5	0	1.0
	Pogonistes rufoaeneus	♂	0.75	0.95	1.0		0.55	1.0	
		♀	0.7	1.0	0.67	0.6	0.17	1.0	1.0
	Pogonistes convexicollis	♂	1.0	1.0	1.0	1.0	0	0	1.0
		♀	1.0	1.0	1.0	1.0	0	0	1.0
	Pogonus cumanus	♂			0	1.0	0.33	0.39	0.5
		♀	0		0	1.0	0.29	0.29	0
Salina on a floodplain terrace	*Pogonus meridionalis*	♂	0	0.35	0.36	0	0.6	1.0	
		♀	0	0.2	0.15	0	1.0		
	Pogonus transfuga	♂	0.02	0.32	0.19	0.02	0.11	0.07	0.15
		♀	0.03	0.26	0	0	0.13	0	0.1
Reedbeds	*Pogonus transfuga*	♂	0.16	0.24	0.71	0.86	0.5	0	0.21
		♀	0.09	0.32	0.33	0.8	0.75		0.21

Table 4. Seasonal changes in infestation rates (in %) *Pogonus transfuga* by Podapolipidae mites in the Lake Elton region (pooled data for 2006-2007).

Habitats	Sex	Months						
		IV	**V**	**VI**	**VII**	**VIII**	**IX**	**X**
Salina on a floodplain terrace	♂ P							33.3
	♀ P				80.2			75.4
	♂ A	53.5	78.5					
	♀ A	40.0	33.3					
Reedbeds	♂ P				72.7	70.0	72.7	63.6
	♀ P				96.1	87.5	85.7	88.8
	♂ A	50.2	70.8					
	♀ A	63.3	76.0					

Notes. P - parental generation; A - ancestral generations.

DISCUSSION

Contradictory observations have been published concerning the life cycles of Pogonini. Turin (2000) found a spring-summer type of seasonal activity in *Pogonus chalceus*, *P. litoralis* and *P. luridipennis*. However Larsson (1939), Heydemann (1962) and Paarmann (1976) considered *P. chalceus* to be a summer breeder. In the Elton region, the maximum reproduction activity in most of the Pogonini species is confined to April to May. This is probably affected by air temperatures and relatively high soil moisture rates during a short spring. Similar observations are available for salt-marshes in Belgium, where the maximal reproduction activity of *P. chalceus* was also recorded in May (Desender, 1985, 2000). Differences in egg-production between old and young females have been reported for *Calathus melanocephalus* (van Dijk, 1979), several *Carabus* spp. (Grüm, 1975) and *Pterostuchus melanarius* (Matalin, 2006). In these species, the average number of eggs in old females was lower than that in young individuals.

Detailed information concerning the duration of pre-imaginal stages is known only for *P. litoralis*. Under laboratory conditions, the larvae developed in 22-24 days and pupae in 15-17 days. Total development from egg to imago lasted 42-48 days (Nekuliseanu, 1987). Such results correspond well with our data on the development of a related species, *P. transfuga* (Table 2).

Gonad maturation has been studied only for *P. chalceus* and *P. litoralis*. Gonad maturation in *P. chalceus* is induced by high air temperatures in northern Africa, with reproduction observed during the summer months. Under such conditions, two generations are produced (Paarmann, 1976). However, in Northern Europe the life cycle of *P. chalceus* is monovoltine (Larsson, 1939; Paarmann, 1979; Desender, 1989; Turin, 2000). A similar life-cycle is known for several species of *Stenolophus* (Matalin, 1997c). In contrast, gonad maturation in *P. litoralis* is completed only after hibernation (Neculisaenu, 1987).

According to our data as well as those of Larsson (1939), six species of Pogonini are "spring" breeders. Already in the first half of April, nearly 90% of specimens of *P. transfuga*, *P. meridionalis* and *C. chloroticus*, as well as 50% of specimens of *P. rufoaeneus* and *P. convexicollis*, can reproduce (Table 1). In these species, maturation of gonads follows the Type 1 pattern described by Thiele (1977). Only *P. cumanus* is an "autumn" breeder. This is surprising, because hitherto all Pogonini have been considered as "spring" breeders or "imaginal" hibernators (Larsson, 1939; Heydemann, 1962; Paarmann, 1976, 1979; Desender, 1985, 1989, 2000; Lindroth, 1992; Turin, 2000; Matalin & Makarov, 2006). Although the new generation of *P. cumanus* emerges in mid May to early June, they reproduce from the end of August to the end of September. Thus, gonad maturation corresponds to the Type 4 pattern described by Thiele (1977).

Like most carabid beetles, all of the studied species of Pogonini are opportunistic predators. As they are similarly sized, we presume that they compete for food. Moreover, six species show the same type of life-cycle and, hence, are potential competitors for living space. To reduce competition, they must partition their habitat in space and/or in time

(Andersen, 1983, 1988; Matalin, 1997a; Brandmayr & Algieri, 2000). This is attained by varying habitat preferences and different life cycle strategies.

Among the studied species, *P. transfuga* and *P. meridionalis* can be regarded as halophiles, because they live in habitats with moderate salinity. To diminish competition, heterogeneous within- and between-habitat distributions (Fig. 5), heterochronous maxima of locomotor activity (Table 1), as well as size differences (*P. transfuga* is larger than *P. meridionalis*) are applied. The other five species are halobiontic because, with only a few exceptions, they occur only in habitats with high salinity (Fig. 5). Among them, only *P. cumanus* is an "autumn-breeder", while the others are "spring-breeders (Table 1). *C. chloroticus* seems to be a more specialized beetle, because it inhabits highly specific microhabitats and possibly has a bivoltine life-cycle. Competition in the three *Pogonistes* species seems to be higher, as they are similar in size and inhabit the same, or similar, habitats. Only *P. rufoaeneus*, the most abundant species, inhabits all of the studied habitats, while *P. angustus* and *P. cinvexicollis* are less abundant and coexist with *P. rufoaeneus* in salt-marshes only.

The discovery of Podapolipidae mites on Pogonini beetles is remarkable. Until now, only species from the tribes Carabini (Eidelberg, 1994; Fain et al., 1995; Regenfuss, 1968), Scaritini (Husband, 2001), Broscini (Eidelberg, 1994), Bembidiini (Eidelberg & Husband, 1993), Platynini (Eidelberg, 1994; Fain et al., 1995; Husband, 1998), Pterostichini (Eidelberg, 1994; Husband, 1998c; Husband & Dastych, 2000), Zabrini (Eidelberg, 1994; Husband & Husband, 1996), Harpalini (Husband, 1998a, b), Panagaeini (Husband, 2000), Callistini (Eidelberg, 1994; Husband & Dastych, 1998), Lebiini (Fain et al., 1995) and Trichognathini (Husband & Eidelberg, 1996) have been known as hosts of Podapolipidae.

The biology of Podapolipidae is still poorly known. These mites are known to show complicated life-histories and to be ectoparasites of different insects (Husband, 2000, Regenfuss, 1968). High specialization is typical of Podapolipidae, because they are usually associated with only one or two host species (Fain et al., 1995; Regenfuss, 1968). Beetles usually become infested during copulation with an infested mate. Apparently, this sharply restricts the mites' dispersal capacity (Regenfuss, 1968) and corresponds well with our observations. Mites have only been found on beetles in copula. Since numerous Carabidae species are known to reveal high mortality rates during hibernation as imago, the capability of some spent beetles to survive winter and to reproduce in the next year together with specimens of the new generation are, we suggest, important preconditions for the successful development of Podapolipidae. On the other hand, this greatly restricts the range of their potential hosts. In September-October, the proportion of already copulated and infested specimens of *P. transfuga* was about 70%, whereas it was 58.5-55.5% during April to May (Table 4).

ACKNOWLEDGEMENTS

We extend our thanks to all our colleagues at the Elton Natural Park (Elton, Volgograd Region, Russia), especially to its director, Mrs Yulia A. Nekrutkina (Volgograd, Russia), and to Mr Artem A. Zaitsev (Moscow State Pedagogical University, Russia) for their assistance in our work, to Dr. Olga L. Makarova (Institute for Problems of Ecology and Evolution, Moscow, Russia) for important remarks concerning the biology of Podapolipidae mites, and to Dr. Sergei I. Golovach (Institute for Problems of Ecology and Evolution, Moscow, Russia) and Dr. Stephen Venn (University of Helsinki, Finland) for a critical review of the text.

This study was financially supported by the Russian Foundation for Basic Research (projects Nos 06-04-49456, 07-04-08381), as well as the Presidental Support Programme for Leading Academic Schools (project No. НШ-2154.2003.4).

REFERENCES

Andersen, J.M. (1983). The life cycles of the riparian species of *Bembidion* (Coleoptera, Carabidae) in Northern Norway. – Notul. Entomol. 63: 195-202.

Andersen, J.M. (1988). Resource partitioning and interspecific interactions among riparian *Bembidion* species (Coleoptera: Carabidae). – Entomol. Gen. 13: 47-60.

Arndt, E. (1991). Carabidae. – In: Die Larven der Käfer Mitteleuropas. Adephaga. (Klausnitzer, B., ed.). Krefeld: Goecke & Evers. 1: 1-141.

Butterfield, J. (1986). Changes in life-cycle strategies of *Carabus problematicus* over a range of altitudes in northern England. – Ecol. Entomol. 11: 17-26.

Bousquet, Y. (2003). Tribe Pogonini Laporte, 1834. – In: Catalogue of Palaearctic Coleoptera. Archostemata – Myxophaga – Adephaga (Löbl, I. & Smetana, A., eds), Stenstrup: Appolo Books. 1: 286-288.

Borovikov, V. (2001). STATISTICA: data analysis on the computer. – St. Petersburg: Piter Press. 1-650 [in Russian].

Brandmayr, P. & Algieri, M.C. (2000). Habitat affinities of chlaeniine species (Coleoptera, Carabidae) in Calabria and the status of *Epomis circumscriptus*, evaluated by the "Cronogeonemie" software. – In: Natural History and Applied Ecology of Carabed Beetles (Brandmayr, P., Lövei, G.L., Zetto-Brandmayr, T., Casale, A. & Vigna Taglianti, A., eds). Sofia-Moscow: Pensoft. 71-78.

den Boer, P.J. (1977). Dispersal power and survival. Carabids in a cultivated countryside (with a mathematical appendix by J. Reddingius). – Miscellaneous Papers Wageningen. 14: 1-190.

Desender, K. (1985). Wing polymorphism and reproductive biology in the halobiont carabid beetle *Pogonus chalceus* (Marsham) (Coleoptera, Carabidae). – Biol. Jb. Dodonaea. 53: 89-100.

Desender, K. (1989). Dispersievermogen en ecologie van loopkevers (Coleoptera, Carabidae) in Belgie: een evolutionaire benadering. – Studiedoc. Kon. Belg. Instit. Natuurwet. Rijksuniv. Gent. 54: 1-136.

Desender, K. (2000). Flight muscles development and dispersal in the life cycle of carabid beetles: patterns and processes. – Bull. Inst. R. Sci. Nat. Belgique. 70: 13-31.

Desender, K. & Vaneeckhoutte, M. (1984). Phoretic associations of carabid beetles (Coleoptera, Carabidae) and mites (Acari). – Rev. Écol. Biol. Sol. 21: 363-371.

Eidelberg, M.M. (1994). Mites of the family Podapolipidae (Heterostigmata, Tarsonemina) of Ukraine and adjacent lands with descriptions of new species. – Vestnik Zoologii (Kiev). 1: 37-43 [in Russian].

Eidelberg, M. & Husband, R.W. (1993). A new species of *Eutarsopolipus* (Acari: Podapolipidae) from *Bembidion saxatile* (Coleoptera: Carabidae). – Int. J. Acarol. 19: 267-272.

Emden, F.I. van (1942). A key to the genera of larval Carabidae. – Trans. Roy. Ent. Soc. London. 92: 1-99.

Fain, A., Noti, M.I. & Dufrêne, M. (1995). Observations on the mites (Acari) associated with Carabidae (Coleoptera) in Belgium. I. Annotated list of the species. – Int. J. Acarol. 21: 107-122.

Grebennikov, V.V. & Bousquet, Y. (1999). Larvae of Pogonini (Coleoptera: Carabidae): Genera *Pogonus*, *Pogonistes*, *Cardiaderus*, and *Thalassotrechus*. – Acta Soc. Zool. Bohem. 63: 427-441.

Grüm, L. (1973). Egg production of some Carabidae species. – Bull. Polish Acad. Sci. 21: 261-268.

Heydemann, B. (1962). Die biozönotische Entwicklung vom Vorland zum Koog. Vergleichend-ökologische Untersuchungen an der Nordseeküsten. II Teil: Käfer (Coleoptera). – Abh. Math-Naturw. Kl. Akad. Wiss. Mainz. 11: 765-964.

Husband, R.W. (1998a). A new species of *Eutarsopolipus* (Acari: Podapolipidae) from *Harpalus pennsylvanicus* (Coleoptera: Carabidae) from East Lansing, Michigan. – Great Lakes Entomol. 31: 141-150.

Husband, R.W. (1998b). New species of *Eutarsopolipus* (Acari: Podapolipidae) from *Harpalus caliginosus* (F.) and *Agonoderus comma* (F.) (Coleoptera: Carabidae) from Kansas and Wyoming, U.S.A. – Entomol. Mitt. Zool. Mus. Hamburg. 12: 255-264.

Husband, R.W. (1998c). Two new species of *Eutarsopolipus* (Acari: Podapolipidae) from *Agonum extensicole* and *Pterostichus lucublandus* (Coleoptera: Carabidae) from Canada, including taxonomic keys to the 13 American species of Podapolipidae from carabid beetles. – Ann. Entomol. Soc. Amer. 91: 279-287.

Husband, R.W. (2000). Two new species of *Dorsipes* (Acari: Podapolipidae) from *Tefflus zebulianus reichardi* (Coleoptera: Carabidae) from the Democratic Republic of Congo, including a key to *Dorsipes* species. – Ann. Entomol. Soc. Amer. 93: 7-14.

Husband, R.W. (2001). A new species of *Eutarsopolipus* (Acari: Podapolipidae) from *Scarites subterraneus* (Coleoptera: Carabidae) from Louisiana, U.S.A. – Int. J. Acarol. 27: 113-117.

Husband, R.W. & Dastych, H. (1998). A new species of *Eutarsopolipus* (Acari: Podalipolipidae) from *Chlaenius sericeus* Frost (Coleoptera: Carabidae) from Athens, Georgia, USA. – Entomol. Mitt. Zool. Mus. Hamburg. 12: 317-326.

Husband, R.W. & Dastych, H. (2000). Two new species of *Dorsipes* (Acari: Podapolipidae) from *Pterostichus niger* (Schall.) (Coleoptera: Carabidae) from Germany, including a key to *Dorsipes* species. – Entomol. Mitt. Zool. Mus. Hamburg. 13: 205-218.

Husband, R.W. & Eidelberg, M. (1996). A new species of *Eutarsopolipus* (Acari: Podapolipidae) from *Trichognathus marginipennis* (Coleoptera: Carabidae) from Brazil. – Int. J. Acarol. 22: 193-197.

Husband, R.W. & Husband, D.O. (1996). A new species of *Eutarsopolipus* (Acari: Podapolipidae) from *Amara californica* Dejean (Coleoptera: Carabidae) from California. – Proc. Entomol. Soc. Wash. 98: 465-470.

Jeannel, R. (1941). Coléoptères Carabiques. I. Faune de France. – Paris: Librairie de la Facuilté des Sciences. 39: 1-571.

Kryzhanovsky, O.L. (1982). Beetles of the suborder Adephaga: Families Rhysodidae, Trachypachidae, Carabidae (introduction and review of the USSR fauna). – Fauna SSSR. Zhestkokrylye (Fauna of the USSR. Coleoptera). Leningrad: Nauka. 1: 1-341 [in Russian].

Larsson, S.G. (1939). Entwicklungstypen und Entwicklungszeiten der dänischen Carabiden. Kommission Hos P. Haase & Søns Forlag, København. – Entomologiske Meddelelser. 20: 277-562.
Larsson, S.G. (1968). Løbebillernes larver. – In: Danmarks Fauna. Sandspringere og løbebiller. 24: 282-243.
Lindroth, C.H. (1992). Ground Beetles (Carabidae) of Fennoscandia. A Zoogeographic Study. Specific Knowledge Regarding the Species. — London: Intercept Ltd. 1: 1-630.
Lorenz, W. (1998). Systematic list of extant ground beetles of the world (Insecta Coleoptera "Geadephaga: Trachypachidae and Carabidae incl. Paussinae, Cicindelinae, Rhysodinae"). First edition. Tutzing: published by the author. 502 pp.
Luff, M.L. (1985). The larvae of the British Carabidae (Coleoptera) VII. Trechini and Pogonini. – Entomologist's Gaz. 36: 301-316.
Luff, M.L. (1993). The Carabidae (Coleoptera) larvae of Fennoscandia and Denmark. – Fauna Ent. Scand. 27: 1-186.
Matalin, A.V. (1997a). Peculiarities of spatial and temporal differentiation of Carabids (Coleoptera, Carabidae) in the steppe zone. – Entomol. Rev. 77: 1155-1166.
Matalin, A.V. (1997b). Specific features of the life cycles of *Pseudoophonus* (s.str.) *rufipes* Deg. (Coleoptera, Carabidae) in southwestern Moldova. – Biol. Bull. 24: 371-381.
Matalin, A.V. (1997c). Life cycles of carabids of the genus *Stenolophus* (Coleoptera, Carabidae) in the steppe zone of Europe. – Entomol. Rev. 77: 1181-1190.
Matalin, A.V. (2003). Variations in flight ability with sex and age in ground beetles (Coleoptera, Carabidae) of south-western Moldova. – Pedobiologia. 47: 311-319.
Matalin, A.V. (2006). Geographic variability of the life cycle in *Pterostichus melanarius* (Coleoptera, Carabidae). – Entomol. Rev. 86: 409-422.
Matalin, A.V. & Makarov K.V. (2006). The life cycle of halophilous ground beetles, *Pogonus* (s. str.) *transfuga* Chaudoir, 1871 (Coleoptera: Carabidae) in the Elton Region. – In: Biodiversity and the Problems of Nature Management in the Elton Region (Kursakova, N.A., Nekrutkina, Yu.A., Sokhina, E.N. & Chernobay, V.F., eds). Volgograd: PrinTerra. 40-46 [in Russian].
Nekuliseanu, Z.Z. (1987). The biology of *Pogonus litoralis* Duft. (Coleoptera, Carabidae). – In: Problems of Soil Zoology (Kurashvili, B.E., ed.). Tbilisi: Metsniereba. 199 [in Russian].
Paarmann, W. (1976). The annual periodicity of the polyvoltine ground-beetle *Pogonus chalceus* Marsh. (Col. Carabidae) and its control by environmental factors. – Zool. Anz. 196: 150-160.
Paarmann, W. (1979). Ideas about the evolution of the various annual reproduction rhythms in carabid beetles of the different climatic zones. – In: On the Evolution of Behaviour in Carabid Beetles (den Boer, P.J., Thiele, H.U. & Weber, F., eds). H. Veeneman & Zonen B.V. Publ., Wageningen. 18: 119-132.
Raynaud, P. (1976). Stades larvaires de Coléoptères carabiques. – L'Entomologiste. 32: 166-174.
Regenfuss, H. (1968). Untersuchungen zur Morphologie, Systematik und Ökologie der Podapolipidae (Acarina, Tarsonemini) (Unter besonderer Berücksichtungen der Parallelevolution der Gattungen *Eutarsopolipus* und *Dorsipes* mit ihren Wirten (Coleoptera, Carabidae). – Zeitschr. wiss. Zool. 177: 183-282.
Regenfuss, H. (1972) Über die Einnischung synhospitaler Parasitenarten aus dem Wirtkorper. – Zeitschr. Zool. Syst. Evol. 10: 44-65.
Safronova, I.N. (2006). Description of the flora of the Pallasovskiy District, Volgograd Region. – In: Biodiversity and the Problems of Nature Management in the Elton Region

(Kursakova, N.A., Nekrutkina, Yu.A., Sokhina, E.N. & Chernobay, V.F., eds). Volgograd: PrinTerra. 5-9 [in Russian].
Sharova, I.Kh. (1958). Carabid larvae beneficial or harmful to agriculture. – In: Uchenye zapiski Moskovskogo gosudarstvennogo pedagogicheskogo institute (Scientific Papers of the Moscow State Pedagogical Institute) (Naumov, S.P., ed.). 12: 4-164 [in Russian].
Sharova, I.Kh. (1964). The family Carabidae. – In: Opredelitel' obitauschikh v pochve lichinok nasekomykh (Identification Book of the Soil-Dwelling Insect Larvae) (Ghilarov, M.S., ed.). Moscow: Nauka. 112-185 [in Russian].
Thiele, H.-U. (1977). Carabid Beetles in their Environments. A Study on Habitat Selection by Adaptations in Physiology and Behaviour. Springer: Berlin-Heidelberg-New York. XVII+369.
Tietze, F. (1963). Untersuchungen über die Beziehungen zwischen Flugelreduction und Ansbidung des Metathorax bei Carabiden, unter besonderer Berücksichtigung der Flugmuskulatur (Coleoptera, Carabidae). – Beitr. Ent. 13: 88-167.
Turin, H. (2000). De Nederlandse Loopkevers: Verspreiding en Oecologie (Coleoptera: Carabidae). – Nat. Natuurhistorisch Mus. Natur., Tekeningen, 666 p.
van Dijk, Th.S. (1979). Reproduction of young and old females in two carabid beetles and the relationship between the number of eggs in the ovaries and number of eggs laid. – In: On the Evolution of Behaviour in Carabid Beetles (den Boer, P.J., Thiele, H.U. & Weber, F., eds). H. Veeneman & Zonen B.V. Publ.: Wageningen. 18: 167-183.
Wallin, H. (1987). Distribution, movements and reproduction of Carabid beetles (Coleoptera, Carabidae) inhabiting cereal fields. – Plant protection reports. Dissertations, Swedish University Agricultural Sciences, Uppsala: SLU/Repro. 15: I-XXV+109.

L. Penev, T. Erwin & T. Assmann (Eds) 2008
Back to the Roots and Back to the Future. Towards a New Synthesis amongst Taxonomic, Ecological and Biogeographical Approaches in Carabidology
Proceedings of the XIII European Carabidologists Meeting, Blagoevgrad, August 20-24, 2007, pp. 325-338.

Defensive strategies against predators in Carabid beetles

Tullia Zetto Brandmayr, Teresa Bonacci,
Antonio Mazzei & Pietro Brandmayr

Dipartimento di Ecologia, Università della Calabria, 87036 Arcavacata di Rende, (CS), Italy.
E-mail: tuzetto@unical.it

SUMMARY

A review of defensive strategies against predators is presented for carabid beetles in *Brachinus* and *Anchomenus* genera. We consider three important topics as unpalatability, gregariousness and aposematism worldwide applied to defence against predators. In the first part of the review we report some of the outstanding works from a rich literature both about the description of these evolutionary aspects and about the relationships among them and their evolutionary dynamics. Secondly we focused on *Anchomenus dorsalis* and *Brachinus sclopeta* referring to our recent works: they are more or less unpalatable, mainly *B. sclopeta*, a bombardier beetle; they aggregate in conspicuous interspecific groups and inside them *A. dorsalis* show an intensive "rubbing behaviour" towards *B. sclopeta*; moreover they share a bright colour pattern that may be thought as aposematism via warning signals. To confirm this hypothesis, laboratory data with *Ocypus olens* (staphyline), shrews and lizards as predators are reported. A highly similar cuticle molecular composition between *B. sclopeta* and *A. dorsalis,* particularly if compared to a non aposematic carabid beetle, may be supposed a mechanism similar to that involved in colour similarity, effective in reducing the predation risk by non visual predators.

Keywords: defence, chemicals, aposematism, insects aggregation

INTRODUCTION

Animals evolved a variety of strategies to avoid or reduce predation; basically, these develop along two different pathways: (i) making animals hard to find (cryptic behaviour) or (ii) enhancing escaping possibility, when detected by predators (Alcock, 1979).

A peculiar drift from the first way may be considered the true opposite behaviour, that is to become very conspicuous to predators. Possible solutions are aposematism and gregariousness, very common among unpalatable and dangerous insects. In this short review we point out whether or not aposematism and gregariousness play a role as defensive strategy in carabid genera *Anchomenus* and *Brachinus*.

First, we synthesise the main steps assessed in a very rich literature by several Authors focusing on the three milestone concepts of unpalatability, aposematism and gregariousness and second, we present and discuss our recent research on these topics in *Anchomenus* and *Brachinus* species.

UNPALATABILITY

This term means to be not good once ingested, is comprehended in the wide classification of the "unprofitability" i.e the character of a prey that gives no net reward to the predator once consumed, leading to learned or evolved avoidance (ex. noxiousness/toxicity, toughness, difficult handling, difficult/costly capture) (Joron, 2002).

Particularly, unpalatability in insects may be caused by chemical defence substances secreted by glandular sources, most of all if their products are sprayed or otherwise applied topically to an enemy or non-glandular defenses as blood or enteric discharging (for a more extended review see Eisner, 1970 and cited list of references).

Many insects are chemically protected by defensive secretions released from exocrine glands located in several parts of their body (Pasteels et al., 1988; Pasteels et al., 1994; Hartmann et al., 2003).

This chemical defense can be autogenous and synthesized from diet (Pasteels et al., 1996) or produced in special pygidial glands as in many beetles (Eisner & Aneshansley, 1999; Eisner et al., 2000, 2001a, b). Alternatively, toxic chemical compounds may be sequestered by host or feeding plants or by animal prey.

APOSEMATISM

Many animals use warning colours to signal their dangerousness to potential predators (Cott, 1940; Guilford, 1990). Aposematic coloration decreases the attack probability in naïve predators, either as effect of the novelty and of aversive colours (Coppinger, 1969, 1970; Roper & Cook, 1989; Gamberale & Tullberg, 1996) or both (Sillén-Tullberg,

1985). It has been more than a century since Wallace (1867) proposed that conspicuous animal colour patterns could evolve to advertise the toxic or unpalatable nature of prey to visually hunting predators. In this time aposematism (the name given by Poulton, 1890, to describe this common association between bright coloration and unpalatability) has become a key system in which to study how receiver psychology can influence the evolution of signals (Rowe & Guilford, 2000).

In their paper "Aposematism: to be red or dead" Row & Guilford (2000) re-discussed how warning coloration evolved; in previous papers (Mallet & Joron, 1999; Lindstrom et al., 1999; Gamberale & Tullberg, 1996), some Authors spoke about an "evolutionary paradox", since there is an initial cost to an aposematic morph; not only will be it more detectable to predators, but also its rarity means that predators are less likely to have learnt the association between the colour and the unpalatability. Therefore, although the pattern might have a selective advantage once it is common, at a low frequency it will suffer from increased predation and remaining cryptic will be strongly favoured (Rowe & Guilford, 2000).

For the question, "how did conspicuousness evolve, since the first pioneers were prone to predation due to increased visibility and encounter rate and, therefore were most probably sampled and killed during the training of predators?" (Lindström et al., 2001) the possible replay according to the Authors is that although learning has been thought as being the main factor facilitating the evolution of aposematism, it is not the only potential aspect of predator psychology that can influence the signals used by the prey. Predators may already have some preferences when they encounter possible new prey types and these preferences may be innate, triggered by novelty or the result of a search image. Reluctance of predators for eating novel prey (neophobia) might balance the initial predation caused by inexperienced predators.

GREGARIOUSNESS

The group formation may also be adaptive in avoiding predation by visually hunting predators and many possible antipredator advantages may be enjoined by animals which clump together (Alcock, 1979). The individual risk to be killed is lowered, depending on the group size, after the predator satiation. Conversely, group formation may be not so advantageous if it makes animals more conspicuous to predators. For example, solitary locusts in transition to the gregarious phase aggregate and, being very active, are more conspicuous to predators as solitary ones, even if they have still a cryptic coloration (Pitt & Ritchie, 2002). This increased predator detection risk may provide conditions under which is adaptive to switch to the aposematic developmental phase accomplished by the moult to the next stage; indeed, this is brightly coloured (yellow-black) to signal unpalatability (Despland & Simpson, 2005).

In insects, aposematism often occurs together with gregariousness (Edmunds, 1974). Some authors suggested that the gregariousness increases the effect of the aposematic signals (Poulton, 1890; Cott, 1940) and this increasing in signal efficiency could influence both the initial unconditioned aversion of naïve predators and the speed and memorability of avoidance learning (Gamberale &Tullberg, 1998). Actually, predator's avoidance learning can be faster and more durable when prey is gregarious because it can see warningly coloured prey items simultaneously or immediately after perceiving the noxious stimulus (Gagliardo & Guilford, 1993; Speed, 2000) and likewise. As result, gregariousness is more common in aposematic distasteful prey than in palatable and cryptic species (Järvi et al., 1981). Moreover, studies on the effect of prey aggregation on phobic reactions in *Gallus gallus domesticus* show good evidence that aggregations of live aposematic prey generate higher levels of avoidance neophobia than solitary prey (Gamberale & Tullberg, 1996; 1998). The gregariousness, which reduces per capita detectability of the prey, is expected to evolve where there is any tendency toward predator satiation; and one of the best ways of satiating predators is to be distasteful (Mallet & Joron, 1999).

A convincing demonstration of the adaptive value of the different combinations of states cryptic, gregarious, solitary and aposematic is given by Hatle et al., (2002) with experimental data about predation on mealworms prey items by bullfrogs. In a trend of increased time to attack the succession of two coupled characters was: cryptic and gregarious, cryptic and solitary, aposematic and solitary, aposematic and gregarious.

EVOLUTION OF UNPALATABILITY, WARNING COLORATION AND GREGARIOUSNESS

As unpalatability, aposematism and gregariousness are strictly connected, they have been an important subject of debate concerning their evolution.

The unpalatability may evolve in kin-grouped prey (Fisher, 1930) and kin selection may favour the evolution of aposematic coloration. On the other hand, the association between gregarious larvae and unpalatability can be explained easily because gregariousness will evolve more readily after unpalatability, rather than before it, as required under the kin selection hypothesis. Moreover, the pattern of unpalatability first, gregariousness thereafter, is well supported in Lepidoptera by phylogenetic analysis (Mallet & Joron, 1999).

Conversely, in locusts gregariousness comes early and thereafter unpalatability is behaviourally acquired by modifying gregarious animal diet, which involves host plants with toxic compounds (Despland & Simpson, 2005).

Alatalo and Mappes (1996) proposed a sequence of events leading from palatability and crypsis to unpalatability and aposematism. This process is founded on three steps: (i) from palatability to unpalatability, (ii) from solitary to aggregated, (iii) from cryptic to aposematic. An alternative evolutionary pathway to aposematism was suggested by Tullberg et al. (2000) from palatable, cryptic, and solitary prey, i.e. (i) from palatability to unpalatability and (ii) from cryptic to aposematic. At this point these Authors state that gregariousness may or may not constitute a third step of the sequence or alternatively it may evolve directly after unpalatability, allowing aposematism to be or not a third step. (see also Sillén-Tullberg & Leimar 1988; Vulinec, 1990).

ARE BRIGHT COLOURFUL CARABID BEETLES APOSEMATIC?

Many cases of carabid aposematism have been reported by Lindroth (1971) for Lebiini, e.g. in the genus *Lebia*, where *L. viridis* is suspected to be related to arboreal Alticinae (flea beetles). Lindroth supposes also that some South African *Lebistina* species exhibit Batesian mimicry versus the poisonous flea beetles *Diamphidia* and *Polyclada*, used by hunting Bushmen in the preparation of an arrow poison that kills warm-blooded animals. A similar black-yellow pattern is shown by the *Eurycoleus* larvae that prey on the pupae of the genus *Amphis* (Endomychidae beetle) (Erwin & Erwin, 1976). One case of mimicry was found by Adis et al. (1997) in carabid beetle *Colliuris batesi* that mimic to avoid attacks of predators (when the individuals are exposed on the soil) groups of ants.

Carabids, in common with all members of the suborder Adephaga, possess pygidial glands which produce defensive secretions. The glands exhibit great variability in structure and variety in the nature of the secretions produced (Thiele, 1977). Some carabid beetles, such as bombardier beetles of the genus *Brachinus* (Weber 1801), are well protected against predators because they are able to release irritating quinones, produced by oxidation of hydroquinones in a double-chambered apparatus (Schildknecht, 1961; Schildknecht et al., 1968); a certain amount of heat and the explosion associated with the reaction reinforce the defensive effect. Predation on these beetles appears to be rare (Juliano, 1985). Unlike most carabid beetles which are homogeneously brown or brown-black, a number of *Brachinus* species are bright orange-red with blue or green elytra; it is likely to be an aposematic signal (Bonacci et al., in press). From literature it is known that *Anchomenus dorsalis* (Pontoppidan,1763) produces toxic methilsalycilate from its pygidial gland (Schildknecht, 1970); like *Brachinus*, it exhibits a bright two-coloured coat (green-blue and red-brown) (see Photo 1).

Photo 1. Interspecific aggregation of *Brachinus sclopeta* (a) and *Anchomenus dorsalis* (b) individuals. In the group frequently we found individuals of *Poecilus cupreus* (c). Scale: 2 mm.

We investigated in laboratory the predatory behaviour of some animals which may prey on carabid beetles. We tested as predator models staphylinid ground beetle *Ocypus olens* (Müller) (Bonacci et al., 2006), shrew, *Crocidura leucodon* (Hermann, 1780) (Bonacci et al., 2004a) and lizard, *Podarcis sicula* (Rafinesque-Schmaltz, 1810) (Bonacci et al., in prep.) and prey were bright coloured or cryptic brown carabid beetles. Among warning coloured models *Brachinus sclopeta* and *Anchomenus dorsalis* were always ranked, together with other aposematic genera. All the experiments revealed a significant predator preference for non visually conspicuous and unprotected prey (Figs 1-2). Our data (Bonacci et al., 2006) suggest that *O. olens* recognises *Brachinus sclopeta* and *Anchomenus dorsalis* as chemically protected prey and has the ability to decide whether or not to attack on the basis of the palatability/dangerousness of the prey.

Since the first two models are not thought to be exclusively visual hunters, possibly also olfactory signals are involved.

From these results we can infer that the coloration of *B. sclopeta* and *A. dorsalis*, very conspicuous to the background, can be actually a warning signal to predators in the sense of aposematism.

CARABID BEETLE AGGREGATIONS

Among Carabid beetles some positive relationships are known as for example the gregariousness, i.e. some individuals associate with others of the same species as well as of different ones at least during some periods of their life cycle (Thiele, 1977). Intraspecific aggregations may be the result of a mutual attraction among conspecific individuals, which is not related to age or sex and depend on olfactory cues perceived by the antennae (Wautier, 1971).

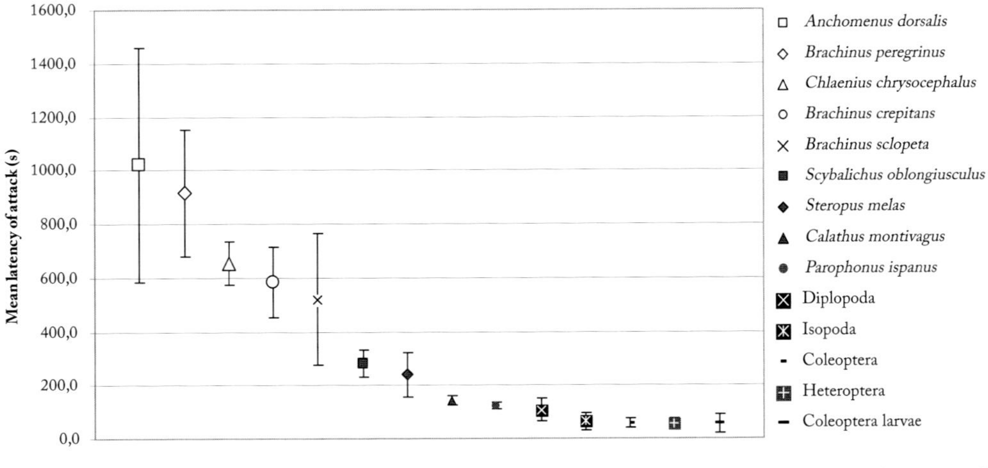

Fig. 1. Mean latency of attack of *Crocidura leucodon* (Hermann, 1780) towards chemical protected (empty symbols) and unprotected species (full symbols) (from Bonacci et al., 2004a).

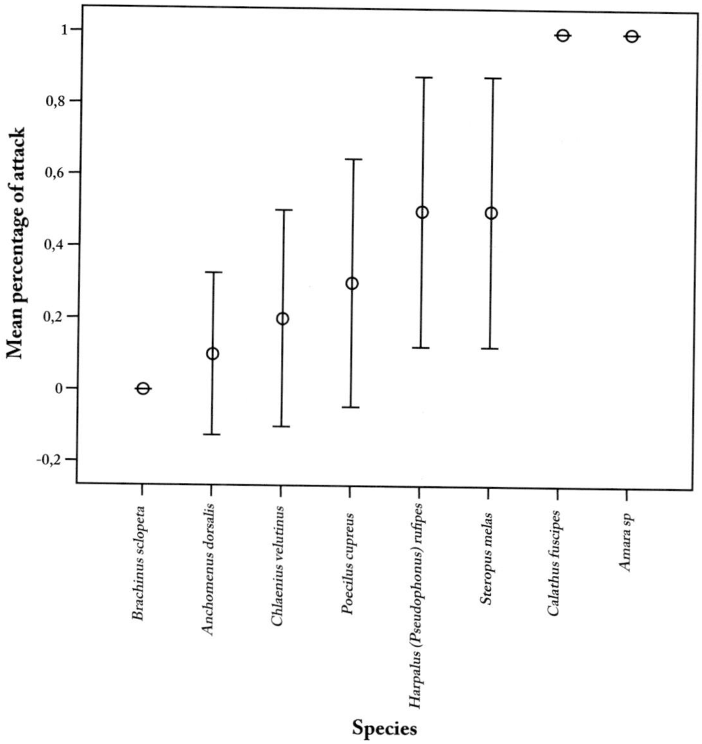

Fig. 2. Percentage of attack of *Ocypus olens* towards some species of carabid beetles. The first three are aposematic and protected species.

Aggregation has been described in adult populations of *Agonum dorsale* (Allen, 1957), *Nebria brevicollis* and *Brachinus crepitans* (Greenslade, 1963), *Brachinus sclopeta* and *Brachinus explodens* (Wautier, 1971), *Brachinus variventris* (Zaballos, 1985), and *Colliuris batesi* Chaudoir, 1862 a carabid beetle inhabiting central amazonian forests (Adis et al., 1997). On the whole, aggregation seems to occur in only a few carabid species and its evolutionary significance may be mainly protection from water loss and keeping together of the sexes (Thiele, 1977) as well as the lowering of individual predation risk (Alcock, 1979).

A. dorsalis adults have been usually found grouping in little amount of individuals inside the aggregations of *Brachinus* species (Juliano, 1985; Zaballos, 1985; Lindroth, 1949; Bonacci et al., 2004b; Mazzei et al., 2005). Lindroth (op. cit) observed also inter-specific interactions among members, recently described in detail as "rubbing behaviour" by Zetto Brandmayr and co-workers, (2006).

In the last years we studied carabid aggregations in Calabria (Southern Italy). The dominant species are aposematic and chemically protected: *Chlaenius chrysocephalus* (60%), *Brachinus brevicollis* (14. 84%), *Brachinus crepitans* (8.63%), *Anchomenus dorsalis* (5.52%), *B. psophia* (4.66%), *B. sclopeta* (2.015%) ($t = -3.646$, $d.f = 1656$, $P < 0,000$). Only one non aposematic species (*Parophonus hispanus*) (4.37%) has been found (Fig. 3).

To the general benefits of gregariousness for aposematic insects (see above at the point "Gregariousness"), further increasing in effectiveness against predators may be postulated:

a) in interspecific aggregations more success depends on the Müllerian mimicry, when two or more protected species share a similar warning pattern; in this form of mimicry the mimics possessing different defence chemicals are better protected than those that share a single defence chemical (Skelhorn & Rowe, 2005).

b) the possibility of repelling predators (also in intraspecific aggregations) is more prolonged in the time.

Indeed, bombardier beetles may discharge their defensive spray a number of times, after which they are temporally unprotected. We observed this in our experiments with lizards and shrews (Bonacci et al., 2004a; and Bonacci et al., in press). Since we tested one beetle at time, after some trials (two or three depending on the models) an experienced predator learnt the dangerousness of the prey, but also the way to escape from the pain, forcing the beetle to discharge its glands before eating it.

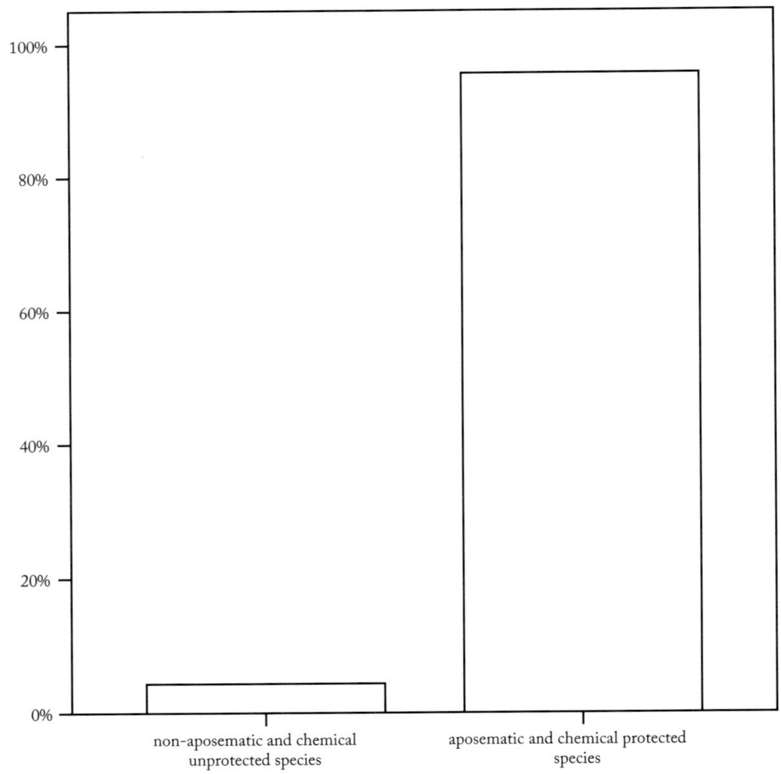

Fig. 3. Aggregation composition (in %) in carabid beetles during the sampling season (Spring, Autumn and Winter) in Calabria (Southern Italy).

The bombardier defensive mechanism is found in at least two different groups of ground beetles. The most well-known bombardiers are the colourful *Brachinus* species we used in our experiments, but *Metrius* and related paussine beetles possess a similar behaviour (Eisner et al., 2000). *Metrius contractus* is not aposematic and discharges its defensive secretion as a froth that clings to its body. In the above cited work the Authors presented a photo of aggregating beetles in laboratory. We can postulate that aggregation may be advantageous also to non-aposematic bombardiers.

DOES THE COMPLEX *BRACHINUS-ANCHOMENUS* REPRESENT A MIMICRY CHAIN?

Mimicry in Carabid beetles is poorly represented. Some cases are reported for genera of Lebiini tribe (Thiele, 1977) known to be ectoparasitoid in their larval stages of certain species of Chrysomelidae (Alticinae) while adults are similar to their host (Balsbaugh, 1967; Lindroth, 1971).

Conversely, in *Eurycoleus macularis*, a lebiine carabid with ectoparaitoid habit, adults and larvae mimic insects other than their host, and this mimetic complex involves three families of beetles (Erotylidae and Tenebrionidae) at the adult stage and two in the larval one (Erotylidae) (Erwin & Erwin, 1976).

A possible mimicry between tiger beetles (Cicindelinae) and three species of Eastern and Northern Africa *Graphipterus* are achieved both in body colouration and in running behaviour (Cassola & Vigna Taglianti, 1988).

We observed that *B. sclopeta* and *A. dorsalis* share a colourful model strongly contrasting to the background (see Photo 1). Although the similarity is not complete, aggregated animals detected by potential visual predators under stones or other diurnal shelters are not easily discriminated owing to their rapidity of movement inside the aggregation, since the outcome is a blinking mixing up of aposematic colours. The conspicuousness of the aggregations increases the effectiveness of the deceit (see Photo 1). The mutual benefit for the members of the aggregation may be thought as effect of Müllerian mimicry.

Aside, a survey by Thiele (1977) of the most common species preying on carabids reported that the main predators are hedgehogs and shrews, moles, birds, owls, frogs and toads, ants, predacious flies, spiders and carnivorous ground beetles. Only some of them are predators that locate prey by visual cues, while the remainder are olfactory/tactile predators. The question arises: how can *B. sclopeta* and *A. dorsalis* avoid the olfactory predators?

Our experiments with *O. olens* (Bonacci et al., 2006), showed that the staphiline has the capacity for learning, which is probably based on the olfactory signals released by prey (possibly alkaloids combined with chemical defences as multimodal signals) (Rothschild et al., 1984; Lindström et al., 2001).

An interesting finding is the similarity in the cuticle molecular composition of *B. sclopeta* and *A. dorsalis* particularly if compared to a non aposematic carabid beetle (Bonacci et al. 2007, in press).

Table 1. Chemical compounds present in *B. sclopeta*, *A. dorsalis* and *P. cupreus* adult individuals (from Bonacci et al., 2007, in press).

Peak No.	Compound	*Brachinus sclopeta*	*Anchomenus dorsalis*	*Poecilus cupreus*
1	p-benzoquinone	x		
2	2-methyl-p-benzoquinone	x		
3	undecane		x	
4	tridecane	x		
5	2.cloro-4-1(1,1-dimethyl)-phenol	x		
6	1-(2-hydroxy-5-methoxyphenyl)-ethanone	x		
7	pentadecane	x		
8	4-methy-1H-indole			x
9	2-tridecanone		x	
10	Unknown compound isomer			x
11	Disoprophylnaftalene isomer			x
12	Disoprophylnaftalene isomer			x
13	Disoprophylnaftalene isomer			x
14	Disoprophylnaftalene isomer			x
15	Disoprophylnaftalene isomer			x
16	Unknown compound isomer			x
17	Eicosene C_{20}			x
18	Eicosane C_{20}	x	x	x
19	7-heneicosene C_{21}	x	x	
20	7-heneicosadiene C_{21}			x
21	Heneicosene C_{21}			x
22	Heneicosane C_{21}	x	x	x
23	8-docosene	x	x	
24	9-docosene C_{22}			x
25	7-docosene C_{22}			x
26	Docosane C_{22}	x	x	x
27	9-tricosene C_{23}	x	x	x
28	7-tricosene C_{23}	x	x	x
29	Tricosane C_{23}	x	x	x
30	Tetracosene C_{24} isomer			x
31	Tetracosene C_{24} isomer			x
32	Tetracosene C_{24} isomer			x
33	8-tetracosene	x	x	
34	Tetracosane C_{24}	x	x	x
35	9-pentacosene	x	x	
36	7-pentacosene	x	x	
37	Pentacosene C_{25} isomer			x
38	Pentacosene C_{25} isomer			x
39	Pentacosene C_{25} isomer			x
40	Pentacosene C_{25} isomer			x
41	Pentacosene C_{25} isomer			x
42	pentacosane	x	x	
43	9-heptacosene	x	x	
44	11-heptacosene C_{27}			x
45	Heptacosane C_{27}	x	x	x
46	Squalene	x	x	
47	11-nonacosene			x
48	Nonacosane C_{29}	x	x	x

Most of the peaks represent hydrocarbons with more than 20 atoms, which in many insects are thought to constitute the basic odorous profile (Table 1). Despite the incomplete chemical similarity, we believe that a mechanism similar to that involved in colour similarity may be effective in reducing the predation risk by non visual predators.

CONCLUSION

To conclude, we postulate highly effective antipredatory strategy in the *Brachinus-Anchomenus* complex. This is realized by unpalatability, aposematism and gregariousness combined with a double mimicry which sends similar signals both to visual and olfactory predators. Multimodal warning displays combine visual signal with components produced in other sensory modalities (Rowe & Guilford, 1999). More work is needed to deeper investigations about the supposed chemical mimicry, but we find this topic very intriguing and possible to be extended to many other carabid taxa.

ACKNOWLEDGEMENTS

Authors are indebted to the organisers of the XIII European Carabidologists Meeting for giving us the possibility to present this short review on an important evolutionary field of carabid biology.

REFERENCES

Adis, J., Amorim, M.A., Erwin, T.L. & Bauer, T. (1997). On ecology, life history and survival strategies of a wing-dimorphic ground beetle (Col., carabidae: Odacanthini: *Colliuris*) inhabiting Central Amazonian inundation forests. – Stud Neotrop Fauna 6 Environm. 32: 174-192.

Alatalo, R.V. & Mappes, J. (1996). Tracking the evolution of warning signals. – Nature 382: 708-710.

Alcock, J. (1979). Animal Behavior. An evolutionary approach. Sixth Sinauer Associates, Inc. Publishers, Sunderland, Massachussets.

Allen, A.A. (1957). The habit of aggregation in *Agonum dorsale* Pont. (Coleoptera, Carabidae). – Entomol. Mon. Mag. 210: 142.

Balsbaugh, E.U. (1967). Possible mimicry between certain Carabidae and Chrysomelidae. – Coleopt. Bull. 2: 139-140.

Bonacci, T., Aloise, G., Brandmayr, P. & Zetto Brandmayr, T. (2004a). Risposte comportamentali di *Crocidura leucodon* (Herrmann, 1780) (Insectivora, Soricidae) ai meccanismi antipredatori di alcuni Artropodi. – Hystrix , Italian Journal of Mammalogy (n.s) 15(1): 73-76.

Bonacci, T., Mazzei, A., Zetto, T. & Brandmayr, P. (2004b). Aposematic aggregation of carabid beetles (Coleoptera: carabidae): preliminary data. – Redia, LXXXVII: 243-245.

Bonacci, T., Massolo, A., Brandmayr, P. & Zetto Brandmayr, T. (2006). Predatory behaviour on ground beetles (Coleoptera: Carabidae) by *Ocypus olens* (Müller) (Coleoptera: Staphylinidae) under laboratory conditions. – Entomological News 117 (5): 545-551.

Bonacci, T., Brandmayr, P., Dalpozzo, R., De Nino, A., Massolo, A., Tagarelli, A. & Zetto Brandmayr T. (2007). Odour and colour similarity in two species of gregarious carabid beetles: one case of Müllerian mimicry? – Entomological News (in press).

Bonacci, T., Aloise, G., Brandmayr, P., Zetto Brandmayr, T. & Capula, M. (2007). Testing the predatory behaviour of Podarcis sicula (Reptilia:Lacertidae) towards aposematic and non-aposematic preys. – Amphibia-Reptilia (in press).

Cassola, F. & Vigna Taglianti, A. (1988). Mimicry in Cicindelini e Graphipterini africani (Coleoptera, Caraboidea). – Biogeographia XIV: 229- 233.

Coppinger, R.P. (1969). The effects of experience and novelty on avian feeding behaviour with reference to the evolution of warning coloration in butterflies. I: reactions of wild-caught adult blue jays to novel insects. – Behaviour 35: 4-60.

Coppinger, R.P. (1970). The effect of experience and novelty on avian feeding behavior with reference to the evolution of warning coloration in butterflies. II: reactions of naïve birds to novel insects. – Am. Nat. 104: 323-335.

Cott, H.B. (1940). Adaptive coloration in animals. Oxford Press, London.

Despland, E. & Simpson, S.J. (2005). Surviving the change to warning colouration: density-dependent polyphenism suggests a route for the evolution of aposematism. – Chemoecology 15: 69-75.

Edmunds, M. (1974). Defence in animals: A survey of anti-predator defences. Longman, Harlow, Essex.

Eisner, M. (1970). Chemical defense against predation in arthropods. – In: Chemical Ecology (Sondheimer, E., Sondheimer, E. & Someone J.B., eds). Academic Press, New York.

Eisner, T. & Aneshansley, D. (1999). Spray aiming in the bombardier beetle: photographic evidence. -Proc. Natl. Acad. Sci. USA, 96: 9705-9709.

Eisner, T., Aneshansley, D.J., Eisner, M., Yack, J., Attygalle, A.B., Alsop, D.W. & Meinwald, J. (2000). Spray mechanism of the most primitive bombardier beetle (*Metrius contractus*). – The Journal of Experimental Biology 203: 1265-1275.

Eisner, T., Yack, J. & Aneshansley D. J. (2001a). Acoustic concomitants of the defensive discharges of a primitive bombardier beetle (*Metrius contractus*). – Chemoecology 11: 221–223.

Eisner, T., Aneshansley, D.J., Yack, J., Attygalle, A.B. & Eisner, M. (2001b). Spray mechanism of crepitogastrine bombardier beetles (Carabidae ; Crepitogastrini). – Chemoecology 11: 209-219.

Erwin, T.L. & Erwin, J.M. (1976). Relationships of predaceous beetles to tropical forest wood decay. Part II. The natural history of neotropical *Eurycoleus macularis* Chevrolat (Carabidae: Lebiini) and its implications in the evolution of ectoparasitoidism. – Biotropica 8: 215-224.

Fisher, R.A. (1930). The genetical theory of natural selection. The Clarendon Press, Oxford.

Gagliardo A. & Guilford, T. (1993). Why do warning-colored prey live gregariously? – Proceedings of the Royal Society of London B-Biological Sciences 251: 69-74

Gamberale, G. & Tullberg, B.S. (1996). Evidence for a peak-shift in predator generalization among aposematic prey. – Proceedings of the Royal Society of London(B) London. 263: 1329-1334.

Gamberale, G. Tullberg, B.S. (1998). Aposematism and gregariousness: the combined effect of group size and coloration on signal repellence. – Proc R Soc Lond B 265: 889-894.

Greenslade, P. J. M. (1963). Further notes on aggregation in Carabidae (Coleoptera), with especial reference to *Nebria brevicollis* (F.). – Entomol. Mon. Mag. 99: 109-114.

Guilford, T. (1990). The evolution of aposematism. – In: Insects defenses: adaptative mechanisms and strategies of prey and predators (Evans, L.D. & Schmidt, J.O. eds). State University of New York Press, NewYork.

Hartmann, T., Häggström, H., Theuring, C., Lindigkeit, R. & Rahier, M. (2003). Detoxification of pyrrolizidine alkaloids by the harvestman *Mitopus morio* (Phalangidae) a predator of alkaloid defended leaf beetles. – Chemoecology 13: 123–127.

Hatle, J.D., Salazar B.A. & Whitman, D.W. (2002) Survival advantage of sluggish individuals in aggregations of aposematic prey, during encounters with ambush predators. – Evolutionary Ecology 16: 415-431.

Järvi T., Sillén-Tullberg B. & Wiklund, C. (1981). The cost of being aposematic. An experimental study of predation on larvae of *Papilio machaon* by the great tit *Parus major*. – Oikos 36: 267-272.

Joron, M. (2002). Aposematic coloration. – In: Encyclopedia of Insects (Cardé, R.T. & Resh, V.H., eds). Academic Press, New York.

Juliano, S.A. (1985). Habitat associations, resources, and predators of an assemblage of *Brachinus* (Coleoptera: Carabidae) from southeastern Arizona. – Canadian Journal of Zoology 63: 1683-1691.

Lindroth, C.H. (1949). Die Fennoskandischen Carabidae. Göteborgs Kungliga Vetenskapsoch Vitterhetø-SamhällesHandinglar,Stockholmhälles Handinglar, Stockholm (B4) 1, Spezieller Teil, 1-709 (1945); 3, Allgemeiner Teil. 1-911.

Lindroth, C.H. (1971). Disappaerance as a protective factor. A supposed case of Batesian mimicry among beetles (Coleoptera: Carabidae and Chrysomelidae). – Entomol. Scand. 2: 41-48.

Lindström, L., Alatalo, R.V., Lyytinen, A. & Mappes, J. (2001). Predator experience on cryptic prey affects the survival of conspicuous aposematic prey. – Proceedings of The Royal Society of London B 268: 1925-1929.

Lindström, L., Alatalo, R.V., Mappes, J., Riipi, M. & Vertainen, L. (1999). Can aposematic signals evolve by gradual change? – Nature 397: 249–251.

Lindström, L., Rowe, C. & Guilford, T. (2001). Pyrazine odour biases food selection in avian predators against conspicuously coloured prey. – Proceedings of the Royal Society of London, Series B 268: 357-361.

Mallet, J. & Joron, M. (1999). Evolution of diversity in warning color and mimicry: polymorphisms, shifting balance, and speciation. – Annu. Rev. Ecol. Syst. 30: 201-233.

Mazzei A., Bonacci, T., Zetto Brandmayr, T. & Brandmayr, P. (2005). Capacità di aggregazione di Coleotteri Geoadefagi, in ambiente ipolitico di suoli argillosi del bioclima mediterraneo arido. – Atti del XV Congresso della Società Italiana di Ecologia. Torino, 12-14 settembre 2005.

Pasteels, J.M, Braekman, J.C. & Daloze, D. (1988). Chemical defense in the Chrysomelidae. – In: Biology of Chrysomelidae (Jolivet P., Petitpierre E. & Hsiao, T.H., eds). Dordrecht: Kluwer Academic Publishers, p. 233–252.

Pasteels, J.M., Rowell-Rahier, M., Braekman, J.C. & Daloze, D., (1994). Chemical defense of adult leaf beetles updated. – In: Novel Aspects of the Biology of Chrysomelidae (Jolivet, P., Cox, M.L. & Petitpierre, E., eds). Kluwer Academic Publishers, Dordrecht, p. 289–301.

Pasteels, J.M., Rowell-Rahier, M., Ehmke, A. & Hartmann, T. (1996). Hostderived pyrrolizidine alkaloids in *Oreina* leaf beetles: physiological, ecological and evolutionary aspects. – In: Chrysomelidae Biology, Ecological Studies. vol 2 (Jolivet, P.H,A. & Cox, M.L., eds), Amsterdam Academic Publishing, Amsterdam, p. 213–225.

Pitt, W.C. & Ritchie M.E. (2002). Influence of prey distribution on the functional response of lizards. – Oikos 96: 157–163.

Poulton, E.B. (1890). The colours of animals: their meaning and use especially considered in the case of insects. Keegan Paul, Trench, Trübner, London.

Roper, T.J. & Cook, S.E. (1989). Responses of chicks to brightly coloured insect prey. – Behaviour 110: 276-293.

Rothschild, M., Moore, B.P. & Brown, W.V. (1984). Pyrazines as warning odour components in the monarch butterfly, *Danaus plexippus*, and in moths of the genera *Zygena* and *Amata* (Lepidoptera). -Biological Journal of the Linnean Society 23: 372-380.

Rowe, C. & Guilford, T. (1999). Novelty effects in a multimodal warning signal. – Anim. Behav. 57: 341-346.

Rowe, C. & Guilford, T. (2000). Aposematism: to be red or dead. – Trends in Ecology & Evolution 15: 261-262.

Schildknecht, H.U. (1961). Die Bombardierkäfer und ihre Explosionschemie. – Angewandte Chemie 73(1): 1-7.

Schildknecht, H.U. (1970). Die Wehrchemie von land- und Wasserkäfern. – Angewandte Chemie 82(1): 16-25.

Schildknecht, H., Maschwitz, U. & Winkler, H. (1968). Zur Evolution der Carabiden-Wehrdrusensekrete. – Naturwissenshaften 55: 112-117.

Sillén-Tullberg B. (1985). The significance of coloration per se, independent of background, for predator avoidance of aposematic prey. – Anim. Behav. 33: 1382-1384.

Sillén-Tullberg, B. & Leimar, O. (1988). The evolution of gregariousness in distasteful insects as a defense against predators. – Am. Nat. 132: 723-734.

Skelhorn, J. & Rowe, C. (2005). Tasting the defence: do multiple defence chemicals interact in Müllerian mimicry ? – Proc. Roy. Soc. London B, 272: 339-345.

Speed, M.P. (2000). Warning signals, receiver psychology and predator memory. – Anim. Behav. 60: 269-278.

Thiele, H.U. (1977). Carabid Beetles in Their Environments. A study on habitat Selection by Adaptations in Physiology and Behaviour. Springer Verlag, Berlin.

Tullberg, B.S., Leimar, O. & Gamberale-Stille, G. (2000). Did aggregation favour the initial evolution of warning coloration? A novel world revisited. – Anim. Behav. 59: 281-287.

Vulinec, K. (1990). Collective security: aggregations by insects as a defense. – In: Insect Defenses. Adaptive Mechanisms and Strategies of Prey and Predators (Evans, D.L. & Schmidt, J.O., eds). State University of New York Press, New York, p. 251-288.

Wallace, A.R. (1867). Proceedings of the Entomological Society of London, London, March 4, lxxx-lxxxi.

Wautier, V. (1971). Un phénomène social chez les Coléoptères: le grégarisme de *Brachinus* (Caraboidea, Brachinidae). – Insectes Sociaux 18: 1-84.

Zaballos, J.P. (1985) Paralelismo fenológico en *Brachinus variventris* Schaufuss, 1862 y *Anchomenus dorsalis* (Pontoppidan, 1963). (Coleoptera Carabidae). – Actas do II Congresso Ibérico de Entomologia: 85-92.

Zetto Brandmayr, T., T. Bonacci, A. Massolo & P. Brandmayr (2006). What is going on between aposematic carabid beetles? The case of *Anchomenus dorsalis* (Pontoppidan 1763) and *Brachinus sclopeta* (Fabricius 1792) (Coleoptera Carabidae). – Ethology Ecology & Evolution 18: 335-348.

L. Penev, T. Erwin & T. Assmann (Eds) 2008
Back to the Roots and Back to the Future. Towards a New Synthesis amongst Taxonomic, Ecological and Biogeographical Approaches in Carabidology
Proceedings of the XIII European Carabidologists Meeting, Blagoevgrad, August 20-24, 2007, pp. 339-351.

Composition of weed community determines carabid assemblage

Pavel Saska

Crop Research Institute, Drnovska 507, Praha 6 – Ruzyne, 161 06, Czech Republic.
E-mail: saska@vurv.cz

SUMMARY

The objective of this study is to determine how composition of carabid assemblage changes with plant (weed) community, hypothesizing that (i) composition of weed assemblage will affect the structure of carabid assemblage, (ii) at least some carabid species will respond positively to occurrence of a preferred weed species, and that (iii) the preferred weed in the field will agree with known food specialization of larvae or adults. Metal enclosures with nested pitfall trap were used to sample carabids in weed patches in Praha – Ruzyně at four sample periods between 9.5.2006 and 23.8.2006. Redundancy Analysis (RDA) revealed that structure of weed community affected composition of carabid assemblage, but presence/absence explanatory variable explained significantly higher portion of variability in carabid density than percentage cover data. Ordination biplots and correlation analysis showed congruent associations of some carabids with particular weeds: *Amara familiaris*, *Brachinus explodens* and *Harpalus luteicornis* were associated with *Stellaria media*, *Acupalpus meridianus* with *Capsella bursa-pastoris* in spring and (together with *Paratachys bistriatus*) with *Tripleurospermum inodorum* in late summer. The results indicate that composition of weed populations translates to the structure of carabid assemblage, and that some associations agree with other biological data of carabids (seed preference, larval requirements etc.). The presence of weed patches is thus crucial for preserving species dependent on weeds in agricultural landscapes.

Keywords: Density, diversity, enclosures, granivory, insect-plant interactions

INTRODUCTION

Current data indicate that besides having negative impact on crops some weed species increase biodiversity of insects and birds in agro-ecosystems (e.g. Marshall et al., 2003; Holland et al., 2006; Storkey, 2006). Birds benefit from increased diversity and amount of seeds, their abundant food (Holland et al., 2006), whereas herbivorous insects depend on weeds more closely since they develop on or inside a plant. The range of herbivores associated to a weed varies markedly (Ward & Spalding, 1993). For example, *Veronica persica* hosts only one species of herbivorous insects, while 71 species associate with *Stellaria media* according to Phytophagous Insects Data Base (Ward & Spalding, 1993). However, this as well as other available databases (Campobasso et al., 1999; www.ecoflora.co.uk) focuses mainly on leaf, stem or root feeding herbivores and pre-dispersal seed predators. Post-dispersal seed predators, which may also be closely connected with a particular plant species too (Schremmer, 1960; Zetto Brandmayr, 1983; Honek et al., 2005; Saska, 2005), are widely neglected.

The truly granivorous carabids mainly belong to the tribes Harpalini (e.g. *Harpalus*, *Ophonus*, *Pseudoophonus*) and Zabrini (*Amara*, *Zabrus*), with a few belonging to the Trechini, Platynini and Pterostichini (e.g. Zetto Brandmayr, 1990; Goldschmidt & Toft, 1997; Honek et al., 2003, 2007). Granivory, typical of adults, is also essential for successful larval development in some species (Zetto Brandmayr, 1976, 1983; Jorgensen & Toft, 1997; Saska & Jarosik, 2001; Saska, 2005). Many species prefer a limited range of seed species (Honek et al., 2007), and these specialists often time their reproduction to seed shed and aggregate at patches with high density of their preferred seed (Zetto Brandmayr, 1983; Honek & Jarosik, 2000; Honek & Martinkova, 2001; Honek et al., 2005).

There are several literature reports on attraction of granivorous carabids to weedy plots. Since the primary observation of Gersdorf (1937), several authors had reported that fields with higher densities of weeds hosted more carabid beetles than those in which weeds were mechanically removed or sprayed by herbicides (Powell et al., 1985; Bosch, 1987; Kokta, 1988; Kromp, 1989, 1990; Altieri et al., 1995; de Snoo et al., 1995; Hawthorne & Hassall, 1995). Speight & Lawton (1976) found more carabids in traps surrounded by *Poa annua* compared to traps surrounded by bare ground in cereal field. Holland et al. (1999) found close association of the occurrence of granivorous *Amara* species with the weed cover in a within-field spatial analysis. Honek & Jarosik (2000) and Honek & Martinkova (2001) showed that granivorous carabids positively responded to increased densities of seeds around pit fall traps.

Although carabids are an intensively studied insect group, data on biology of many common species are lacking. Luff (2002) calls for more investigations on biology of common field carabid species and their requirements for habitats. So far only Honek et al. (2005) refers to the effect of composition of the weed patches on the structure of granivorous carabid assemblage. In their study, *Amara montivaga* Sturm aggregated in stands grown by *Taraxacum officinale*, while in stands only a few meters apart grown by other weeds *A. montivaga* was scarce (Honek et al., 2005).

The principal aim of this paper is to investigate the link between composition of weed and carabid communities. It is expected that (i) composition of weed assemblage will affect the structure of carabid assemblage, (ii) at least some carabid species will respond positively to occurrence of a preferred weed species, and that (iii) the preferred weed in the field will be congruent with known food specialization of larvae or adults.

MATERIAL AND METHODS

Sampling

Sampling was conducted in Praha-Ruzyně, Czech Republic (50° 06´ N 14° 16´E), in May-August 2006. In order to study the relationship of assemblages of weeds and carabids, naturally established weed patches were selected, so weed species typical for sample date and locality were included. List of common species of weeds included in the survey is shown in Table 1. For comparative reasons, plots of bare ground and neighbouring grassy strip were also sampled.

Metal quadrate enclosures (0.16 m^2) were used to delimit the study patches, because they are suitable to estimate true population density (Desender & Maelfait, 1986). They were pressed in soil, so they reached ca 15 cm deep and projected ca 15 cm above ground. This assured that the surface-active arthropods present within the enclosure could not escape. First, the composition of vegetation within each enclosure was recorded and expressed as a relative abundance, i.e. percentage of ground covered by individual species. Then the vegetation was removed and running carabids were collected by hand.

Table 1. The mean percentage cover of weed species/habitat types within the enclosures (%) per sample period. Species that covered less than 10 % of enclosure area not shown. Numbers in square brackets – number of samples (enclosures).
Sample weeks: 9/5/2006 (week 19), 23/5/2006 (21), 9/6/2006 (23), 9/8/2006 (32).

Plant species/ habitat	Mean percentage cover per sample week			
	19 [8]	21 [10]	23 [11]	32 [14]
Capsella bursa-pastoris (L.) Med.	37	23	25	0
Stellaria media (L.) Vill.	37	37	43	3
Cirsium arvense (L.) Scop.	2	1	1	19
Taraxacum officinale sp. agg.	0	10	1	1
Geranium robertianum L.	0	4	7	0
Convolvulus arvensis L.	0	4	8	1
Polygonum aviculare L.	0	0	0	23
Tussilago farfara L.	0	0	0	21
Tripleurospermum inodorum (L.) Schultz-Bip.	0	0	0	20
Bare ground	23	2	10	8
Grassy strip	0	20	6	0

A pitfall trap (two nested plastic cups, diameter 7 cm, depth 12 cm) was placed in one of the corners of the enclosure, flush with soil surface. No bait or preservative was used and a metal square equipped with two nails screened the trap from rain and sunshine. Finally, enclosures were covered by a roof netting (mesh size 0.2 mm) in order to avoid arthropods inside the enclosure flying out and those from outside flying in. The sampling was replicated four times a year (9/5/2006 [week 19], 23/5/2006 [21], 9/6/2006 [23], 9/8/2006 [32]), with 8-14 replications each sample date (Table 1). In total 43 samples were taken. Enclosures were in service for two weeks after which all trapped carabids were identified to species (larvae to adults) (Luff, 1993; Hůrka, 1996) and released.

Data processing

The statistical analysis was made in order to reveal how composition of weed assemblage affected the structure of carabid community in weed patches. A matrix of numbers of individuals per carabid species and sample served as basic data set, each species of carabid representing a dependent variable. Two separate sets of analyses were made according to the type of environmental variables used: (i) percentage cover of particular plant species or bare ground as measured in the enclosures, and (ii) presence/absence data. In the latter set of analyses, the percentage cover data were converted to binary format in the way that percentage cover of a plant/bare ground > 19.9 % of the enclosure area was replaced by "1", and a percentage cover < 20% of the enclosure area was replaced by "0". The threshold of 20 % was set arbitrary.

The data were analysed using a multivariate approach in CANOCO 4.5 for Windows (Ter Braak & Šmilauer, 2002), separately for each sample period. That was in order to control for the temporal shift in the composition of weed and carabid communities between sample periods. Carabid species that occurred in singletons or doubletons on a particular sample date were excluded from analysis. First, the data were subjected to Detrended Canonical Analysis (DCA) in order to test for the character of the relationship between explanatory and dependent variables. Since the lengths of the first canonical axes were in all cases shorter than 3 (i.e. linear response), Redundancy Analysis (RDA) was used in the final analysis. The RDA was followed by the Monte-Carlo permutation test to reveal the significance of individual environmental variable (i.e. plant species/bare ground) on structure of carabid community. In these analyses the weight of rare species (those represented in fewer individuals than there were samples each sample period) was reduced down to 0.1. The results were visualized in biplots, created in CanoDRAW (Ter Braak & Šmilauer, 2002). The affiliation of particular carabid species to a weed species/bare ground was assessed using the significance (at $p<0.05$) of the correlation coefficients estimated by RDA (Ter Braak & Šmilauer, 2002).

RESULTS

The structure of weed community varied with season (Table 1). In May (weeks 19 and 21), *C. bursa-pastoris* and *S. media* predominated (Table 1). In June (week 23), *T. officinale*, *G. robertianum* and *Co. arvensis* became also abundant in the study plots (Table 1). In August (week 32), the structure of the weed assemblages was completely changed, with *P. aviculare*, *Tr. inodorum*, *Tu. farfara* and *Ci. arvense* dominating the assemblage (Table 1).

Within the enclosures, 33 adult carabid species and three genera in larval stage were found in 501 individuals total (Table 2). *A. familiaris*, *A. meridianus* and *B. explodens* were the most abundant carabid species in the enclosures (Table 2). Nineteen singleton and doubleton species (regarding to their abundance each sample week) were omitted from the analysis. Among the species recorded, 22 were previously reported as granivorous; these summed up 378 individuals in total.

Table 2. The abundance of carabid beetles within the enclosures (0.16 m^2), Praha – Ruzyně. Sample weeks: 9/5/2006 (week 19), 23/5/2006 (21), 9/6/2006 (23), 9/8/2006 (32).

Species	Abundance				
	Total	Sample week			
		19	21	23	32
Acupalpus meridianus (Linnaeus)	121	23	46	22	30
Amara aenea (DeGeer)	31	9	13	7	2
Amara apricaria (Paykull)	1	0	0	0	1
Amara aulica (Panzer)	3	0	0	0	3
Amara bifrons (Gyllenhal)	1	0	0	0	1
Amara convexiuscula (Marsham)	1	0	0	0	1
Amara familiaris (Duftschmid)	130	46	31	52	1
Amara littorea C.G. Thomson	1	0	0	0	1
Amara montivaga Sturm	1	0	1	0	0
Amara ovata (Fabricius)	1	0	0	1	0
Amara sabulosa (Audinet-Serville)	1	0	0	0	1
Amara similata (Gyllenhal)	5	2	2	0	1
Anchomenus dorsalis (Pontoppidan)	8	6	0	2	0
Badister bullatus (Schrank)	1	0	0	0	1
Bembidion lampros (Herbst)	11	2	3	3	3
Bembidion quadrimaculatum (Linnaeus)	1	0	0	0	1
Brachinus crepitans (Linnaeus)	2	2	0	0	0
Brachinus explodens Duftschmid	34	8	5	21	0
Calathus melanocephalus (Linnaeus)	1	0	0	0	1
Harpalus affinis (Schrank)	30	5	7	14	4
Harpalus distinguendus (Duftschmid)	5	0	0	5	0
Harpalus honestus (Duftschmid)	1	0	0	1	0

Species	Abundance				
	Total	Sample week			
		19	21	23	32
Harpalus luteicornis (Duftschmid)	18	7	7	3	1
Harpalus tardus (Panzer)	1	0	0	1	0
Leistus ferrugineus (Linnaeus)	2	0	0	2	0
Microlestes minutulus (Goeze)	9	4	3	1	1
Notiophilus palustris (Duftschmid)	7	1	1	4	1
Ophonus azureus (Fabricius)	8	1	3	4	0
Paratachys bistriatus (Duftschmid)	10	0	0	0	10
Poecilus cupreus (Linnaeus)	2	0	0	0	2
Pseudoophonus rufipes (DeGeer)	15	0	2	0	13
Stomis pumicatus (Panzer)	2	2	0	0	0
Trechus quadristriatus (Schrank)	2	0	0	1	1
larva *Amara*	21	3	8	9	1
larva Harpalini	2	0	0	1	1
larva Platynini	11	3	3	5	0
Total	501	124	135	159	83

Both types of Redundancy Analyses (RDA) indicated an effect of weed community on composition of carabid assemblage. The percentage of variance explained by all canonical axes is summarized separately for all analyses made in Table 3. In the percentage cover analysis, Monte-Carlo permutation tests revealed that weed community significantly affected the structure of carabid assemblage in weeks 19 and 32 but not in weeks 21 and 23 (Table 3). That was due to *S. media* that explained significant amount of variation in carabid distribution (Monte-Carlo permutation test: F=7.12, p=0.002) in week 19 (9/5/2006), and by *Tr. inodorum* (Monte-Carlo permutation test: F=33.27, p=0.002) in week 32 (9/8/2006). Other weed species did not contribute significantly to the overall variability in the data (Table 3), although biplots based on RDA indicated an

Table 3. Comparison of the Redundancy analyses performed with percentage cover or presence/absence (binary) explanatory variables.
Sample weeks: 9/5/2006 (week 19), 23/5/2006 (21), 9/6/2006 (23), 9/8/2006 (32).

	Explanatory variable							
	Percentage cover				Binary			
	19	21	23	32	19	21	23	32
Explained variance [%]	60.8	43.5	41.5	85.8	64.3	77.2	58.3	87.7
N weeds included in analysis [1]	3	3	3	5	3	7	6	6
N weeds significant [1]	1	0	0	1	1	1	2	2
N significantly correlated pairs of weeds and carabids	3	4	2	2	3	3	3	4

[1] Monte-Carlo permutation test (p<0.05)

affiliation of carabids to more weed species than shown by Monte-Carlo analysis (Fig. 1). This indication was confirmed by the values of the correlation coefficients estimated by the RDA for each weed-carabid pair, which revealed significant correlations of 8 carabid species with some of the weeds or habitat types (Table 4).

The presence/absence analysis explained more variance in the data than percentage cover analysis (Table 3). Monte-Carlo permutation tests revealed that weed community significantly affected the structure of carabid assemblage in all sample weeks (Table 3). In weeks 19 and 21, only *S. media* explained significant amount of variation in carabid distribution (Monte-Carlo permutation test: F=7.12, p=0.002 [week 19], F=3.13, p=0.01 [week 21]). Two species, *G. robertianum* (Monte-Carlo permutation test: F=2.31, p=0.046) and *S. media* (Monte-Carlo permutation test: F=2.99, p=0.016) significantly affected the carabid assemblage in week 23. Similarly, significant effects of *Tr. inodorum* (Monte-Carlo permutation test: F=12.39, p=0.002) and *S. media* (Monte-Carlo permutation test: F=2.81, p=0.034) were found in week 32. Other weed species did not contribute significantly to the overall variability in the data. Correlation

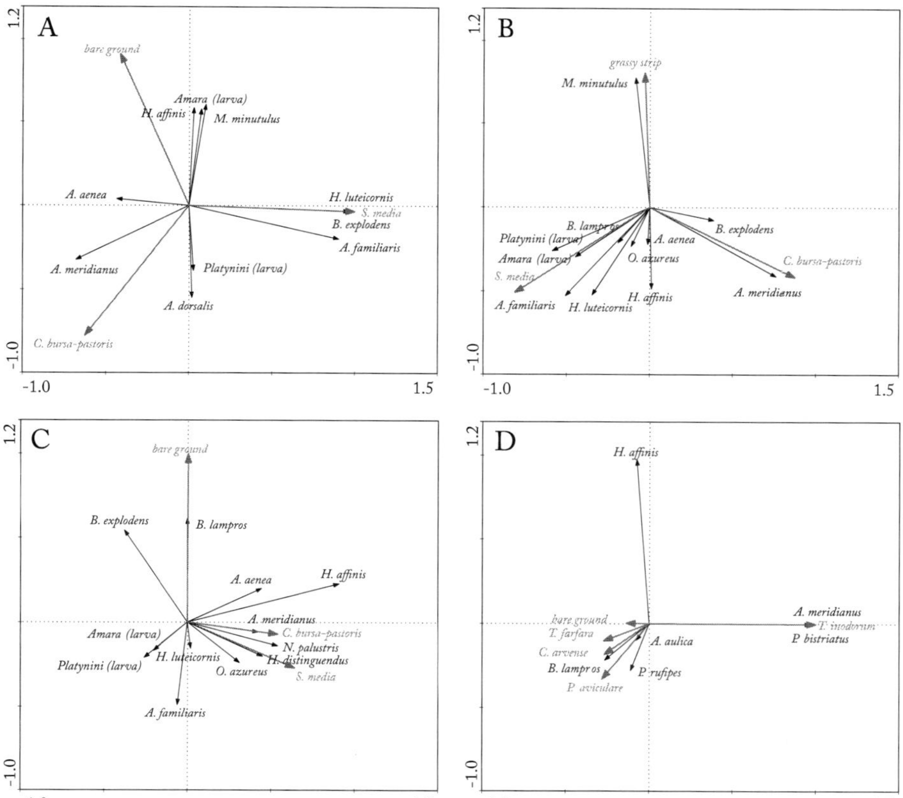

Fig. 1. Percentage cover analysis of the vegetation effects on abundance of carabid beetles (RDA). A – week 19; B – week 21; C – week 23; D – week 32.

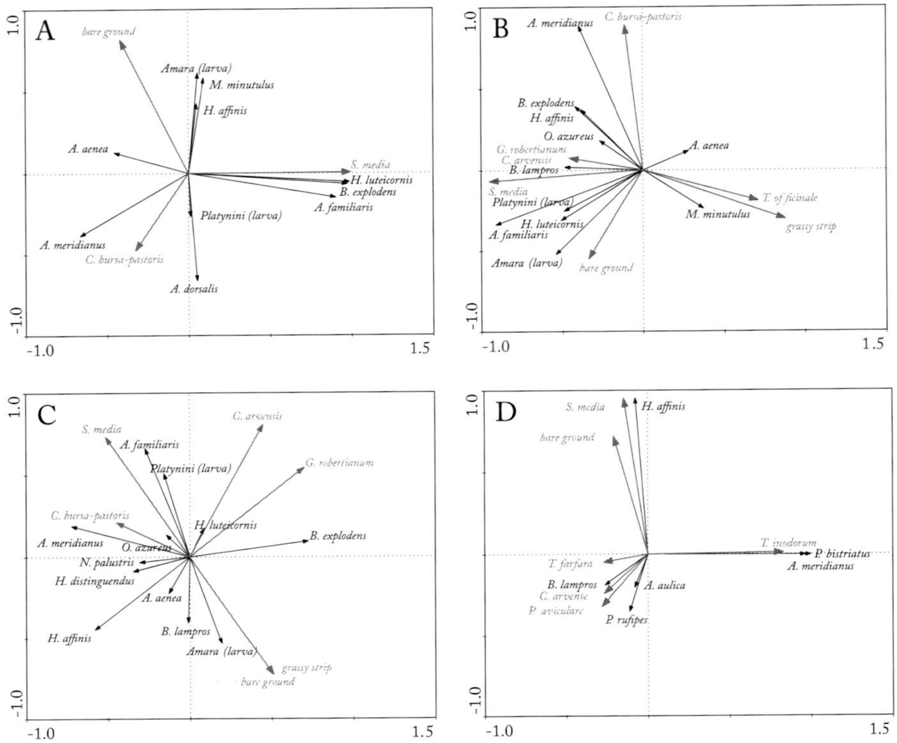

Fig. 2. Presence/absence analysis of the vegetation effects on abundance of carabid beetles (RDA). A – week 19; B – week 21; C – week 23; D – week 32.

Table 4. Association of carabid beetles with particular weed species or habitat (RDA, $p<0.05$). Sample weeks: 9/5/2006 (week 19), 23/5/2006 (21), 9/6/2006 (23), 9/8/2006 (32).

Species	**Association [week]** **Relative cover [%]**	**Presence/absence**
Acupalpus meridianus	*C. bursa-pastoris* [21]	*C. bursa-pastoris* [21][23]
	T. inodorum [32]	*T. inodorum* [32]
Amara familiaris	*S. media* [19][21]	*S. media* [19][21][23]
Bembidion lampros	bare ground [23]	
Brachinus explodens	*S. media* [19]	*S. media* [19]
		G. robertianum [23]
Harpalus affinis		*S. media* [32]
		bare ground [32]
Harpalus luteicornis	*S. media* [19][21]	*S. media* [19]
Microlestes minutulus	grassy strip [21]	
Ophonus azureus	*S. media* [23]	
Paratachys bistriatus	*T. inodorum* [32]	*T. inodorum* [32]
larva *Amara*		bare ground [21]

coefficients estimated by the RDA for each combination of weed and carabid revealed significant correlations of 7 carabid species with 10 weeds or bare ground (Tables 3, 4; Fig. 2). All significant pairs of weeds and carabids estimated by both types of analyses are summarized in Table 4.

DISCUSSION

In this paper, the link between composition of naturally established weed and carabid assemblages at micro-habitat scale was studied using quadrat enclosures. Although many species of carabids seemed not to be affected by the vegetation composition, density of some carabid species correlated with presence or percentage cover of a particular plant species. Thus, the first two hypotheses expecting that composition of weed assemblage would affect the structure of carabid assemblage, since at least some carabid species would respond to occurrence of a preferred plant species, were supported.

This study used quadrat enclosures, which were found to be good estimator for carabid density (Desender & Maelfait, 1986). However, rather low individual numbers were found in the study habitats: 501 individuals of 36 taxa were recorded, with mean density of 12 individuals per sample. Density of many species was insufficient for statistical analysis, which is the major constraint of this study: only 17 species could have been included in analysis in at least one of the sample date. But, average of 12 individuals per sample of 0.16 m^2 gives a mean density of 75 individuals per m^2, which agrees with densities reported in the literature (Desender & Pollet, 1986; Desender & Alderweireldt, 1988; Ulber & Wolf-Schwerin, 1995).

In order to find the most suitable method for analyzing environmental effects on the assemblage composition data, the effect of vegetation composition on carabid abundance was tested using two different approaches: (i) percentage cover or (ii) presence/absence of weeds. In general, analysis in which presence/absence data were used explained more variance every time compared to the percentage cover analysis. This is especially remarkable in weeks 21 and 23, when explanatory power of the percentage cover model was below 50%. Supposedly low explanatory power of that model was due to automatic omitting (provided by the software) less abundant weed species (explanatory variables) from the Monte-Carlo tests due to their negligible variance, whereas these weeds were retained in the binary models since their values were equalled to the pre-dominant weed species (all weeds with > 20 % cover had even weight). This approach is ecologically sensible seeing that the density of carabids within a patch is not proportionate to patch size (P. Saska pers. obs.). In other words, even a small patch of a preferred weed may host as many individuals of a specialized species of carabid as a large patch.

Beside increase in explained variance, including more explanatory variables in the binary analyses brought more weed species significantly explaining variability in the carabid data in the permutation tests, and revealed higher number of significant associations of carabids with weeds compared to percentage cover analysis. Thus, binary expression of vegetation data seems to be a better predictor of the vegetation effects on carabid community structure, especially in case of small samples.

Results of both analyses gave congruent results with respect to observed significant weed-carabid associations, and in this way provided strong support for their establishment. For example, *A. familiaris* was found more often in patches of *S. media*, or *A. meridianus* in patches of *C. bursa-pastoris* and *Tr. inodorum*. In contrast, *B. lampros*, *M. minutulus* and *O. azureus* (to name a few examples) associated with bare ground, grassy strip and *S. media* according to the percentage cover analysis, but did not in the presence/absence analysis. Supposedly these discrepancies came from different numbers of environmental variables included in the model. As discussed above, binary analysis had higher predictive power than percentage cover analysis, so associations shown by percentage cover analysis but not substantiated by the binary analysis should be regarded with caution.

The last hypothesis tested in this paper was that plant-carabid association in the field would correspond with known food preference/requirements of the species. Data for some species support this hypothesis. The occurrence of *A. familiaris* correlated both with percentage cover and presence of *S. media* in most sample dates. Decrease in abundance of this species in the study fields was explained by a decrease in presence of weeds, mainly *S. media* in a study of Ulber & Wolf-Schwerin (1995), but their observation was not supported by any statistical test. *A. familiaris* has been reported to be specialized on *S. media* seed as adult and larva (Aubrook, 1949; Saska & Jarosik, 2001; Honek et al., 2007), and now its close association to that weed is strongly supported. Additionally, *H. luteicornis* associated with the same weed in some weeks, supporting the adult preference for small seeds (including *S. media*) in a cafeteria experiment (Honek et al., 2007). Or, *A. meridianus* preferred seeds of small Brassicaceae including *C. bursa-pastoris* in the same experiment of Honek et al. (2007) and occurred more likely in patches containing *C. bursa-pastoris* in spring in this study. The summer shift in habitat preference to *Tr. inodorum* patches shown by RDA may be well explained by a change in weed community (spring/summer seed-producing *C. bursa-pastoris* replaced by autumn seed-producing *T. inodorum*) during larval development. This hypothesis is supported by the prevalence of teneral adults of this species in August.

There are two species of carnivorous carabids that have been assigned to a particular weed species in both types of analyses. Since larva of *B. explodens* is a parasitoid of *Amara* pupae (Saska & Honek, 2004), its association to *S. media* patches where *A. familiaris* (one of the host species) predominated may be regarded as a support for that relationship, so far not unobserved in the field. *P. bistriatus* associated with *Tr. inodorum*, but this is very likely a result of microclimatic conditions or presence of a prey that was related to the plant rather than direct association with this plant.

In conclusions, this paper shows how knowledge on plant-carabid association helps us to understand carabid distribution at micro-habitat scale. Some species were consistently found to associate their presence/abundance with particular weeds; within their geographical range, their occurrence can now be predicted based on the presence of preferred plant species. Due to having vital role for many carabids, the presence of weeds in agricultural landscapes is thus crucial for preserving species directly or indirectly dependent on weeds; the knowledge on plant-carabid interaction also has conservation concern. To finish, data presented in this paper may be used to complement the lists of herbivorous insects related to particular herb species (Ward & Spalding, 1993; Campobasso et al., 1999; www.ecoflora.co.uk).

ACKNOWLEDGEMENTS

The work was supported by the grant no. 522/06/P366 of the Grant Agency of the Czech Republic.

REFERENCES

Altieri, M.A., Wilson, R.C. & Schmidt, L.L. (1995). The effect of living mulches and weed cover on the dynamics of foliage- and soil- arthropod communities in the three crop systems. – Crop Protection 4: 201-213.

Aubrook, E.W. (1949). *Amara familiaris*. – Entomologist's Monthly Magazine 85: 44.

Bosch, J. (1987). Der Einfluss einiger dominanter Ackerunkräuter auf Nutz- und Achadarthropoden in einem Zuckerrübenfeld. – Zeitschrift fuer Pflanzenkrankheiten und Pflanzenschutz 94: 398-408.

Campobasso, C., Colonnelli, E., Knutson, L., Terragitti, G. & Cristofaro, M. (1999). Wild Plants and Their Associated Insects in the Palearctic Region, Primarily Europe and the Middle East. U.S. Department of Agriculture, Agricultural Research Service.

de Snoo, G.R., van der Poll, R.J. & de Leeuw, J. (1995). Carabids in sprayed and unsprayed crop odges of winter wheat, sugar beet and potatoes. – In: Arthropod natural enemies in arable land I. Density, spatial heterogeneity and dispersal. (Toft, S. & Riedel, W., eds), Acta Jutlandica 70: 199-211.

Desender, K. & Alderweireldt, M. (1988). Population dynamics of adult and larval Carabid beetles in a maize field and its boundary. – Journal of Applied Entomology 106: 13-19.

Desender, K. & Maelfait, J.P. (1986). Pitfall trapping within enclosures: a method for estimating the relationship between the abundances of coexisting carabid species (Coleoptera: Carabidae). – Holoarctic Ecology 9: 245-250.

Desender, K. & Pollet, M. (1986). Adult and larval abundance from carabid beetles (Col., Carabidae) in a pasture under changing grazing management. – Mededelingen van de Faculteit Landbouwwetenschappen, Rijksuniversiteit Gent 51: 943-955.

Gersdorf, E. (1937). Ökologisch-faunistische Untersuchungen über die Carabiden der mecklenburgischen Landschaft. – Zoologische Jahrbücher, Abteilung für Systematik, Ökologie und Geographie der Tiere 70: 17-86.

Goldschmidt, H. & Toft, S. (1997). Variable degrees of granivory and phytophagy in insectivorous carabid beetles. – Pedobiologia 41: 521-525.

Hawthorne, A. & Hassall, M. (1995). The effect of cereal headland treatments on carabid communities. – In: Arthropod natural enemies in arable land I. Density, spatial heterogeneity and dispersal. (Toft, S. & Riedel, W., eds), Acta Jutlandica 70: 185-198.

Holland, J.M., Hutchison, M.A.S., Smith, B. & Aebischer, N.J. (2006). A review of invertebrates and seed-bearing plants as food for farmland birds in Europe. – Annals of Applied Biology 148: 49-71.

Holland, J.M., Perry, J.N. & Winder, L. (1999). The within-field spatial and temporal distribution of arthropods in winter wheat. – Bulletin of Entomological Research 89: 499-513.

Honek, A. & Jarosik, V. (2000). The role of crop density, seed and aphid presence in diversification of field communities of Carabidae (Coleoptera). – European Journal of Entomology 97: 517-525.

Honek, A. & Martinkova, Z. (2001). Aggregation of ground beetles (Carabidae, Coleoptera) on winter rape seeds dispersed on the ground. – Plant Protection Science 37: 97-102.

Honek, A., Martinkova, Z. & Jarosik, V. (2003). Ground beetles (Carabidae) as seed predators. – European Journal of Entomology 100: 531-544.

Honek, A., Martinkova, Z. & Saska, P. (2005). Post-dispersal predation of *Taraxacum officinale* (dandelion) seed. – Journal of Ecology 93: 345-352.

Honek, A., Martinkova, Z., Saska, P. & Pekar, S. (2007). Size and taxonomic constraints determine the seed preferences of Carabidae (Coleoptera). – Basic and Applied Ecology 8: 343-353.

Hůrka, K. (1996). Carabidae of the Czech and Slovak Republics. Carabidae České a Slovenské Republiky. Kabourek, Zlín.

Jorgensen, H.B. & Toft, S. (1997). Role of granivory and insectivory in the life cycle of the carabid beetle *Amara similata*. – Ecological Entomology 22: 7-15.

Kokta, C. (1988). Bezeihungen zwischen der verunkratung und phytophagen laufkäfern der gattung *Amara*. – Mitteilungen aus der Biologischen Bundesanstalt für Land und Forstwirtschaft Berlin Dahlem 247: 139-145.

Kromp, B. (1989). Carabid beetle communities (Carabidae, Coleoptera) in biologically and conventionally farmed agroecosystems. – Agriculture, Ecosystems and Environment 27: 241-251.

Kromp, B. (1990). Carabid beetles (Coleoptera, Carabidae) as bioindicators in biological and conventional farming in Austrian potato fields. – Biology and Fertility of Soils 9: 182-187.

Luff, M.L. (1993). The Carabidae (Coleoptera) Larvae of Fennoscandia and Denmark. Brill, Leiden.

Luff, M.L. (2002). Carabid assemblage organization and species composition. – In: The Agroecology of Carabid Beetles (Holland, J.M., ed.). Intercept, Andover, pp. 41-79.

Marshall, E.J.P., Brown, V.K., Boatman, N.D., Lutman, P.J.W., Squire, G.R. & Ward, L.K. (2003). The role of weeds in supporting biological diversity within crop fields. – Weed Research 43: 77-89.

Powell, W., Dean, G.J. & Dewar, A. (1985). The influence of weeds on polyphagous arthropod predators in winter wheat. – Crop Protection 4: 298-312.

Saska, P. (2005). Contrary food requirements of the larvae of two *Curtonotus* (Coleoptera: Carabidae: *Amara*) species. – Annals of Applied Biology 147: 139-144.

Saska, P. & Honek, A. (2004). Development of the beetle parasitoids, *Brachinus explodens* and *B. crepitans* (Coleoptera: Carabidae). – Journal of Zoology, London 262: 29-36.

Saska, P. & Jarosik, V. (2001). Laboratory study of larval food requirements in nine species of *Amara* (Coleoptera: Carabidae). – Plant Protection Science 37: 103-110.

Schremmer, F. (1960). Beitrag zur biologie von *Ditomus clypeatus* Rossi, eines körnersammeln den Carabiden. – Zeitschrift der Arbeitsgemeinschaft österreichisches Entomologen 12: 140-145.

Speight, M.R. & Lawton, J.H. (1976). The influence of weed cover on the mortality imposed on artificial prey by predatory ground beetles in cereal fields. – Oecologia 23: 211-223.

Storkey, J. (2006). A functional group approach to the management of UK arable weeds to support biological diversity. – Weed Research 46: 513-522.

Ter Braak, C.J.F. & Šmilauer, P. (2002). CANOCO Reference Manual and CanoDraw for Windows User's Guide. Software for Canonical Community Ordination (Version 4.5). Biometris, Wageningen and Ceske Budejovice.

Ulber, B. & Wolf-Schwerin, G. (1995). A comparison of pitfall trap cathes and absolute density estimates of carabid beetles in oilseed rape fields. – In: Arthropod natural enemies in arable land I. Density, spatial heterogeneity and dispersal. (Toft, S. & Riedel, W., eds), Acta Jutlandica 70: 77-86.

Ward, L.K. & Spalding, D.F. (1993). Phytophagous British insects and mites and their food-plant families – total numbers and polyphagy. – Biological Journal of the Linnean Society 49: 257-276.

Zetto Brandmayr, T. (1976). Nutrizione e allevamento di Carabidi esclusivamente fitofagi: spermofagia larvale di *Ophonus ardosiacus* Lutsh. – Redia 59: 197-206.

Zetto Brandmayr, T. (1983). Life cycle, control and propagation rhythm and fecundity of *Ophonus rotundicollis* Fairm. et Lab. (Coleoptera, Carabidae, Harpalini) as an adaptation to the main feeding plant *Daucus carota* L. (Umbelliferae). – In: Ecology of Carabids: the synthesis of field study and laboratory experiment. (Brandmayr, P., Den Boer P.J. & Weber, F., eds). University of Münster, Münster, pp. 93-103.

Zetto Brandmayr, T. (1990). Spermophagous (seed-eating) ground beetles: first comparison of the diet and ecology of the harpaline genera *Harpalus* and *Ophonus* (Col., Carabidae). – In: The Role of Ground Beetles in Ecological and Environmental Studies (Stork, N., ed.). Intercept, Andover, pp. 307-316.

L. Penev, T. Erwin & T. Assmann (Eds) 2008
Back to the Roots and Back to the Future. Towards a New Synthesis amongst Taxonomic, Ecological and Biogeographical Approaches in Carabidology
Proceedings of the XIII European Carabidologists Meeting, Blagoevgrad, August 20-24, 2007, pp. 353-362.

Sofia–Moscow

A real time extinction: the case of *Carabus clatratus* in Italy (Coleoptera, Carabidae)

Achille Casale[1] & Enrico Busato[2]

[1] Università di Sassari, Dipartimento di Zoologia e Genetica Evoluzionistica, Via Muroni 25, 07100 Sassari, Italy. E-mail: casale@uniss.it
[2] Università di Torino, Di.Va.P.R.A. – Entomologia e Zoologia applicate all'Ambiente "C. Vidano", Via Leonardo da Vinci 44, 10095 Grugliasco (TO), Italy.
E-mail: enrico.busato@unito.it

SUMMARY

In Europe, highly hygrophylic carabids in lowlands are markedly exposed to the risk of extinction due to the loss or modification of their habitat. Among these, *Carabus clatratus*, one of the few species of the genus *Carabus* characterized by a semi-aquatic way of life, is reported as threatened or disappeared in several areas in all European countries. In Italy, the historical distribution of this species is reported from some 25 localities. A chrono-geonemic survey, however, shows that in the last decade of the past century it was confined to not more than three localities in Tuscany, reduced to one only at the beginning of this century. We supposed that the rapid and unexpected extinction of *C. clatratus* in some undisturbed and not modified biotopes, where it was very abundant until a few years ago, was due to the massive colonization of an alien, very invasive species, the red swamp crayfish *Procambarus clarkii*, imported into Italy for aquaculture from the southeastern United States. An original experiment and a video, made under laboratory conditions, prove for the first time that this crayfish can be a very able predator on adult individuals of *C. clatratus*, and strongly supports the hypothesis that it is at present the primary factor of extinction of *C. clatratus* in Italy.

Keywords: Coleoptera, *Carabus clatratus*, Decapoda, *Procambarus clarkii*, predation, chronogeonemy, extinction

INTRODUCTION

As postulated by authors such as Wilson (1992), the extinction of invertebrate organisms - at a global or local scale – is an event difficult to prove on the ground of objective data: inadequate knowledge of population sizes, scarcity of investigations in the field, and the loss of information on the life histories of many so called "rare" taxa, make the assumption that a species has really and definitively disappeared from a given area difficult, and sometimes impossible.

Carabus (*Limnocarabus*) *clatratus* Linné, 1761 is species of nice large-sized carabids in which several peculiarities quite unusual among the congeneric species are summarized. At first, it presents - with several subspecies - a (virtually) very wide area of distribution, from southern France to Siberia (see map in Turin et al., 2003). The eastern vicariant and close adelphotaxon *C. maacki* Morawitz, 1862, of the Far East and Japan, previously treated as a subspecies of *C. clatratus*, is now recognized as a distinct species (Deuve, 2004). Secondly, it is one of the few *Carabus* species (not more than six among the over 800 described species) with a semi-aquatic way of life, able to dive voluntary and stay under water for several minutes (Sturani, 1962; Busato, personal observations) (Fig. 7). Finally, it is one of the very few *Carabus* species with a pteridimorphic state of wings, being some individuals fully winged and able to fly (Lindroth, 1985).

In spite of these morphological, chorological, and ecological features, *C. clatratus* is presently one of the most localized and endangered species on the European continent, and is reported as threatened or disappeared in several areas in all European countries (Assmann, 2003). This is a rather common situation: in all Europe, highly hygrophylic carabids in lowlands are very exposed to the risk of extinction due to the loss or modification of their habitat.

In Italy, the "historical" distribution of this species, represented by the subspecies *antonellii* Luigioni, 1921 (Fig. 1), was reported from 26 localities (Bucciarelli, 1963; Magistretti, 1965; Casale et al., 1982, 2006) (Fig. 2).

A chrono-geonemic survey, however, shows that in the last decade of the past century it was confined to not more than five localities in Tuscany, reduced to two only at the beginning of this century (Brandmayr et al., 2006) (Figs 3-5).

Recently, several amateur collectors informed us that the species had disappeared in some undisturbed and non-modified biotopes (among others, the Fucecchio marsh in Tuscany, where it was very abundant until a few years ago: Bordoni, 1995).

One of the authors of the present contribution (A.C.) supposed that this rapid and unexpected extinction was due to the massive colonization of an alien and very invasive species, the red swamp crayfish *Procambarus clarkii* (Girard, 1852) (Crustacea, Decapoda), imported into Italy for aquaculture from the southeastern United States in the eighties of the last century. The distribution of this species is now increasing, with incredible densities of population and a deep impact on both aquatic vegetation and fauna, being the young individuals predators, and the adults mostly vegetarian (see, for a review, Delmastro, 1999; Acquapace et al., 2006). This species is also able to move on the ground, in wet

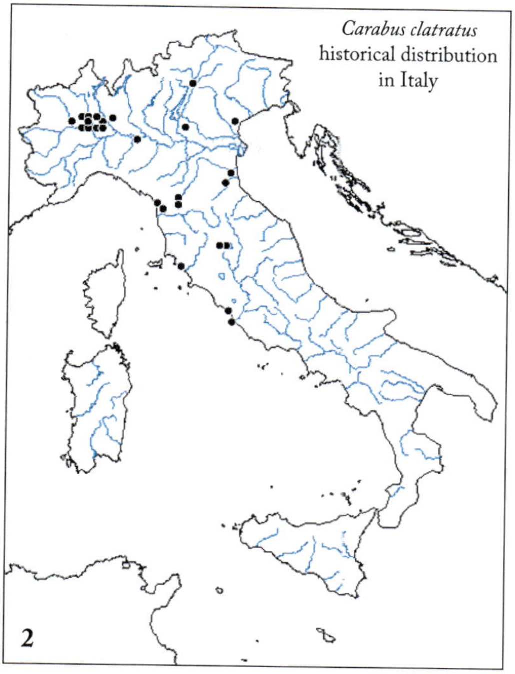

Figs 1-2. Fig. 1: *Carabus clatratus antonellii* Luigioni, 1921, a hibernating individual in rotten wood, dorsal aspect (Tuscany, Lago di Chiusi) (photo A. Vigna Taglianti). Fig. 2: *C. clatratus antonellii,* historical distribution in Italy (data from Bucciarelli, 1963; Magistretti, 1965; Casale et al., 1982, 2006).

conditions; surprisingly, its distribution in Italy (Froglia, 2006) is presently overlapping on that of the former distribution of *C. clatratus* in this country (Figs 2, 6).

An original experiment and a video, made under laboratory conditions by E.B., proves for the first time that this is a real fact, not a theory.

MATERIAL AND METHODS

The data reported in the current paper concern some adult individuals of *Carabus* (*Limnocarabus*) *clatratus antonellii* collected on 5.III.2006 (3 ♂♂: Tuscany, Siena province, Lago di Montepulciano m 248, A. Petrioli legit), 13.II.2007 (2 ♂♂, 3 ♀♀: Tuscany, Siena province, Lago di Montepulciano m 248 and Lago di Chiusi m 251, P. Cavazzuti and A. Casale legit) and different instars of *Procambarus clarkii* (10 young stages and 5 adults) collected in Piedmont, Carmagnola (hamlet San Michele, locality Martinetto, Canale di San Grato m 235, 1.VI.2006, G. B. Delmastro legit) and of *Austropotamobius pallipes* (Lereboullet, 1858) collected in Piedmont, Giaveno (bridge of the provincial road no. 193, "Rio della Colletta" m 550, 2.VI.2006, G. B. Delmastro legit).

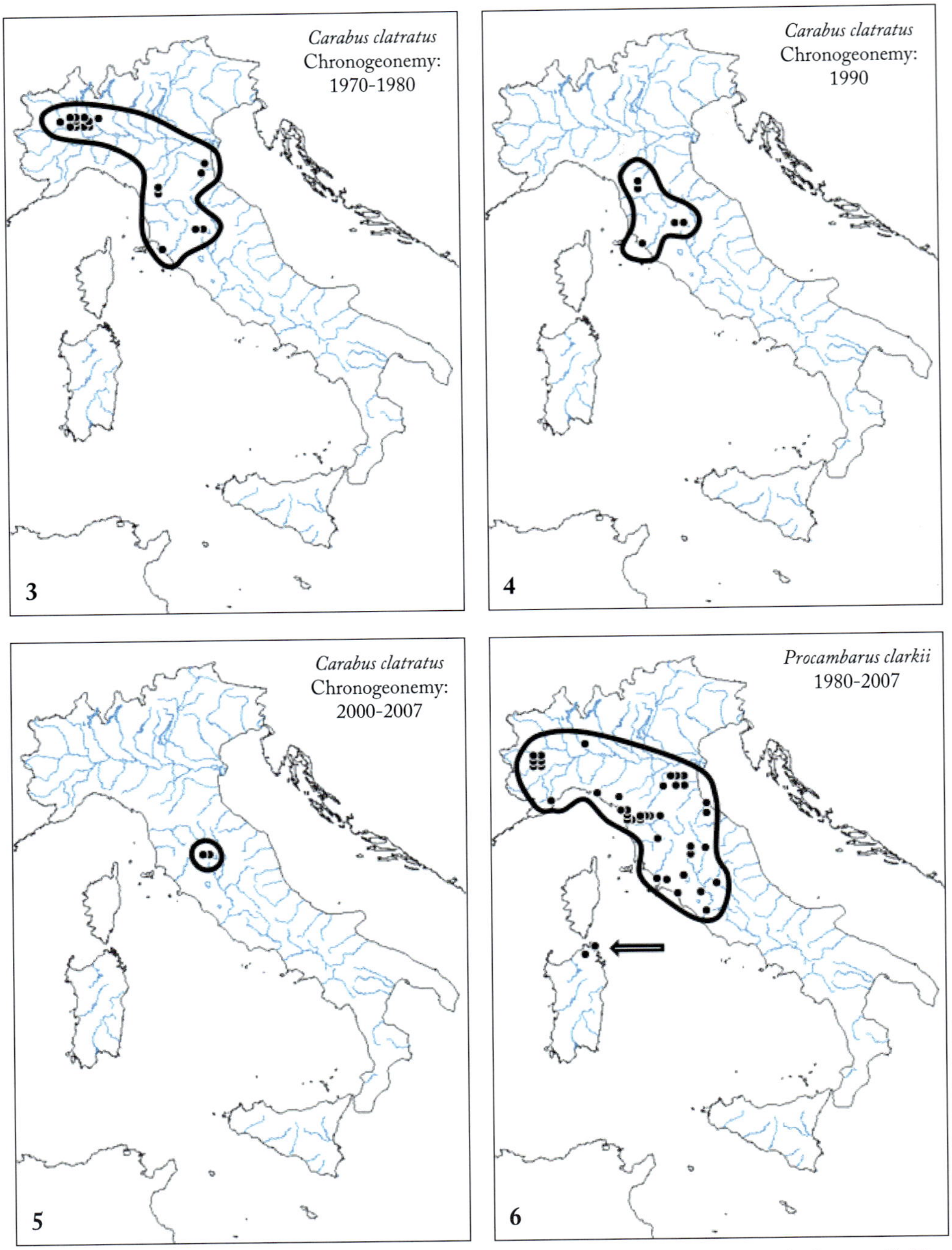

Figs 3-6. Figs 3-5: Chronogeonemy and decline of *C. clatratus antonellii* in Italy (original). Fig. 6: Present distribution of the red swamp crayfish *Procambarus clarkii* in Italy (data from Froglia, 2006, modified).

Breeding of *C. clatratus* was carried out in a 240 L aquarium-terrarium (cm 99.5 x 45.0 x 55) characterized by a 14 cm layer of volcanic soil on the bottom (Fig. 8). Some aquatic plants (*Microsorum pteropus*, *Anubias barteri*, *Echinodorus osiris*) were introduced to simulate a natural surrounding and to help the beetles reach the surface of the water when the air reserves were low. A plastic container (cm 38.0 x 28.5) containing a 5 cm layer of soil was introduced into the aquarium-terrarium to let the beetles live outside the water and as an ovipositional site for the females. Every two days, food (apple and beef) was provided to *C. clatratus* on a piece of polystyrene to avoid moulds, and changed every two days.

Three quarters of the soil was covered with moss, and pieces of bark were placed to allow the movements of the beetles between the soil and the water. The terrarium was supported by a wooden framework covered by bark to simulate a hollow stump immersed in the water for almost 2/3 of its length (30 cm), half of which were beneath the bottom substrate.

A pump connected with a biological filter kept the water clean. To allow the growth of the aquatic plants, the aquarium-terrarium had, on the lid, a 25 watt (1000° Kelvin) neon lamp regulated by a timer according to the photoperiod of the rearing period. The aquarium-terrarium was located in a place with constant temperature (20±1°C), while the lid was raised to 45° to avoid moisture.

The two crayfish species were separately bred in two different aquariums of 48 L each (cm 49.5 x 29.7 x 33.0) with 8 cm of quartz sand on the bottom.

Predation trials were carried out moving two specimens of *C. clatratus* from the rearing aquarium to the aquarium with crayfishes. The remaining specimens of *C. clatratus* (2 ♂♂ e 3 ♀♀) reared in 2007 are still alive.

Preventively the water level was reduced to 15 cm and peaces of bark were placed on the bottom and partially immersed in the water to create a platform where the beetles could walk.

In the aquarium-terrarium just one or few specimens of crayfish were introduced for the time necessary for the observations.

Figs 7-8. Fig. 7: *C. clatratus antonellii*, an adult individual staying in water in breeding terrarium-aquarium (photo E. Busato). Fig. 8: Terrarium-aquarium for adult individuals of *C. clatratus antonellii* (photo E. Busato).

RESULTS

In the 48 L aquarium, the predation trials carried out with the local crayfish (*Austropotamobius pallipes*) assessed that this species is completely harmless towards *Carabus clatratus*. To further confirm this result, two specimens of *A. pallipes* were kept for 10 days in the aquarium-terrarium used for beetle breeding (240 L) and no predatory activity was observed.

Conversely, in the 48 L aquarium containing the red swamp crayfish (*Procambarus clarkii*), a predation activity was observed every time the beetles moved into the water. *P. clarkii* made an attempt to *C. clatratus*' life also when the beetles were outside the water, *P. clarkii* crouched under the surface of the water for many minutes moving closer to prey slowly and progressively. Then they jumped forward with the claws open to capture the insects dragging them under water and eating them (Figs 9-10).

The trials were carried out with two *C. clatratus* males. Both of them, when captured, floundered until the air reserves stored under the elytra were over. *C. clatratus* can stay immersed for quite a long time, until 17'30" (Sturani, 1962, and our personal observations). During this time the crayfish kept *C. clatratus* with a claw between the thorax and the abdomen, while the other claw was used to cut the antennae and the legs used by the insect to anchor itself to the substrate and to reach the water surface. When the beetle was deprived of the appendages but was still alive, the crayfish devoured it after ripping the cuticle. In both cases the beetle was eaten starting from the head; each part of the insect was emptied. The abdomen was devoured starting from the connection with the thorax (Figs 10-11), then after the elytra were discarded, the beetle was eaten thanks to the laceration of the cuticle.

Each meal, starting from the attack, lasted 2h 30' finally leaving only the elytra (nibbled at the attachment with the abdomen), some pieces of legs, and few remains of the integument (Fig. 12). During the trials, *P. clarkii* individuals, confirming their amphibious character, often went outside the water climbing onto the bark platform where the beetles rested.

DISCUSSION AND CONCLUSIONS

Our experiment proves for the first time that the crayfish *Procambarus clarkii* is a very efficient predator on the carabid *Carabus clatratus*.

In fact, in areas where the crayfish is more abundant, the rarity or the disappearance of the aquatic vegetation and several aquatic organisms, including Gastropoda, Odonata, Coleoptera Hydradephaga, local crayfish species, amphibians and smaller sized fishes, is reported (Diamond, 1996; Gamradt et al., 1997; Gutiérrez-Yurrita et al., 1998; Delmastro, 1999; Gherardi et al., 1999; Acquistapace et al., 2006; Fabbri, 2007). Furthermore, the present distribution of the crayfish (Fig. 6) is perfectly overlapping on the former distribution of *C. clatratus* in Italy (Fig. 2), the highest density of its populations is reported really in the last localities where this carabid was present and sometimes abundant (Padanian plain, Tuscany, Adriatic coast close to Ravenna) and, finally, the increasing spread of the crayfish in the last three decades is markedly coincident with the decline of *C. clatratus* in Italy, as showed by the chronogeonemy analysis of this species (Figs 3-5).

We can also propose that the crayfish, a thermophilic species, has over the past few years taken advantage of the increasing temperatures induced by the global change. These environmental conditions, on the contrary, probably induced a negative impact on *C. clatratus*, which is confined to wet biotopes and mesophilic forests in Northern and Central Italy, as a relict element of northern of Pleistocene origin in the Italian fauna (see Vigna Taglianti, 1998).

Unfortunately, this situation could be soon replicated in Southern France, where *P. clarkii* is spread in some localities (see Laurent et al., 1991), in which *C. clatratus* is represented by the endemic subspecies *arelatensis* Lapouge, 1903.

Another interesting datum, that we may anticipate, is that the predation seems to occur only on adult individuals, which are highly hygrophilous, stay for a long time in the water (Fig. 7), and have a long life span (more than one year, in laboratory conditions:

Figs 9-12. Four phases of predation by *P. clarkii* on *C. clatratus antonellii* (photo E. Busato).

Busato, personal observations on *C. clatratus antonellii*). On the contrary, the impact on larval stages has been probably moderate, or absent. In fact, larvae of *C. clatratus*, unlike those of other *Carabus* species with a semi-aquatic way of life (as *C. nodulosus* Creutzer, 1799: Sturani, 1964), are mostly terricolous, with fossorial behaviour; furthermore, they display a peculiarly short development (30-60 days), without having a diapause (Sturani, 1962; Huk, 1998; Huk & Kühne, 2000; in *C. clatratus antonellii*, bred for this experiment, duration larva-prepupa: 42-50 days: Busato, pers. obs.).

In spite of the fact that some of the biotopes where *C. clatratus* was, or is still present in Italy, are protected areas, currently no action has been realized to oppose the increasing distribution of the crayfish, which can be both active on the continent, and introduced by man on islands (see the recent introduction into Sardinia in last years: Fig. 6).

The extinction of the Italian populations of *C. clatratus* is much more dramatic, since these populations were attributed to an endemic, well characterized subspecies, highly interesting from the biogeographic point of view, on which a genetic survey is now in progress (Mossakowski, personal communication).

ACKNOWLEDGEMENTS

For the gift of living individuals of *Carabus clatratus antonellii*, *Procambarus clarkii* and *Austropotamobius pallipes*, we are particularly indebted to Andrea Petrioli (Florence), Pierfranco Cavazzuti (Pagno), and to Gianni Delmastro (Carmagnola), respectively.

Our sincere thanks are also extended to Augusto Vigna Taglianti (Rome), who authorized the reproduction of his nice photograph of *C. clatratus*, Dietrich Mossakowski (Bremen) for instructive carabidological discussions both in the field and laboratory, Peter John Mazzoglio (Torino) for the revision of the English text, and Ivo Manca (Sassari) for his valuable help in cartographic elaboration of maps.

We are also very grateful to Barbara Anselmi (WWF, Siena) and Carlo Accame (WWF-oasis of Lago di Chiusi), for their kindness and valuable information.

REFERENCES

Acquistapace, P., Cacchiani, A. & Gherardi, F. (2006). The impact of the introcuded crayfish, *Procambarus clarkii*, on a lake community in Tuscany. Biologia Ambientale 20 (1): 45-47.

Assmann, T. (2003). Conservation biology, pp. 427-437. – In: Turin, H., Penev, L. & Casale, A., eds. The genus *Carabus* in Europe – A Synthesis. Pensoft, Sofia-Moscow, 511 pp.

Bordoni, A. (1995). I Coleotteri del Padule di Fucecchio (Coleotterofauna di una biocenosi palustre dell'Italia centrale, Toscana). Tipografia Artigiana, Pistoia, 229 pp.

Brandmayr, P., Casale, A., Puzzo, F. & Scalercio, S. (2006). Chronogeonemy analysis: some examples regarding species of the Italian fauna. – In: Ruffo, S. & Stoch, F., eds. Checklist and distribution of the Italian fauna. Memorie del Museo civico di Storia naturale, Verona, 2.serie, Sezione Scienze della Vita 17: 41-45.

Bucciarelli, I. (1963). Un interessante reperto faunistico nei dintorni di Milano: il *Carabus clathratus antonellii* Luigioni. Bollettino della Società entomologica italiana 93: 43-50.

Casale, A., Sturani, M. & Vigna Taglianti, A. (1982). Coleoptera, Carabidae. I. Introduzione, Paussinae, Carabinae. Fauna d'Italia, 18, Calderini Ed., Bologna, 499 pp.

Casale, A., Vigna Taglianti, A., Brandmayr, P. & Colombetta, G. (2006). Insecta Coleoptera Carabidae. – In: Ruffo, S. & Stoch, F., eds. Checklist and distribution of the Italian fauna. Memorie del Museo civico di Storia naturale, Verona, 2.serie, Sezione Scienze della Vita 17: 159-163 (with maps on CD-ROM).

Delmastro, G.B. (1999). Annotazioni sulla storia naturale del gambero della Louisiana *Procambarus clarkii* (Girard, 1852) in Piemonte centrale e prima segnalazione regionale del gambero americano *Orconectes limosus* (Rafinesque, 1817) (Crustacea: Decapoda: Astacidea: Cambaridae). Rivista piemontese di Storia naturale 20: 65-92.

Deuve, T. (2004). Illustrated Catalogue of the Genus *Carabus* of the World (Coleoptera: Carabidae). Pensoft, Sofia-Moscow, 461 pp., 24 colour plates.

Diamond, J.M. (1996). A-bombs against amphibians. Nature 383: 386-387.

Fabbri, R. (2007). Modificazioni nella comunità odonatologica nell'oasi di Punte Alberete, Parco del Delta del Po. Riassunti del Convegno: Le libellule in Italia - Ricerche e conservazione. Parco Naturale Valle del Ticino, Cameri, 10-11 febbraio 2007: 12.

Froglia, C. (2006). Crustacea Malacostraca Decapoda. – In: Ruffo, S. & Stoch, F., eds. Checklist and distribution of the Italian fauna. Memorie del Museo civico di Storia naturale, Verona, 2.serie, Sezione Scienze della Vita 17: 113-114 (with maps on CD-ROM).

Gamradt, S.C., Kats, L.B. & Anzalone, C.B. (1997). Aggression by Non-Native Crayfish Deters Breeding in California Newts. Conservation Biology 11 (3): 793-796.

Gherardi, F, Baldacchini, G.N., Barbaresi, S., Ercolini, P., De Luise, G., Mazzoni, D. & Mori, M. (1999). The situation in Italy, pp. 107-128. – In: Gherardi F. & Holdich D. M. eds. Crayfish in Europe as alien species. How to make the best of a bad situation?. Crustacean Issues 11, Balkema, Rotterdam, x + 299 pp.

Gutiérrez-Yurrita, P.J., Sancho, G., Bravo, M.A., Baltanás, A. & Montes C. (1998). Diet of the red swamp crayfish *Procambarus clarkii* in natural ecosystems of the Doñana National Park temporary fresh-water marsh (Spain). Journal of Crustacean Biology 18 (1): 120-127.

Huk, T. (1998). Ausbreitungsvermögen, Lebenszyklus, Larvalökologie und Habitatwahl von *Carabus clatratus* (Coleoptera, Carabidae). Angewandte Carabidologie 1: 41-50.

Huk, T. & Kühne, B. (2000). Egg laying strategy and aspects of larval biology of two *Carabus* species (Coleoptera, Carabidae), pp. 161-168. – In: Brandmayr, P., Lövei, G., Zetto Brandmayr, T., Casale, A. & Vigna Taglianti, A., eds. Natural History and Applied Ecology of Carabid Beetles, Proceedings of the IX European Carabidologists' Meeting, Pensoft, Sofia-Moscow.

Laurent, P.-J., Leloirn, H. & Neveu, A. (1991). Remarques sur l'acclimatation en France de *Procambarus clarkii* (Decapoda Cambaridae). Bulletin mensuel de la Société Linnéenne de Lyon 60 (5): 166-173.

Lindroth, C.H. (1985). The Carabidae (Coleoptera) of Fennoscandia and Denmark. I. Fauna Entomologica Scandinavica 15 (1), 226 pp.

Magistretti, M. (1965). Coleoptera Cicindelidae, Carabidae. Catalogo topografico. Fauna d'Italia, 8, Calderini Ed., Bologna, 512 pp.

Sturani, M. (1962). Osservazioni e ricerche biologiche sul genere *Carabus* Linnaeus (sensu lato) (Coleoptera Carabidae). Memorie della Società entomologica italiana 41: 85-202.

Sturani, M. (1964). Osservazioni biologiche e morfologiche sul *Carabus* (*Hygrocarabus*) *variolosus* Fabricius (Coleoptera Carabidae). Atti Accademia nazionale italiana di Entomologia, Rendiconti 11 (1963): 182-184.

Turin, H., Penev, L. & Casale, A. (Eds) (2003). The genus *Carabus* in Europe – A Synthesis. Pensoft, Sofia-Moscow, 511 pp.

Vigna Taglianti, A. (1998). I Carabidi nella Faunistica e Biogeografia. Atti Accademia nazionale italiana di Entomologia, Rendiconti 46: 245-276.

Wilson, O.L. (1992). The Diversity of Life. Harvard University Press.

L. Penev, T. Erwin & T. Assmann (Eds) 2008
BACK TO THE ROOTS AND BACK TO THE FUTURE. TOWARDS A NEW SYNTHESIS AMONGST TAXONOMIC, ECOLOGICAL AND BIOGEOGRAPHICAL APPROACHES IN CARABIDOLOGY
Proceedings of the XIII European Carabidologists Meeting, Blagoevgrad, August 20-24, 2007, pp. 363-372.

Sofia–Moscow

The distribution, habitat, and the nature conservation value of a Natura 2000 beetle, *Carabus hungaricus* Fabricius, 1792 in Hungary

Sándor Bérces[1], Győző Szél[2], Viktor Ködöböcz[3] & Csaba Kutasi[4]

[1]Duna-Ipoly National Park Directorate. H-1021 Budapest, Hűvösvölgyi út 52.
E-mail: bercess@gmail.com
[2] Hungarian Natural History Museum, H-1088 Budapest, Baross u. 13.
[3]Hortobágy National Park Directorate, H-4024 Debrecen, Sumen u. 2.
[4]Bakony Natural History Museum, H-8420 Zirc, Rákóczi tér 3-5.

SUMMARY

Carabus hungaricus Fabricius, 1792 usually inhabits sandy grasslands and dolomitic grasslands in Hungary. It is listed in the Habitat Directive and it is a characteristic species of the Pannonian biogeographic region. This paper summarizes all available data (literature data, personal communications, all available museum specimens, original research) on the current distribution of *Carabus hungaricus* in Hungary making use of GIS. The most numerous populations of this carabid beetle live in Pannonic sand steppe biotopes, the most vulnerable of the dolomitic grasslands. In Hungary, *Carabus hungaricus* is a vulnerable species according to the IUCN criteria. Known habitat types, habitat preferences, co-occurring ground beetle species, and endangering environmental factors are discussed.

Keywords: Natura 2000, *Carabus hungaricus*, nature conservation, distribution, Hungary

INTRODUCTION

In the Pannonian biogeographical region, *Carabus hungaricus* Fabricius, 1792 is a species of community interest, whose conservation requires the designation of special areas of conservation. Hungary has enlarged the EU with its entry not only with one more country, but also with a whole biogeographic region – the Pannonian biogeographic region. Thus the Hungarian government has greater responsibility to preserve this biogeographic region than other member states. When the designation process of the Natura 2000 network started, relatively little was known about the distribution and population size of *Carabus hungaricus,* which is a strictly protected beetle in Hungary.

Carabus hungaricus Fabricius, 1792 usually inhabits sandy grasslands and dolomitic grasslands in Hungary. The type locality "Hungaria" in Fabricius's description from 1792 refers most likely to the Buda Mountains, which was undoubtedly a place frequently visited by collectors and naturalists of that time. The first exact locality recorded for this species should be attributed to T. Koy, who published in 1800 a specimen catalogue (of the Buda Mts.), which is regarded as the first Hungarian faunistic list (Koy, 1800). In spite of this being the first record of the mentioned ground-beetle species, today *Carabus hungaricus* seems to be more frequent in the sandy grasslands of plains than in the dolomitic regions of the Buda Mts.

All subspecies of *Carabus hungaricus* are restricted to the Palearctic region. The European distribution of this species is disjunctive – three major distribution areas can be distinguished: 1) Ukrainian and Russian steppes, 2) Bulgaria, and 3) the Carpathian Basin. In the whole distribution area of the beetle, the habitats where this species occurs are fragmented, and as a result often isolated (Fig. 1).

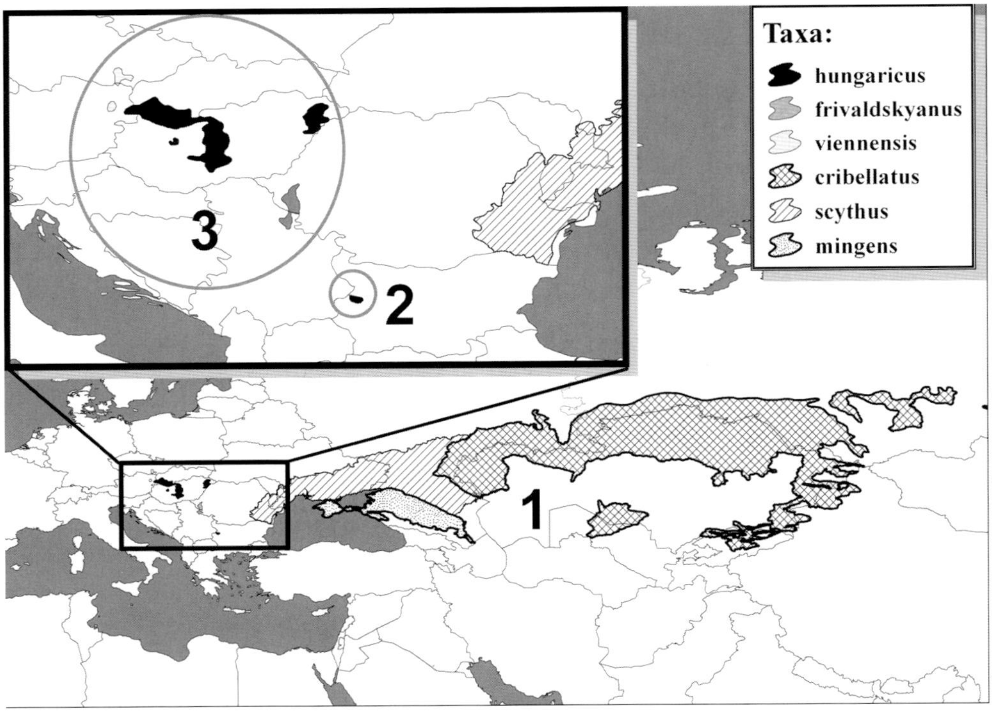

Fig. 1. The distribution area of subspecific taxa of *Carabus hungaricus*
1) Ukrainian and Russian steppes; 2) Bulgaria; 3) Carpathian Basin

Two further subspecies – *mingens* Quensel, 1806, and *cribellatus* M. F. Adams, 1812 live in the Russian and Ukrainian steppes. Some authors distinguish two further subspecies in this area: *scytus* Motschulsky, 1847 (Kryzhanovskij, 1953; Turin et al., 2003), and *gastridulus* Fischer von Waldheim 1823 (Löbl & Smetana, 2003). Furthermore, *cribellatus* M.F.Adams, 1812 is regarded as a distinct species and not a subspecies of *hungaricus* (Turin et al., 2003).

In the Carpathian Basin, two taxa have been described besides the nominotypic subspecies: *hungaricus viennensis* Kraatz, 1877, living in the Vienna Basin (Austria) and in South Moravia (Czech Republic), and *frivaldskyanus* Breuning, 1933 occurring in the Banat Region (Romania, Serbia). Many authors (Freude, 1976; Turin et al., 2003) question the subspecies rank of both forementioned taxa.

Carabus hungaricus is a typical steppe species, inhabiting dry calcareous and acidic sandy grasslands (in lowlands), and dolomitic grasslands (in mountains) in the Carpathian Basin. The majority of its populations inhabit calcareous sandy grasslands from the Deliblat (Serbia) throughout the Banat (Serbia, Romania) and sandy areas along the Danube River all the way to Vienna (Austria) and South Moravia (Czech Republic). Numerous populations occur on acidic types of sand grasslands in the Nyírség area, near the city of Debrecen (Hungary). A classical collecting place for this beetle was in the dolomitic grasslands of Buda Mountains, near Budapest. It also occurs on the same type of vegetation on dolomitic grasslands in the Eastern Bakony Mountains, north of Lake Balaton (Hungary).

Little is known about the populations living in the southern part of the Carpathian Basin in Romania. There are only three old records from Romania – Temesvár (Timişoara), Máslak (Masloc), and Németremete (Remetea Mica) (Turin et al., 2003). *Carabus hungaricus* was newly discovered in Temes County near Nagyzsám (Jamu Mare) by Lie (Lie, 1994), who caught 13 and 105 specimens near a black locust (*Robinia pseudoacacia)* forest in 1994 and 1995, respectively (Lie, 1994, 1995). In Serbia, only the Deliblat Region is known as a suitable habitat for *Carabus hungaricus* (Breuning, 1933; Pavićević & Mesaroš, 1997).

There are many published localities from South Slovakia: Pozsony (Bratislava) (Csiki, 1905-08; Frivaldsky, 1874; Majzlan, 1998), Trencsén (Trenčín), Peréd (Tešedíkovo) (Breuning, 1933); Peres (Pereš), Szentgyörgyhalma (Jurský Chlm) (Majzlan, 1998); Bassóc Hill in Marcelháza (Marcelová), Búcs (Búč), Köbölkút (Gbelce), and Helemba (Chl'aba) (Majzlan, 2005).

In the Czech Republic, only two recent localities are known in South Moravia (Lukáš Čížek, personal communication).

In Austria, *Carabus hungaricus* is almost extinct and occurs in the region of Lake Fertő (Neuseedler See) (Müller-Motzfeld, 2004; Turin et al., 2003). Published data on *Carabus hungaricus* localities in Austria: Niederösterreich (Lower Austria): Lajta Mountain (I. Frivaldsky, 1865); Vienna (X. and XI. Bezirk (districts 10 and 11)), Hennersdorf, Deutsch-Altenburg (Breuning 1933); Burgenland: Bruck an der Leitha; Joiszer Trift; Neusiedel am See (Breuning 1933).

Carabus hungaricus is also listed in the Russian (Ivanenko, 1999) and in the Ukrainian Red Book (Serbaka, 1994), while in Moldavia it is critically endangered (Neculiseanu et al., 1999).

Our goal was to clarify the recent distribution, habitat choice, and nature conservation value of *Carabus hungaricus* in Hungary. To achieve this goal we processed and analyzed our data (see Materials and methods) with GIS and used the standardized criteria published by IUCN (IUCN 2001). We also provide information about the typical coexisting ground beetle species.

MATERIAL AND METHODS

We summarised all available data from the collection of the Hungarian Natural History Museum, where probably the most numerous material of *Carabus hungaricus* specimens are preserved. A database was built on the basis of museum data, and of all available literature data (Ádám & Merkl, 1986; Csiki, 1905-08, 1946; Hůrka, 1973; Kaszab & Székessy, 1953; Kempelen, 1868; Kutasi, 1998; Kutasi et al., 2004; Kutasi et al., 2005; Kutasi & Szél, 2006; Kuthy, 1896; Merkl, 1991; Nározsny, 1938; Szél, 1985; Szél & Ádám, 1992; Szél & Bérces, 2002; Tóth, 1973). Aditionally, amateur collectors (Imre Retezár, József Muskovits, György Rozner, László Somay), shared their unpublished data with us. All these data were processed with GIS. In some cases it was impossible to identify the current location of an old collection place. In such cases, the centre of the corresponding geographical entity from the description was geo-referenced. From 2004 to 2006 60 habitat paches known form the literature, personal communication, or just supposed to be a potential habitat of *Carabus hungaricus* were trapped, and marked in the field using GPS (type: eTrex Legend by Garmin). These data was summarized in a GIS database (ESRI shape file), which contained a total of 189 records.

To locate new localities for *C. hungaricus*, we analyzed orthophotos, and contacted several biologists who could point out such places mainly upon botanical characteristics.

RESULTS

Our research proved the presence of *Carabus hungaricus* in a total of 42 localities, out of which 17 were new records in Hungary. On many locations near Budapest (Tétényi-fennsík, Hármashatár-hegy) we were unable to find *Carabus hungaricus* again in places where this species has been extensively collected just 50 years ago. New records were made in North-West Hungary in Kisalföld region (Csép, Mocsa: Bélapuszta); along the Danube River (Duna-Tisza Mid-Region sand area: Sződliget, Erdőkertes, Tatárszentgyörgy), and in the North-East Hungary (Nyírség area: Bánk, Nyíradony, Nyírbéltek, Nyírgelse, Vámospércs). New localities were also found in the in dolomitic grasslands of Buda Mountains near Budapest (Budaőrs: Farkas-hegy; Biatorbágy: Bolha-hegy).

According to the data gathered, *Carabus hungaricus* occurs mainly in sandy and dolomitic habitat types in Hungary. These habitats are heavily fragmented and are scattered in over 200 habitat patches. On the distribution map (Fig. 2), five main distribution areas of *Carabus hungaricus* in Hungary can be distinguished .

The first of these distribution areas in the eastern part of Hungary is the so called Nyírség area, where sand steppe habitats are of acidic type. The second one is located along the Danube, south of Budapest, mainly between the rivers Danube and Tisza, where *Carabus hungaricus* lives in calcareous sandy grasslands, while in the most southern part of this distribution area, a loess habitat was also found. West of Budapest, along the Danube River, calcareous sandy habitats, where the sand is more or less mixed with loess are found. Dolomitic type of habitats are located in the Buda Mountains and in the East Bakony Mountains. *Carabus hungaricus* was also collected on Sarmatian (Upper Miocenic age) limestones, near Budapest on the Érd-Tétényi table.

On two occasions *Carabus hungaricus* was found in not natural habitat: in the Nyírség area, near Újfehértó (Kutasi et al., 2004), one specimen was collected from an abandoned apple plantation; and near Nyíradony one more specimen has been collected in an abandoned arable land.

Carabus hungaricus lives in habitats where mainly no other *Carabus* species occur, however, some other *Carabus* species can also be found in its habitat (Fig. 3). Most often we find *Carabus scabriusculus* Olivier, 1795, which is a typical steppe species. In dolomitic

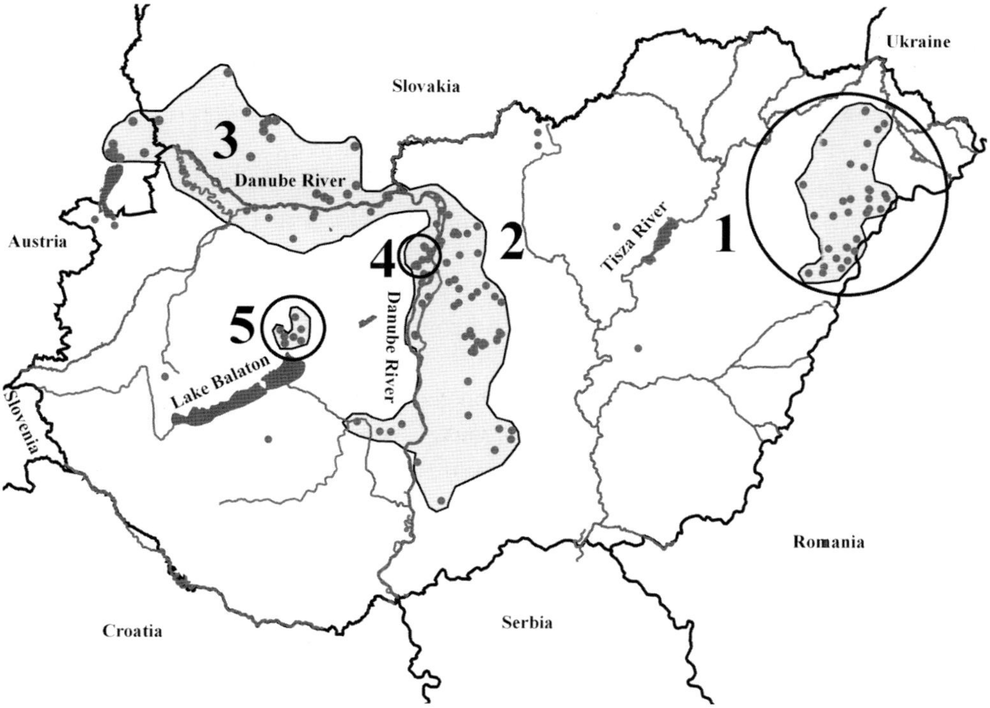

Fig. 2. The distribution of *Carabus hingaricus* in Hungary.
1) Nyírség; 2) Area between Duna and Tisza Rivers; 3) West of Budapest along the Danube River; 4) Buda Mountains; 5) East-Bakony Mountains

Fig. 3. *Carabus* species occurring with *C. hungaricus* (Photo: Imre Retezár).

grassland habitats we find other numerous species like *Carabus scheidleri* Panzer, 1799, or *Carabus convexus* Fabricius, 1775. *Carabus coriaceus* Linneaus, 1758, another accompanying carabid species, is a species inhabiting mainly forests and forest edges, but has a great dispersal power and can easily enter the habitats of *Carabus hungaricus* (in lowlands, but especially in the hills).

Other typical ground beetles found in the habitats of *Carabus hungaricus* were *Calathus erratus* (C. R. Sahlberg, 1827), *Calathus ambiguus* (Paykull, 1790), *Calathus fuscipes* (Goeze, 1777) and *Zabrus spinipes* (Fabricius, 1798). *Licinus cassideus* (Fabricius, 1792) is considered to be rare in Hungary and lives in the same habitats as *Carabus hungaricus* both in hills and lowlands. In the open sandy steppe habitat, *Osimus ammophilus* Dejean, 1829, (a Hungarian Red Data Book species) can be found occasionally.

DISCUSSION

In its whole distribution area *Carabus hungaricus* is endangered, in some countries almost extinct or critically endangered (Austria, Müller-Motzfeld, 2004; Moldova, Neculiseanu et al., 1999). The populations in Czech Republic and Slovakia are local (Turin et al., 2003).On the basis of this investigation, it can be summarized that in Hungary *Carabus hungaricus* can be found in many places, but the habitat of this species is extremely fragmented. The further fragmentation of its habitat is would cause a decrease in *C. hungaricus* population size, especially in the agglomeration of the capital city Budapest or near larger cities like Győr, Komárom, Ócsa.

As endangering factors could be regarded the ongoing urbanization near big cities like Budapest, Győr or Komárom. The habitats of *Carabus hungaricus* – due to our experiences – could be harmed by waste deposition, building of industrial areas and roads, mining practises, sports (quad, motorcycling), the afforestation of its habitat with black locust (*Robinia pseudoacacia*), scots pine (*Pinus sylvestris*) or European black pine (*Pinus nigra*), and cultivated hybrid poplar (*Populus x hybrida*). It seems that the populations in dolomitic habitats are more vulnerable, mainly because they are more fragmented, and smaller in size. In the last 50 years over 10, 000 hectares of dolomitic grasslands were forested with European black pine plantations (Tamás, 2001) in Hungary.

Without maintaining the open habitats by grazing, these could become spontaneously reforested with bushes or trees, causing the extinction of *Carabus hungaricus*.

To preserve this Pannonian beetle, which is of community importance, the proper designation of the Natura 2000 network is needed. The responsibility to preserve the populations of *Carabus hungaricus* is higher for the Hungarian decision makers, because the majority of the populations of this beetle are situated in Hungary, where it is a vulnerable species (Bérces et al., 2007). The further habitat loss can lead to local extinctions, which endanger the survival of *Carabus hungaricus* in Hungary.

ACKNOWLEDGEMENTS

We are grateful to the staff of the Duna–Ipoly National Park for helping with the field work among others to Sándor Bíró, Péter Csáky, Péter Csonka, Antal Halász, Róbert Grósz, István Staudinger, Valentin Szénási and Tamás Vidra. We would also like to thank Zoltán Barina who works in the Hungarian Natural History Museum for giving us advises on trapping in the Gerecse area. We would also like to extend our gratitude to those people who shared their records with us: Tamás Kovács, Tibor Magura, József Muskovits, Imre Retezár, László Somay and György Rozner. The project was supported by the Hungarian National R&D Programme "Faunagenesis project" (NKFP 3B023-04).

REFERENCES

Ádám, L. & Merkl, O. (1986). Adephaga of the Kiskunság National Park, I.: Carabidae (Coleoptera). – In: The Fauna of the Kiskunság National Park, I. (Mahunka, S., ed.), Akadémiai Kiadó, Budapest, pp. 119-142.

Bérces, S., Szél, Gy., Ködöböcz, V., Kutasi, Cs., Szabó, K., Fülöp, D., Pénzes, Zs., Peregovits, L. (2007) A magyar futrinka [*Carabus hungaricus*], in (Forró, L., ed.): A Kárpát-medence állatvilágának kialakulása [Faunagenesis of the Carpathian-Basin]

Breuning, S. (1933). Monographie der Gattung *Carabus* L. (IV. Teil). Bestimmungs-Tabellen der europäischen Coleopteren (Troppau) 106. Heft pp. 705-704.

Csiki, E. (1905-08). Magyarorszag bogarfaunaja 1. kotet Adephaga 1. Caraboidea.[The beetle fauna of Hungary 1. volume Adephaga 1. Caraboidea.], Budapest. (in Hungarian)

Freude, H. (1976). Familie: Carabidae (Laufkäfer). – In: Freude, H., Harde, K. W. és Lohse, G. A. (eds): Die Käfer Mitteleuropas Band. 2. Goecke & Evers Verlag, Krefeld.
Frivaldsky, I. (1865). Jellemzo adatok Magyarorszag faunajahoz. [Typical faubistic data for Hungary] A Magyar Tudomanyos Akademia Evkonyvei 11 (4): 1-275, 13 pls. (in Hungarian)
Frivaldsky, J. (1874). Magyarorszag tehelyropuinek futonczfelei. [Hungarian ground beetels] (Carabidae) F. Eggenberger Magyar Akademiai Konyvarusnal, Budapest. (in Hungarian)
Guéorguiev V.B. (1989). Prinos kum izuchavaneto na predstavitelite na semeistvo Carabiade (Coleoptera) ot Bulgaria. II. Acta Zoologica Bulgarica 38: 82-84 (in Bulgarian with English summary).
Guéorguiev V.B. & Guéorguiev B.V. 1995. Catalogue of the ground-beetles of Bulgaria (Coleoptera: Carabidae). Pensoft, Sofia-Moscow, 279 pp.
Guéorguiev B.V. & Sakalian V.P. 1997. Vertical distribution of Carabidae (Coleoptera, Carabidae) in Bulgaria. Acta Zoologica Bulgarica 49: 52-57.
Hůrka, K. (1973). Fortpflanzung und Entwicklung der mitteleuropäischen *Carabus*- und *Procerus*-Arten. – Studie CSAV, 9, Academia, Praha.
IUCN. (2001). IUCN Red List Categories and Criteria: Version 3.1. IUCN Species Survival Commission. IUCN, Gland (Switzerland) & Cambridge (UK).
Kaszab, Z. & Szekessy, V. (1953). Batorliget bogar-faunaja, Coleoptera. – In: Batorliget elovilaga [The beetle fauna of Batorliget – In: The flora and fauna of Batorliget] (in Hungarian) (Szekessy, V., ed.). Akademiai Kiado, Budapest, pp. 194-285.
Kempelen, R. (1868). III. Heves es kulso Szolnok torvenyesen egyesult varmegyek allattani leirasa. – In: Heves- es Kulso Szolnok torvenyesen egyesult varmegyek leirasa [Faunistic desription of the counties Heves and Szolnok in: The decription he counties Heves and Szolnok] (Albert, F., ed.): A Magyar Orvosok es Termeszetvizsgalok XIII. nagygyulese [Congress of the Hungarian doctors and naturalists], Eger, pp. 175-226. (in Hungarian)
Koy T. (1800). Alphabetisches Verzeichniss meiner Insecten-Sammlung. Gewidmet Seinen Entomologischen Freuden von Tobias Koy. Ofen, 64 pp.
A Kornyezetvedelmi Miniszter 13/2001. V. 9 KoM rendelete. (2001). A vedett es fokozottan vedett noveny-es allatfajokrol, a fokozottan vedett barlangok korerol, valamint az Europai Kozossegben termeszetvedelmi szempontbol jelentos noveny-es allatfajok kozzetetelerol. [Order of the Minister for environment 13/2001. V. 9 KoM About the protected and strictly protected plant and animal species, as well as about the stictly protected caves, and about species of Community interests] – Magyar Kozlony 53: 3446-3511. (in Hungarian)
Kryzhanovskij, O.L. (1953). Zhuki-zhuzhelitsy roda *Carabus* Srednei Asii (The ground-beetles of the genus *Carabus* in Middle Asia). – Opredeliteli po faune SSSR, Moskow-Leningrad: Zool. inst. Akad. auk SSSR 52: 79-80pp.
Kutasi, Cs. (1998). Futóbogarak (Coleoptera, Carabidae) Litér környékéről. (Ground beetles (Coleptera: Carabidae) of the environment of Litér (West Hungary).– Folia Musei Historico-Naturalis Bakonyiensis 13 (1994): 73-87.
Kutasi, Cs., Markó, V. & Balog, A. (2004). Species Composition of Carabid (Coleoptera: Carabidae) Communities in Apple and Pear Orchards in Hungary. – Acta Phytopatologica et Entomologica Hungarica 3 (1-3): 71-78.
Kutasi, Cs., Szél, Gy., & Retezár, I. (2005). Species composition of ground beetle assemblages of dolomitic grasslands in Hungary. – In: XII European Carabidologists Meeting.

Ground beetles as a key group for biodiversity conservation studies in Europe. Murcia, Spain, 2005. 19-22 September. Abstracts of talks et posters (Serrano, J., Gómez-Zurita, J. et Ruiz, C., eds). Murcia, Spain, pp. 289-293.
Kutasi, Cs. & Szél, Gy. (2006). Ground beetle assemblages of dolomitic grasslands in Hungary. – Entomologica Fennica 17: 253-257.
Kuthy, D. (1896 [1897]). Coleoptera. – In: A Magyar Birodalom Allatvilaga (Fauna Regni Hungariae). [Fauna of the Hungarian Empire.] A K. M. Termeszettudomanyi Tarsulat, Budapest, pp. 1-213. (in Hungarian)
Lie, P. (1994). Neue Beiträge zur Kenntniss der Carabofauna des Rumänischen Banates für das Jahr 1993 (Coleoptera, Carabidae). – Folia entomologica hungarica 55: 225-232.
Lie, P. (1995). Beiträge zur Kenntniss des *Carabus hungaricus* frivaldskyanus Breuning neuentdeckt im Banat, Rumänien (Coleoptera, Carabidae). – Folia entomologica hungarica 56: 85-88.
Löbl, I. & Smetana, A. (Eds) 2003: Catalogue of Palaearctic Coleoptera. Volume 1. Archostemata – Myxophaga – Adephaga. Stenstrup, Denmark, Apollo Books 182 p
Majzlan, O. (1998). Chrobaky (Coleoptera) diluvia Peresa a Jurskeho Chlmu na juhu Slovenska. (The Beetle Species (Coleoptera) of the peres locality and Jurskeho Chlmu in South Slovakia.) Rosalia (Nitra) 13: 179-206 (in Slovak).
Majzlan, O. (2005). Bystruska juzna (*Carabus hungaricus*). p. 341. – In: Priaznivy stav biotopov a duhov europskeho vyznamu. Manual k programom starostlivosti o uzemia NATURA 2000 [The favurable state of the animal species and the biotopes of communitie interests. Manual for the favurable treatment programme of the Natura 2000 sites.] (Polak P. & Saxa A., eds). Statna ochrana prirody SR, Banska Bystrica [State protection of Nature in Banska Bystrica, Slovak Republic], p. 341. (in Slovak).
Merkl, O. (1991). Reassessment of the beetle fauna of Batorliget, NE Hungary (Coleoptera). – In: The Batorliget Nature Reserves after forty years, I. (Mahunka, S., ed.). Akademiai Kiado, Budapest, pp. 381-498.
Müller-Motzfeld, G. (Ed.) (2004). Band 2. Adephaga 1: Carabidae (Laufkäfer). – In: Die Käfer Mitteleuropas (Freude, H., Harde, K. W., Lohse, G. A., & Klausnitzer, B., eds). Spektrum-Verlag, Heidelberg-Berlin, 521 pp.
Narozsny, Z. (1938). Adatok Magyarorszag nagyfuto feleihez (Carabini). – Doktori ertekezes. [Data for Hungaries Carabini. Doctor thesis.] Debreceni szemle: 1-19. (in Hungarian)
Pavićević, D., & Mesaroš, G. (1997). Carabini of Jugoslavia and adjacent regions (Coleoptera: Carabidae). – Catalogus Fauna Jugoslaviae. Encyclopedia. Ecolibri-Bionet, Belgrade,CD-ROM.
Szel, Gy. (1985). A Carabus-genus Karpat-medenceben elo fajainak elterjedese es alfaji tagozodasa (Coleoptera: Carabidae). – Doktori ertekezes. [The distribution and subspecieses of the genus *Carabus* in the Carpathian Basin. Doctor thesis.] Budapest, 77 pp. (in Hungarian)
Szel, Gy., & Adam, L. (1992). Bogarkozossegek vizsgalata dolomitgyepekben (Coleoptera). (Examinations on beetle-communities in dolomitic grasslands) – Folia entomologica hungarica 52: 232-236. (in Hungarian)
Szel, Gy. & Berces, S. (2002). Carabidae (Coleoptera) from the Ferto-Hansag National Park. – In: The Fauna of the Ferto-Hansag National Park, II. (Mahunka, S., ed.). Magyar Termeszettudomanyi Muzeum, Budapest, pp. 379-399.

Tamas Julia (2001). A feketefenyvesek telepitese Magyarorszagon, kulonos tekintettel a dolomitkoparokra (Austrian pine plantations in Hungary with special attention to dolomite hills) Termeszetvedelmi Kozlemenyek 9: 75-85, 2001. (in Hungarian)

Toth, L. (1973). A Bakony hegyseg futobogar-alkatu faunajanak alapvetese (Coleoptera: Cicindelidae et Carabidae). [A monographie of the cabaoid beetles of the Bakony MoutainsG – A Veszprem megyei Mueumok Kozlemenyei 12: 275-351. (in Hungarian)

Turin, H., Penev, L. & Casale, A. (eds.) (2003).The Genus *Carabus* in Europe. – A Synthesis. Pensoft Publishers, Sofia-Moscow & European Invertebrate Survey, Leiden.

Neculiseanu, Z., Danila A, Cilipic G. (1999): Lista insectelor rare si amenintate cu disparitia din Republica Moldova. [List of rare insects in Moldova] http://www.salvaeco.org/insecte/page/carabus_pachistus_hungaricus.php (visited: 2007.12.19.)

Ivanenko, V. N. (Ed.) (1999): Endangered Animals of Russia: from knowledge to action http://www.nature.ok.ru/doc/nasekom/7_4.htm, (visited: 2007.12.19).

Serbaka, M.M. (1994). Червона книга України. Тваринний та рослинний світ [Red Book of Ukraine. Part I. Fauna]. Ministry of Ecology and Natural resources of Ukraine, Kiyv, Ukrainehttp://mail.menr.gov.ua/publ/redbook/redbook.php?lang=ukr&kingdom=1&class=5&ordo=25&fam=38&num=87, (visited: 2007.12.19).

L. Penev, T. Erwin & T. Assmann (Eds) 2008
BACK TO THE ROOTS AND BACK TO THE FUTURE. TOWARDS A NEW SYNTHESIS AMONGST TAXONOMIC, ECOLOGICAL AND BIOGEOGRAPHICAL APPROACHES IN CARABIDOLOGY
Proceedings of the XIII European Carabidologists Meeting, Blagoevgrad, August 20-24, 2007, pp. 373-384.

Carabidae as monitoring subject in the light of EU Natura 2000 (Habitats Directive)

Erik Arndt

Anhalt University of Applied Sciences, Department LOEL, Strenzfelder Allee 28, D-06406 Bernburg, Germany. E-mail: earndt@loel.hs-anhalt.de

SUMMARY

The European nature conservation network Natura 2000 consists of the Habitats and Birds Directives. The importance of the Habitats Directive for applied carabidology is demonstrated here. The Habitats Directive contains annexes listing protected habitat types as well as protected species. Both habitat types of Annex I and species of Annex II require the designation of special areas of conservation and special surveillance and/ or management programs. The work with ground beetles concerns on one hand the surveillance and conservation of 9 European carabid species listed in Annex II. On the other hand ground beetles can be used as indicator group in the monitoring program of habitats listed in Annex I. Finally, an example is given of habitat monitoring using Carabidae as indicator group in Germany. Criteria for evaluation of the species' conservation status are demonstrated.

Keywords: Natura 2000, Habitats Directive, *Carabus menetriesi*, monitoring, conservation status

INTRODUCTION

The loss of natural environment, destruction of habitats, fragmentation of populations and decrease of biodiversity are serious ecological problems of our time. Carabidologists are directly concerned by these processes as increasing Red Lists reflect.

The European Union created the Natura 2000 network to counteract the loss of biodiversity and natural habitats. Natura 2000 forms the main legal framework to protect nature and biodiversity at the EU level. It is the legal basis for

- the establishment of an European network of protected areas
- provisions for the identification and designation of individual sites
- the use of measures to promote connectivity between sites and overall coherence (= network).

Natura 2000 consists of the Habitats and Birds directives. The Birds directive (EC 1979) regulates the conservation of wild living birds as already its name implies. Entomologists are especially concerned with the Habitats Directive (Directive 92/43/EEC, EC 1992)*.

It is the aim of this contribution to demonstrate the importance of the EU Natura 2000 network for applied carabidology. The subject should be of special interest for colleagues who are residents of the "new" EU Member States. Following questions are in the main focus of the contribution: What is the aim of the Habitats Directive? Why is Annex II of Habitats Directive of any interest for carabidologists? Is there any intersection point between ground beetles and protected habitat types (listed in Annex I of Habitats Directive)?

AIM AND CONTENTS OF THE HABITATS DIRECTIVE

The main aims of the Directive 92/43/EEC are:

- to ensure biodiversity through the conservation of natural habitats and of wild fauna and flora in the European territory of the Member States
- to maintain or restore natural habitats and species of community interest at favourable conservation status
- to take account of economic, social and cultural requirements and regional and local characteristics.

Therefore the Habitats Directive should lead to a coherent European ecological network of special areas of conservation under the title "Natura 2000". This network is composed of sites hosting "species of community interest" as well as habitat types. These sites were proposed by the EU Member States as so-called pSCI (proposed Sites of Community Interest). The proposals are evaluated by the EU and are confirmed if applicable as SCIs. After this procedure the Member States must declare the sites as special areas of conservation (SAC). The total area of SACs should come to around 10% of the area of each Member State. At the moment it is between 7.4 % (France) and more than 20% (Greece, Denmark).

* Further information is actually available under: http://ec.europa.eu/environment/nature/nature_conservation/index_en.htm

In preparation of the Habitats Directive EU Member States proposed a variety of natural habitats and a number of endangered species that are listed in several annexes of the Habitats Directive*:

- *Annex I* contains *Natural Habitat Types* of community interest whose conservation requires the designation of special areas of conservation.
- *Annex II* contains *Animal and Plant Species* of community interest whose conservation requires the designation of special areas of conservation.

Species are also listed in Annexes IV (species in need of strict protection) and V (species whose exploitation may be subject to management measures) of the Habitats Directive. But whether these are not carabids or they are already included in Annex II. While working with ground beetles we have two points of contact with the directive: threatened species *and* threatened habitats. The Habitats Directive defines a "favourable conservation status" (FCS) both for *habitats and species.*

SPECIES OF ANNEX II

Annex II contains a comparably small number of species. By far not all endangered species of community interest are included. There were listed species which are not only of conservational importance but play also an important role as bio-indicators. Record of a species listed in Annex II in a certain EU Member State means that the country is obligated to declare special areas of conservation and to conduct a surveillance of each species.

The "Conservation status" of a species means the sum of influences acting on the species concerned that may affect the long-term distribution and abundance of its populations within the territory. The conservation status will be taken as "favourable" (FCS) when:

- population dynamics data on the species concerned indicate that it is maintaining itself on a long-term basis as a viable component of its natural habitats, and
- the natural range of the species is neither being reduced nor is likely to be reduced for the foreseeable future, and
- there is, and will probably continue to be, a sufficiently large habitat to maintain its populations on a long-term basis (Article 1(i) of Habitats Directive, EC 1992).

Based of proposals of the 27 EU Member States, 869 species are listed in Annex II, including 297 animal species and 38 Coleoptera with the following Carabidae (in the version of February 2007):

***Carabus (Morphocarabus) hampei* Küster, 1846**

- Distribution: NE Hungary, Romania; outside EU: Ukraine.
- Habitat: Various open and half open habitats.
- Remarks: Almost no information available about ecology and endangerment.

* Further information and complete lists are available under: http://ec.europa.eu/environment/nature/nature_conservation/eu_nature_legislation/habitats_directive/index_en.htm

***Carabus (Morphocarabus) zawadszkii* Kraatz, 1854**
- Distribution: Slovakia, Hungary, Romania; outside EU: Ukraine.
- Habitat: Forests from lowland to mountain level.
- Remarks: Almost no information available about ecology and endangerment.

****Carabus (s.str.) menetriesi pacholei* Sokolar, 1911***
- Distribution: Subspecies in South-East Germany, Austria, Czechia, Slovakia (see Farkač & Hůrka, 2005; Müller-Kroehling, 2005; Pawłowski, 2005; Zulka & Paill, 2005, and other contributions in "Angewandte Carabidologie", supplement vol. IV, 2005).
- Habitat: Character species of wooded *Sphagnum* bogs (of Central European mountains).
- Remarks: Threatened by habitat loss, isolation of populations, entomologists. Taxonomical status of some populations (e.g. in NE Germany) still unclear. These populations do not fall into the scope of the directive, if regarded as nominate subspecies!

***Carabus (Pachystus) hungaricus* Fabricius, 1792**
- Distribution: Disjunct areal, Austria; Czechia, Slovakia, Hungary, Romania; also in Eastern Europe between Ukraine and Caspian Sea.
- Habitat (in southeastern part of Central Europe): Character species of xerophilous steppes.
- Remarks: Endangered by ploughing or loss of pasture habitats and replacement by woodland (Franz, 1983; Kutasi & Szel, 2006; Turin et al., 2003).

***Carabus (Hygrocarabus) variolosus* Fabricius, 1787**
- Distribution: Czechia, Slovakia, Hungary, Romania; Bulgaria; also in some Eastern European and Balkan states; the subspecies [or species?] *nodulosus* Creutzer, 1799 in Germany, Austria; extremely endangered in France, probably extinct in Switzerland and Italy (Turin et al., 2003).
- Habitat: Character species of native forests, e.g. beach forests with dead wood, small ponds and brooks; hygrophilous and semiaquatic (Matern et al., 2007; Turin et al., 2003).
- Remarks: Indicator of undisturbed forest brooks; endangered by habitat destruction. Partly threatened by entomologists (over-collecting). The species was proposed for Annex II by the new EU Member States. However, *C. nodulosus* is even stronger endangered in Central Europe, than *C. variolosus* in Eastern Europe. If both taxa regarded as separate species (as actually done, e.g. by Turin et al., 2003; Arndt & Trautner, 2006) *C. nodulosus* is excluded from Habitats Directive conservation efforts. This is an actual political problem in the discussion concerning the Habitats Directive in Central European countries (Matern & Assmann, 2005; Müller-Kroehling, 2006).

****Carabus (Chrysocarabus) olympiae* Sella, 1855**
- Distribution: Endemic in western Italian Alps.
- Habitat: *Fagus* forests, *Rhododendron* shrubs and subalpine pastures (Negro et al., 2007).

* Priority species, which means "species of which the EC has particular responsibility in view of the proportion of their natural range which falls within the territory of EU" (EC, 1992).

- Remarks: Probably a glacial relict, extremely vulnerable. Threatened by forest cleaning and entomologists (over-collecting). Object of reintroduction projects in western Alps (Turin et al., 2003).

***Duvalius gebhardti* (Bokor, 1926)**
- Distribution: Hungary
- Habitat: Caves.
- Remarks: Very rare species, occurring very locally.

***Duvalius hungaricus* (Csiki, 1903)**
- Distribution: Slovakia, Hungary
- Habitat: Caves.
- Remarks: Very rare species, occurring very locally.

***Rhysodes sulcatus* (Fabricius, 1787)**
- Distribution: From France to Romania in 7 EU Member States, also on Balkan, in Ukraine and Russia.
- Habitat: Character species of natural broad leaved forests. Occurring in rotten wood.
- Remarks: Surprisingly less known ecology for a Central and West European species. Decline of populations in whole distribution area probably because of forest cleaning.

The following species are probably proposed by Turkey for Annexes II or IV (Schnitter pers. comm.):

***Megacephala euphratica* Latreille & Dejean, 1822**
- Distribution: Spain, Greece (Rhodes, Crete), Cyprus, Turkey to Central Asia; also in parts of North Africa and Arabian Peninsula.
- Habitat: Character species of salt marshes and salt meadows.
- Remarks: Threatened by agriculture (loss of saline habitats).

***Cephalota (Taenidia) eiselti* (Mandl, 1967)**
- Distribution: Endemic species of Turkey.
- Habitat: Inland salt meadows.
- Remarks: Endangered by cattle grazing and loss of salt marshes.

Turkish species of ***Carabus*** subgenus ***Procerus*** (there are differing classifications concerning subspecies and species, see Cavazutti, 1989; Turin et al., 2003):

***C. (Procerus) scabrosus* Olivier, 1795**
- Distribution: Greece, Bulgaria, Turkey, Ukraine and Black Sea coast of Russia, Caucaus region.

- Habitat: Various habitats, e.g. forests and macchia, from lowland to montane regions (1.300m).
- Remarks: Factors of endangerment in Turkey less known.

***C. (Procerus) syriacus* Kollar, 1843**
- Distribution: Turkey and states in Near East.
- Habitat: Various habitats, e.g. beech forests, macchia, pastures; from lowland to alpine regions (2.000m).
- Remarks: Factors of endangerment in Turkey less known.

Monitoring of Annex II species

Articles 1 and 11 of Habitats Directive obligate Member States to undertake surveillance of the conservation status of the species with particular regard to priority species.

The conservation status can be defined favourable when

a) population dynamics data of the species concerned indicate that it is maintaining itself on a long-term basis as a viable component of its natural habitats;
b) the natural range of the species is neither being reduced nor is likely to be reduced for the foreseeable future;
c) there is, and will probably continue to be, a sufficiently large habitat to maintain its populations on a long-term basis.

"This definition contains the main parameters (population dynamics, range, sufficient habitat, prospects of long-term viability) for defining and assessing both the current and target conservation status. It also provides a framework for more specific definitions on a species-by-species basis. All these parameters therefore need to be considered thoroughly when designing measures for a certain species. It is important to note that the assessment of conservation status not only includes an element of 'diagnosis' based on current conditions, but also an important element of 'prognosis' (foreseeable future) based on influences. Such foreseeable future influences could be specific or general threats, positive or negative, medium- to long-term impacts, etc.

The concept of FCS is not limited to the Natura 2000 network or to the species protected by this network (i.e. Annex II species). It applies to the *overall situation* of all species of community interest (Annexes II, IV and V), which needs to be assessed and surveyed in order to judge whether it is favourable or not. Assessing and evaluating the conservation status of habitats and species *within* the Natura 2000 network is therefore not always enough, especially when the occurrences of habitats or species are only partly covered by the network, maybe even in some cases only to a relatively small extent." (EC, 2007)

Methods or extent of the surveillance (respectively assessment, monitoring) are given by EC only in a very wide frame (EC, 2005a). Therefore each Member State must develop its own detailed program.

Example case: Monitoring of Annex II species in Germany

Carabus menetriesi pacholei is the only Annex II ground beetle species occurring in Germany. The methods of examination as well as criteria for the conservation status of *C. menetriesi* were defined by Schnitter et al. (2006). There are four possible levels of the conservation status: A (excellent), B (good), C (not favourable), and D (extremely low; habitat destroyed or population extinct). Pitfall traps shall be used as sample method for first examination of habitats and traps for living specimens in succeeding examinations. The criteria for the conservation status of *C. menetriesi* are given Table 1.

Table 1. Criteria for evaluation of the conservation status – the example *Carabus menetriesi* (acc. to Schnitter et al., 2006)

Population	**A (excellent)**	**B (good)**	**C (medium to bad)**
Size of habitat complex	>100 ha	20-100 ha	<20 ha
Abundance	high (records in nearly all research sites)	medium (records in majority of sites)	low (records in few sites)
Habitat			
Naturalness of habitat	mostly natural	slightly changed (fenn character still distinct)	strongly changed towards heathland or high forest
Insolation	halfshadow, tree coverage 0.2-0.7	halfshadow, tree coverage 0.2-0.7	open areas or dark shadow
Size of favourable habitat sites	large (>40 ha)	medium	low (<20 ha)
Impairments			
Water budget (primary factor!)	wet to extremely wet	moist	moderately moist or surface dry
Drainage ditches	lacking	old ditches, moderate incline	active ditches
Nutrient budget (indicator plants)	extremely low nutrient content	low nutrient content	moderately low nutrient content
Occurrence of other *Carabus* spp (indicating habitat changes)	less than 5 other Carabidae, no *Carabus* spp	other *Carabus* spp occur regularly	ubiquitous *Carabus* spp occur in larger abundances
Endangerments (by collecting specimens)	no indication	illegal pitfall traps recorded first time	illegal pitfall traps several times/often

Every six years Member States shall draw up a report including in particular (acc. Article 17 of Habitats Directive):

- the main results of the surveillance,
- information concerning the conservation status of the species in Annex II.

HABITATS OF ANNEX I AND CHANCES OF HABITAT MONITORING FOR APPLIED CARABIDOLOGY

There are listed 217 natural and semi-natural habitat types in Annex I from the Boreal biogeographic region in the North to the Mediterranean in the South and from the Macaronesian in the West to the Pannonian in the East*.

Analogous to species of Annex II, there is required a surveillance of habitats and a report on their conservation status.

Monitoring of Annex I habitats

Member States shall undertake surveillance (= monitoring) of the conservation status of the natural habitats with particular regard to priority natural habitat types. Priority natural habitat types means natural habitat types in danger of disappearance, which are present on the territory referred to in Article 2 and for the conservation of which the community has particular responsibility in view of the proportion of their natural range which falls within the EU territory. These priority natural habitat types are indicated by an asterisk (*) in Annex I (EC, 2007).

The conservation status of a natural habitat is considered "favourable" when

a) its natural range and areas it covers within that range are stable or increasing;
b) the specific structure and functions which are necessary for its long-term maintenance exist and are likely to continue to exist for the foreseeable future;
c) the conservation status of its typical species is favourable (EC, 1992).

The monitoring of habitats and their management should guarantee that their conservation status does not decrease. Type and intensity of surveillance (= monitoring) of the protected habitats shall be determined by each Member State. As for species, requirements for assessment and monitoring are given by EC only in a very wide frame (EC, 2005a).

Every six years Member States shall draw up a report including in particular:

- the main results of the surveillance,
- information concerning the conservation status of natural habitats.

Possibilities to include ground beetles in the habitat monitoring: The example of Saxony Free State in Germany

The monitoring concept of Annex I habitats in Saxony (Germany) was developed by Richter & Arndt (unpubl. research report 2002). The vegetation, but also animal groups

* Further information and complete lists are available under: http://ec.europa.eu/environment/nature/nature_ conservation/eu_nature_legislation/habitats_directive/index_en.htm

like birds, ground beetles, and in some habitats butterflies, dragonflies or Orthoptera were included as indicators for the conservation status. Carabidae was accepted as an indicator group for 30 terrestrial habitat types by the Saxonian government (examples are given in Table 2).

All in all there are 270 SCI covering 47 different habitat types in Saxony (total area of 509 km^2). Ground beetles were examined in ca. 230 monitoring sites covering 30 habitat types.

For the evaluation of ground beetle monitoring there was developed a specific evaluation system based on German Industrial Standard DIN 38410 including 'positive-' and 'negative indicators'. The indicator species were classified using the species indicator system (Dufrêne & Legendre, 1997) evaluating samples of more than 300 sites of conservation status A (excellent), B (good) and C (not favourable). Habitat surveillance as well as establishment of management plans for the SACs were financed by EU. Several entomologists could profit by these measurements.

Table 2. Examples of habitat types in the Saxony Free State (Germany) listed in Annex I. Ground beetles were included in the monitoring program as well as management plans of these habitats. Dominance of "character species" and lack of "negative indicators" may indicate a favourable conservation status (Richter & Arndt, unpubl. research report, 2005). Note that the indicator value may be restricted to the region of Saxony and not equally valid in other European regions.

EU-Code	Habitat type	Area in Saxony [ha]	Number of SCIs	Sites in monitoring	Character species (selection)	Negative indicators (selection)
2310	Dry sand heaths with *Calluna* and *Genista*	194	8	**9**	*Calathus erratus, Harpalus rufipalpis, Cymindis angularis*	*Anchomenus dorsalis, Carabus nemoralis, Pseudophonus rufipes*
4010	Northern Atlantic wet heaths with *Erica tetralix*	29	12	**9**	*Carabus nitens, Pterostichus rhaeticus, Bembidion mannerheimi*	*Amara bifrons , A. similata, Anchomenus dorsalis, Anisodactylus binotatus*
7110	* Active raised bogs	13	7	**6**	*Carabus menetriesi, Agonum ericeti, Agonum gracile*	*Amara aenea, A. communis, A. familiaris, Poecilus cupreus*
9110	*Luzulo-Fagetum* beech forests	4957	131	**60**	*Cychrus attenuatus, Pterostichus burmeisteri*	*Amara* spp *(except A. convexior), Poecilus versicolor*
9170	*Galio-Carpinetum* oak-hornbeam forests	2602	90	**38**	*Calosoma inquisitor, Abax ovalis*	*Amara aulica, A. similata, Badister bullatus, Poecilus versicolor, Pseudophonus rufipes*

9180	* *Tilio-Acerion* forests of slopes, screes and ravines	1445	157	**48**	*Carabus intricatus*	
91D4	* *Picea* Bog woodland	143	20	**10**	*Bembidion humerale, Pterostichus diligens, Carabus auronitens*	*Amara aulica, A. similata, Badister bullatus, Calathus fuscipes, Poecilus versicolor, Pseudophonus rufipes*
91F0	Riparian mixed forests of *Quercus, Ulmus, Fraxinus* along the great rivers (*Ulmenion minoris*)	1074	24	**24**	*Patrobus atrorufus, Limodromus assimilis, Pterostichus anthracinus*	*Amara aenea, A. lunicollis, A. similata, Anisodactylus binotatus, Calathus fuscipes, Poecilus versicolor, Pseudophonus rufipes*
9410	Acidophilous *Picea* forests of the montane to alpine levels (*Vaccinio-Piceetea*)	668	19	**7**	*Carabus linnei, C. silvestris; Pterostichus niger, Trechus splendens*	*Amara* spp, *Harpalus* spp

* priority habitat type!; SCI: Sites of Community Interest;

CONCLUSIONS

The Habitats Directive is a complex European conservation measurement. Its realisation brings money and jobs also for entomologists, especially because the measures are co-financed by the EU. Every Member State is obligated to realise a surveillance of Annex II species and their populations as well as Annex I habitats. The EU states must select suitable indicators for this monitoring. The discussion about type and extant of the monitoring has been started in several or all Member States yet. Every 6 years the Member States must send a report about the results of the surveillance to the EU.

Member States and the Commission encouraged the necessary research and scientific work having regard to these species (Article 18 of Habitats Directive, EC, 1992). Financial support is distributed by the competent authorities of the Member States and co-financed by the EU up to 85% for realisation of Natura 2000 tasks and projects (EC, 2005b). There are several possibilities including projects in the scope of 7th EU framework program or life projects at international level as well as co-financed projects (e.g. management plans) at regional level*.

* Several documents regarding financing are provided on the EU web site: http://ec.europa.eu/environment/ nature/nature_conservation/natura_2000_network/financing_natura_2000/index_en.htm.

ACKNOWLEDGEMENTS

I want to thank all the colleagues who provided me information about monitoring standards, endangered species and habitats, in particular Peer Schnitter (Halle, Germany) and Wolfgang Paill (Vienna, Austria). P. Schnitter kindly took over the proof reading of the MS.

REFERENCES

Arndt, E. & Trautner, J. (2006). Carabini, Cychrini. – In: Die Käfer Mitteleuropas. Band 2, Adephaga 1. Carabidae (Laufkäfer) (Freude, H., Harde K.W., Lohse, G.A. & Klausnitzer, B., eds) – Spektrum-Verlag, Heidelberg/Berlin, p. 28-63.

Cavazzuti, P. (1989). Monografia del genere *Procerus* (Coleoptera, Carabidae, Carabini). – Edizione L'Artistica Savigliano, Mem. Ass. Natur. Piem. I, 200pp.

Dufrêne, M. & Legendre, P. (1997). Species assemblages and indicator species: the need for a flexible asymmetrical approach. – Ecological monographs 67: 345-366.

European Commission (1979). Council Directive 79/409/EEC of 2 April 1979 on the conservation of wild birds. – Official Journal L 103, 25.4.1979: 1–18.

European Commission (1992). Council Directive 92/43/EEC of 21 May 1992 on the conservation of natural habitats and of wild fauna and flora. – Official Journal L 206, 22/07/1992: 7–50.

European Commission (2000). Managing NATURA 2000 Sites. The provisions of Article 6 of the 'Habitats' Directive 92/43/CEE. – Office for Official Publications of the European Communities, Luxembourg .

European Commission (2005a). Assessment, monitoring and reporting of conservation status. Preparing the 2001-2007 report under Article 17 of the Habitats Directive (DocHab-04-03/03 rev. 3). http://forum.europa.eu.int/Public/irc/env/monnat/library?1/=/committeesswokingsgroup/habitatsscommitteessswg/reporting_framework&vm=detailed&sb=Title. May 05, 2005. [COULDN'T FIND THE RESOURCE CITED]

European Commission (2005b). Financing Natura 2000. Guidance Handbook. Official Journal reference: S 73-070009.

European Commission (2007). Guidance document on the strict protection of animal species of Community interest under the Habitats Directive 92/43/EEC. February 2007, 88pp. http://circa.europa.eu/Public/irc/env/species_protection/library?l=/commission_guidance/final-completepdf/_EN_1.0_&a=d. [COULD NOT LOCATE RESOURCE: "Current path '/commission_guidance/final-completepdf/_EN_1.0_' does not correspond to an existing library item"]

Farkač, J. & Hůrka, K. (2005). *Carabus menetriesi* in der Tschechischen und in der Slowakischen Republik. – Angewandte Carabidologie", suppl. IV: 29-34.

Franz, H. (1983). Rote Liste der in Österreich gefährdeten Käferarten (Coleoptera). Hauptteil. – In: Rote Liste gefährdeter Tiere Österreichs (Gepp, J., ed.). – Bundesministerium für Gesundheit und Umweltschutz, Wien, p. 9-92.

Kutasi, C. & Szel, G. (2006). Ground beetle assemblages of dolomitic grasslands in Hungary. – Entomologica Fennica 17: 253-257.

Matern, A., Drees, C., Kleinwächter, M. & Assmann, T. (2007): Habitat modelling for the conservation of the rare ground beetle species *Carabus variolosus* (Coleoptera, Carabidae) in the riparian zones of headwaters. – Biological Conservation 136: 618-627.

Matern, A. & Assmann, T. (2005). Nationale Verantwortlichkeit und Rote Listen – *Carabus nodulosus* als Fallbeispiel für die Zusammenführung von Verbreitungsdaten und Gefährdungssituation und die damit verbundenen Probleme. – Nat.schutz Biol. Vielfalt 8: 235-254.

Müller-Kroehling, S. (2005). Verbreitung, Habitatbindung und Lebensraumansprüche der prioritären FFH-Anhang II-Art *Carabus menetriesi pacholei* Sokolar, 1911 (*bohemicus* Tanzer 1934) (Böhmischer Hochschmoorlaufkäfer) in Ostbayern und Überlegungen zu ihrem Schutz. – Angewandte Carabidologie, suppl. IV: 65-86.

Müller-Kroehling, S. (2006). Ist der Gruben-Laufkäfer *Carabus* (*variolosus*) *nodulosus* ein Taxon des Anhangs II der FFH-Richtlinie in Deutschland? – Waldökologie online 3: 57-62.

Negro, M., Casale, A., Migliore, L., Palestrini, C. & Rolando, A. (2007). The effect of local anthropogenic habitat heterogeneity on assemblages of carabids (Coleoptera, Caraboidea) endemic to the Alps. – Biodivers. Conserv. 16: 3919–3932.

Pawłowski, J. (2005). The *Carabus menetriesi* Hummel, 1827 (Coleoptera, Carabidae), a postglacial (and glacial or may be preglacial?) relic in Poland and in adjancent countries in Central and Eastern Europe. – Angewandte Carabidologie, suppl. IV: 97-100.

Schnitter, P., Eichen, C., Ellwanger, G., Neukirchen, M. & Schröder, E. (2006). Empfehlungen für die Erfassung und Bewertung von Arten als Basis für das Monitoring nach Artikel 11 und 17 der FFH-Richtlinie in Deutschland. – Berichte des Landesamtes für Umweltschutz Sachsen-Anhalt (Halle), Sonderheft 2: 1-370pp.

Turin, H., Penev, L., Casale, A., Arndt, E., Assmann, T., Makarov, K., Mossakowski, D., Szél, G. & Weber, F. (2003). Chapter 5. Species accounts. – In: Turin, H., Penev, L. & Casale, A. (Eds). The genus *Carabus* in Europe. A synthesis. Pensoft, Sofia-Moscow, p. 151-286.

Zulka, K.-P. & Paill, W. (2005). *Carabus menetriesi pacholei* Sokolar, 1911 in Österreich (Coleoptera, Carabidae). – Angewandte Carabidologie, suppl. IV: 87-92.

L. Penev, T. Erwin & T. Assmann (Eds) 2008
BACK TO THE ROOTS AND BACK TO THE FUTURE. TOWARDS A NEW SYNTHESIS AMONGST TAXONOMIC, ECOLOGICAL AND BIOGEOGRAPHICAL APPROACHES IN CARABIDOLOGY
Proceedings of the XIII European Carabidologists Meeting, Blagoevgrad, August 20-24, 2007, pp. 385-395.

Correcting for detection biases in the pitfall trapping of ground beetles (Coleopetera, Carabidae)

Evan D. Esch, Joshua M. Jacobs, Colin Bergeron & John R. Spence

Department of Renewable Resources, University of Alberta. Edmonton, Alberta, Canada. 751 General Services Bldg. T6G 2H1. E-mail: eesch@ualberta.ca

SUMMARY

Effects of body size, temperature, sex and habitat complexity on the probability of capturing ground beetles (Coleoptera: Carabidae) in pitfall traps were assessed for 11 boreal forest carabid species using laboratory experiments. Body size and temperature were positively correlated with the probability of catch. The probability of capture was used to develop correction factors to reduce biases associated with these variables and improve the ability of pitfall traps to estimate abundance from activity measured in the field. Use of the correction factors increased the correlation between pitfall trap samples and known or estimated abundance of forest carabid populations in two sets of field data.

Keywords: pitfall traps, probability of capture, size, temperature, correction factor

INTRODUCTION

Since their first reported uses by Hertz (1927) and Barber (1931), pitfall traps have become the most widely used method of collecting ground beetles (Coleoptera: Carabidae). The relationship between activity as measured directly by pitfall trap samples and the underlying abundance of carabid populations has been studied extensively (Briggs, 1960; Mitchell, 1963; Adis, 1979; Müller, 1984; den Boer, 1986; Topping & Sunderland, 1992; Spence & Neimelä, 1994). It is well established that the probability of capturing any species in a pitfall trap is a function of both its abundance and activity (Tretzel, 1955; Thiele, 1977; Luff, 1982, 1986; Franke et al., 1988). The probability of capture varies with species (Halsall & Wratten, 1988) and probably depends on numerous factors. For example, size and sex of individuals will influence activity measured for any species (Halsall & Wratten, 1988; Benest, 1989). Environmental factors such as air temperature (Honêk, 1997; Raworth & Choi, 2001) and habitat structure (Greenslade, 1964) have also been demonstrated to influence the probability of capture.

Many authors have focused on the shortcomings of pitfall trapping (e.g., Adis, 1979) and these are well understood. Nonetheless and despite objections from some quarters, a huge literature about carabids and other arthropods is based on data from pitfall trapping (Dunn, 1989; Spence & Niemelä, 1994). It is probably legitimate to use activity data to compare relative abundances of carabids within species and habitats of similar complexity, but the common comparisons of activity among species can be less rigorously justified as reflections of abundance. A few authors have attempted to understand the shortcomings of pitfall data more deeply and correct for the problems (e.g., Raworth & Choi, 2001; Perner & Schueler, 2004). One way to address the detection biases associated with a trapping technique is to adjust the raw activity data for the differential probability of capture (MacKenzie & Kendall, 2002).

In this study we present a correction factor that appears to adjust pitfall trap data for 11 boreal carabid species to more accurately represent abundances. Our species-specific correction factors are based on manipulative laboratory experiments. We focused on four main factors: 1) size of individuals of a species; 2) their gender; 3) temperature of the environment, and 4) habitat complexity. The correction factors were tested using both field experiments and published comparisons of activity and density for the carabid species studied here (Spence & Niemelä 1994).

MATERIALS AND METHODS

Laboratory experiment. Ground beetles were collected in unbaited pitfall traps in late July and early August 2006 from the aspen-dominated forest at the George Lake Field Station (53°57'N, 114°06'W) operated by the University of Alberta ca. 100 km northwest of Edmonton, Alberta, Canada. One species, *Sericoda quadripunctata* (Degreer) was hand collected at the Ecosystem Management Emulating Natural Disturbance (EMEND) research site in Northern Alberta (56°46'N, 118°22'W). Beetles were stored in an incubator at ca. 9 °C under constant darkness for five to ten days before experiments in an attempt to counteract seasonal light and temperature cues that could alter species-specific activity patterns (Thiele, 1977). Different species and sexes were kept in separate waxed paper cups. Beetles were provided water, but not food.

Thirty-two mesocosms were created using white plastic tubs (30 x 60 x 30 cm), each fitted with a screened top. Each mesocosm was filled with a layer of clay (ca. 15 cm) topped by a layer of 30 year-old composted soil (ca. 3 cm). One plastic pitfall trap, 11.2 cm in diameter and 8.0 cm deep, was installed in the center of each mesocosm with its rim flush with the compost. Mesocosms were held on shelves in a climate-controlled room with a 12L:12D photoperiod.

To determine the probability of capture, trials of 5 individuals of the same species and sex were placed in a mesocosm with a small quantity of dry dog food (Alpo™ Chicken, Liver & Rice). Beetles were allowed to acclimatize in the mesocosms for 1 hr. before data were collected (Olvido, 1995). During the acclimatization period, the pitfall traps were sealed with a cardboard disk. After this period, the cardboard was removed and the screen lids were clamped firmly in place on the tubs. Mesocosms were left undisturbed for 24 hrs., after which the number of beetles in the pitfall trap was recorded.

Eleven species were used in the mesocosms: *S. quadripunctata*, *Agonum retractum* Leconte, *Calathus ingratus* Dejean, Synuchus impunctatus Say, *Pterostichus adstrictus* Eschsholtz, *Scaphinotus marginatus* Fischer, *Pterostichus melanarius* Illiger, *Carabus chamissonis* Fischer, *Carabus nemoralis* Mueller, *Carabus taedatus* Fabricius, and *Calosoma frigidum* Kirby. Mean body length ranged from 4.9 to 21.3 mm and was determined by averaging the body lengths of 5 individuals collected at George Lake or EMEND. To determine the effect of temperature, trials were conducted comparing capture rates among species at 5 temperatures: 15 °C, 17 °C, 19 °C, 21 °C and 26 °C. Temperature was measured using the median value of a max/min mercury thermometer (Taylor) in the climate-controlled room. The range did not fluctuate more than 4 °C and was clearly centered on the desired mean temperature. Two levels of habitat complexity were compared for 4 of the 11 species (*C. ingratus*, *S. impunctatus*, *P. adstrictus*, *P. melanarius*). Habitat complexity was increased by randomly adding 20 branches (ca. 15 cm long and ca. 2 cm in diameter) of trembling aspen (*Populus tremuloides* Michaux.) into the mesocosms. Equal replication for each treatment combination was not always possible (see Appendix 1 for details about the experimental combinations).

The probability of capture for a species was determined from the proportion of beetles in a mesocosm that were captured in repeated trials under a given set of conditions. Body size, temperature, sex, and habitat complexity were regressed against probability of capture using forward stepwise regression (SPSS, 2005).

Field Experiment. Carabids for the field experiment were collected from the boreal mixed-wood forest adjacent to the experimental enclosures. Unbaited pitfall traps were used to collect the beetles. The carabids were stored at ca. 9 °C under constant darkness in waxed paper cups. Beetles were given water and a small piece of moss for shelter.

Nine enclosures (1.0 x 1.0 x 0.4m) were constructed at the EMEND research site in summer 2005. Enclosures were made from oriented strand board (OSB) and penetrated the forest floor to at least the depth of the mineral soil. A screened top was placed over each enclosure using Velcro strips. Four pitfall traps were installed ca. 30cm from each corner and were filled with ca. 100mL of silicone-free ethylene gycol (GM Dex-Cool®). Traps were left operational with the covers on the enclosures for a year prior to the experiment to trap out resident beetle populations and allow the plants to recover from disturbance during construction.

In the summer of 2006, a total of 26 individuals representing 6 species (*A. retractum*, *C. ingratus*, *Stereocerus haematopus* (Dejean), *P. adstrictus*, *Pterostichus punctatissimus* Randall, and *C. chamissonis*; for abundances see Table 1) were simultaneously observed

Table 1. The number of beetles released into forest floor enclosures and recaptured 10 days later by pitfall traps. Probability of capture was calculated by Probability = 0.043 (mean body length (mm)) + 0.022 (14°C) – 0.468. Catch was corrected using N=I/P, where *N* is corrected catch, *I* is Beetles recaptured, and *P* is probability of capture.

Species	Body length (mm)	Beetles released	Beetles recaptured	Probability of capture	Corrected catch
A. retractum	6.66	27	0	0.13	0
C. ingratus	7.99	27	1	0.18	5
S. haematopus	10.26	81	10	0.28	36
P. adstrictus	11.80	45	5	0.35	14
P. punctatissimus	15.66	45	16	0.51	31
C. chamissonis	16.70	9	4	0.56	7

in each enclosure. Beetles were marked on the elytra with a small abrasion (Winder, 2004) and released into the enclosures by transferring them with the moss they were taking shelter in to the center of the enclosure. Pitfall traps were collected 10 days later and marked beetles were counted and identified. Beetle size was regressed against probability of capture (SPSS, 2005). The regression constants observed in the field were calculated from raw abundance and compared to those observed in the laboratory using a t-test (Zar, 1999).

Correction factors. The number of individuals in a sample *I* can be equated to the population size *N* multiplied by the probability of capture *P* (MacKenzie & Kendall, 2002).

$$I=NP$$

Therefore, the size of the population *N* can be estimated from the number of individuals in sample *I* divided by the probability of capture.

$$N = I/P$$

With pitfall traps, the probability of capture is a function of both activity and abundance (Tretzel, 1955; Thiele, 1977; Luff, 1982, 1986; Franke et al., 1988). If the probability of capture associated with species-specific activity patterns can be identified and predicted, it can be incorporated into the above equation to remove the biases normally associated with activity patterns. This correction factor could be used to increase the accuracy of estimates of abundance calculated from activity data collected with pitfall traps.

The probability of capture established in the mesocosms was substituted into the above expression to create a correction factor. This correction factor was applied to pitfall traps samples taken in the forest floor enclosures. Corrected and uncorrected pitfall trap samples were compared to the known abundances in the enclosures using a χ^2 test (SPSS, 2005). Corrected and uncorrected pitfall trap samples were then correlated with the known abundance of beetles of each species in the enclosures (Spearman's ρ, SPSS, 2005).

To further assess the efficacy of this correction factor it was applied to data from Spence & Neimelä (1994). Corrected and uncorrected data from pitfall trap samples were correlated with quadrat samples sorted by a litter washing technique described by these authors.

RESULTS

Probability of capture. Probability of capture increased with average beetle size for the 11 species observed. The relationship between probability of capture and body size is described by the regression equation: Probability = 0.044 (mean body length (mm)) – 0.022 (r^2=0.27, p<0.0001). The probability of capture also increased with temperature. When temperature was included into the stepwise regression the probability of capture was predicted more accurately: Probability = 0.043 (mean body length (mm)) + 0.022 (temperature (°C)) – 0.468 (r^2 = 0.35, p<0.0001). Habitat complexity (p=0.21) and sex (p=0.90) had no significant effect on the probability of capture.

Field Experiment. Pitfall traps in the forest floor enclosures recaptured 37 individuals at a mean temperature of 14 °C (range of variation during the experiment: 6-25 °C). The smallest of the six species (*A. retractum*) was the only species not recaptured. Pitfall trap catches significantly underestimated beetle abundance in the enclosures (DF=53, χ^2=169, α= 0.001).

The probability of capture in the forest floor enclosures increased with beetle size. The relationship between probability of capture in the field and body size is described by the regression equation: Probability = 0.043 (mean body length (mm))-0.32 (r^2= 0.30, p<0.0001). This regression equation has the same slope (t_{233}=0.33, p<0.01) but the y-intercept differed significantly (t_{233}=7.175, p$\geq$ 0.5) from that observed in the laboratory-generated relationship between length and probability of capture.

Correction factors. The expression used to correct pitfall trap samples in the forest floor enclosures for biases associated with body size and temperature was calculated as: N=I/0.043 (mean body length (mm)) + 0.022 (temperature (°C)) – 0.468. Abundances in the forest floor enclosures were more strongly correlated with the corrected catch (ρ=0.567, p<0.01) than the uncorrected catch (ρ =0.480, p<0.01) (Table 1). However, the corrected pitfall trap catch still underestimated the beetle abundances in the enclosures (DF=53, χ^2=137, p= 0.001).

Temperature data were not available for the re-analysis of data from Spence & Neimelä (1994). Therefore, the correction factor used to remove bias associated with beetle size was calculated as: N = I/0.044 (mean body length (mm)) – 0.022. The number of carabids collected from quadrat samples were more strongly correlated with the corrected pitfall trap catch (ρ=0.515, p=0.04) than the uncorrected pitfall trap catches (ρ=0.458, p=0.07). However, the correction factor did not always reduce the difference between pitfall trap catch and litter washes (Table 2).

Table 2. The number of beetles collected from litter samples and pitfall traps. Probability of capture was calculated by Probability = 0.043 (mean body length (mm)) – 0.022. Catch was corrected using N=I/P, where N is corrected catch, I is pitfall trap catch, and P is probability of capture. Body length, litter wash and pitfall trap catch data was taken from Spence & Neimelä (1994).

Species	Body length (mm)	Litter wash	Pitfall trap catch	Probability of capture	Corrected catch
T. apicalis	4.3	7	1	0.16	6
B. dorsale	5.1	1	0	0.20	0
B. nigrinus	5.3	1	0	0.21	0
A. sordens	5.9	0	1	0.23	4
A. retractum	6.7	138	20	0.27	75
A. gratiosum	7.6	0	1	0.30	3
E. americanus	7.8	0	1	0.31	3
C. ingratus	8.8	51	66	0.36	185
H. fulvilabris	9.7	9	2	0.40	5
P. foveocolis	10.1	1	0	0.41	0
P. adstrictus	10.9	41	309	0.45	692
P. decentis	11.2	45	79	0.46	172
S. marginatus	14.8	0	9	0.61	15
C. chammissonis	14.3	0	6	0.59	10
P. punctatantisimus	16.4	1	0	0.68	0
C. fridgidum	20.2	1	49	0.85	58

DISCUSSION

Correction factors. Pitfall trap samples do not always compare favorably with estimates of carabid densities (Briggs, 1960; Desender & Maelfait, 1986; Spence & Neimelä, 1994) and thus activity data must be used carefully when comparing species across habitats (Greenslade, 1964; Neimelä et al., 1990). Upon encountering a pitfall trap all species do not have an equal likelihood of being captured (Halsall & Wratten, 1988). This can have serious consequences for ecological studies that attempt to quantify beetle populations or communities (MacKenzie & Kendall, 2002). These proposed correction factors are extremely valuable tools because they correct for a substantial component of the bias associated with differential probability of capture and increase the utility of pitfall traps as a means to measure beetle abundances.

Both correction factors used in this experiment increased the correlation between pitfall trap samples and density of beetles in the field. The correction factor for body size alone explained less variance that the correction factor for both body size and temperature. However, a correction factor that includes fewer parameters may introduce less error into the estimate of abundance. Balancing the potential bias associated with pitfall trapping against the reduced precision associated with increasing the number of parameters considered is vital to the selection of most appropriate correction model (Skalski et al.,

1983). In addition, temperature may vary across microhabitats and change in relation to dial and seasonal activity patterns making it difficult to determine the appropriate temperature for adjustments. A further advantage of using the simpler of the two correction factors is that no additional data need be collected from the field and, thus, this could be applied to many existing data sets. Determining which correction factor is most appropriate will clearly depend on the objectives of the study.

Our correction factors are based on the assumption that carabid behavior observed in the laboratory accurately reflects carabid behavior in the wild. The differences in experimental design between laboratory and field experiments, such as density of beetles, species composition, trap number, trap arrangement, and temperature compromise our ability to compare rigorously the data between laboratory and field experiments. However, high congruence between probability regressions observed in the laboratory and field support the assumption that behaviors in the laboratory and field are comparable. The discrepancy between the y-intercepts the regressions observed in the laboratory and field are most likely due to lower temperature in the field, particularly at night. Nonetheless, we are confident in applying these correction factors to pitfall activity data for this assemblage of boreal carabids. Further work to assess the generality of these factors for boreal species from across the Palearctic would be most useful.

Body size. The increased likelihood of finding larger beetles in pitfall traps samples has been previously reported (Franke et al., 1988; Spence & Neimelä, 1994). Yet the exact mechanism underlying this relationship has not been established. It may result from the generally increased mobility of larger species (Luff, 1975; Thiele, 1977; den Boer, 1981; Frank et al., 1988) leading to higher probability of individuals encountering a trap. Others suggest that individuals of smaller-bodied species are better at escaping traps (Luff, 1975; Wagge, 1985). However, since the relationship between size and probability of capture observed in the laboratory was similar to that observed in the forest floor enclosures, where ethylene glycol was used as a preservative in the traps, it is unlikely that escape from the traps significantly affected our results. Halsall and Wratten (1988) claim that probability of capture is not strictly related to body size but is alternatively a function of the beetle's ability to perceive and regain balance once it has encountered the edge of a trap. Wallin and Ekbom (1994) also suggested that the likelihood of capture is unrelated to body size but is correlated with hunger and feeding behavior in carabids. Nonetheless, the value of a regression analysis can validly be assessed in terms of its ability to predict an outcome given certain criteria (Underwood, 1997), regardless of whether or not the underlying functional linkages have been firmly established.

Temperature. The relationship between temperature and probability of capture is likely a result of increased activity levels at warmer temperatures (Honêk, 1997). It is likely that this relationship will vary seasonally with the physiological state of the beetles (Spence & Niemelä, 1994; Raworth & Choi, 2001) and it deserves further study.

Sex. Sex had no effect on capture of carabids by pitfall traps. This is contrary to the findings of Benest (1989) and Raworth and Choi (2001). Our inability to detect a difference between the sexes in this study may have been influenced by the time of year carabids were captured and/or the use of a 12D:12L photoperiod. As a result, differences between sexes such as presence/absence of eggs (Grüm, 1984) or movement associated with mating strategies may have less importance than at other times of year when reproductive activity would be greatest. Further work to develop more effective correction factors must consider these possible relationships.

Habitat complexity. Interestingly, habitat complexity had no effect on the capture of carabids in pitfall traps. Previously, interactive effects between habitat and species have been observed to affect beetle movement (Greenslade, 1964; Evans, 1986). However, the simple manipulation of habitat complexity in the mesocosms may not have altered the behavior of the species considered here. Nonetheless, variation in the amount of fine woody debris from aspen is a common feature of the forest habitats used by the species that we considered and it is somewhat comforting to know that such variation does not significantly affect carabid catch.

Unexplained variance. A large component of the variance observed in the mesocosms was not explained by body size, temperature, sex, or habitat complexity. Factors such as hunger (Wallin & Ekbom, 1994), humidity (Honêk, 1997), seasonal activity patterns (Nelemans et al., 1989; Niemelä & Spence, 1992), and habitat preferences (Lindroth, 1961-67; Thiele, 1977) most likely account for the majority of this variance. Thus, there is plenty of scope for further studies following this general approach.

As with any data transformation or correction, how the transformation works and what it does to the data must be understood before it is applied. These corrections may be a valuable tool for removing a component of the bias associated with pitfall trapping. However, this does not mean that cautions, considerations, and recommendations developed over years of rigorous research should be forgotten. However, failure to explain everything satisfactorily should also not be used to suppress or abandon innovative experiments to improve our ability to interpret pitfall trap data. As expressed by the old Chinese proverb anglicized by Adlai Stevenson in praise of Eleanor Roosevelt, it is perhaps more useful to “light a candle than curse the darkness”.

ACKNOWLEDGEMENTS

We would like to thank S.E. Abele, M. Koivula, C.J.K. MacQuarrie, and C.M. Wood for their helpful contributions.

REFERENCES

Adis, J. (1979). Problems of interpreting arthropod sampling with pitfall traps. – Zoologischer Anzeiger 202: 177-184.

Benest, G. (1989). The sampling of a carabid community.I: The behavior of a carabid when facing a trap. – Revue d'écologie et de biologie du sol. 26: 205-211.

Barber, H.S. (1931). Traps for cave-inhabiting insects. – Journal of Elisha Mitchell Science Society 46: 259-266.

Briggs, J.B. (1960). A comparison of trapping and soil sampling in assessing populations of two species of ground beetles (Col.: Carabidae). – East Malling Research Station Annual Report 48: 108-112.

den Boer, P.J. (1981). On the survival of populations in a heterogeneous and variable environmental. – Oecologia (Berlin) 50: 39-53.

Desender, K. & Maelfait, J.-P. (1986). Pitfall trapping within enclosures: A method for estimating the relationship between the abundances of coexisting carabid species (Coleoptera: Carabidae). – Holarctic Ecology 9: 245-250.

Dunn, M. (1989). Bibliography of information on pitfall trapping. – Young Entomologists Association Quarterly 6: 41-42.

Evans, M.E.G. (1986). Carabid locomotor habits and adaptations. – In: Carabid Beeltes, their Adaptations and Dynamics (den Boer, P.J., M.L. Luff, D. Mossakowski & Weber, F., eds). Gustav Fischer, Stuttgart. p. 57-77.

Franke, U., Friebe, B. & Beck, L. (1988). Methodisches zur Ermittlund der Siedlungsdichte von Bodentiere aus Quadraproben und Barberfallen. – Pedobiologia 32: 253-264.

Greenslade, P.J.M. (1964). Pitfall trapping as a method for studying populations of Carabidae (Coleoptera). – Journal of Animal Ecology 13: 301-310.

Grüm, L. (1984). Carabid fecundity as affected by extrinsic and intrinsic factors. -Oecologia 65: 1432-1939.

Halsall, N.B. & Wratten, S.D. (1988). The efficiency of pitfall trapping of polyphagous predatory Carabidae. – Ecological Entomology 13: 293-299.

Hertz, M. (1927). Huomiota petokuorianisten olinpaikoista. – Luonnon Ystävä 31: 218-222.

Honêk, A. (1997). The effect of temperature on activity Carabidae (Coleoptera) in a fallow field. – European Journal of Entomology. 94: 97-104.

Lindroth, C.H. (1969). The ground-beetles of Canada and Alaska (Carabidae excl. Cicindelinae). – Opuscula Entomologica Supplenta. 20, 24, 26, 29, 33, 34, 35: 1-1192.

Luff, M.L. (1975). Some features influencing the efficiency of pitfall traps. – Oecologia (Berlin) 19: 345-357.

Luff, M.L. (1982). Population dynamics of Carabidae. – Annals of Applied Biology 101: 164-172.

Luff, M.L. (1986). Aggregation of some Carabidae in pitfall traps. – In: Carabid Beeltes, their Adaptations and Dynamics (den Boer, P.J., Luff, M.L., Mossakowski, D., and Weber, F., eds). Gustav Fischer, Stuttgart. p. 385-397.

MacKenzie, D.I. & Kendall, W.L. (2002). How should detection probability be incorporated into estimates of relative abundance? – Ecology 83: 2387-2393.

Mitchell, B. (1963). Ecology of two carabid beetles, *Bembidion lampros* (Herbst) and *Trechus quadristriatus* (Schrank). II. Studies on populations of adults in the field, with species references to the technique of pitfall trapping. – Journal of Animal Ecology 32: 377-392.

Müller, L.K. (1984). Die Beduetung der Fallenfang-Methode für die Lösung ökologischer Fragestellungen – Zoologische Jahrbücher Abteilung für Systematik, Ökologie und Geographie der Tiere 111: 281-305.

Nelemans, M.N.E., den Boer, P.J. & Spee, A. (1989). Recruitment and summer diapause in the dynamics of a population of *Nebria brevicolis* (Coleoptera: Carabidae) – Oikos 56: 157-169.

Neimelä, J., Halme, E., Pajunen, T. & Haila, Y. (1990). Balancing sampling effort in pitfall trapping of carabid beetles. – Entomologica Fennica 1: 233-238.

Olvido, A.E. & Mousseau, T.A. (1995). Effects of rearing environment on calling-song plasticity in the striped ground cricket. – Evolution 49: 1271-1277.

Perner, J. & Schueler, S. (2004). Estimating the density of ground-dwelling arthropods with pitfall traps using a nested-cross array. – Journal of Animal Ecology 73: 469-477.

Raworth, D.A. & Choi, M.-Y. (2001). Determining numbers of active carabid beetles per unit area from pitfall-trap data. – Entomologia Experimentalis et Applicata 98: 95-108.

Skalski, J.R., Robson, D.S. & Simmons, M.A. (1983). Comparative census procedures using single mark-recapture methods. – Ecolocgy 64: 752-760.

Spence, J.R. & Neimelä, J. (1994). Sampling carabid assemblages with pitfall traps: The madness and the method – Canadian Entomologist 126: 881-894.

SPSS Inc. (2005). SPSS® 14.0 for Windows. Chicago.

Thiele, H.-U. (1977). Carabid Beetles in their Environments – Springer – Verlag, Berlin.

Topping, C.J. & Sunderland, K.D. (1992). Limitations to the use of pitfall traps in ecological studies exemplified by a study of spiders in a field of winter wheat – Journal of Applied Ecology 29: 485-491.

Tretzel, E. (1955). Technik und Bedeutung des Fallenfanges für ökologische Untersuchungen – Zoologischer Anzeiger 155: 276-287.

Underwood, A.J. 1997: Experiments in Ecology: Their Logical Design and Interpretation Statistical Principals – Cambridge University Press, Cambridge.

Wallin, H. & Ekbom, B. (1994). Influences of hunger and prey densities on movement patterns in three species of *Pterostichus* beetles (Coleoptera: Carabidae). – Environmental Entomology 23: 1171-1181.

Wagge, B.E. (1985). Trapping efficiency of carabid beetles in glass and plastic pitfall traps containing different solutions – Fauna Norvegicum, Series 32: 33-36.

Winder, L. (2004). Marking by abrasion or branding and recapturing carabid beetles in studies of their movement – International Journal of Pest Management 50: 161-164.

Zar, J.H. (1999) Biostatistical Analysis: Prentice-Hall Inc., New Jersey.

Appendix 1. Number of trials run in the mesocosm for each treatment combination.

Species	Sex	Temperature °C					Total	Habitat complexity	
		15	17	19	21	26		Simple	Complex
S. quadripunctata	F	0	0	0	8	0	8	0	0
	M	0	0	0	4	0	4	0	0
A. retractum	F	3	6	9	0	0	18	0	0
	M	1	2	3	0	2	8	0	0
C. ingratus	F	3	5	12	0	8	28	5	6
	M	3	6	12	0	9	30	0	4
S. impunctatus	F	2	4	9	0	3	18	4	2
	M	2	4	3	0	2	11	0	2
P. adstrictus	F	4	8	15	0	6	33	4	4
	M	1	2	6	0	4	13	2	2
S. marginatus	F	1	2	3	0	4	10	0	0
	M	2	4	9	0	4	19	0	0
P. melanarius	F	3	6	0	0	6	15	6	6
	M	3	6	0	0	6	15	6	6
C. chamissonis	F	0	0	0	5	0	5	0	0
	M	0	0	0	4	0	4	0	0
C. nemoralis	F	0	0	3	0	2	5	0	0
	M	1	2	0	0	0	3	0	0
C. taedatus	F	2	4	3	0	2	11	0	0
	M	0	0	0	0	0	0	0	0
C. frigidum	F	0	0	0	0	0	0	0	0
	M	0	0	0	0	0	0	0	0
Total		31	61	87	21	58	258	27	32

L. Penev, T. Erwin & T. Assmann (Eds) 2008
BACK TO THE ROOTS AND BACK TO THE FUTURE. TOWARDS A NEW SYNTHESIS AMONGST TAXONOMIC, ECOLOGICAL AND BIOGEOGRAPHICAL APPROACHES IN CARABIDOLOGY
Proceedings of the XIII European Carabidologists Meeting, Blagoevgrad, August 20-24, 2007, pp. 397-408.

Sofia–Moscow

Towards combined methods for recording ground beetles: Pitfall traps, hand picking and sifting in Mediterranean habitats of Israel

Anika Timm[1], Tamar Dayan[2], Tal Levanony[2], David W. Wrase[3] & Thorsten Assmann[1]

[1] Institute of Ecology and Environmental Chemistry, Leuphana University Lueneburg, Scharnhorststr.1, 21314 Lueneburg, Germany. E-mail: timm@leuphana.de
[2] Department of Zoology, Tel Aviv University, Tel Aviv 69978, Israel
[3] Dunckerstr. 78, 10437 Berlin, Germany

SUMMARY

Three different methods (pitfall traps, hand picking and sifting) were used to sample ground beetles in a woodland site and a batha site in Upper Galilee, Israel. Better knowledge about the detectability of ground beetle species in Mediterranean habitats and the most suitable methods of collecting are important for studying the taxon in landscape planning or nature conservation research approaches. Rarefaction procedures show that pitfall traps alone catch only half of the species detected by a combination of methods. Some guilds are not (or only rarely) found in pitfall traps. Among these are the myrmecophile (e.g. *Paussus turcicus*, *Macrocheilus saulcyi*), inhabitants of the superficial underground compartment (*Zuphium numidicum* and a microphthalmic *Parazuphium* species), and some litter inhabiting species (e.g. *Metadromius carmelitanus*). The reasons for the different catchabilities are discussed. Generally, we recommend a combination of recording techniques to obtain a comprehensive overview of the diverse fauna of Mediterranean habitats.

Keywords: Upper Galilee, Israel, pitfall traps, hand picking, sifting

INTRODUCTION

Pitfall traps were first described by Barber (1931) and are today one of the standard methods used to study ground beetles. Since this initial description, the use of pitfall traps as a sampling method has been discussed extensively. Baars (1979) described the use of continuous pitfall sampling as an important method to measure the size of carabid populations. Other studies have proved that the number of animals trapped also depends on their epigeic activity (Andersen, 1995; Perner & Schueler, 2004). Many other factors that also influence the effectiveness of pitfall traps have been described in the literature (e.g. Adis, 1979; Heydemann, 1955). Halsall & Wratten (1988) even state that pitfall trap catches are only poorly related to population densities and that the size or speed of movement of the beetle plays no role in capture efficiency. See also Esch, et al., this volume.

Methods such as sifting, hand picking, light trapping or net sweeping can also be used to study ground beetle assemblages (e.g. Freude et al., 1965; White, 1983). Some of these techniques seem appropriate if ground beetles of the given habitats do not show locomotor activity on the surface. For this reason, we compare the efficiency of pitfall trapping, sifting and hand picking. Most previous studies on the efficiency of different methods were conducted in temperate climate regions of Europe and North America (e.g. Andersen, 1995; Prasifka et al., 2007). For the Mediterranean region, to date, there have been no studies which deal with the efficiency of pitfall traps and other methods of collecting ground beetle assemblages. We chose two different East-Mediterranean habitats to determine the efficiency of three different methods. One study site is an evergreen oak woodland and the other is an open grazed habitat, a so-called batha. Knowledge of the detectability of ground beetle species in Mediterranean habitats and the most suitable methods for collecting are important for studying the significance of this animal group in landscape planning and biological conservation research approaches. Ground beetles are used increasingly for nature conservation strategies. The Convention on the Conservation of European Wildlife and Natural Habitats of the European Union (e.g. Guenther & Assmann, 2004; Matern et al., 2007; Ssymank, 1998) or the Endangered Species Act of the United States of America (e.g. Mello, 2005; Talley et al., 2007).), for instance, explicitly protect habitats where ground beetle species are found. Prerequisites for nature conservation strategies are methods for obtaining comprehensive knowledge on the existing fauna. This is crucial for identifying changes and threats. Therefore, the main focus of our contribution is (1) to compare the efficiency of the methods in two widely distributed habitats of the East Mediterranean and (2) to derive from these results recommendations regarding monitoring of those types of habitats.

MATERIALS AND METHODS

The two study sites are located in Upper Galilee, North Israel, near the Lebanese border and close to the villages of Bar'am and Ziv'on. The sites are on terra rossa soils on hard limestone. Here, we studied the ground beetle fauna for one year using 10 pitfall traps per site (first opened 14 March 2005; closed 17 March 2006). The traps were filled with approximately 2 cm Renner (1980) liquid (30% ethanol, 20 % glycerol, 10% acetic acid) and emptied every second week. The sites were located on an open meadow (batha; geographic coordinates: N 033°01', E 035°25') with a dominance of *Sarcopotherium spinosum* and two *Cistus* species, and in a woodland (geographic coordinates: N 033°02', E 035°25') dominated by *Quercus calliprinos*. In both habitats, ground beetles were also collected by hand (mainly beneath stones). The collection activities took the two collectors seven hours in the woodland and six hours in the open meadow. In the woodland, ground beetles were also collected by sifting the litter layer and the top mineral horizons of the soil down to a depth of approximately 8 mm. In most cases, the sifted litter was examined on the same day (while it was still relatively humid), but in a few cases it was not examined until the next day. For examination, the sifted litter was sifted again with a mesh size of 4 mm onto a white sheet. Most of the beetles in the sifted litter were found as a result of their running activity. The last step was to examine the rest of the sifted litter for inactive and larger specimens (cf. Freude et al., 1965; White, 1983).

Rarefaction was used to study the efficiency of the different collection methods. This was performed using the online calculator by Brzustowski (http://www2.biology.ualberta.ca/jbrzusto/rarefact.php#Inputs) which is based on the program RAREFACT.FOR written by Charles J. Krebs. The nomenclature of the ground beetles follows Löbl & Smetana (2003) and Deuve (2004). A few species are either not known to science or could not be determined (see Table 1 for further details).

Table 1. Catches in woodland and batha

	Woodland			Batha		
	Pitfall Traps	**Sifting**	**Hand Picking**	**Pitfall Traps**	**Hand Picking**	**Total specimens**
Amara pumilio	.	.	.	.	13	13
Amblystomus cephalotes	.	.	2	.	.	2
Apotomus clypeonitens	.	.	.	.	3	3
Bembidion leucoscelis	.	.	1	.	.	1
Bembidion liliputanum	.	14	.	.	.	14
Bembidion phoeniceum	6	2	.	.	.	8
Broscus laevigatus	18	.	.	22	18	58
Broscus nobilis	6	.	.	3	6	15
Calathus cinctus	110	.	.	1	10	121
Calathus longicollis	658	.	6	50	42	756
Calathus mollis	.	.	5	.	.	5
Carabus impressus	8	.	2	58	12	80
Carabus phoenix	2	.	.	.	.	2
Carabus piochardi	2	.	.	37	1	40
Carabus sidonius	18	.	.	110	20	148

	Woodland			Batha		
	Pitfall Traps	**Sifting**	**Hand Picking**	**Pitfall Traps**	**Hand Picking**	**Total specimens**
Carabus syrus	11	.	.	83.	11	105
Carterus cribratus	.	.	2	.	9	11
Cymindis pallida	7	.	2	5	.	14
Cymindis spec.[1]	6	.	2	2	.	10
Eucarterus sparsutus	.	.	.	15	.	15
Harpalus caiphus	.	.	.	1	4	5
Laemostenus quadricollis	60	.	.	1	1	62
Leistus caucasicus	108	.	1	.	.	109
Macrocheilus saulcyi	.	.	.	.	5	5
Metadromius carmelitanus	.	89	.	.	.	89
Microdaccus pulchellus	.	.	.	.	4	4
Microlestes cf. apterus	.	.	.	.	2	2
Microlestes baudii	.	.	.	.	2	2
Microlestes maurus	5	.	.	15	21	41
Nebria hemprichi	66	.	7	.	.	73
Notiophilus danieli	3	.	.	3	2	8
Odotoncarus asiaticus	.	.	5	65	9	79
Odotoncarus samson	.	.	1	.	2	3
Olisthopus glabricollis	2	.	.	.	1	3
Ophonus puncticeps	.	.	.	.	2	2
Ophonus rufibarbis	.	.	.	.	1	1
Orthomus berytensis	.	.	.	6	.	6
Orthomus sidonicus	1614	.	16	.	.	1630
Parazuphium spec.[2]	.	.	.	.	1	1
Paussus turcicus	.	.	.	.	8	8
Philorhizus notatus	.	.	.	1	.	1
Platyderus spec.[3]	1	.	3	.	.	4
Platytarus reichei	.	.	.	.	7	7
Polyderis cardioderus	.	.	.	.	2	2
Pseudaristus punctatissimus	.	.	.	1	33	34
Scarites saxicola	.	.	.	73	5	78
Trechus crucifer	1	9	.	.	.	10
Trechus quadristriatus	.	.	.	.	3	3
Trechus saulcyanus	266	27	.	.	.	293
Zuphium numidium	.	.	.	.	2	2
Total	2978	141	55	552	262	3988
Total species	22	5	14	20	32	

[1] The determination of these beetles was impossible. We agree with Mateu 1956 that in the systematics and taxonomy of the *Cymindis axillaris* group only chaos exists.

[2] It is a microphthalmic *Parazuphium* species which has not been described yet.

[3] The determination of the Israeli *Platyderus* species is not possible for us. A revision of the south-European and southwest-Asian species is urgently necessary.

RESULTS

During the year cycle, 2978 specimens were caught in the woodland site and 552 ground beetles in the open meadow with pitfall traps; 196 and 226 specimens were captured by hand picking in the woodland and in the batha, respectively.

In the batha, 30 species were detected by hand picking and 16 by pitfall traps. Almost two-thirds of the species found by hand picking were not found in the pitfall traps, while five species were only caught in pitfall traps. On the woodland site, however, 22 species were found in pitfall traps, 14 by hand picking and five by sifting. Nearly half of the species found by sifting and hand picking were not caught in pitfall traps. Two species were found in the woodland exclusively by means of pitfall traps. At both sites, some species were caught using only one of the methods (Table 1).

The ant nest beetle *Paussus turcicus*, the helluonine ground beetle *Macrocheilus saulcyi*, the zuphiine ground beetles *Zuphium numidicum, Parazuphium spec.*, and some other species (e.g. *Apotomus clypeonitens, Microdaccus pulchellus*) are examples of species found exclusively by hand picking. While only a few specimens of these species were recorded, litter sifting revealed high densities of one species which was not found in pitfall traps: the lebiine *Metadromius carmelitanus*. We also found this litter layer inhabiting ground beetle in high population densities in other woodlands of northern Israel (e.g. several woodlands in Upper Galilee, the Carmel Mountains and the Golan Heights), but not in pitfall traps. *Carabus phoenix* and *Eucarterus sparsutus* are examples of species recorded exclusively from pitfall trapping.

The rarefaction curves (Figs 1a-b) as well as Table 1 indicate that the efficiency of the methods used varies, especially in relation to the number of species and specimens recorded and in time required. Hand picking was carried out for a few hours, and the number of specimens collected was relatively low, but the number of species detected by this method was high in comparison to pitfall trapping. Hand picking and sifting raised the total number of species considerably for each site.

DISCUSSION

Comparison of the methods

Using pitfall traps we caught only ground beetles which move on the ground. Species which inhabit the lower horizons of the soil or on the vegetation are only very seldom recorded. Moreover, the catch rate for epigeic species depends on the "environmental resistance" caused by the vegetation structure which obstructs the beetles in their locomotor activity (Heydemann, 1957). The physiological condition of the animals also has an effect on the intensity of locomotor activity and thus also on the catch rates (Chiverton, 1984). The most important methodological aspect of pitfall traps is that

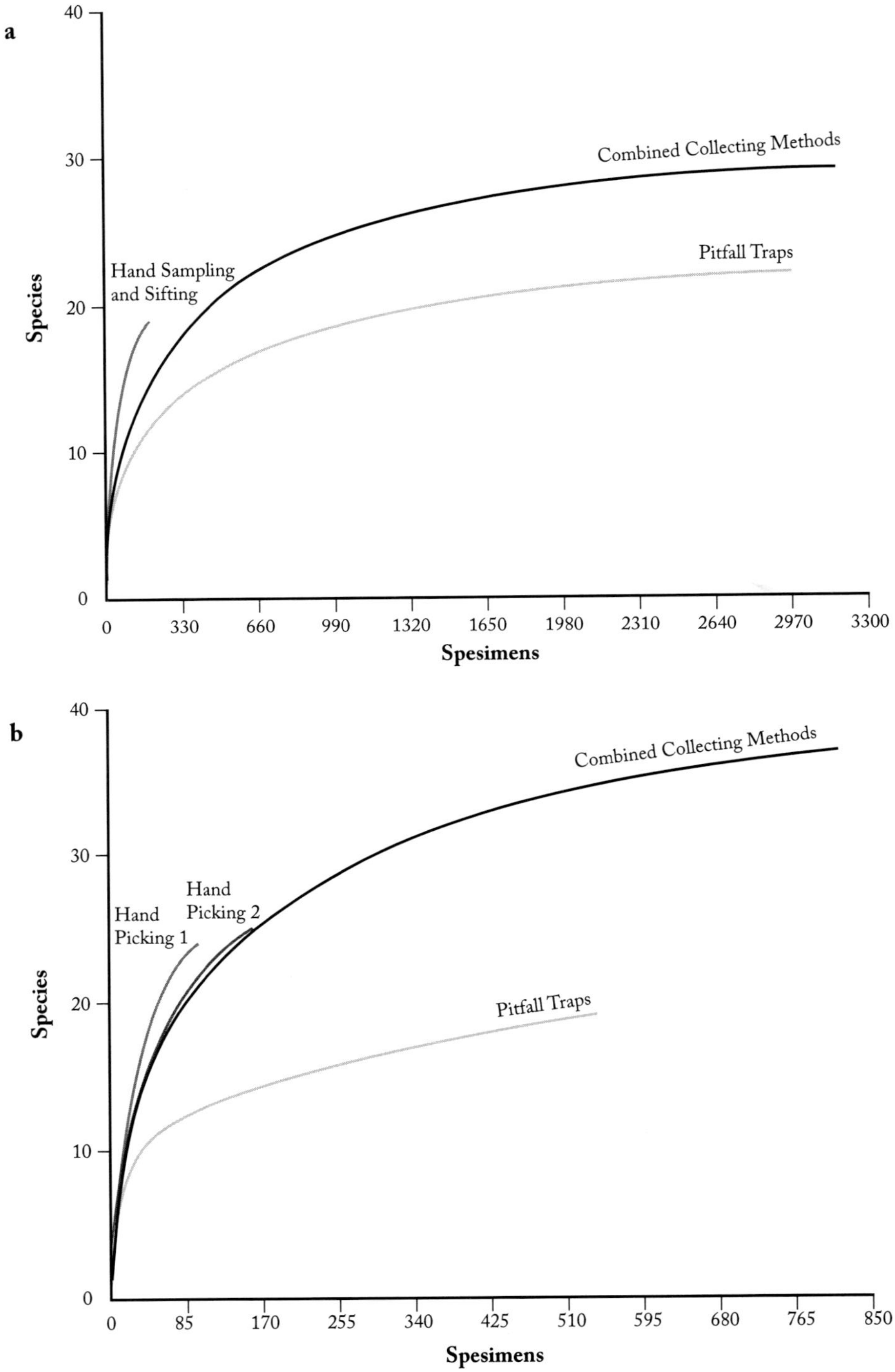

Fig. 1. Species richness (rarefaction) in the woodland site (1a; above) and the batha sites (1b; below). 'Hand Picking 1' and 'Hand Picking 2' refer to the results of two collectors.

they sample the carabid assemblage independent from the scientist. This is the central aspect which has to be questioned for the other methods we used (hand picking and litter sifting). The success of hand picking depends strongly on the experience of the collector. The species that were caught only by pitfall traps could, with more intensive work, also have been caught by hand. These species are mostly rare species, some of which occur only during specific seasons (e.g. *Eucarterus sparsutus* in summer) when no hand picking took place. Furthermore, the beetle findings depend on the collector's subjective view and also on the point of collection. *Ophonus puncticeps* and *O. rufibarbis* as well as *Parazuphium spec.* from the batha site are good examples: each species was recorded by a different collector (but none by pitfall traps). Given the few hours that were invested to obtain the additional samples by hand picking or sifting, the efficiency of these "methods of beetle collecting" is substantial.

Nevertheless, using a variety of methods rather than only pitfall traps gives a different impression of the carabid assemblages in a study area. Renner (1980) showed the difference between pitfall traps and some other catching methods (e.g. taking soil samples, checking tree trunks and mushrooms). He indicated that with a combination of different methods 25% more species were caught as if only pitfall traps were used. In our case, the combination of different methods doubled the species number in the batha and enabled us to catch nearly 50% more species in the woodland. In some other sclerophyllic woodlands of the Mediterranean region, species scarcity has also been demonstrated within the ground beetle assemblage (e.g. four species for the karst formation of Trieste, Brandmayr et al., 1983). In other regions of the Mediterranean the species number clearly seems to be higher (e.g. in Calabria: 26 species following Pizzolotto et al., 2005; in Central Spain: 20 species following Serrano et al., 2005). We found 30 species in the scerophyllous oak woodland we studied in Upper Galilee. This number exceeds even those of some beech dominated stands of Mediterranean mountains (e.g. Brandmayr et al., 1983; Pizzolotto et al., 2005). In general, species richness of our sclerophyllous woodland site is comparable to other habitats of the Mediterranean.

Composition of the carabid assemblages

Digging ground beetles, such as harpalines and, to a lesser extent, zabrines, are characteristic for Mediterranean habitats. They carry seeds and other plant parts to underground chambers to supply their offspring with food (Brandmayr et al., 1983; Brandmayr & Zetto Brandmayr, 1987; Zetto Brandmayr, 1990). In Israel, we found several species from this guild (e.g. *Odontocarus samson, O. asiaticus, Pseudaristus punctatissimus*). These species exhibit epigeic activities when foraging and are therefore well represented in pitfall traps. In the Mediterranean habitat, some species of the subfamily Harpalinae can be found, especially during the brood care period, inside the soil and under large stones, as described by Brandmayr & Zetto Brandmayr (1987).

Predominantly epigeic active species can be divided by their functional morphology into two groups: pushers and runners. These include, for example, species of the genera *Carabus, Orthomus* and *Nebria* (Evans & Forsythe, 1984) or visually hunting species e.g. *Notiophilus danieli* (cf. Bauer, 1975). These species are not only known from Mediterranean habitats, but also from temperate and boreal climate regions (e.g. Guenther & Assmann, 2004; Spence et al., 1996; Sroka & Finch, 2006). Species that are associated with ant nests are not found at all in cooler regions but are very common in warmer areas, and we found two species of this group, *Paussus turcicus* and *Macrocheilus saulcyi*, on our batha sampling site. The relationship of ant nest beetles to ants is well documented. *Paussus turcicus,* for example, lives, like the common west Mediterreanen *Paussus favieri*, in the ant nests of *Pheidole pallidula* (Escherich, 1898). We observed *Macrocheilus saulcyi* several times in Israel: in most cases the specimens were associated with ants. A trophic relationship between army ants (*Neivamyrmex nigrescens*) and adults of two Nearctic helluonine species (*Helluomorphoides latitarsis* and *H. ferrugineus*) has already been documented by Topoff (1969). Reichardt (1974) dissected many South American species of this group and found fragments of ants in the gut. He took this as evidence that members of Helluomorphina are predators of ants. We assume that the representatives of this group in Africa and Asia also feed on ants. In addition to the Paussinae and the helluomorphines species, *Pseudotrechus mutilatus* is also found living in ant nests; in the Mediterrranean this species lives in the nests of the ant *Messor barbarus* (cf. Antoine, 1963).

We found this species frequently in southern Spain and in the Maghreb; it was not, however, caught in pitfall traps. The low epigeic activity of this ecological group could be the reason for their under-representation in pitfall trap catches.

Another group that is very characteristic for the Mediterranean region is that of the endogeic and hypogean species (cf. Casale et al., 1998). The superficial underground compartment, described by Juberthie et al. (1981), harbours not only blind species (Drovenik et al., 2008) but also species with reduced eyes (e.g. *Limnastis galilaeus* and *Parazuphium chevrolati*, Nitzu & Decu, 1998). Both zuphiine species that were found in the batha site belong to these microphtalmic ground beetles. While only few species of the latter group can be found using non-trapping methods, the high densities of *Metadromius carmelitanus* detected this way are noteworthy. This lebiine species was found more frequently in the sifting samples than the *Trechus* species, but never occurred in pitfall traps, while this latter method served well to detect the Trechini. The tarsae of most lebiine beetles are equipped with a large number of adhesive setae, in contrast to the trechine beetles, which have only a few (cf. Schürstedt et al., 2000). It is possible that, with the help of these setae, *M. carmelitanus* is able to escape from the traps. This would, in view of the fact that no other carabid beetle has a higher density than *M. carmelitanus* in these woodlands, at least explain the absence of this beetle in pitfall traps. Although we recorded a large variety of species and employed different capture methods, our data do not present a complete picture of ground beetle species in the region: species living in the canopy of the woodlands were not included. Several species of *Calosoma* (Löbl & Smetana, 2003) have been recorded for Israel and we also know *Lebia rutilicollis* from trees in Upper Galilee. Therefore, it is very likely that the real number of species at the study sites is still undocumented, in spite of the fact that various catching methods were used.

Monitoring carabid beetles: methodology

In Europe, but also in North America, beetles are increasingly being considered in conservation biology (e.g Red Lists, the Habitat and Species Directive of the European Union). Ground beetles are also very important for the description of changes in habitats (Butterfield et al., 1995; Pearce & Venier, 2006; Rainio & Niemelä, 2003; Ssymank, 1994), because they appear in almost all terrestrial habitats and show some remarkable trends in their population-dynamic development (Lindroth, 1972). If Carabidae should receive comparable importance in Israel, we should not only use pitfall traps as a tracking method, but must also take other methods into consideration. If the results of these "alternative" sampling methods are to be compared satisfactorily, standards have to be developed for them. This might be possible for sifting if defined areas are sampled (for example, through screening and selection of 10 times 0.25 m^2, cf. Spence & Niemelä, 1994 and a defined depth of 5 cm). We doubt whether comparable standardization is possible for hand picking. However, we suggest that this method nonetheless should be carried out, as it can at least prove the presence of some species of certain ecological groups or guilds (e.g., myrmicophilic, endogeic species).

REFERENCES

Adis, J. (1979). Problems of interpreting arthropod sampling with pitfall traps. – Zoologischer Anzeiger 202: 177-184.

Andersen, J. (1995). A comparison of pitfall trapping and quadrat sampling of Carabidae (Coleoptera) on river banks. – Entomologica Fennica 6: 65-77.

Antoine, M. (1963). Coléoptères carabiques d'Maroc. Cinquiéme partie. – Mémoires de la Société des Sciences naturelles et physiques du Maroc, Zoologie (Nouvelle Série, Zoologie) 9: 359-692.

Baars, M.A. (1979). Catches in pitfall traps in relation to mean densities of carabid beetles. – Oecologia 41: 25-46.

Barber, H.S. (1931). Traps for cave-inhabiting insects. – Journal of the Elisha Mitchell Scientific Society 46: 259-266.

Bauer, T. (1975). Zur Biologie und Autökologie von *Notiophilus biguttatus* F. und *Bembidion foraminosum* Strm. (Coleopt., Carabidae) als Bewohner ökologisch extremer Standorte. – Zoologischer Anzeiger Jena 194: 305-318.

Brandmayr, P., Colombetta, G. & Polli, S. (1983). Waldcarabiden des Triester Karstes als Indikatoren des makroklimatischen Überganges vom kontinentalen Europa zur Mediterraneis (Coleoptera, Carabidae). – Zoologisches Jahrbuch der Systematik 110: 201-220.

Brandmayr, P. & Zetto Brandmayr, T. (1987). The problem of presocial behaviour in ditomine ground beetles. – Pubblicazioni dell 'istituto di entomologia dell 'universita di Pavia 36: 15-18.

Brzustowski, J. Rarefaction Calculator. – http://www2.biology.ualberta.ca/jbrzusto/rarefact.php#Inputs, date accessed: 9th December 2007.

Butterfield, J., Luff, M.L., Baines, M. & Eyre, M.D. (1995). Carabid beetle communities as indicators of conservation potential in upland forests. – Forest Ecology and Management 79: 63-77.

Casale, A., Taglianti, A.V. & Juberthie, C. (1998). Coleoptera Carabidae. – In: Encyclopaedia Biospeologica (Juberthie, C. & Decu, V., eds). – Société de Biospéologie, Bucarest. p. 1047-1081.

Chiverton, P.A. (1984). Pitfall-trap catches of the carabid beetle *Pterostichus melanarius*, in relation to gut contents and prey densities, in insecticide treated and untreated spring barley. – Entomologia Experimentalis et Applicata 36: 23-30.

Deuve, T. (2004). *Carabus (Lamprostus) sidonius* Lapouge, 1907, bona species, et note sur les Carabes du Liban (Coleoptera, Carabidae). – Coléopteres 10: 91-105.

Drovenik, B., Weber, F., Paill, W. & Assmann, T. (2008). *Aphaenopidius kamnikensis* Drovenik 1987 in Kärnten. – Angewandte Carabidologie 8: 63-67 (in press).

Escherich, K. (1898). Zur Anatomie und Biologie von *Paussus turcicus* Friv – Zugleich ein Beitrag zur Kenntniss der Myrmecophilie. – Habilitationschrift, Technische Hochschule Karlsruhe, Jena. p. 48.

Evans, M.E.G. & Forsythe, T.G. (1984). A comparison of adaptations to running, pushing and burrowing in some adult Coleoptera: especially Carabidae. – Journal of Zoology 202: 513-534.

Freude, H., Harde, K.W. & Lohse, G.A. (1965). Die Käfer Mitteleuropas. – Goecke & Evers, Krefeld. p. 214.

Guenther, J. & Assmann, T. (2004). Fluctuations of carabid populations inhabiting an ancient woodland (Coleoptera, Carabidae). – Pedobiologia 48: 159-164.

Halsall, N.B. & Wratten, S.D. (1988). The efficiency of pitfall trapping for polyphagous predatory Carabidae. – Ecological Entomology 13: 293-299.

Heydemann, D.B. (1955). Carabiden der Kulturfelder als ökologische Indikatoren. – In: Bericht über die 7. Wandersammlung Deutscher Entomologen (Sachtleben, H., ed.). – Deutsche Akademie der Landwirtschaftswissenschaften, Berlin. p. 172-185.

Heydemann, D.B. (1957). Die Biotopstruktur als Raumwiderstand und Raumfülle für die Tierwelt. – Verhandlungen der deutschen Zoologischen Gesellschaft 1956: 332-347.

Juberthie, C., Delay, B. & Bouillon, M. (1981). Extension du milieu souterrain superficiel en zone non calcaire. Description d'un nouveau milieu et de son peuplement par les Coléoptères troglobies. – Mémoires de biospéologie 7: 19-52.

Lindroth, C.H. (1972). Changes in the Fennoscandian ground-beetle fauna (Coleoptera, Carabidae) during the twentieth century. – Annales Zoologici Fennici 9: 49-64.

Löbl, I. & Smetana, A. (2003). Catalogue of Palaearctic Coleoptera. – Apollo Books, Stenstrup. p. 819.

Matern, A., Drees, C., Kleinwächter, M. & Assmann, T. (2007). Habitat modelling for the conservation of the rare ground beetle species *Carabus variolosus* (Coleoptera, Carabidae) in the riparian zones of headwaters. – Biological Conservation 136: 618-627.

Mateu, J. (1956). Misión H. Coiffait al Libano Lebiidae y Brachinidae (Col. Carabidos). – Archivos del Instituto de Aclimatación, Almeria 5: 33-49.

Mello, M.J. (2005). Inventory of macrolepidoptera and other insects in the Boston Harbor Islands national park area. – Northeastern Naturalist 12: 99-144.

Nitzu, E. & Decu, V. (1998). First record of *Limnastis galilaeus* Brullé, 1875 and *Parazuphium chevrolati* (Castelnau, 1833) (Coleoptera, Caraboidaea) in the subterranean habitat from Southern Dobrudja, Romania. – Entomologische Mitteilungen aus dem Zoologischen Museum Hamburg 12: 229-236.

Pearce, J.L. & Venier, L.A. (2006). The use of ground beetles (Coleoptera: Carabidae) and spiders (Araneae) as bioindicators of sustainable forest management: a review. – Ecological Indicatores 6: 780-793.

Perner, J. & Schueler, S. (2004). Estimating the density of ground-dwelling arthropods with pitfall traps using a nested-cross array. – Journal of Animal Ecology 73: 469-477.

Pizzolotto, R., Brandmayr, P. & Mazzei, A. (2005). Carabid beetles in mediterranean region: biogeographical and ecological features. – DIAS Report 114: 243-254.

Prasifka, J.R., Lopez, M.D., Hellmich, R.L., Lewis, L.C. & Dively, G.P. (2007). Comparison of pitfall traps and litter bags for sampling ground-dwelling arthropods. – Journal of Applied Entomology 131: 115-120.

Rainio, J. & Niemelä, J. (2003). Ground beetles (Coleoptera, Carabidae) as bioindicators. – Biodiversity and Conservation 12: 487-506.

Reichardt, H. (1974). Monograph of the Neotropical Helluonini, with notes and discussions on old world forms (Coleoptera: Carabidae). – Studia Entomologica 17: 211-301.

Renner, K. (1980). Faunistisch-ökologische Untersuchungen der Käferfauna pflanzensoziologisch unterschiedlicher Biotope im Evessell-Bruch bei Bielefeld-Sennestadt. – Berichte des naturwissenschaftlichen Vereins Bielefeld Sonderheft 2: 145-176.

Schürstedt, H., Rossbach, A. & Assmann, T. (2000). Morphological differentiations of tarsal structures in ground beetles living in reedbed habitats (Coleoptera, Carabidae). – In: Natural history and applied ecology of carabid beetles (Brandmayr, P., Lövei, G., Brandmayr, T., Casale, A. & Vigna Taglianti, A., eds). – Pensoft, Sofia-Moscow. p. 81-87.

Serrano, J., Ruiz, C., Andújar, C. & Lencina, J.L. (2005). Land use and ground beetle assemblages in the national park of Cabañeros, Central Spain (Coleoptera: Carabidae). – DIAS Report 114: 275-289.

Spence, J.R., Langor, D.W., Niemelä, J., Carcamo, H.A. & Currie, C.R. (1996). Northern forestry and carabids: the case for concern about old-growth species. – Annales Zoologici Fennici 33: 173-184.

Spence, J.R. & Niemelä, J.K. (1994). Sampling carabid assemblages with pitfall traps: The madness and the method. – The Canadian Entomologist 126: 881-894.

Sroka, K. & Finch, O.-D. (2006). Ground beetle diversity in ancient woodland remnants in north-western Germany (Coleoptera, Carabidae). – Journal of Insect Conservation 10: 335-350.

Ssymank, A. (1994). Indikatorarten der Fauna für historisch alte Wälder. – NNA-Berichte 3/94: 134-141.

Ssymank, A. (1998). Das europäische Schutzgebietssystem NATURA 2000: BfN-Handbuch zur Umsetzung der Fauna-Flora-Habitat-Richtlinie (92/43/EWG) und der Vogelschutzrichtlinie (79/409/EWG). – Bundesamt für Naturschutz, Bonn-Bad Godesberg. p. 560.

Talley, T.S., Fleishman, E., Holyoak, M., Murphy, D.D. & Ballard, A. (2007). Rethinking a rare-species conservation strategy in an urban landscape: The case of the valley elderberry longhorn beetle. – Biological Conservation 135: 21-32.

Topoff, H.R. (1969). A unique predatory association between carabid beetles of the genus Helluomorphoides and colonies of the army ant *Neivamyrmex nigrescens*. – Psyche 76: 375-381.

White, R.E. (1983). A field guide to the beetles of North America. – Houghton Mifflin Company, Boston, New York. p. 384.

Zetto Brandmayr, T. (1990). Spermophagous (seed-eating) ground beetles: first comparison of the diet and ecology of the harpaline genera *Harpalus* and *Ophonus* (Col., Carabidae). – In: The role of ground beetles in ecological and environmental studies (Stork, N. E., eds). – Intercept, Andover (Hampshire). p. 307-316.

L. Penev, T. Erwin & T. Assmann (Eds) 2008
BACK TO THE ROOTS AND BACK TO THE FUTURE. TOWARDS A NEW SYNTHESIS AMONGST TAXONOMIC, ECOLOGICAL AND BIOGEOGRAPHICAL APPROACHES IN CARABIDOLOGY
Proceedings of the XIII European Carabidologists Meeting, Blagoevgrad, August 20-24, 2007, pp. 409-423.

Sofia–Moscow

Behavioural patterns of nocturnal carabid beetles determined by direct observations under red-light conditions

Claudia Drees, Andrea Matern & Thorsten Assmann

Institute of Ecology and Environmental Chemistry, University of Lüneburg, D – 21335 Lüneburg, Germany

SUMMARY

Information on the locomotory activity of carabid beetles in their habitats has to date mostly been obtained by pitfall trapping or by telemetric methods (e.g. harmonic radar). However, both methods have certain shortcomings, such as the dependency on running activity in the case of pitfall traps or the restricted applicability for smaller species due to the weight of transponders (telemetric methods). Both pitfall trapping and telemetric methods allow only several observations per hour or day and, consequently, an estimation of merely the minimum distance covered. A continuous observation of a beetle in the field, however, can reveal important behavioural traits, such as feeding or mating and dependency on habitat characteristics. It can also provide information about the behaviour of a beetle in different habitats. We propose a method of direct observation of nocturnal beetles under red-light conditions in order to gain insight into the biology of ground beetle species.

We investigated parameters of habitat choice, hunting behaviour and feeding of two carabid beetle species. 28 individuals of the endangered semi-aquatic *Carabus variolosus* and 13 individuals of the ripicolous *Omophron limbatum* were continuously observed during their main activity period in the first half of the night and were statistically analysed. We saw running behaviour in favourite and non-favourite habitats, foraging, hunting in the water, and feeding. Our findings complemented previous knowledge for both species: The home range of *O. limbatum* includes densely vegetated areas as hunting habitats. *C. variolosus* was observed to hunt for small invertebrates in or close to the brook in our study area, but only on humid soil. Our method apparently did not have any influence on the beetles and was found to be efficient at least for epigaeic animals and in habitat types with herb layers which are not too dense to constrain individual tracking. The suggested technique has the potential to provide interesting data on the microhabitat use of other little known nocturnal insect species.

Keywords: *Carabus variolosus*, continuous tracing, individual tracking, *in situ* observation, microhabitat use, *Omophron limbatum*

INTRODUCTION

The knowledge of behavioural patterns of animals, especially when these are endangered, is of great importance for efficient conservation (Sutherland, 1998). In the attempt to preserve populations and species, special focus should be placed on all behavioural traits that may directly and/or indirectly decrease effective population size (Anthony & Blumstein, 2000). In the face of environmental changes and human impact, a thorough understanding of mobility, dispersal, mating systems and feeding habits is of particular importance in achieving this aim (Ulfstrand, 1996). One important issue with respect to (small) invertebrates is the connection between behavioural patterns, such as feeding or mating, and habitat features. Observing animals in their habitats, i.e. *in situ*, is the best means of enhancing knowledge on the biology of the study species. Autecological knowledge is fundamental for the management and conservation in particular of rare species and those which only occur locally in small areas (Negro et al., 2008).

Different methods have been proposed and applied to gather information about behavioural patterns of individuals in the field (e.g. fluorescent powder tracking (Nicolas & Colyn, 2007), tracing by harmonic radar (Mascanzoni & Wallin, 1986; Riley et al., 1996) or radiotelemetry (Riecken & Raths, 1996; Negro et al., 2008)). In contrast to direct observations in the field where individuals are traced continuously (Schtickzelle et al., 2007), these methods do not completely reveal behavioural traits and their dependence on habitat characteristics. However, it is important to consider not only the dispersal capabilities of an organism but also the complex interaction between the organism's behaviour and landscape pattern and use (Vuilleumier & Metzger, 2006). Knowledge obtained by mark-recapture experiments and census data on habitat preferences can be verified and enhanced by continuous observation (Haynes & Cronin, 2006).

Information on the locomotory activity of carabid beetles in their habitats has so far mostly been obtained by pitfall trapping (Lövei & Sunderland, 1996) or by telemetric methods such as harmonic radar (Mascanzoni & Wallin, 1986; Wallin, 1988; Hockmann et al., 1989, 1992; Niehues et al., 1996; Charrier et al., 1997) or radio-telemetry (Riecken & Raths, 1996; Negro et al., 2008). However, both methods have certain shortcomings, such as the dependency on running activity (pitfall traps) or the restricted applicability for smaller species due to the weight of transponders (telemetric methods). Though the direct observation method does not have these disadvantages its use is complicated by the fact that many species are nocturnal (Thiele & Weber, 1968). One method of continuous observation of beetles in the field was proposed by Baars (1979) who traced radioactively marked beetles in their habitat. Chemical light tags have also been used in combination with radiotelemetry (Beaudoin-Ollivier et al., 2003).

We introduce another, easier-to-use method of continuous observation of nocturnal beetles in the field, making use of the fact that photoreception is not uniform among different insect taxa. While bees were, for example, shown to perceive green, blue and ultraviolet, which has often been generalised, butterflies were shown to have tetrachromatic vision with an additional sensitivity maximum at long-wave red (Scharstein & Stommel, 1999). Spectral range sensitivity of carabids was studied by Hasselmann (1962), who demonstrated a low sensitivity above 550 nm for two *Carabus* species, irrespective of their nocturnal or diurnal activity. Thus, we observed beetles under red-light conditions in order to gain insight into their biology. We present the results obtained by this method for two ground beetle species differing in both body size and type of preferred habitat.

MATERIAL AND METHODS

Study species

Carabus variolosus Fabricius, 1787 is an extremely hygrophilous and stenotopic flightless European ground beetle of woodlands which inhabits the immediate surroundings of spring-brooks, swamps and small rivers (Turin et al., 2003; Matern et al., 2007a, 2007b). Inhabiting a sensitive and anthropogenically frequently influenced ecotone, the beetle is only locally distributed all over its range, where it is mostly rare and endangered or threatened.

Omophron limbatum Fabricius, 1776 has a Palaearctic distribution (Lindroth, 1945). The stenotopic species occurs on sparsely vegetated river banks and lakesides (Turin, 2000). It has a preference for sun-exposed and sandy soils, in which it rests during the day. Like other typical riparian carabid species *O. limbatum* is capable of flight (Desender, 1989; Günther et al., 2004).

Observation under red-light conditions

Based on the findings of Hasselmann (1962), our observations took place under red-light conditions (590-680 nm) which were obtained by covering a lamp with transparent red foil (Heyda, handicraft supply). Preparatory to observation, a beetle was individually marked with a white line or white dots (Edding, 751) at the edge of its elytra. Beetles were placed in their habitat fenced in by a tube (10 cm diameter) for several hours prior to observation. Each individual was used only once for observation.

Beetles were observed during their main activity period. Observation started with the removal of the fence. The beetles' behaviour was recorded continuously by taping time, position and behavioural patterns of the observed beetle on a voice recorder. The observing person stayed one to two metres away from the beetle in order to avoid any disturbance. One person observed one beetle. The positions of the beetles were marked with numbered sticks (Fig. 1) at regular time intervals which were – if necessary – placed with a certain delay, again to avoid disturbance. Observation ended either after a phase of non-activity of at least 30 minutes or after losing sight of the observed individual. If one individual was lost early in the night we started observing another beetle which was kept as substitute.

Investigations of *C. variolosus* took place in the forest of Arnsberg, NW Germany (see Matern et al. (2007a) for a habitat description). Information on the diurnal activity of *Carabus variolosus* was gathered using 40 dry pitfall traps placed in the habitat. We checked these traps every four hours on five days during the main reproductive period of *C. variolosus* (Matern et al., 2007b) and thus determined its diurnal activity period. Beetles for observation were captured in dry pitfall traps placed in the same habitat patch. 38 *C. variolosus* individuals, both reproductive and newly hatched, were observed during spring- (May-June 2003 and 2004) and late-summer-season (August 2003), respectively (Table A1). Beetles were observed at 2-minute intervals. Movement and behaviour of 28 beetles were observed at a minimum of 6 different positions and for at least 54 minutes. These were analysed, adding up to a total observation time of 69.4 hours.

Investigations of *O. limbatum* took place at a restitution area in the city of Osnabrück (NW Germany) with a mosaic of pioneer vegetation, open sandy areas, and small ponds surrounded by sandy banks. *O. limbatum* individuals were captured by bloating them with water from their daytime-resting site. Observations of *O. limbatum* took place in July and August 2000. As the activity diurnal period of this species starts with dusk, observation started at sunset. Observations were carried out as for *C. variolosus* but with 5-minute recording intervals. Altogether 17 beetles were recorded, 13 of which were found at more than 5 different positions and were thus analysed. These 13 individuals were observed for a total of 21.8 hours.

Distances covered and habitat choice

Each morning after observation, the routes of the beetle(s) (cf. Fig. 1) were recorded by measuring the distances between the position-marking sticks and the turning angles between the sections. The habitat at each position was characterised within its immediate surrounding.

For *C. variolosus* the habitat at each position was characterised within a diameter of 20 cm. Following Matern et al. (2007a) we estimated the cover (in %) of the following habitat parameters: (lentic and running) water, bushes and trees, moss, herbaceous plants, bare soil, leaf litter, needle litter, woody debris and grass. For *O. limbatum* the vegetation cover (in %) at each position was estimated within a diameter of 10 cm.

In order to compare the characteristics of the habitat chosen by the beetle with the habitat structure of the whole investigation area, we estimated the habitat parameters at 100 randomly chosen spots within the investigation area. Mean individual habitat

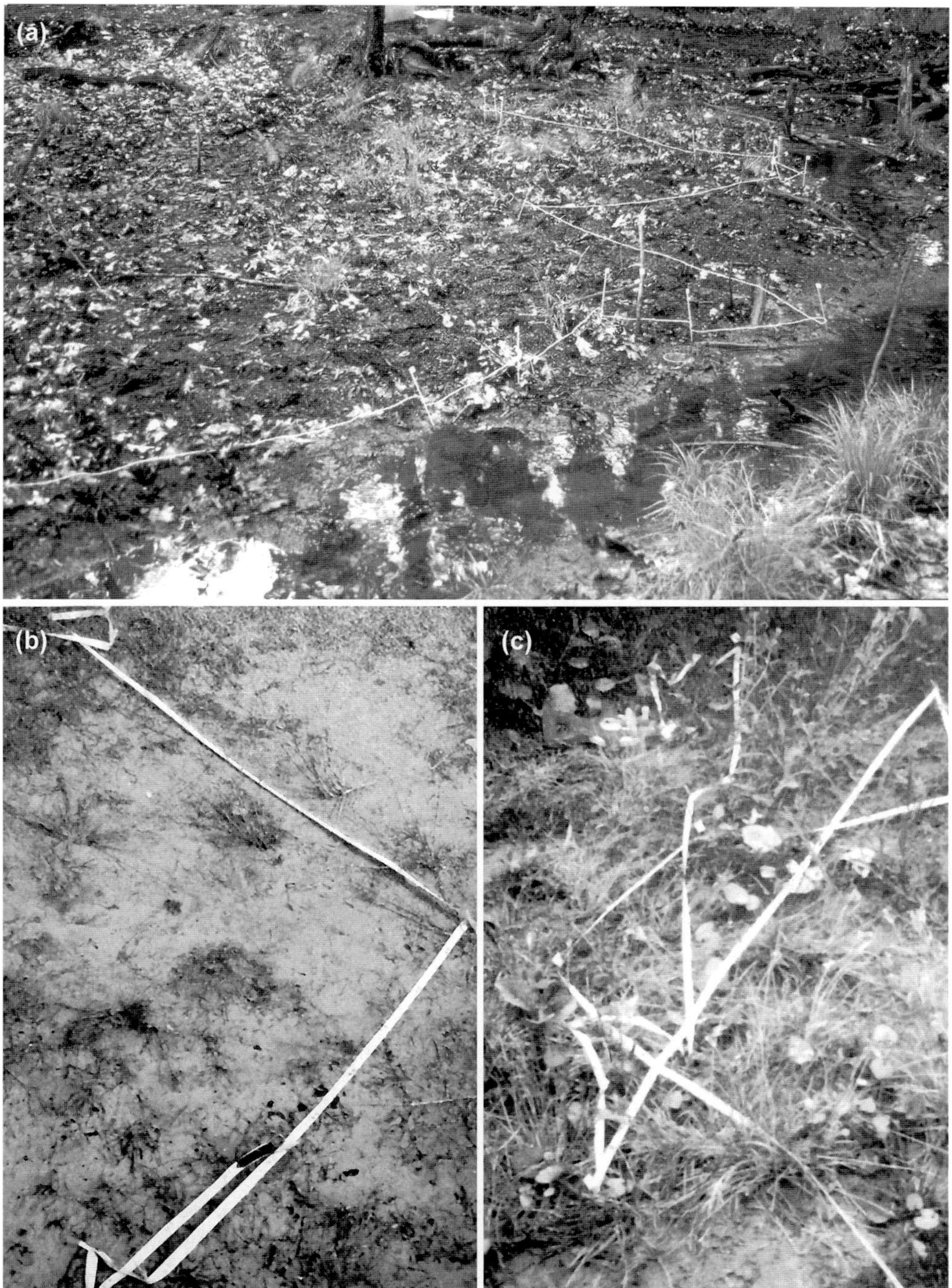

Fig. 1. Parts of observed routes covered by *C. variolosus* (a, individual no. 1230, cf. Table 1A), distance shown: approximately 19m) and *O. limbatum* (b, 2.20m and c, 5.40m) within one observation.

preferences of the observed beetles were compared to the habitat parameters assessed at these 100 spots. Mean individual habitat preference was determined by weighting habitat parameters at each position-marking stick by residence time, i.e. longer residence at a certain spot gives a larger contribution of the habitat parameters at this spot to the mean individual habitat preference of the beetle. For *C. variolosus*, we considered only the beetles which were traced in spring (15 beetles), as the habitat parameters for the randomly chosen spots were also estimated during spring. Results of the pairwise tests for each parameter (Mann-Whitney *U*-test, calculated with Statistica, Ver. 7.1, StatSoft Inc.) were corrected for multiple testing by means of the FDR-method (Benjamini et al., 2001).

Behavioural patterns

For *C. variolosus*, we observed and distinguished six different behavioural patterns: resting, running, foraging, feeding, swimming and diving. *Resting* describes a rest for at least 20 seconds. *Foraging* is characterised by movements where the beetle showed forwards and backwards movements and frequent turns while keeping its head close to the ground or pushing it into the mud. *Feeding* means a beetle obviously feeding on prey, e.g. chewing with parts of prey visible. *Diving* describes a beetle running under water surface while being completely submerged.

For *O. limbatum*, only resting and walking behaviour was differentiated the latter of which was divided in 'directed movement' and 'random walk' in the sense of Baars (1979).

Behaviour and habitat

Habitat preferences were investigated in more detail by analysing the habitat parameters around the sites where a beetle displayed certain behavioural patterns. For *C. variolosus* we investigated a possible relationship of the behavioural patterns resting, running and foraging to environmental structures. The habitat parameters at all residence spots at which a beetle displayed a certain behavioural pattern were averaged (weighted mean over residence time) and, again, compared to the habitat parameters at the randomly chosen spots by a pairwise test (Mann-Whitney *U*-test, results FDR-corrected).

RESULTS

Distances covered

Altogether, 24 *C. variolosus* individuals were captured while assessing the diurnal acitivity, showing that the first half of the night was the main diurnal activity period (χ^2 *goodness of fit*; *d.f.*=1; *p*=0.042, Fig. 2). Therefore, observations of this species started at sunset.

28 *C. variolosus* individuals were observed and the results analysed; mean individual observation time was 149 minutes in which the beetles covered an average of 853 cm (Table 1). Although the mean observation time of males and females did not differ significantly (Mann Whitney *U*-test: p=0.160), male beetles were observed at significantly fewer different positions (13) than females (22 positions on average) (Mann Whitney *U*-test: p=0.0014). Interestingly, males covered significantly shorter distances (mean distance: 598 cm) than females (1074 cm) (Mann Whitney *U*-test: p=0.025), whereas the velocity did not differ significantly (males: 5.4 cm/minute, females: 8.2 cm/minute; Mann Whitney *U*-test: p=0.345) (Table A1). The maximum distance covered within a 2-minute interval was 427 cm.

The 13 *O. limbatum* individuals were observed for a mean time of 101 minutes in which the beetles covered 645 cm on average (Table 1). The mean velocity was 7.6 cm/min, the maximum distance covered within a 5-minute interval was 290 cm.

Table 1. Overview of the number of *C. variolosus* and *O. limbatum* observed and analysed. Mean observation times, distances covered and velocities and their ranges (min-max) are given.

Species	Beetles analysed (beetles observed)	Male/ female	Observation time [min]	Distance covered [cm]	Velocity [cm/min]
Carabus variolosus	28	13/15	149	853	6.9
	(38)		(54-294)	(195-2,248)	(0.7-22.8)
Omophron limbatum	13	6/7	101	646	7.6
	(17)		(35-150)	(128-1,701)	(1.1-25.6)

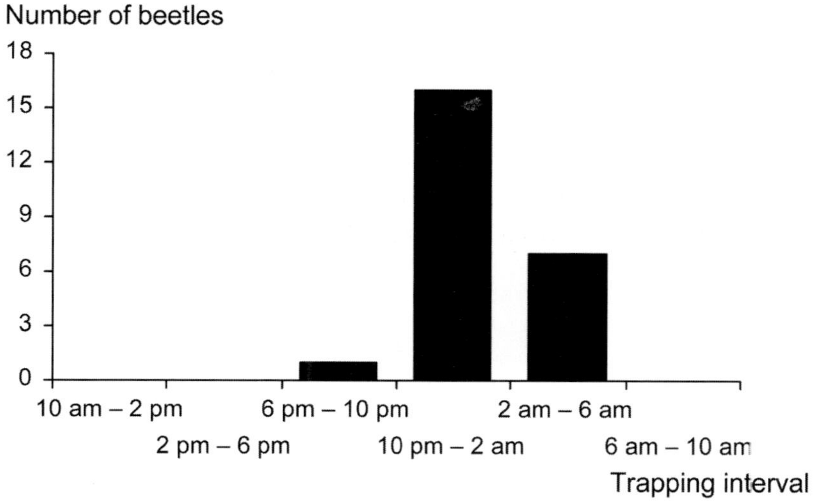

Fig. 2. Diurnal activity of *C. variolosus* in the Arnsberg forest (NW Germany) as shown by controls of dry pitfall traps every 4 hours on 5 days in spring 2004.

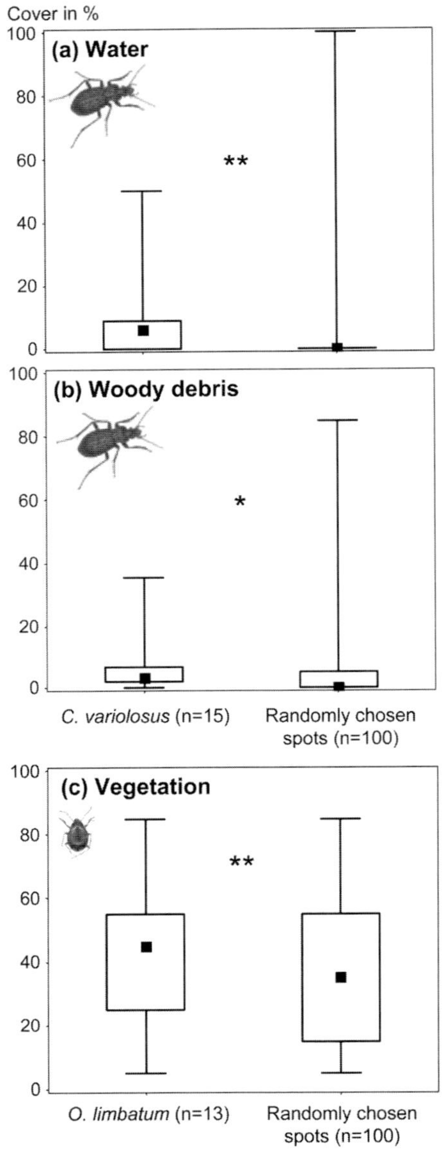

Fig. 3. Habitat characteristics at the observation spots (weighted means) of *C. variolosus* and *O. limbatum* compared with 100 randomly chosen spots within the respective investigation area. Boxplots indicate median (■), 25-75% percentiles (box) and ranges (whisker). Asterisks indicate significant differences between the groups: ***: p<0.001; **: p<0.01; *: p<0.05. Only significant results are shown. Drawing of *C. variolosus* by P. Schüle.

Habitat choice

A typical walking pattern of *C. variolosus* is shown in Fig. 1a. It is characterised by frequent direction changes and short walking distances, which fits the pattern of 'random walk' described by Baars (1979). Usually the beetles stayed in the bare-soil dominated area close to the shallow brook (bottom right in Fig. 1a) which they followed at a certain distance.

The analysis of the habitat structure at the observation spots of the 15 beetles observed in spring compared to the 100 randomly chosen spots (reference) in the respective study areas confirmed a preference for special habitat characteristics: *C. variolosus* prefers a greater amount of woody debris and a certain amount of water in the immediate surrounding of its residence (Fig. 3a-b). Other parameters describing the habitat did not differ significantly from the reference points. The habitat structure at the observation points did not differ significantly between the sexes for *C. variolosus* (Mann-Whitney *U*-test, $p<0.05$).

O. limbatum individuals were observed in both open areas (Fig. 1b) and areas with denser vegetation (Fig. 1c). Evidently *O. limbatum* showed a particular preference for areas with higher vegetation density (Fig. 3c).

Behavioural patterns

In *C. variolosus*, six different behavioural patterns were observed: resting (28 beetles), running (28), foraging (23), feeding (12), diving (9) and swimming (7). Resting, running and foraging were observed in 55, 48 and 34% of the observation intervals, respectively (several activities per interval allowed). Feeding (5%), diving (1%) and swimming (0,3%) were found less often. In areas with stony ground, beetles were observed searching the ground along the edge of stones (foraging). Furthermore, we observed beetles that came out of the water and chewed; in some cases we found the beetles with gammarids between their mandibles (feeding). Gammarids were very abundant in the brooks crossing the study area. *C. variolosus* was even observed feeding on an adult crane fly (Diptera: Tipulidae). These were laying their eggs in the shallow water of the brook and therefore dipping their abdomina into the water.

For *O. limbatum*, resting and random walk was found in all the individuals observed (13 beetles each), whereas only 8 individuals displayed directed movement. The two walking patterns are illustrated in Figs 1b-c, respectively.

Behaviour and habitat

The positions chosen by the *C. variolosus* individuals for the activities resting, running and foraging varied significantly in some cover parameters from randomly chosen spots (Figs 4a-c). Interestingly, beetles that showed the behaviour 'running' preferred significantly more bare soil, grass, and woody debris compared to the reference and areas with hardly any herb cover (Fig. 4a). Resting beetles preferred areas with a higher cover of woody debris but less bare soil (Fig. 4b). We observed that beetles ran in straight lines over long distances when beetles crossed stretches of dry forest floor which was not part of their foraging habitat. Foraging beetles favoured areas with greater amounts of woody debris, grass, water, and bare soil in their immediate surroundings compared to the reference (Fig. 4c).

O. limbatum mostly displayed directed movement in sparsely vegetated areas (Fig. 1b). It also moved more or less non-directionally (random walk) in adjacent areas with dense vegetation (random walk, Fig. 1c).

DISCUSSION

Red-light conditions proved a valuable tool with which to track carabid beetles in their habitat, observe their behaviour and movement patterns, gain an impression of their places of residence, and analyse the connection between behavioural patterns and habitat variables. The observed beetles showed no noticeable movements towards or away from the observer that could have indicated any influence. Our observation of *C. variolosus* and *O. limbatum* suggests that the method we used is suitable for epigaeic species unable or reluctant to fly and, at any rate, applicable in habitat types with herb layers which are not too dense to constrain individual tracking. In contrast to radio or radar tracking experiments, beetles are not handicapped by heavy and unwieldy tags, the influence of which is difficult to estimate (Negro et al., 2008).

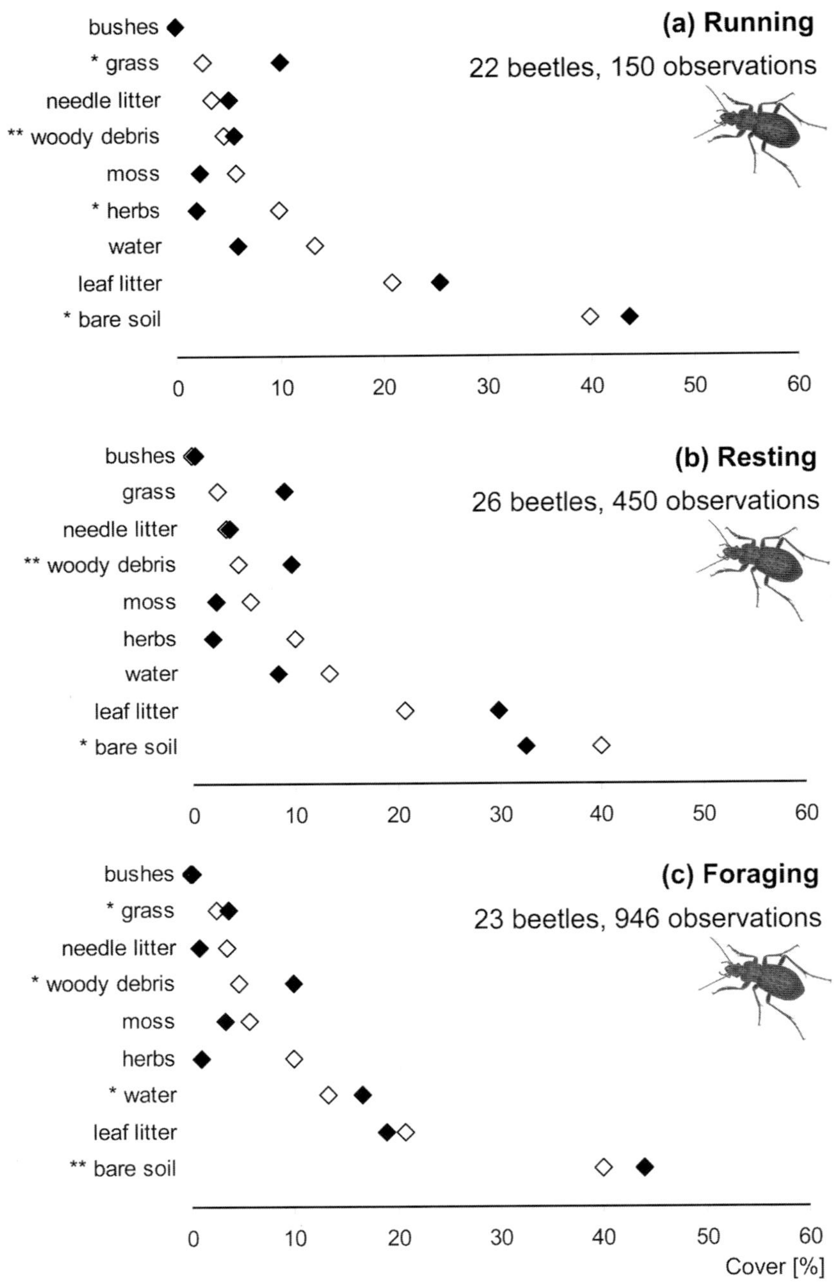

Fig. 4. Comparison of habitat structure at the positions chosen by *C. variolosus* (black diamonds) for the behavioural patterns *running*, *resting* and *foraging* with the habitat structure at randomly chosen spots (open diamonds). Asterisks indicate significant differences between the groups: ***: $p<0.001$; **: $p<0.01$; *: $p<0.05$. In contrast to *foraging*, *running* and *resting* were only analysed when they occurred as only behavioural pattern within one observation interval. Drawing of *C. variolosus* by P. Schüle.

In the case of *C. variolosus*, the preference for bare soil and the importance of water fits well with the results of Matern et al. (2007a) obtained by means of habitat suitability models. However, it is by direct observation that a reason for this habitat selection can be inferred: It was precisely on humid soil and in the water where foraging behaviour was observed. These results complement the detailed analyses of the habitat preferences and population structure of *C. variolosus* which have been published only recently (Matern et al., 2007a, 2007b) and correspond well to some aspects of the biology of the species, e.g. the hunting behaviour of both larvae and adults have been known for decades (Sturani, 1963a, b). The remarkable behaviour of this ground beetle species (both larvae and adults hunt for prey by swimming and diving) was confirmed by our observations for adult beetles. This close connection to water, the reason why *C. variolosus* is described as semi-aquatic, occurs in only a few other carabid species (*C. clatratus, C. galicianus, C. alysidotus, C. melancholicus*, Turin et al., 2003).

Our observation of *O. limbatum* expands on earlier descriptions of the habitat that have so far focussed on the diurnal resting place of the species. In addition to sparsely vegetated sandy soils, the home range apparently includes adjacent areas with dense vegetation. Apparently, random walk is linked to hunting activity; this observation is supported by crop analyses where remains of aphids were found (F. Prüßner, pers. comm.). Patches of sandy soils along river banks, which are rapidly (re)colonised by flight activity of the beetles, according to observations of Günther et al. (2004) and Günther & Assmann (2005), are therefore only a part of the habitat of the beetles.

As the example of these two species already shows, direct observations contribute essentially to the knowledge of the biology of species. They can, on the one hand, increase knowledge where it is lacking, e.g. for poorly known species, and can, on the other hand, serve to verify knowledge obtained by laboratory experiments (Yamazaki, 2006). In addition to observations of behaviour in laboratories and artificial landscapes (e.g. Olden et al., 2004; Bonacci et al., 2006; Dorosheva & Reznikova, 2006), *in situ* observations offer explanations for certain habitat associations that could also be found by other methods (e.g. habitat models) which, however, merely indicate correlations and not causal relationships. Using *in situ* observation methods Newbold (2007), for example, was able to attribute the attraction of lizards to dung heaps to better visibility of prey. As far as habitat use is concerned, direct observation is also more insightful than other tracking techniques such as mark-recapture, telemetry and harmonic radar. The latter are well suitable for determining (diurnal) activity periods and for estimating home ranges (Hockmann et al., 1989, 1992; Riecken & Raths, 1996; Hedin & Ranius, 2002) but allow less detailed information on (micro-)habitat preferences. Moreover, *in situ* observations are suitable for studying the dependency of (movement) behaviour on different microhabitat types or at habitat boundaries (Goodwin & Fahrig, 2002; Haynes & Cronin, 2006). This is especially important for developing realistic scenarios and models of dispersal processes (Goodwin & Fahrig, 2002; Chapman et al., 2007; Tyre et al., 2007). Information about the use of certain structural habitat elements, and consequently their importance, can also considerably help to improve habitat quality (Newbold, 2007, this study). Consequently, knowledge of species behaviour acquired by observation can be crucial for the conservation of rare or endangered species (Yamazaki, 2006; Schtickzelle et al., 2007). The proposed method will help to expand the possibilities in this field.

ACKNOWLEDGEMENTS

We thank Fritzi Prüßner for collecting the data on *Omophron limbatum* and Karsten Schlumbohm for collecting the data on the late-summer 2003 generation of *Carabus variolosus*. The local authorities in Soest and Arnsberg are thanked for issuing permits. This research was supported by a grant from the Federal Environmental Foundation (DBU) to A.M.

REFERENCES

Anthony, L.L. & Blumstein, D.T. (2000). Integrating behaviour into wildlife conservation: the multiple ways that behaviour can reduce N_e. – Biological Conservation 95: 303-315.

Baars, M.A. (1979). Patterns of movement of radioactive carabid beetles. – Oecologia 44: 125-140.

Beaudoin-Ollivier, L., Bonaccorso F., Aloysius, M. & Kasiki, M. (2003). Flight movement of *Scapanes australis australis* (Boisduval) (Coleoptera: Scarabaeidae: Dynastinae) in Papua New Guinea: a radiotelemetry study. – Australian Journal of Entomology 42: 367-372.

Benjamini, Y., Drai, D., Elmer, G., Kafkafi, N. & Golani, I. (2001). Controlling the false discovery rate in behavior genetics research. – Behavioural Brain Research 125: 279-284.

Bonacci, T., Brandmayr, P., Giglio, A., Massolo, A., Mazzei, A., Odoguardi, R., Romeo, M., Talarico, F.F. & Zetto Brandmayr, T. (2006). Agonistic behaviour of *Scarites buparius* (Forster, 1771) (Coleoptera: Carabidae) in relation to body size. – Entomologica Fennica 17: 340-344.

Chapman, D.S., Dytham, C. & Oxford, G.S. (2007). Landscape and fine-scale movements of a leaf beetle: The importance of boundary behaviour. – Oecologia 154: 55-64.

Charrier, S., Petit, S. & Burel, F. (1997). Movements of *Abax parallelepipedus* (Coleoptera, Carabidae) in woody habitats of a hedgerow network landscape: a radio-tracing study. – Agriculture, Ecosystems and Environment 61: 133-144.

Desender, K. (1989). Ecomorphological adaptations of riparian carabid beetles. – Verhandelingen van het Symposium "Invertebraten van Belgie": 309-314.

Dorosheva, E. & Reznikova, Z. (2006). An experimental study of behavioral mechanisms of relationships between red wood ants and ground beetles. – Entomological Review 86: 139-148.

Goodwin, B.J. & Fahrig, L. (2002). Effect of landscape structure on the movement behaviour of a specialized goldenrod beetle, *Trirhabda borealis*. – Canadian Journal of Zoology 80: 24-35.

Günther, J. & Assmann, T. (2005). Restoration ecology meets carabidology: effects of floodplain restitution on ground beetles (Coleoptera, Carabidae). – Biodiversity and Conservation 14: 1583-1606.

Günther, J., Hölscher, B., Prüßner, F. & Assmann, T. (2004). Survival at river banks – power of dispersal and population structure of *Elaphrus aureus* and *Omophron limbatum* in north-western Germany (Coleoptera, Carabidae). – Mitteilungen der Deutschen Gesellschaft für Allgemeine und Angewandte Entomologie 14: 517-520.

Hasselmann, E.-M. (1962). Über die relative spektrale Empfindlichkeit von Käfer- und Schmetterlingsaugen bei verschiedenen Helligkeiten. – Zoologische Jahrbücher Physiologie 69: 537-576.

Haynes, K.J. & Cronin, J.T. (2006). Interpatch movement and edge effects: the role of behavioral responses to the landscape matrix. – Oikos 113: 43-54.

Hedin, J. & Ranius, T. (2002). Using radio telemetry to study dispersal of the beetle *Osmoderma eremita*, an inhabitant of tree hollows. – Computers and Electronics in Agriculture 35: 171-180.

Hockmann, P., Menke, K., Schlomberg, P. & Weber, F. (1992). Untersuchungen zum individuellen Verhalten (Orientierung und Aktivität) des Laufkäfers *Carabus nemoralis* im natürlichen Habitat. – Abhandlungen aus dem Westfälischen Museum für Naturkunde 54: 65-98.

Hockmann, P., Schlomberg, P., Wallin, H. & Weber, F. (1989). Bewegungsmuster und Orientierung des Laufkäfers *Carabus auronitens* in einem westfälischen Eichen-Hainbuchen-Wald (Radar-Beobachtungen und Rückfang-Experimente). – Abhandlungen aus dem Westfälischen Museum für Naturkunde 51: 1-71.

Lindroth, C.H. (1945). Die fennoskandischen Carabidae: I Spezieller Teil. – Göteborgs Kungliga Vetenskaps – och Vitterhets-Samhälles Handlingar Sjätte följiden. Series B 4: 1-709.

Lövei, G.L. & Sunderland, K.D. (1996). Ecology and behavior of ground beetles (Coleoptera: Carabidae). – Annual Review of Entomology 41: 231-256.

Mascanzoni, D. & Wallin, H. (1986). The harmonic radar – a new method of tracing insects in the field. – Ecological Entomology 11: 387-390.

Matern, A., Drees, C., Kleinwächter, M. & Assmann, T. (2007a). Habitat modelling for the conservation of the rare ground beetle species *Carabus variolosus* (Coleoptera, Carabidae) in the riparian zones of headwaters. – Biological Conservation 136: 618-627.

Matern, A., Drees, C., Meyer, H. & Assmann, T. (2007b). Population ecology of the rare carabid beetle *Carabus variolosus* (Coleoptera: Carabidae) in north-west Germany. – Journal of Insect Conservation DOI 10.1007/s10841-007-9096-3.

Negro, M., Casale, A., Migliore, L., Palestrini, C. & Rolando, A. (2008). Habitat use and movement patterns in the endangered ground beetle species, *Carabus olympiae* (Coleoptera: Carabidae). – European Journal of Entomology 105: 105-112.

Newbold, T.A.S. (2007). Use of dung piles by the side-blotched lizard (*Uta stansburiana*). – Southwestern Naturalist 52: 616-619.

Nicolas, V. & Colyn, M. (2007). Efficiency of fluorescent powder tracking for studying use of space by small mammals in an African rainforest. – African Journal of Ecology 45: 577-580.

Niehues, F.J., Hockmann, P. & Weber, F. (1996). Genetics and dynamics of a *Carabus auronitens* metapopulation in the Westphalian Lowlands. – Annales Zoologici Fennici 33: 85-96.

Olden, J.D., Hoffman, A.L., Monroe, J.B. & Poff, N.L. (2004). Movement behaviour and dynamics of an aquatic insect in a stream benthic landscape. – Canadian Journal of Zoology 82: 1135-1146.

Riecken, U. & Raths, U. (1996). Use of radio telemetry for studying dispersal and habitat use of *Carabus coriaceus* L. – Annales Zoologici Fennici 33: 109-116.
Riley, J.R., Smith, A.D., Reynolds, D.R., Edwards, A.S., Osborne, J.L., Williams, I.H., Carreck, N.L. & Poppy, G.M. (1996). Tracking bees with harmonic radar. – Nature 379: 29-30.
Scharstein, H. & Stommel, G. (1999). Photorezeption. – In: Dettner, K. & Peters, W. (eds), Lehrbuch der Entomologie. – Fischer, Stuttgart. 320-345.
Schtickzelle, N., Joiris, A., van Dyck, H. & Baguette, M. (2007). Quantitative analysis of changes in movement behaviour within and outside habitat in a specialist butterfly. – BMC Evolutionary Biology 7: DOI 10.1186/1471-2148-7-4.
Sturani, M. (1963a). Nuove ricerche biologiche e morfologiche sul *Carabus* (*Hygrocarabus*) *variolosus* Fabricius (Coleoptera Carabidae). – Bollettino di Zoologia agraria a di Bachicoltura Serie II: 25-34.
Sturani, M. (1963b). Osservazioni biologiche e morfologiche sul *Carabus* (*Hygrocarabus*) *variolosus* Fabricius (Coleoptera Carabidae). Nota preliminare. – Atti della Accademia Nazionale Italiana di Entomologia 11: 182-184.
Sutherland, W.J. (1998). The importance of behavioural studies in conservation biology. – Animal Behaviour 56: 801-809.
Thiele, H.U. & Weber, F. (1968). Tagesrhythmen der Aktivität bei Carabiden. – Oecologia 1: 315-355.
Turin, H. (2000). De Nederlandse Loopkevers – Verspreiding en oecologie. – Nationaal Natuurhistorisch Museum Naturalis, Leiden.
Turin, H., Penev, L. & Casale, A. (Eds) (2003). The Genus *Carabus* in Europe. – Pensoft, Sofia-Moscow.
Tyre, A., Kerr, G.D., Tenhumberg, B. & Bull, C.M. (2007). Identifying mechanistic models of spatial behaviour using pattern-based modelling: An example from lizard home ranges. – Ecological Modelling 208: 307-316.
Ulfstrand, S. (1996). Behavioural ecology and conservation biology. – Oikos 77: 183.
Vuilleumier, S. & Metzger, R. (2006). Animal dispersal modelling: Handling landscape features and related animal choices. – Ecological Modelling 190: 159-170.
Wallin, H. (1988). Movements of carabid beetles (Coleoptera, Carabidae) inhabiting cereal fields: a field tracing study. – Oecologia 77: 39-43.
Yamazaki, K. (2006). Feeding of a shore-inhabiting ground beetle, *Scarites aterrimus* (Coleoptera: Carabidae). – Coleopterists Bulletin 60: 75-79.

APPENDIX

Table A1. Results of individual tracing of *Carabus variolosus* in the Arnsberg forest (NW Germany) grouped by the sex of the beetles.

Beetle number	Start of observation	End of observation	Observation duration [min.]	No. of positions	Distance covered [cm]	Velocity [cm/min]
females						
1230	21.5.03 21:18	21.5.03 23:23	125	25	2,087	16.70
2110	28.5.03 20:30	28.5.03 23:39	189	15	1,675	8.86
0005	18.6.03 21:40	18.6.03 23:20	100	19	1,632	16.32
0112	31.7.03 21:14	1.8.03 0:56	222	46	1,222	5.50
0311	13.8.03 21:14	13.8.03 22:42	88	19	1,547	17.58
0140	14.8.03 20:08	14.8.03 22:24	136	13	449	3.30
4022	19.8.03 21:36	19.8.03 23:54	138	14	483	3.50
4006	27.8.03 20:00	28.8.03 0:54	294	41	1,433	4.87
0062	27.8.03 20:42	27.8.03 21:58	76	23	1,730	22.76
5110	28.8.03 20:40	28.8.03 23:30	170	16	339	1.99
3060	30.8.03 21:18	31.8.03 0:24	186	25	867	4.66
0321	16.5.04 21:03	17.5.04 1:11	248	14	196	0.79
603	27.5.04 23:10	28.5.04 1:40	150	29	1,252	8.35
0501	8.6.04 22:04	9.6.04 0:44	160	22	685	4.28
3600	8.6.04 22:04	9.6.04 0:44	160	14	518	3.24
		mean	162.8	22.3	1,074	8.18
males						
3004	4.8.03 21:48	4.8.03 23:32	104	14	390	3.75
0520	5.8.03 21:22	5.8.03 23:38	136	14	493	3.63
0525	7.8.03 21:56	7.8.03 23:36	100	14	402	4.02
0041	8.8.03 21:10	8.8.03 22:26	76	11	465	6.12
0111	21.8.03 21:24	22.8.03 0:14	170	22	877	5.16
0602	10.5.04 19:40	11.5.04 0:04	264	18	585	2.22
0610	16.5.04 22:20	17.5.04 1:06	166	10	240	1.45
1360	18.5.04 21:31	19.5.04 1:35	244	12	337	1.38
1222	18.5.04 23:18	19.5.04 0:12	54	10	775	14.35
1630	31.5.04 21:18	31.5.04 23:54	156	19	2,248	14.41
1320	31.5.04 23:00	31.5.04 23:55	55	10	310	5.64
5200	1.6.04 21:50	2.6.04 0:04	134	11	318	2.37
2500	1.6.04 21:52	1.6.04 22:54	62	6	329	5.31
		mean	132.4	13.2	598	5.37
		Total mean	148.7	18.1	853	6.87

L. Penev, T. Erwin & T. Assmann (Eds) 2008
BACK TO THE ROOTS AND BACK TO THE FUTURE. TOWARDS A NEW SYNTHESIS AMONGST TAXONOMIC, ECOLOGICAL AND BIOGEOGRAPHICAL APPROACHES IN CARABIDOLOGY
Proceedings of the XIII European Carabidologists Meeting, Blagoevgrad, August 20-24, 2007, pp. 425-450.

Sofia–Moscow

Influences of succession and harvest intensity on ground beetle (Coleoptera, Carabidae) populations in the boreal mixed-wood forests of Alberta, Canada: species matter

Joshua M. Jacobs[1,2], Timothy T. Work[2] & John R. Spence[1]

[1] Department of Renewable Resources, University of Alberta, Alberta, Canada.
E-mail: email@joshuajacobs.org
[2] Département des Sciences Biologiques, Université du Québec à Montréal, Quebec, Canada

SUMMARY

We studied responses of carabid beetles to variable retention harvest and succession of canopy-tree species in stands of natural fire origin in NW Alberta, Canada. We sampled carabids using pitfall traps, 1-, 2- and 5-years following harvest from four cover-types along a boreal successional gradient (deciduous-dominated, deciduous-dominated with a conifer understory, mixed-wood, and coniferous-dominated) and five levels (75%, 50%, 20% 10% and 2% residual) of variable retention harvesting in each successional stage. We also sampled never cut stands within each successional stage. The ten most abundant species made up ca. 95% of the 32 976 individuals collected. Two species were distributed across all successional cover-types, and the remaining eight species were restricted to either early- or late-successional forests. Five species (*Calathus ingratus*, *Platynus decentis*, *Stereocerus haematopus*, *Agonum retractum* and *Calathus advena*) responded negatively to forest harvest, and their population sizes dropped below detection in many of the higher-intensity harvest treatments. However, responses to harvest vary with species and, thus, the overall consequences for carabid assemblages will likely depend on the mix of species involved.

Keywords: disturbance, sustainable forest management, ecosystem management

INTRODUCTION

The role of succession and disturbance in determining composition of epigaeic assemblages has been the focus of intense study in the boreal forest (Spence et al., 1996; Niemelä, 1999; Buddle et al., 2000; Heliölä et al., 2001; Koivula et al., 2002; Work et al., 2004; Cobb et al., 2007). Most studies have considered the significance of post-disturbance recovery on actively reforested landscapes (e.g., Niemelä et al., 1993) or, alternatively, the role of succession in forests that do not recover from disturbance through broad changes in stand characteristics as observed in the mixed-wood region of the boreal forest (e.g., Buddle et al., 2000). However, joint influence of disturbance and succession on distribution and abundance of individual species could potentially explain much of the variation observed in epigaeic assemblages across the boreal forest.

The Canadian boreal forest covers about ca. 340 million ha (Natural Resources Canada, 2003). Composition and structure of this forest is generally thought to reflect patterns generated by a wildfire regime on boreal landscapes (Bergeron et al., 1999). Large, 'stand-replacing' fires initiate secondary succession in these landscapes, and stand composition is thought to flow from the combined results of variable burn severity (Koivula & Spence, 2006) and differential growth rates among tree species (Viereck, 1983). Typically, these areas are first colonized and dominated by the fast-growing species, Trembling Aspen (*Populus tremuloides* Michx.) and Balsam Poplar (*Populus balsamifera* L.), which can sucker from residual root masses not killed by fires. More slowly growing, shade-tolerant tree species, mainly conifers, will establish in the understory of these forests, depending on seed production (Bergeron & Dubuc, 1988) and proximity to potential seed sources, such as neighbouring intact stands (McClanahan, 1986). Over most of this region, White Spruce (*Picea glauca* [Moench] Voss) is most abundant in the understory of deciduous stands, although Black Spruce (*Picea mariana* [Mill.] B.S.P.) can dominate in wetter areas or on north-facing slopes. As gaps in the deciduous canopy develop – usually as a result of blowdown resulting from decay fungi, insects, diseases or meteorological vectors (McCarthy, 2001), this initial cohort of coniferous trees enters the canopy, creating a mixed-wood forest. In the long-term (100-300 yrs) absence of stand-replacing disturbance or significant episodes of gap dynamics, these stands will mature into old-growth stands dominated by conifer species (Rowe, 1994).

With introduction of industrial-scale forestry, harvesting has become the major disturbance factor in the boreal forests of Canada (Pratt & Urquhart, 1994). Like natural disturbances, harvesting can also reinitiate the successional process by returning late successional forest to early successional stages. Clear-cut harvesting resembles the action of severe forest fires, and variable retention and selective harvesting may alter spatial patterns of succession in ways somewhat similar to more patchy burns and processes of gap dynamics (Bergeron et al., 1999). By removing canopy trees in early successional stages, stand development may be accelerated to later successional stages (Harvey et al., 2002). However, conversely, removing trees in late-successional stands moves stands or patches within stands towards earlier successional stages, as fast growing deciduous tress colonize the newly disturbed areas. The specific effects of such changes on ground beetle assemblages have not been well studied.

Distribution of insect species tends to follow gradients that result from both succession and disturbance. Numerous studies have identified species that are restricted to, or are significantly more dominant, in a particular stage of the successional gradient (Spence et al., 1996; Work et al., 2004; Jacobs et al., 2007). Although changes in canopy tree species are the most apparent successional change, Work et al. (2004) found that moss cover, forb cover and coarse woody debris (CWD) composition co-vary with canopy changes and, moreover, are potentially important determinants of ground beetle distributions. Many disturbance-adapted species flourish for short periods following disturbance, and some are specifically adapted, for example, to short time periods following fire (Evans, 1971; Wikars, 1992; Koivula et al., 2006; Cobb et al., 2007).

In this paper we assess the relative influence of succession and anthropogenic disturbances resulting from large-scale forest harvesting, on ground beetles in north-western Alberta, Canada. Furthermore, we investigate interactions between these two habitat-altering factors on boreal carabids and ask how their species-specific influences change over the first five years following harvest. Because succession and disturbance play such major roles in controlling stand characteristics, we predicted that these processes and their interactions would significantly influence distribution and abundance of ground beetles in a forested landscape. Here, we seek especially to understand how much generality exists in responses of species normally thought of as 'forest specialists'.

METHODS

Study Site

This study was conducted at the EMEND (Ecosystem Management by Emulating Natural Disturbance) research site, a large-scale experiment comparing alternative silvicultural strategies using replicated whole-stand treatments. The site is ca. 24 km^2 and is located ca. 90 km north-west of Peace River, Alberta (56°46' 13" N, 118°22' 28" W). Elevation at EMEND ranges from 677-880 m a.s.l., and the soils are fine-textured lacustrian (Work et al., 2004).

The forests included in the EMEND experiment are divided into four locally common cover-types, representing major successional stages of the boreal mixed-wood forest. The initial successional forest cover-type is deciduous dominated (DDOM) and defined as having a canopy comprised of ≥70% deciduous trees, mainly Trembling Aspen and Balsam Poplar, with minor elements of Paper Birch (*Betula papyrifera* Marshall). Trees in DDOM stands are generally 90-120 years old. The second successional cover-type (DDOMU) has a deciduous canopy as in DDOM, but also includes a significant conifer understory. The main coniferous species is White Spruce but the forest at EMEND also includes Black Spruce, Balsam Fir (*Abies balsamea* [L.] P. Mill.) and Lodgepole Pine (*Pinus contorta* Dougl.). The mid-successional cover-type (MIX) is a mix of deciduous and coniferous canopy with neither comprising more than 70% of the canopy. The oldest successional cover-type (CDOM) has ≥70% coniferous trees in the canopy and these are ca. 200-240 years old. The dominant shrubs at EMEND are *Viburnum edule* (Michx.), *Rosa acicularis* Lindl., *Shepherdia canadensis* (L.), *Alnus crispa* ([Aiton.] Pursh.) and *A. tenuifolia* Nutt.

Relatively homogeneous forest stands representing each of the four cover-types were divided into 10-ha compartments. Compartments in each cover-type received one of five retention-harvest treatments (2%, 10%, 20%, 50% or 75% of the original volume), one of two experimental burn treatments (not further considered here), or were left as uncut 'controls'. Treatments were replicated in different forest stands three times across the experiment, each forest stand received all treatments in a different configuration, for a total of 96 experimental compartments (72 considered here).

Harvest treatments were applied using mechanical harvesting machines ('Feller Bunchers') during the winter of 1998-1999. The harvest pattern was a uniform shelter-wood harvest system, with residuals left by cutting a portion of trees in 15-m wide strips of forest defined by 5-m wide machine corridors cut 20-m apart (center-center). In the 75% retention compartments, only machine corridors were cut. In strip-cut treatments retaining less than 75%, retention trees were removed from forested strips in the following ratios: 50% 1 (cut) : 2 (left); 20%, 3 : 1; and 10% 7 : 1. Clear-cuts (2% residual) were done as generally applied in western Canada, i.e., ca. 1-2% standing residual, and machinery was not restricted to corridors. Two aggregated, elliptical residual patches (ca. 0.4 ha and 0.2 ha) were left unharvested within all harvested compartments. Beetles in these aggregated residual patches were not considered in this study but are being considered elsewhere.

Carabid samples

We sampled ground beetles in all compartments from May to August (the frost-free season) in 1999, 2000 and 2004 using pitfall traps. Traps were 12 cm in diameter and consisted of an outer permanent cup (1000 mL) and a removable inner cup (500 mL) covered by a 20-cm^2 suspended rain cover (Spence & Niemelä, 1994). Traps were filled with ethylene glycol as a killing agent and preservative, and were serviced and emptied approximately every three weeks. Samples were stored in 70% ethanol until the ground beetles were picked from the samples and identified to species with nomenclature following Bousquet (1991).

Within each compartment, six 20-m transects were randomly established as permanent plots for the collection of environmental data and variables pertaining to stand structure. Pitfall traps were placed at either end of three of these transects chosen randomly within each compartment. Thus, carabid samples used for this study were collected using

432 traps (72 compartments x 6 traps) in each of thee three years resulting in 40 482, 45 584, and 41 367 total trap days in 1999, 2000, and 2004, respectively.

Data analysis

We sought to determine whether catches of individual species varied among successional stages and differing intensities of forest harvest. Furthermore, we investigated the relative influence of these two gradients (succession and disturbance), if they were interacting with each other, and how their effects and possible interaction varied among years. Here, we consider only the ten most abundant carabid species from the pooled three-year catch of 32 709 individual carabids. Less abundant species were not included in this analysis but are reported elsewhere in a comparison of assemblage-level changes to harvesting treatments (Work et al., in prep.).

Raw catch-rate data were adjusted before analysis as follows. To account for trap disturbances, the annual overall abundance of each species was totalled for each compartment and divided by the total number of total trap days accumulated by the six pitfall traps within that compartment. Square-root or a modified logarithmic transformation (McCune & Grace, 2002) were applied to improve normality of catch rates based on QQ plots (Zar, 1996) for each species.

Standardized and appropriately transformed data were analyzed using a repeated-measures analysis of variance (RM-ANOVA), with % retention (disturbance) and cover-type (succession) as independent variables and year as the repeated measure. For these analyses, the level of replication was the compartment, i.e., the pooled catch from all six traps within the compartment. P-values of within subject factors were corrected for departures in sphericity using the Greenhouse-Geisser epsilon value (Scheiner & Gurevitch, 1993). All statistical analyses were performed using R 2.1.1 (R Development Core Team, 2005).

RESULTS

The ten most abundant species represented 95% of the overall catch of 59 species. These were *Pterostichus adstrictus* Eschscholtz, 1823 (Fig. 1, n=9743); *Calathus ingratus* Dejean, 1828, (Fig. 2, n=4813); *Platynus decentis* (Say, 1823), (Fig. 3, n=4233); *Stereocerus haematopus* (Dejean, 1831) (Fig. 4, n=3760); *Agonum retractum* LeConte, 1848 (Fig. 5, n=2977); *Calathus advena* (LeConte, 1848) (Fig. 6, n=2228); *Patrobus foveocollis* (Eschscholtz, 1823) (Fig. 7, n=1124); *Pterostichus pensylvanicus* LeConte, 1873) (Fig. 8, n=918); *Carabus chamissonis* Fischer von Waldheim, 1820) (Fig. 9, n=894); and *Calosoma frigidum* Kirby, 1837 (Fig. 10, n=623).

Responses across the successional cover-type gradient

Forest successional stage (cover-type) significantly affected the catch of eight of the ten abundant species. Furthermore, the effect of successional stage varied significantly with year for all but one, viz. *C. chamissonis*, of these eight species (significant cover-type x year interaction; Table 1). Although the ANOVA did not demonstrate a significant effect of forest succession for *C. ingratus*, there was a significant interaction between succession and year for this species.

Three predominant patterns in catch were observed among the responses of individual species to forest succession. Catches of the two most abundant species, viz. *P. adstrictus* and *C. ingratus*, did not differ significantly among the successional cover-types (Figs 1-2). However, *P. decentis, A. retractum, P. foveocollis, P. pensylvanicus, C. chamissonis* and *C. frigidum* were caught most commonly in early successional deciduous forests (Figs 3, 5, 7-10). *Stereocerus haematopus* and *C. advena*, on the other hand, were more commonly caught in late successional conifer forests (Figs 4, 6).

Except for *C. frigidum*, no species showed consistent interaction between successional stage and disturbance (harvest) intensity (Table 1). However, year of sample was significant for all species, and year significantly interacted with successional stage for eight of the abundant species (Table 1). High intensities of harvesting tends to homogenize stands in terms of regenerating tree-species and age-structure resulting in a loss of initial successional stage effect in these stands five years following harvesting. For example, the response to successional stage by the late-successional species *C. advena* (Fig. 6) was consistent across disturbance levels in 1999 and 2000, but in 2004 the overall catch was much lower in all harvesting intensities and successional stages, and the strong effect of successional stage observed immediately after harvest was no longer apparent. The early successional species *P. decentis* responded similarly; i.e., we identified a strong effect of successional stage in 1999 and 2000 but not in 2004.

Response to increasing intensities of forest harvesting

All ten abundant species were affected by the intensity of disturbance (forest harvest), either as a main effect (7 species) or interacting with year (8 species) (Table 1). We detected three main types of response to the disturbance of forest harvest. Firstly, catch of *Pterostichus adstrictus* initially increased in the first and second post-harvest summers in disturbed compartments across all successional stages (Fig. 1). Five years following harvest, however, the catch rate of *P. adstrictus* declined in these disturbed compartments. Secondly, five mature-forest species, viz, *P. decentis, S. haematopus, A. retractum* and *C. advena* (Figs 3-6), responded only slightly to disturbance in the first and second post-harvest year, but displayed a strong negative response to disturbance intensity five years following harvest. Thirdly, *C. ingratus* was more generally distributed than the mature forest species and was only strongly negatively affected by disturbance intensity in the early successional clear- cuts (Fig 2).

Table 1. Results from the RM-ANOVA on the standardized abundances of the 10 most abundant species of ground beetles collected at the EMEND research site for between subject comparisons Cover-Type (Successional Stage), Harvest (Disturbance) Intensity and their interaction, and within subject comparisons of Year (time since harvest), and all interactions of Year with the 2 main factors.

	df	SS	MS	F	P	P_{GG}[†]
Cover-Type						
Pterostichus adstrictus[a]	3	0.004	0.001	0.129	0.943	
Calathus ingratus[b]	3	0.392	0.131	1.887	0.144	
Platynus decentis[b]	3	6.598	2.199	20.776	<0.001***	
Stereocerus haematopus[a]	3	0.400	0.133	8.862	<0.001***	
Agonum retractum[b]	3	7.645	2.548	35.343	<0.001***	
Calathus advena[a]	3	0.440	0.147	17.889	<0.001***	
Patrobus foveocollis[b]	3	0.363	0.121	3.094	0.036*	
Pterostichus pensylvanicus[a]	3	0.203	0.068	17.658	<0.001***	
Carabus chamissonis[a]	3	0.077	0.026	4.929	0.005**	
Calosoma frigidum[a]	3	0.167	0.056	13.629	<0.001***	
Harvest Intensity						
Pterostichus adstrictus	5	0.227	0.045	4.524	0.002**	
Calathus ingratus	5	0.736	0.147	2.126	0.078	
Platynus decentis	5	1.758	0.352	3.322	0.012*	
Stereocerus haematopus	5	0.108	0.022	1.444	0.226	
Agonum retractum	5	0.934	0.187	2.590	0.037*	
Calathus advena	5	0.166	0.033	4.054	0.004**	
Patrobus foveocollis	5	0.602	0.120	3.075	0.017*	
Pterostichus pensylvanicus	5	0.053	0.011	2.767	0.028*	
Carabus chamissonis	5	0.073	0.015	2.800	0.027*	
Calosoma frigidum	5	0.047	0.009	2.318	0.058	
Cover-Type x Harvest Intensity						
Pterostichus adstrictus	15	0.108	0.007	0.719	0.753	
Calathus ingratus	15	0.734	0.049	0.707	0.765	
Platynus decentis	15	0.865	0.058	0.545	0.901	
Stereocerus haematopus	15	0.136	0.009	0.605	0.856	
Agonum retractum	15	1.160	0.077	1.072	0.405	
Calathus advena	15	0.060	0.004	0.491	0.934	
Patrobus foveocollis	15	0.266	0.018	0.453	0.952	
Pterostichus pensylvanicus	15	0.058	0.004	1.003	0.468	
Carabus chamissonis	15	0.051	0.003	0.652	0.816	
Calosoma frigidum	15	0.088	0.006	1.444	0.166	
Year						
Pterostichus adstrictus	2	0.700	0.350	109.163	<0.001***	<0.001***
Calathus ingratus	2	1.663	0.832	33.560	<0.001***	<0.001***
Platynus decentis	2	6.882	3.441	128.564	<0.001***	<0.001***
Stereocerus haematopus	2	0.251	0.125	54.217	<0.001***	<0.001***

	df	SS	MS	F	P	P_{GG}[†]
Agonum retractum	2	3.164	1.582	64.956	<0.001***	<0.001***
Calathus advena	2	0.248	0.124	49.749	<0.001***	<0.001***
Patrobus foveocollis	2	0.151	0.076	4.379	0.015*	0.020*
Pterostichus pensylvanicus	2	0.065	0.033	30.879	<0.001***	<0.001***
Carabus chamissonis	2	0.012	0.006	5.324	0.006**	0.006**
Calosoma frigidum	2	0.071	0.035	24.444	<0.001***	<0.001***
Year x Cover-Type						
Pterostichus adstrictus	6	0.029	0.005	1.504	0.185	0.188
Calathus ingratus	6	0.575	0.096	3.869	0.002**	0.002**
Platynus decentis	6	1.227	0.205	7.640	<0.001***	<0.001***
Stereocerus haematopus	6	0.034	0.006	2.471	0.029*	0.033*
Agonum retractum	6	0.569	0.095	3.891	0.002**	0.002**
Calathus advena	6	0.086	0.014	5.743	<0.001***	<0.001***
Patrobus foveocollis	6	0.315	0.053	3.043	0.009**	0.013*
Pterostichus pensylvanicus	6	0.018	0.003	2.915	0.012*	0.012*
Carabus chamissonis	6	0.010	0.002	1.558	0.168	0.168
Calosoma frigidum	6	0.065	0.011	7.508	<0.001***	<0.001***
Year x Harvest Intensity						
Pterostichus adstrictus	10	0.290	0.029	9.051	<0.001***	<0.001***
Calathus ingratus	10	1.580	0.158	6.377	<0.001***	<0.001***
Platynus decentis	10	0.591	0.059	2.206	0.024*	0.030*
Stereocerus haematopus	10	0.098	0.010	4.228	<0.001***	<0.001***
Agonum retractum	10	0.592	0.059	2.431	0.013*	0.015*
Calathus advena	10	0.104	0.010	4.169	<0.001***	<0.001***
Patrobus foveocollis	10	0.147	0.015	0.851	0.581	0.569
Pterostichus pensylvanicus	10	0.016	0.002	1.473	0.161	0.162
Carabus chamissonis	10	0.034	0.003	2.994	0.002**	0.003**
Calosoma frigidum	10	0.044	0.004	3.053	0.002**	0.002**
Year x Cover-Type x Harvest Intensity						
Pterostichus adstrictus	30	0.118	0.004	1.225	0.228	0.232
Calathus ingratus	30	0.938	0.031	1.262	0.197	0.207
Platynus decentis	30	0.956	0.032	1.191	0.258	0.269
Stereocerus haematopus	30	0.078	0.003	1.128	0.323	0.328
Agonum retractum	30	0.863	0.029	1.182	0.267	0.272
Calathus advena	30	0.042	0.001	0.569	0.960	0.957
Patrobus foveocollis	30	0.577	0.019	1.112	0.340	0.347
Pterostichus pensylvanicus	30	0.029	0.001	0.907	0.608	0.608
Carabus chamissonis	30	0.029	0.001	0.862	0.670	0.670
Calosoma frigidum	30	0.089	0.003	2.037	0.005**	0.005**

[a] Catch-rates were square root transformed prior to ANOVA calculations.

[b] Catch-rates were log transformed prior to ANOVA calculations.

[†] Greenhouse-Geisser corrected P-values used to correct for departures from sphericity for within subject comparisons.

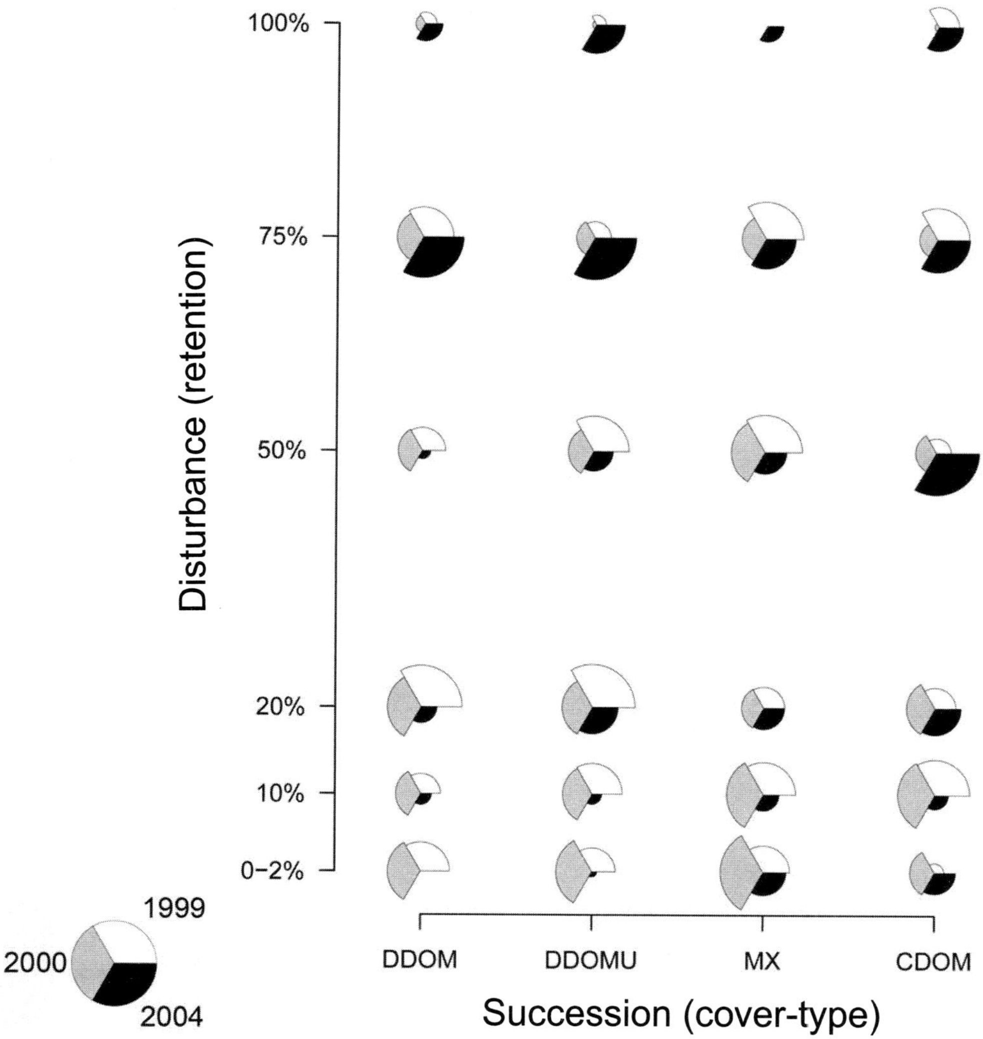

Fig. 1. Star plots of the abundance of *Pterostichus adstrictus* across a disturbance gradient (y-axis) and successional gradient (x-axis) over 3 years (1999=white, 2000= grey, 2004=black) n=9743. The size (area) of each 'pie piece' represents the proportional mean abundance relative to the maximum mean abundance of all treatment combinations. DDOM- deciduous dominated, DDOMU- deciduous dominated with a coniferous understory, MIX- mixed canopy, CDOM- coniferous dominated.

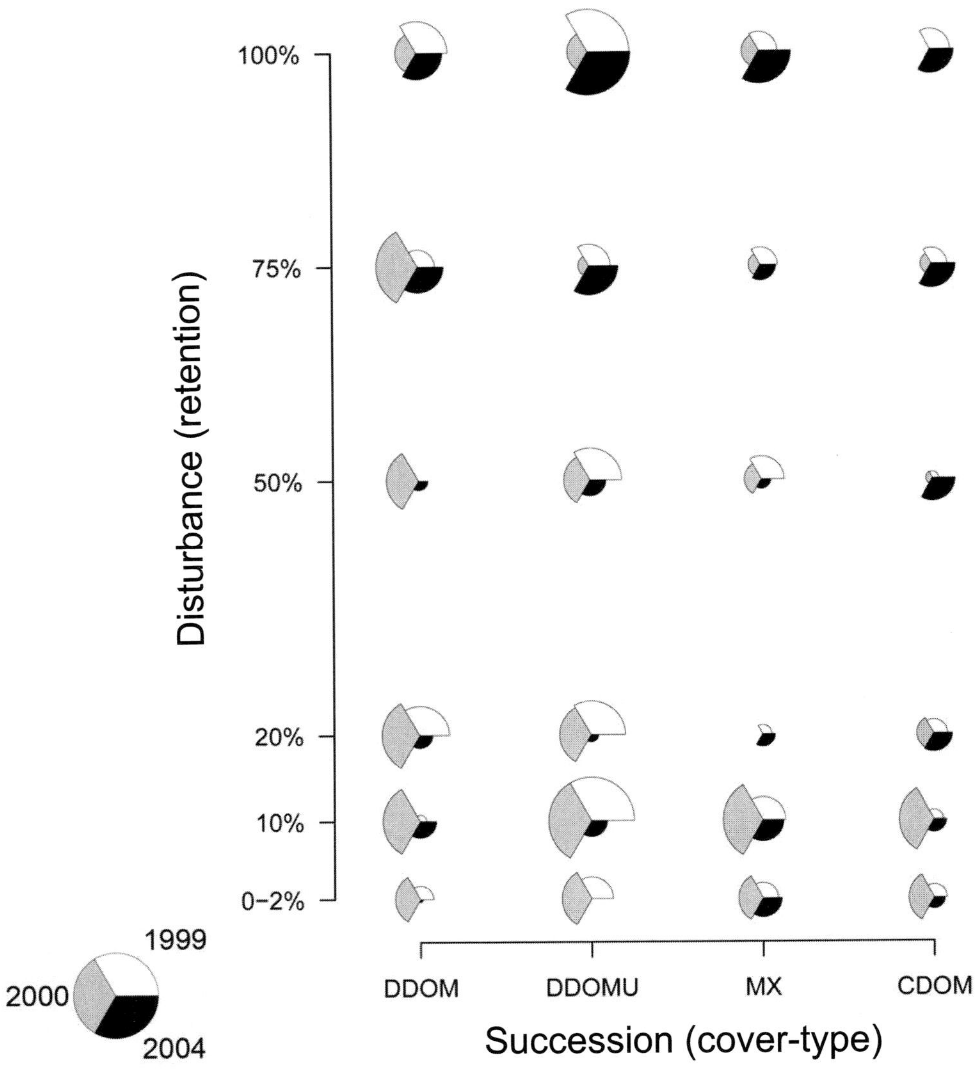

Fig. 2. Star plots of the abundance of *Calathus ingratus* across a disturbance gradient (y-axis) and successional gradient (x-axis) over 3 years (1999=white, 2000= grey, 2004=black) n=4813. The size (area) of each 'pie piece' represents the proportional mean abundance relative to the maximum mean abundance of all treatment combinations. DDOM- deciduous dominated, DDOMU- deciduous dominated with a coniferous understory, MIX- mixed canopy, CDOM- coniferous dominated.

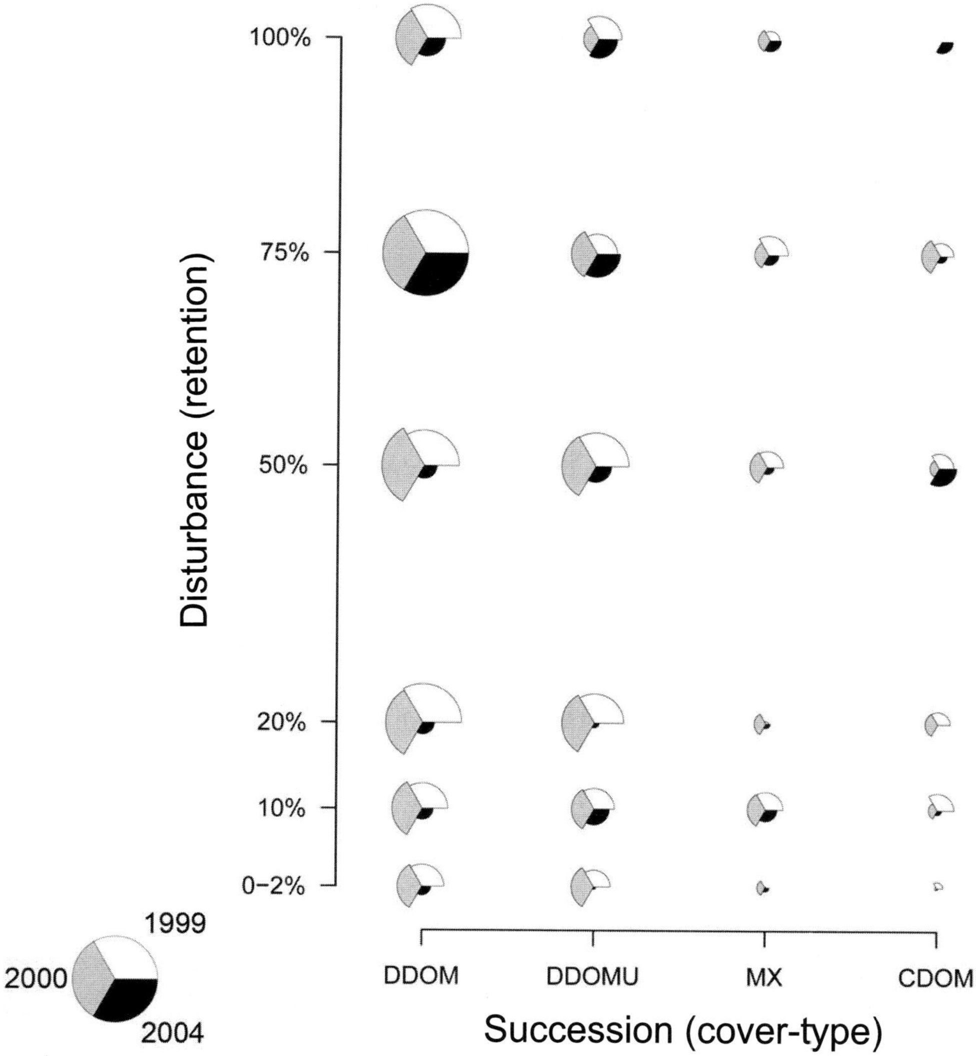

Fig. 3. Star plots of the abundance of *Platynus decentis* across a disturbance gradient (y-axis) and successional gradient (x-axis) over 3 years (1999=white, 2000= grey, 2004=black) n=4233. The size (area) of each 'pie piece' represents the proportional mean abundance relative to the maximum mean abundance of all treatment combinations. DDOM- deciduous dominated, DDOMU- deciduous dominated with a coniferous understory, MIX- mixed canopy, CDOM- coniferous dominated.

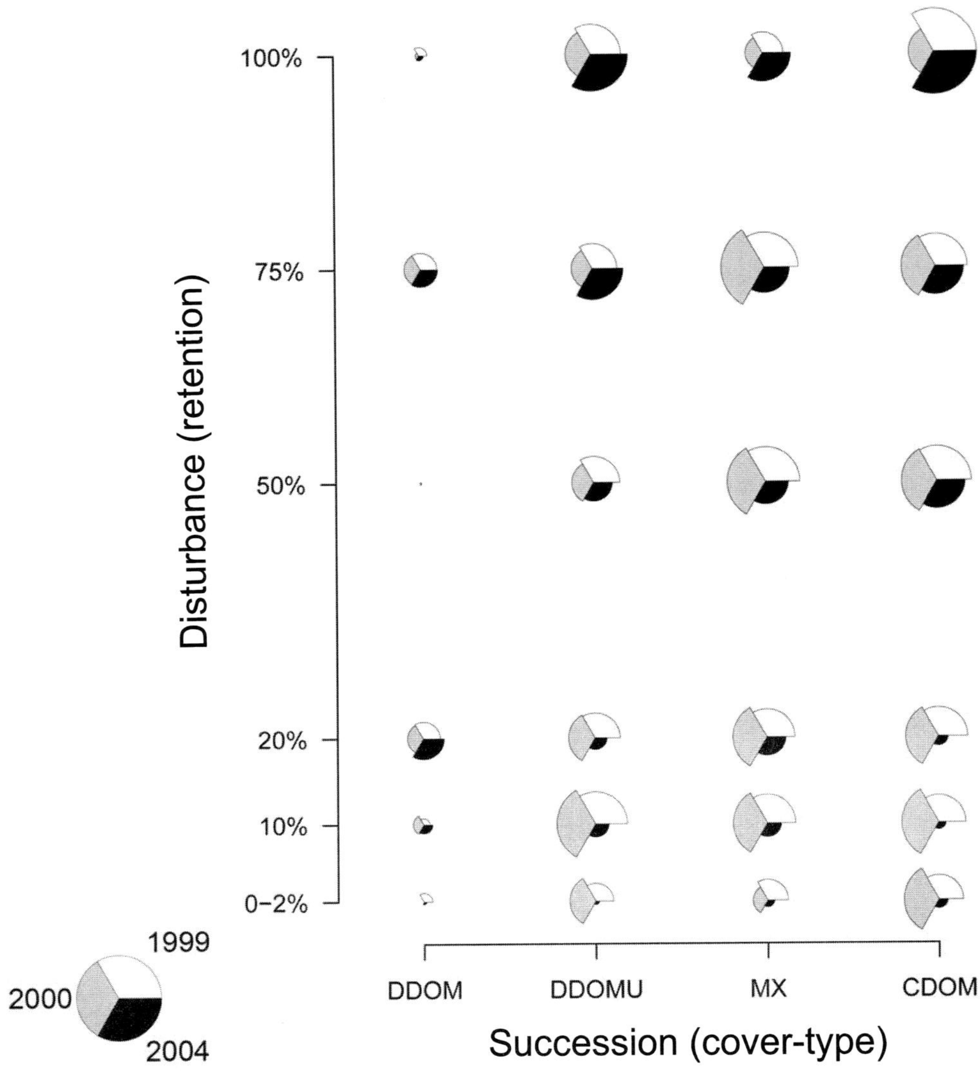

Fig. 4. Star plots of the abundance of *Stereocerus haematopus* across a disturbance gradient (y-axis) and successional gradient (x-axis) over 3 years (1999=white, 2000= grey, 2004=black) n=3760. The size (area) of each 'pie piece' represents the proportional mean abundance relative to the maximum mean abundance of all treatment combinations. DDOM- deciduous dominated, DDOMU- deciduous dominated with a coniferous understory, MIX- mixed canopy, CDOM- coniferous dominated.

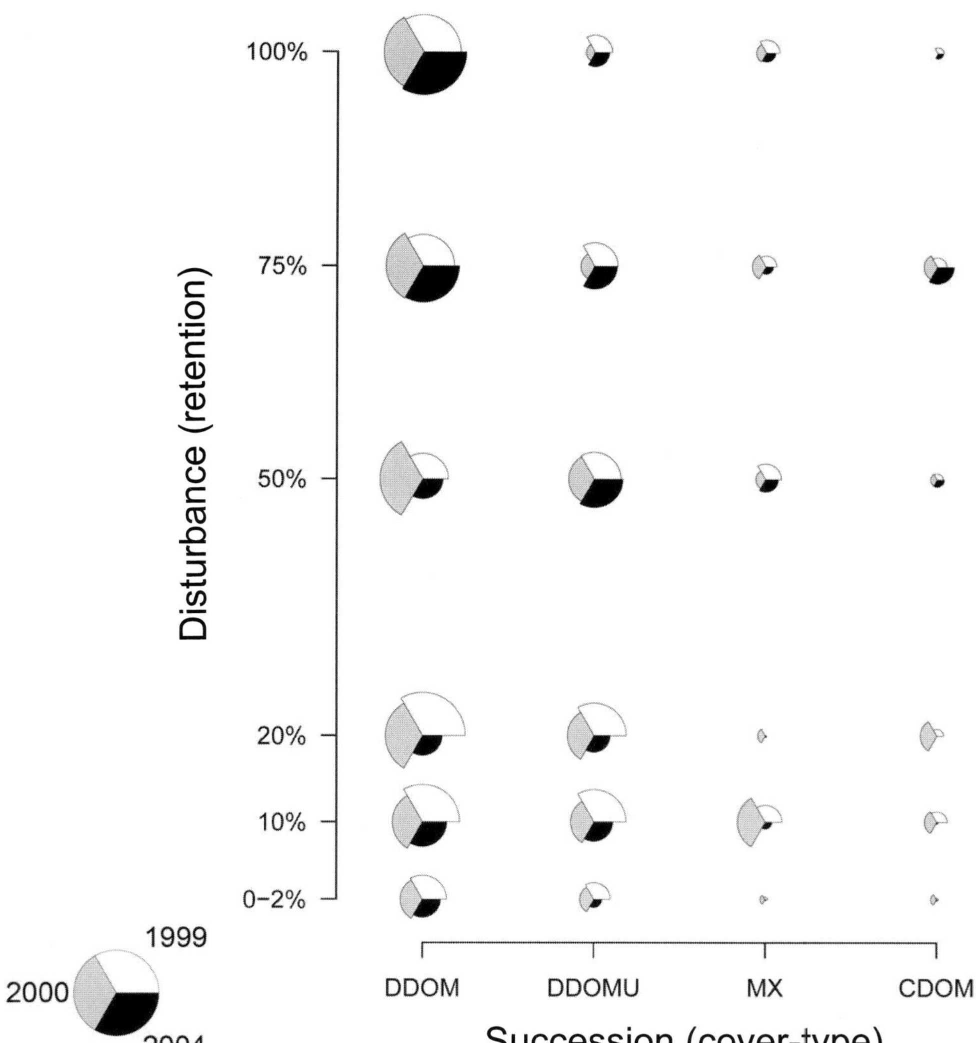

Fig. 5. Star plots of the abundance of *Agonum retractum* across a disturbance gradient (y-axis) and successional gradient (x-axis) over 3 years (1999=white, 2000= grey, 2004=black) n=2977. The size (area) of each 'pie piece' represents the proportional mean abundance relative to the maximum mean abundance of all treatment combinations. DDOM- deciduous dominated, DDOMU- deciduous dominated with a coniferous understory, MIX- mixed canopy, CDOM- coniferous dominated.

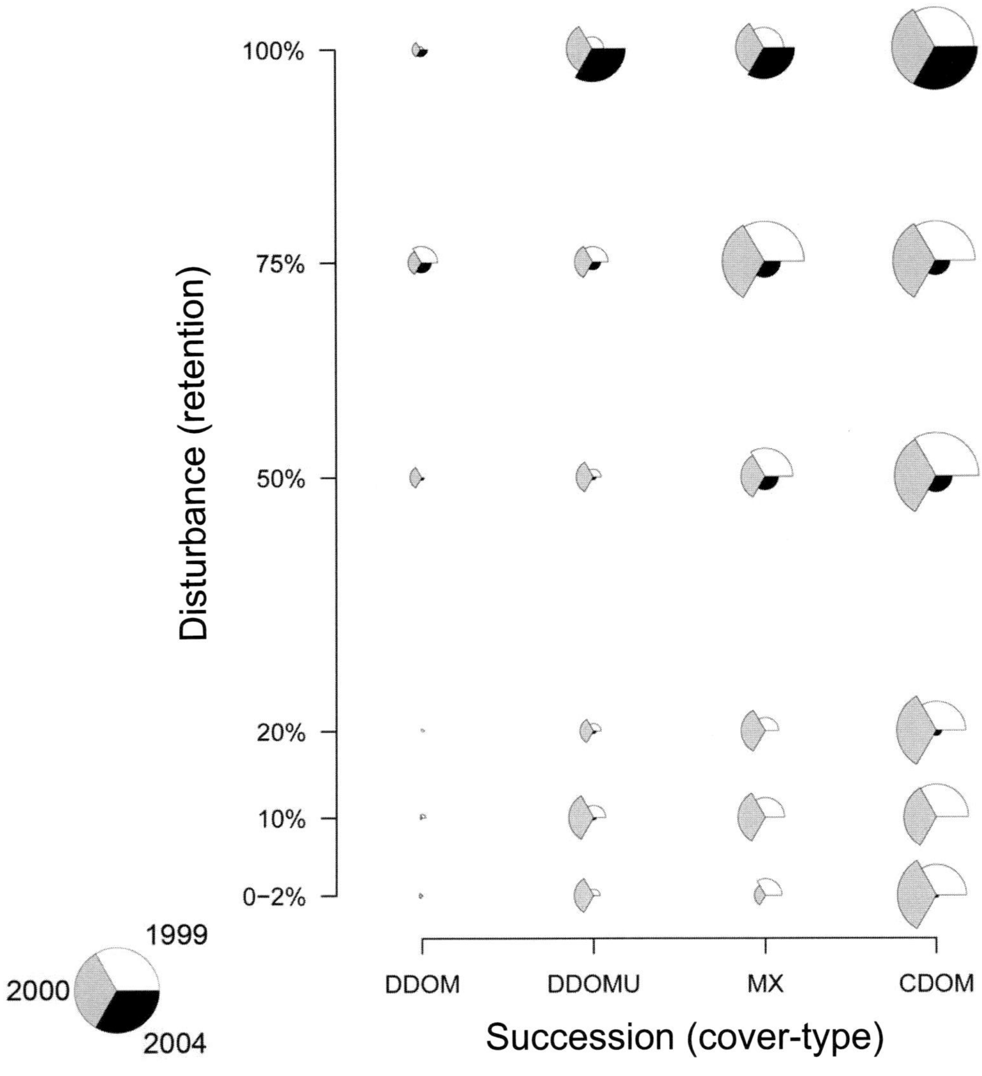

Fig. 6. Star plots of the abundance of *Calathus advena* across a disturbance gradient (y-axis) and successional gradient (x-axis) over 3 years (1999=white, 2000= grey, 2004=black) n=2228. The size (area) of each 'pie piece' represents the proportional mean abundance relative to the maximum mean abundance of all treatment combinations. DDOM- deciduous dominated, DDOMU- deciduous dominated with a coniferous understory, MIX- mixed canopy, CDOM- coniferous dominated.

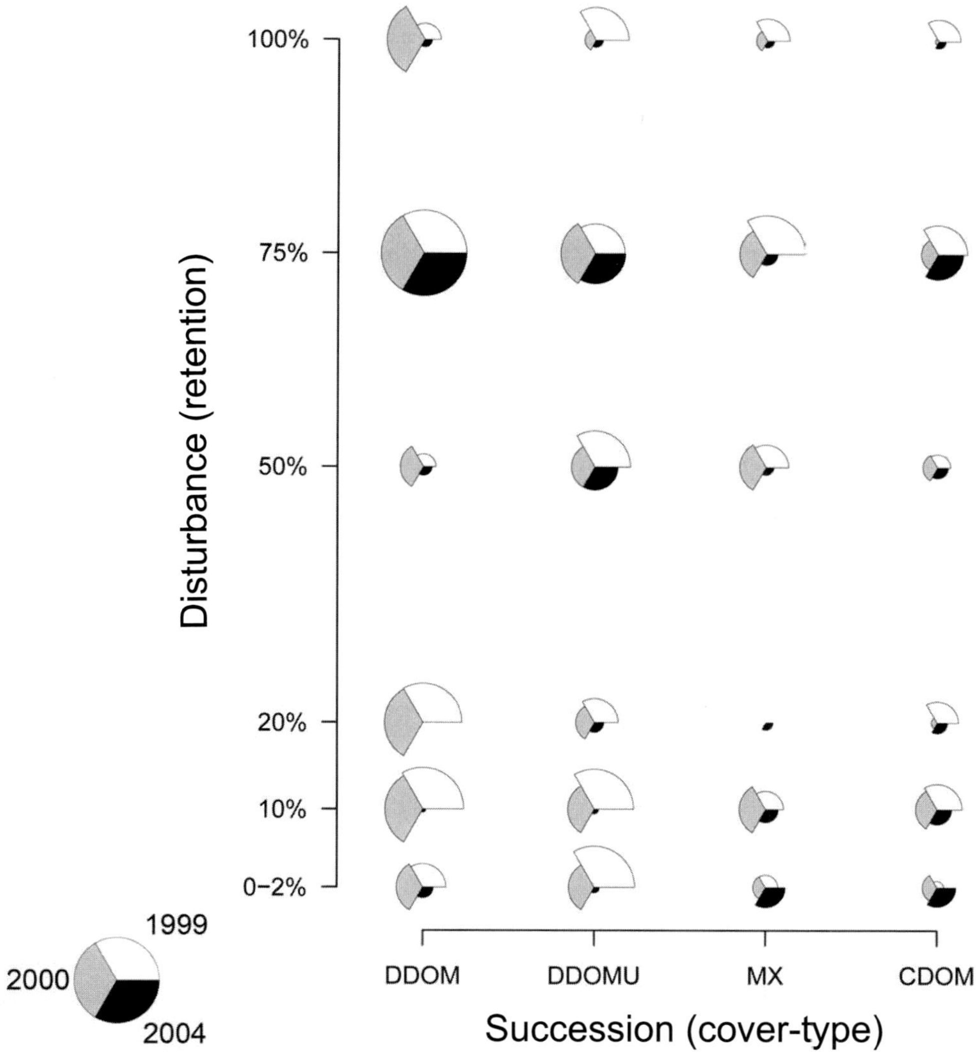

Fig. 7. Star plots of the abundance of *Patrobus foveocollis* across a disturbance gradient (y-axis) and successional gradient (x-axis) over 3 years (1999=white, 2000= grey, 2004=black) n=1124. The size (area) of each 'pie piece' represents the proportional mean abundance relative to the maximum mean abundance of all treatment combinations. DDOM- deciduous dominated, DDOMU- deciduous dominated with a coniferous understory, MIX- mixed canopy, CDOM- coniferous dominated.

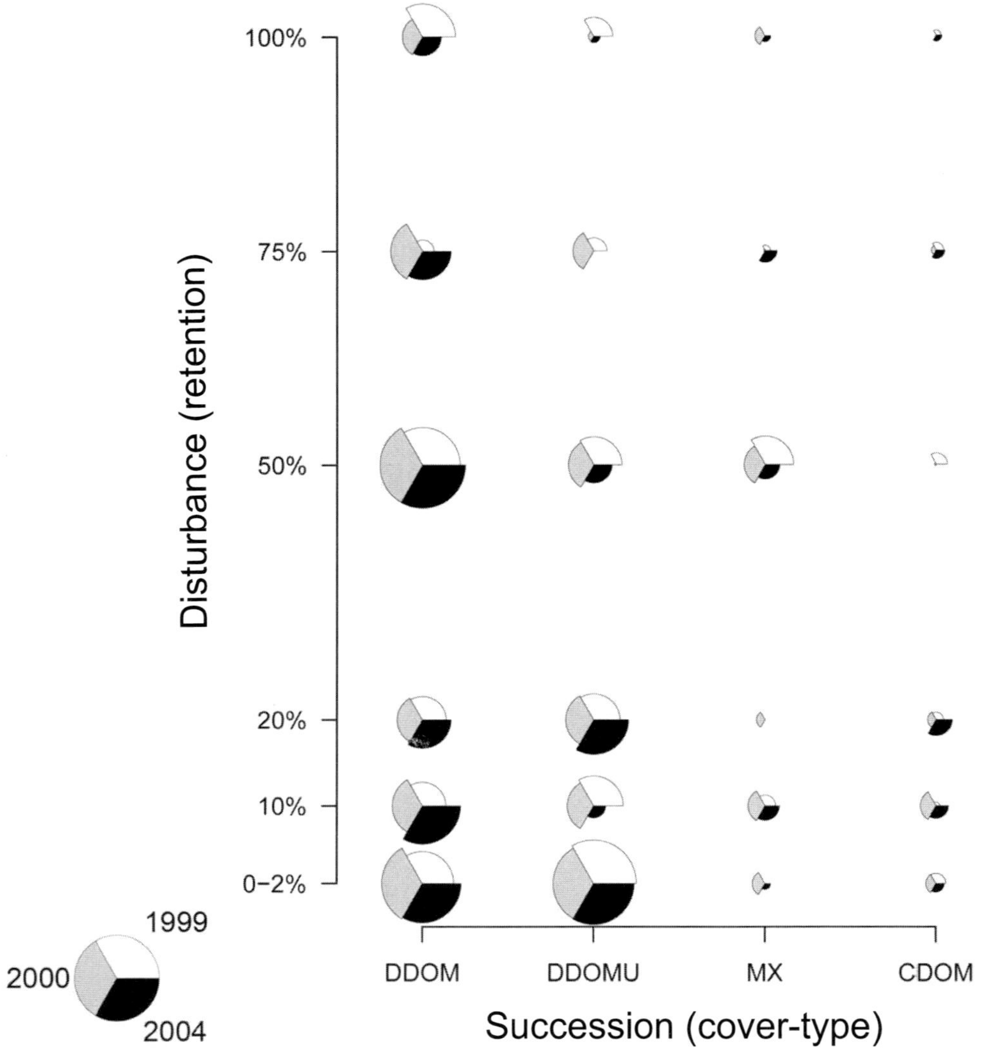

Fig. 8. Star plots of the abundance of *Pterostichus pensylvanicus* across a disturbance gradient (y-axis) and successional gradient (x-axis) over 3 years (1999=white, 2000= grey, 2004=black) n=918. The size (area) of each 'pie piece' represents the proportional mean abundance relative to the maximum mean abundance of all treatment combinations. DDOM- deciduous dominated, DDOMU- deciduous dominated with a coniferous understory, MIX- mixed canopy, CDOM- coniferous dominated.

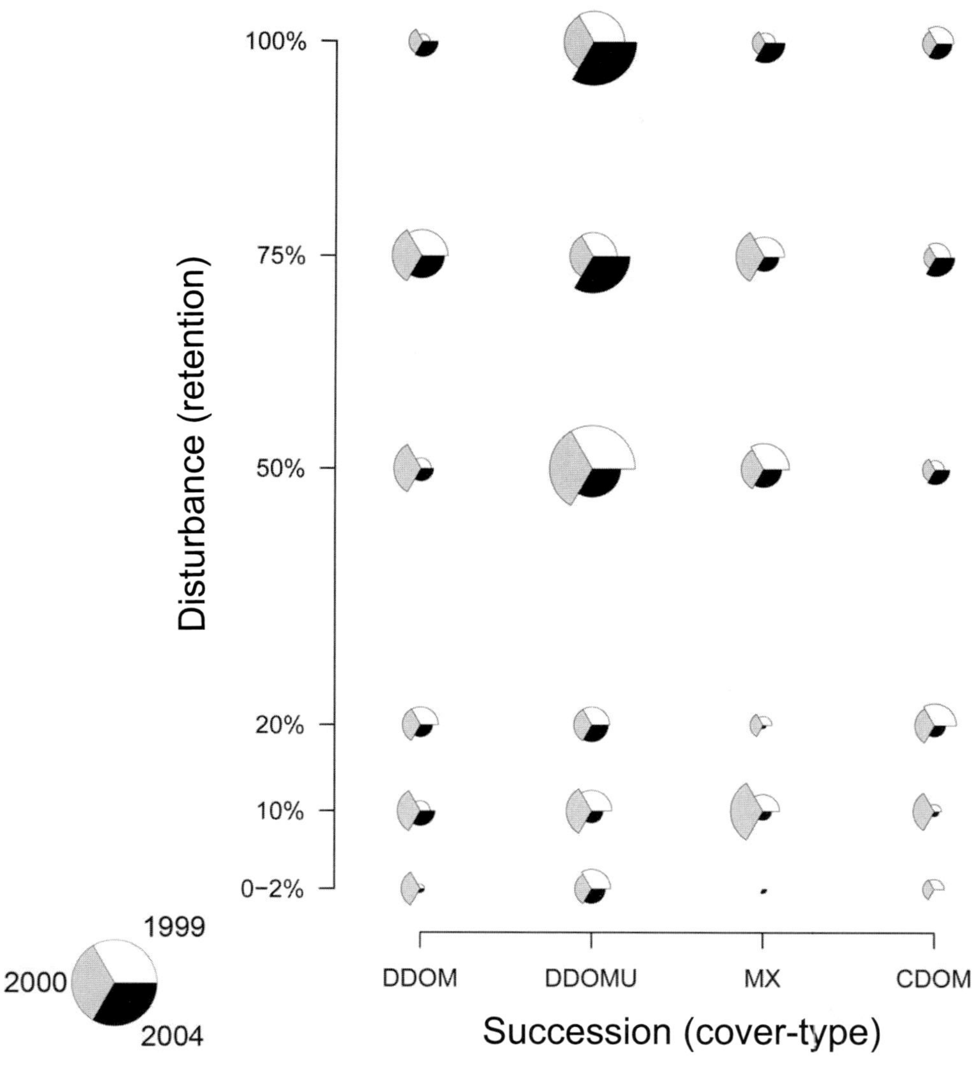

Fig. 9. Star plots of the abundance of *Carabus chamissonis* across a disturbance gradient (y-axis) and successional gradient (x-axis) over 3 years (1999=white, 2000= grey, 2004=black) n=894. The size (area) of each 'pie piece' represents the proportional mean abundance relative to the maximum mean abundance of all treatment combinations. DDOM- deciduous dominated, DDOMU- deciduous dominated with a coniferous understory, MIX- mixed canopy, CDOM- coniferous dominated.

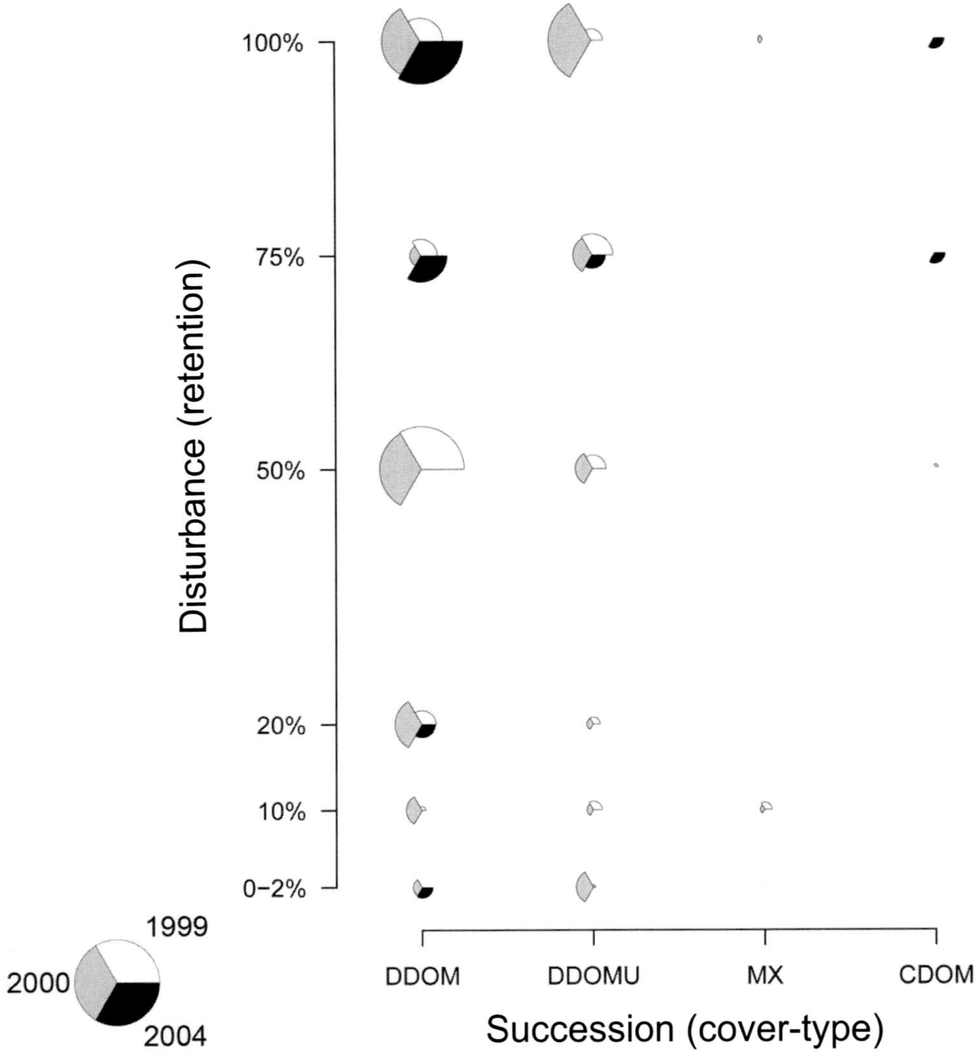

Fig. 10. Star plots of the abundance of *Calosoma frigidum across* a disturbance gradient (y-axis) and successional gradient (x-axis) over 3 years (1999=white, 2000= grey, 2004=black) n=623. The size (area) of each 'pie piece' represents the proportional mean abundance relative to the maximum mean abundance of all treatment combinations. DDOM- deciduous dominated, DDOMU- deciduous dominated with a coniferous understory, MIX- mixed canopy, CDOM- coniferous dominated.

Catch rates for *P. pensylvanicus* were high in the disturbed habitats shortly following harvesting but, in contrast to *P. adstrictus*, *P. pensylvanicus* persisted in the disturbed areas at least five years following harvest (Fig. 8). Catches of *Carabus chamissonis* seemed to increase or be stable under moderate levels of disturbance but at harvest intensities of 20% and greater catch rates fell so that the species was almost undetectable in the clear-cut treatments (Fig. 9). Response of *C. frigidum* also differed from that of the other abundant species (Fig. 10). The overall significant interaction between year, disturbance and successional stage is difficult to explain. Catch of *C. frigidum* was high in undisturbed early successional compartments and in low to moderate across disturbance treatments, but fell as disturbance increased and fell to zero in mid- to late-successional stands.

DISCUSSION

In this study we demonstrated strong influence of both successional stage, expressed in terms of mixed-wood forest cover-types, and anthropogenic disturbance (harvest) intensity on the spatial distribution and catch rates (activity-abundance) of boreal ground beetles. Interestingly, RM-ANOVAs on catches of all abundant species, except *C. frigidum,* showed either significant main effects or interactions with year for both 'successional stage' and 'disturbance' without significantly interacting with each other. Thus, for nine of the ten species tested, the effects of succession and disturbance appeared to be additive in any given year. Species-specific responses to successional stage, however, varied over the five years of the study, likely reflecting the variable but ecologically significant effects of forest regeneration across compartments.

The response of *C. frigidum* differed from the rest of the abundant ground beetle species due to its unique life-history strategy. This species is known to feed on caterpillars (Cameron & Reeves, 1990), and can be a significant agent of caterpillar mortality during outbreaks (Wagner & Leonard, 1980). During the first years of this study, large populations of Large Aspen Tortrix (*Choristoneura conflictana* [Walker]) were reported to have defoliated ca. 2.3 and 3.5 million ha in Alberta in 2000 and 2001, respectively, but these out-breaking populations collapsed by 2004 (Alberta Sustainable Resource Development, 2000, 2001, 2004). The catch-rate for *C. frigidum* is probably directly linked to temporal variation in the availability of prey in these stands.

Forest Successional Stage

The ten 'core' carabid species that were the focus of this study are well adapted to live in boreal mixed-wood forests. Populations of particular species flourish notably in each successional stage, although all species may be found in all stages, presumably because legacy elements (characteristics from earlier successional stages) persist across the successional gradient (Lindenmayer & Franklin, 2002). These elements may include, for example, small numbers of deciduous trees or expanding clones that exist in gaps of largely coniferous stands or small aggregates of conifers that persist as 'fire skips' within deciduous stands.

Patterns observed in ground beetle catch as stands recover reflect the biological characteristics of particular species. For example, populations of open-habitat associated species such as *P. adstrictus* rapidly increase in the more open habitats created by disturbance but then seem to decline as the forest regenerates. Other species, such as *P. decentis, A. retractum*, and *P. foveocollis*, all flourish in early successional deciduous forests but decrease in abundance as succession proceeds. As the coniferous component of these stands increases, species like *S. haematopus* and *C. advena* begin to dominate. The abundance of *C. ingratus*, however, does not appear to be influenced by successional stage; hence this species may be adapted to other features that do not change with succession of canopy trees, but which may instead reflect local disturbance level as determined by gap-dynamic processes.

Changes in vegetation can affect animal communities (Siemann et al., 1998), but direct effects of specific vegetation characteristics have been notoriously difficult to demonstrate for ground beetles (e.g., Niemelä & Spence, 1994). Temperature and humidity in the forest floor are partly controlled by canopy composition (Koivula et al., 1999), and these effects taken together with soil and topographic variability generate a myriad of different microhabitats for Carabids. Koivula et al. (1999) identified deciduous leaf litter as a crucial component for many early-successional species. Many late-successional species seem to be associated with shady habitats with little vegetation, characteristic of forest floors under the closed canopy of coniferous forests (Lindroth, 1966).

Disturbance

Niemelä et al. (1993) categorized carabids in terms of population responses to forest harvest into groups of 1) forest generalists, 2) open-habitat specialists and 3) mature-forest species. All these categories can be identified among the ten 'core' species of this study. The general carabid response to harvesting is an initial increase of 'open-habitat' species (e.g., *P. adstrictus*), followed by a decrease of forest generalists (e.g., *C. ingratus*) and the extirpation of mature-forest specialists (e.g., *P. decentis, C. advena*) (Niemelä et al., 1993). However, Niemelä et al. (1993) study was based on only samples from coniferous stands; in the present paper we further generalize the response to disturbance to include the successional affinities of ground beetles as follows (Fig. 11):

Class I – *Forest generalists* occur in all forest types across the successional gradient (Fig. 11). Generally, these species are not detrimentally affected by harvesting, however high intensities of logging (e.g., clear-cuts) have the potential to extirpate some of these species. The best example of Class I response is that observed here for *C. ingratus*. This species occurred in almost every successional stage under all disturbance intensities in the 6 years of study. However, it was absent from clear-cut areas of the early successional deciduous stands. *Pterostichus pensylvanicus* is another species that could be considered a forest generalist, however this species occurs mainly in deciduous forests. Populations of these species could be maintained by high intensities of partial cutting.

Class II – *Open-habitat generalists* may be found locally across the successional and disturbance gradients, but are typically promoted by forest harvesting (Fig. 11). *Pterostichus adstrictus* clearly displays such a response. Interestingly, this species became virtually absent from the clear-cut areas of the deciduous-dominated sites five years following harvest. Aspen regeneration in these stands was dense and approximately 2-3 m in height, possibly creating forest floor conditions more similar to closed-canopy stands. *Pterostichus adstrictus* also inhabits other open areas, including agricultural lands (Ribera et al., 1999), orchards (Epstein et al., 1999) and grasslands (Rushton et al., 1989). Generally, Class II species are not of conservation concern in forest management.

Class III – *Species of mature deciduous forests* presented the most common harvesting response observed in this study, with four species exhibiting this response (*P. decentis*, *A. retractum*, *P. foveocollis* and *C. chamissonis*). These species maintain their highest populations in mature deciduous stands but have the ability to maintain smaller populations in other successional stages (Fig. 11). They are typically tolerant to moderate levels of logging (ca. 50%) and have the ability to maintain populations in disturbed stands 2-4 years following disturbance. Populations of *Carabus chamissonis* crashed immediately after logging in lodgepole pine forests (Niemelä et al., 1993), but this was not observed in our study. Catch rates of this species fell below detectable levels in the conifer-dominated

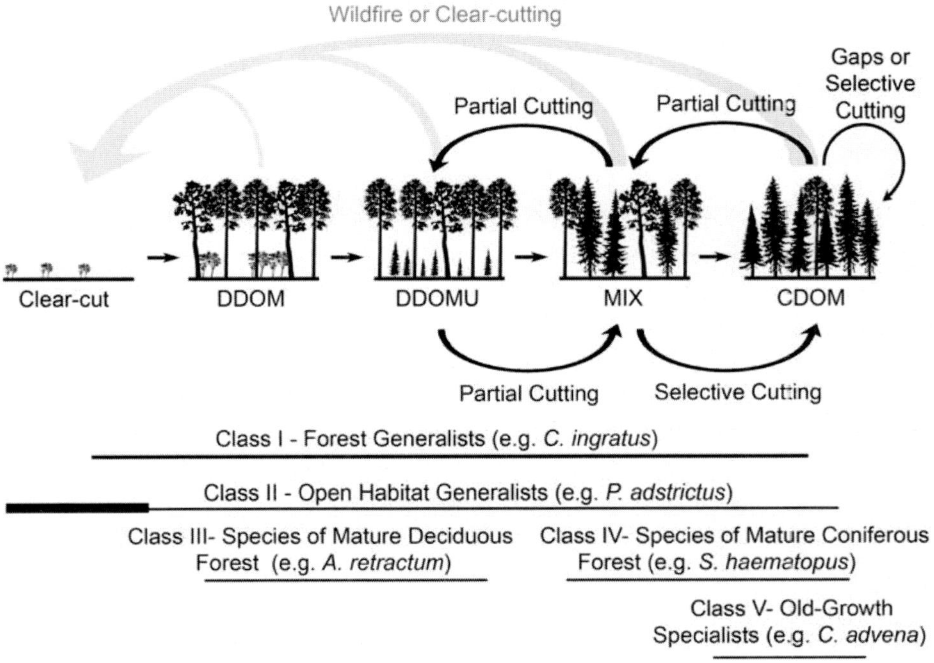

Fig. 11. Graphic model of species' habitat affinities and responses to harvesting (successional pathway modified from Bergeron et al. (2002)).

areas, but apparently the species was able to better maintain populations in the early successional stands, especially under more moderate levels of disturbance.

Class IV – *Species of mature coniferous forests* responded to disturbance similarly to Class III species but tended to be more abundant in conifer-dominated stands (e.g., *S. haematopus*). These species are apparently tolerant to moderate levels of disturbance and can maintain populations in late successional forests of selectively-cut Aspen and older Spruce trees (Fig. 11). As with Class III species, Class IV species have the potential to become extirpated in intensively-logged stands, although this effect may not be observable for 2-4 years following harvest.

Class V – *Old-growth specialists* are adapted to the unique characteristics of old-growth stands (e.g., *C. advena*). Any level of disturbance will greatly reduce the population levels of these species, although this response can take 2-4 years to manifest itself. Perhaps the only management option for these species is to maintain old stands (>200 years) and the associated characteristics of those stands (i.e., high volumes of deadwood, mixed composition, irregular structure) on the forested landscape. Seymour & Hunter (1999) proposes a system using different rotation lengths to maintain natural forest age structure (ie. 100 year cycle across 63% of landbase, 200 year cycles across 23% and 300 year cycles across 14%). Bergeron et al. (2002) suggest the same age structure could be reached using selective cutting while maintaining higher harvesting levels. However, placing a minimum of 20% of the forested land base, representing the diversity of forest types found in the boreal, into long-term ecological reserves may afford the best protection against loosing these old-growth species (Senate Subcommittee on the Boreal Forest, 1999).

Temporal dynamics

The mature forest species (Classes III-V) persisted for two years post-harvest, but disappeared soon after that, i.e., they were absent five years post harvest from the high-intensity treatments. Other studies from Canada (Niemelä et al., 1993; Spence et al., 1996) and Europe (Szyszko, 1990; Koivula & Niemelä, 2002) have documented similar persistence by mature-forest species in clear-cuts. The persistence of these species for at least two years following harvest could represent individuals wandering from neighbouring mature forest stands, or reflect individuals that were present before the disturbance but were unable to successfully reproduce in the altered environment (Koivula et al., 2002). It remains unclear how long it will take these mature-forest species to return to logged environments. Niemelä et al. (1993) found that after 27 years many of these species still had not returned. In Finland, however, some mature-forest specialists begin to return after canopy closure (20-30 years) (Koivula et al., 2002) . It may take hundreds of years for some ground beetle species to recover from large-scale logging (Desender et al., 1999).

The proper duration of a study is essential to understanding ecological processes that occur over extended periods of time. Most studies of ground beetles have focussed on the short-term impacts of harvesting or natural disturbance (e.g. Abildsnes & Tømmeros,

2000; Saint-Germain et al., 2005). Some studies (e.g. Buddle et al., 2000; Gandhi et al., 2001) have investigated ground beetles on larger temporal scales through adopting a chronosequence approach. However, the results of such studies are confounded by uncontrollable differences in habitats and local ecosystem history. Long term experiments, such as EMEND, provide a more accurate understanding of long-term ecological processes. As we saw, sampling 1 and 2 years post harvest would have been insufficient to detect the decline of mature forest species (Response Classes III-V) seen 5 years post harvest. Hence, we submit that any recommendations for forest management directed towards conservation of biodiversity based short-term windows of observation will be at best incomplete and could, at worst, be catastrophic for maintaining species in managed landscapes.

We have demonstrated that the relative abundance of particular ground beetle species in the boreal forest is linked to both successional stage of the forest and the relative intensity of disturbance at any given stage. Although the initial impact of disturbance was similar across successional stages, post-disturbance forest recovery seems to unleash a diverse and more species-specific response. This response apparently depends on the details of local habitat variation associated with succession from different starting points in a landscape that varies in edaphic characteristics. Nonetheless, the main themes of the response of abundant carabid species seem to be interpretable in terms of overall successional stage and intensity of disturbance.

ACKNOWLEDGEMENTS

The EMEND study is a partnership among scientists from various institutions (universities, the Canadian Forest Service, etc.), the Government of Alberta, the Sustainable Forest Management Network (SFMN) and the forest industry, in particular Canadian Forest Products (CANFOR), Daishowa-Marubeni International (DMI), Manning Diversified Forest Products and the Weyerhaeuser Company. This study was specifically funded by the Forest Resource Improvement Association of Alberta through funds available to CANFOR and DMI, the Alberta Forest Research Institute and the Natural Sciences, the Sustainable Forest Management Network and the Natural Sciences and Engineering Research Council of Canada through grants to J. R. Spence. We thank Jason Edwards, Charlene Hahn, Dustin Hartley, David Langor, Tyler Cobb, Karen Cryer, Evan Esch, Colin Bergeron, Matti Koivula and Jan Volney for ideas and assistance, an anonymous referee whose comments improved the clarity of this manuscript, and the many students who capably assisted with field collection of the data.

REFERENCES

Abildsnes, J. & Tømmeros, B.Å. (2000). Impacts of experimental habitat fragmentation on ground beetles (Coleoptera, Carabidae) in a boreal spruce forest. Annales Zoologici Fennici 37: 201-212.

Alberta Sustainable Resource Development. (2000). 2000 Annual Report: Forest Health in Alberta. Department of Sustainable Resource Development, Land and Forest Division, Forest Management Branch, Forest Health Section., Edmonton, Alberta.

Alberta Sustainable Resource Development. (2001). 2001 Annual Report: Forest Health in Alberta. Department of Sustainable Resource Development, Land and Forest Division, Forest Management Branch, Forest Health Section., Edmonton, Alberta.

Alberta Sustainable Resource Development. (2004). 2004 Annual Report: Forest Health in Alberta. Department of Sustainable Resource Development, Land and Forest Division, Forest Management Branch, Forest Health Section., Edmonton, Alberta.

Bergeron, Y. & Dubuc, M. (1988). Succession in the southern part of the Canadian boreal forest. Plant Ecology 79: 51-63.

Bergeron, Y., Harvey, B., Leduc, A. & Gauthier, S. (1999). Forest management guidelines based on natural disturbance dynamics: Stand- and forest-level considerations. Forestry Chronicle 75: 49-54.

Bergeron, Y., Leduc, A., Harvey, B.D. & Gauthier, S. (2002). Natural Fire Regime: A Guide for Sustainable Management of the Canadian Boreal Forest. Silva Fennica 36: 81-95.

Bousquet, Y. (1991). Checklist of beetles of Canada and Alaska. Agriculture Canada, Ottawa, Ont.

Buddle, C.M., Spence, J.R. & Langor, D.W. (2000). Succession of boreal forest spider assemblages following wildfire and harvesting. Ecography 23: 424-436.

Cameron, E.A. & Reeves, R.M. (1990). Carabidae (Coleoptera) associated with gypsy moth, *Lymantria dispar* (L.) (Lepidoptera: Lymantriidae), populations subjected to Bacillus thuringiensis Berliner treatments in Pennsylvania. Canadian Entomologist 122: 123-129.

Cobb, T.P., Langor, D.W. & Spence, J.R. (2007). Biodiversity and multiple disturbances: boreal forest ground beetle (Coleoptera: Carabidae) responses to wildfire, harvesting, and herbicide. Canadian Journal of Forest Research 37: 1310-1323.

Desender, K., Ervynck, A. & Tack, G. (1999). Beetle diversity and historical ecology of woodlands in Flanders. Belgian Journal of Zoology 129.

Epstein, D.L., Zack, R.S., Brunner, J.F., Gut, L. & Brown, J.J. (1999). Effects of Broad-Spectrum Insecticides on Epigeal Arthropod Biodiversity in Pacific Northwest Apple Orchards. Environmental Entomology 29: 340-348.

Evans, W.G. (1971). The attraction of insects to forest fires. – In: Proceedings of the Tall Timbers Research Station, Tallahasse, Fl.

Gandhi, K.J.K., Spence, J.R., Langor, D.W. & Morgantini, L.E. (2001). Fire residuals as habitat reserves for epigaeic beetles (Coleoptera: Carabidae and Staphylinidae). Biological Conservation 102: 131-141.

Harvey, B.D., Leduc, A., Gauthier, S. & Bergeron, Y. (2002). Stand-landscape integration in natural disturbance-based management of the southern boreal forest. Forest Ecology and Management 155: 369-385.

Heliölä, J., Koivula, M. & Niemelä, J. (2001). Distribution of Carabid Beetles (Coleoptera, Carabidae) across a Boreal Forest–Clearcut Ecotone. Conservation Biology 15: 370-377.

Jacobs, J., Spence, J.R. & Langor, D. (2007). Influence of boreal forest succession and dead wood qualities on saproxylic beetles. Agricultural and Forest Entomology 9: 2-15.

Koivula, M., Cobb, T., Dechene, A.D., Jacobs, J. & Spence, J.R. (2006). Responses of two Sericoda Kirby, 1837 (Coleoptera: Carabidae) species to forest harvesting, wildfire, and burn severity. Entomologica Fennica 17: 315-324.

Koivula, M., Kukkonen, J. & Niemelä, J. (2002). Boreal carabid-beetle (Coleoptera, Carabidae) assemblages along the clear-cut originated succession gradient. Biodiversity and Conservation 11: 1269-1288.

Koivula, M. & Niemelä, J. (2002). Boreal Carabid Beetles (Coleoptera, Carabidae) in Managed Spruce Forests – a Summary of Finnish Case Studies. Silva Fennica 36: 423-436.

Koivula, M., Punttila, P., Haila, Y. & Niemelä, J. (1999). Leaf litter and the small-scale distribution of carabid beetles(Coleoptera, Carabidae) in the boreal forest. Ecography 22: 424-435.

Koivula, M. & Spence, J.R. (2006). Effects of post-fire salvage logging on boreal mixed-wood ground beetle assemblages (Coleoptera, Carabidae). Forest Ecology and Management 236: 102-112.

Lindenmayer, D. & Franklin, J.F. (2002). Conserving Forest Biodiversity: A Comprehensive Multiscaled Approach. Island Press, Washington, D.C.

Lindroth, C.H. (1966). The ground beetles of Canada and Alaska: Part 4. Opuscula entomologica, supplementa.

McCarthy, J. (2001). Gap dynamics of forest trees: A review with particular attention to boreal forests. Environmental Reviews 9: 1-59.

McClanahan, T.R. (1986). The effect of a seed source on primary succession in a forest ecosystem. Plant Ecology 65: 175-178.

McCune, B. & Grace, J.B. (2002). Analysis of Ecological Communities. MJM Software Design, Gleneden Beach, OR.

Natural Resources Canada. (2003). The state of Canada's forests 2002-2003. Natural Resources Canada, Canadian Forest Service, Ottawa, Ontario.

Niemelä, J. (1999). Management in relation to disturbance in the boreal forest. Forest Ecology and Management 115: 127-134.

Niemelä, J., Langor, D. & Spence, J.R. (1993). Effects of Clear-Cut Harvesting on Boreal Ground-Beetle Assemblages (Coleoptera, Carabidae) in Western Canada. Conservation Biology 7: 551-561.

Niemelä, J.K. & Spence, J.R. (1994). Distribution of Forest Dwelling Carabids (Coleoptera) – Spatial Scale and the Concept of Communities. Ecography 17: 166-175.

Pratt, L. & Urquhart, I. (1994). The Last Great Forest. NeWest Press, Edmonton, Alberta, Canada.

R Development Core Team. (2005). R: A language and environment for statistical computing. R Foundation for Statistical Computing, V., Austria. ISBN 3-900051-07-0, URL http://www.R-project.org.

Ribera, I., McCracken, D.I., Foster, G.N., Downie, I.S. & Abernethy, V.J. (1999). Morphological diversity of ground beetles (Coleoptera: Carabidae) in Scottish agricultural land. Journal of Zoology 247: 1-18.

Rowe, J.S. (1994). A New Paradigm for Forestry. Forestry Chronicle 70:565-568.

Rushton, S.P., Luff, M.L. & Eyre, M.D. (1989). Effects of Pasture Improvement and Management on the Ground Beetle and Spider Communities of Upland Grasslands. The Journal of Applied Ecology 26: 489-503.

Saint-Germain, M., Larrivée, M., Drapeau, P., Fahrig, L. & Buddle, C.M. (2005). Short-term response of ground beetles (Coleoptera: Carabidae) to fire and logging in a spruce-dominated boreal landscape. Forest Ecology and Management 212: 118-126.

Scheiner, S.M. & Gurevitch, J. (1993). Design and analysis of ecological experiments. Chapman and Hall, New York, New York, USA.

Senate Subcommittee on the Boreal Forest. (1999). Standing Senate Committee on Agriculture and Forestry. Competing Realities: the Boreal Forest at Risk. Report of the Sub-Committee on Boreal Forest of the Standing Senate Committee on Agriculture and Forestry. Ottawa, Ontario.

Seymour, R.S. & Hunter, M.L.J. (1999). Principles of ecological forestry. Cambridge University Press, Cambridge, UK.

Siemann, E., Tilman, D., Haarstad, J. & Ritchie, M. (1998). Experimental tests of the dependence of arthropod diversity on plant diversity. American Naturalist 152: 738-750.

Spence, J.R., Langor, D.W., Niemelä, J., Carcamo, H.A. & Currie, C.R. (1996). Northern forestry and carabids: The case for concern about old-growth species. Annales Zoologici Fennici 33: 173-184.

Spence, J.R. & Niemelä, J.K. (1994). Sampling Carabid Assemblages with Pitfall Traps – the Madness and the Method. Canadian Entomologist 126: 881-894.

Szyszko, J. (1990). Planning of prophylaxis in threatened pine forest biocenoses based on an analysis of the fauna of epigeic Carabidae. Warsaw Agricultural University Press, Warsaw, Poland.

Viereck, L.A. (1983). The effects of fire in black spruce ecosystems of Alaska and northern Canada. JohnWiley & Sons, New York, New York, USA.

Wagner, T.L. & Leonard, D.E. (1980). Mortality factors of satin moth, *Leucoma salicis* [Lep.: Lymantriidae], in aspen forests in Maine. BioControl 25: 1573-8248.

Wikars, L.O. (1992). Forest fires and insects. Entomologisk Tidskrift 113: 1-11.

Work, T.T., Shorthouse, D.P., Spence, J.R., Volney, W.J.A. & Langor, D. (2004). Stand composition and structure of the boreal mixed-wood and epigaeic arthropods of the Ecosystem Management Emulating Natural Disturbance (EMEND) landbase in northwestern Alberta. Canadian Journal of Forest Research 34: 417-430.

Zar, J. (1996). Biostatistical Analysis. Prentice Hall, Upper Saddle River, NJ.

L. Penev, T. Erwin & T. Assmann (Eds) 2008
BACK TO THE ROOTS AND BACK TO THE FUTURE. TOWARDS A NEW SYNTHESIS AMONGST TAXONOMIC, ECOLOGICAL AND BIOGEOGRAPHICAL APPROACHES IN CARABIDOLOGY
Proceedings of the XIII European Carabidologists Meeting, Blagoevgrad, August 20-24, 2007, pp. 451-467.

Sofia–Moscow

Carabid beetle movements in a clear-cut area with retention groups of trees

Jarosław Skłodowski

Dept. Forest Protection and Ecology, Forestry Faculty, SGGW - Warsaw University of Life Sciences (former Warsaw Agricultural University), Nowoursynowska 159, 02-776 Warszawa, Poland, E-mail: sklodowski@wl.sggw.pl

SUMMARY

Using the capture-mark-recapture (CMR) method the movements of forest carabids were studied in a clear-cut with two retention groups of live trees (sizes 400 and 700 m^2). Three exploratory questions were evaluated: (1) do large-sized forest species utilize the retention groups more often than the clear-cut sections, while small-sized non-forest species are expected to avoid the retention groups? (2) does ploughing in the clear-cut sections stimulate the activity of forest species if furrows are oriented perpendicularly to clear-cut edge, or does it stimulate non-forest species if the furrows are oriented parallel to clear-cut edge? (3) are forest and non-forest species expected to experience the micro-habitat differentiation within the clear-cut differently? The results supported the exploratory question (1) for non-forest species, while question (2) was confirmed for the forest species *Carabus arcensis*, and the non-forest species *Pterostichus caerulescens*, *Amara lunicollis* and *Harpalus rufipalpis*. Question (3) was confirmed by showing that small-sized non-forest species experience the conditions of a clear-cut in a different way compared with the forest species.

Keywords: retention groups, clear-cut area, beetle movement, Carabidae

INTRODUCTION

Modern forestry, and clear-cutting in particular, has strongly affected the macrofauna, including ground-dwelling carabid beetles, as reported by Szyszko (1983, 2002), Niemelä et al. (1993), Szujecki et al. (1983), Skłodowski (1995, 2006a), Atlegrim et al. (1997), du Bus de Warnaffe & Lebrun (2004), Koivula (2001, 2002), Pontégnie et al. (2005) and Matveinen-Huju (2006, 2007). These papers have revealed that clear cutting results in an increase in species diversity, mostly as a result of an influx or increased abundance of species associated with open habitats and forest-habitat generalists; forest specialists, however, may drastically decrease or even locally disappear.

Carabid assemblages are subject to changes following clear-cutting of a stand and the often subsequently-performed top-soil preparation (hereafter ploughing) to aid tree regeneration (Koivula & Niemelä, 2003; Pihlaja et al., 2006). Within the time span of the last two decades, substantial changes in forest management practices have been introduced to European forestry, including efforts on minimizing the impact of forestry (Koivula, 2002; Matveinen-Huju et al., 2006). For example, in Poland the largest allowable clear-cut size has been reduced down to 6 ha, and the maximum width of a clear-cut opening has been fixed to be no wider than two heights of a stand, that is up to 80 m. Moreover, in Poland small groups of trees are obligatorily retained within every clear cut area. Such innovative techniques may theoretically facilitate the resilience and recovery of forest species in an area subjected to logging, and the retention groups as such are intended not only to serve as "life boats" but also as "stepping stones" for forest-associated organisms (Lindenmayer & Franklin, 2006).

The importance of old-growth stand islands for carabid was studied by Matveinen-Huju et al. (2006). The above mentioned authors have checked the correctness of the hypothesis stating that the invasion into rather large-size tree retention groups is realized by those forest species proffering the medium and moist habitat, and the invasion into smaller-size tree retention groups by the species requiring open and dry habitats. They have no unambiguous evidence to support this hypothesis. Therefore, when further studying the question, one may attempt describing the role of tree retention groups in the movement of invertebrates (eg. carabids), by observing their routes between particular tree retention groups.

The general pattern of movement of carabids in the environment has been studied by Baars (1979), Mools (1979, 1987), and Rijnsdorp (1980). The effect of logging on carabids has been previously studied by analyzing the movement patterns of carabids within a recently clear-cut area. The movements of carabids have also been studied in "mosaic" landscape and in different types of forest, such as within clear-cut areas that include retention patches of trees, using a variety of methods: the catch-mark-recapture technique (CMR; Petit & Burel, 1993; Joyce et al., 1999; Skłodowski, 1999), portable radar (Charrier et al., 1996), harmonic radar (Wallin & Ekbom, 1988; Lövei et al., 1997) and radio telemetry (Riecken & Raths. 1996).

In the present study I used CMR technique to study movements of carabids in a "mosaic" of clear-cut and forested (retention patch) sections within a single stand. The advantages of the method are (1) the possibility of obtaining simultaneous observations of as much as several hundreds of marked beetles, (2) the often-applied use of dry (non-killing) pitfall traps can be considered nearly harmless for the carabids themselves, and

(3) the method is relatively low in costs as compared to radar and transmitter techniques. The drawback, however, is the lack of information concerning the movements of the marked specimens between the catches.

The analysis of movements of forest carabids in a stand with clear-cut and forested sections has to be based on a representative sample. Therefore, I selected the CMR technique to follow the catch rates of carabids in a harvested stand by placing dry pitfall traps regularly in squire gird all over the clear-cut and, retention-patch, and forest-edge sections of one stand. By placing the traps regularly within the clear-cut sections I aimed at evaluating the effect of ploughing (soil scarification) on carabid movements.

Large-sized forest carabid species disappear in clear cut open area, following the ploughing of soil. It can be assumed that those specimens of forest species present in a forest section will move towards the clear-cut area. If so, the furrows ploughed perpendicularly to the stand edge, should favor the carabid movement while those made parallel to forest edge should hamper the beetle free movement. On the other hand, the species of open areas and eurytopic species (that is the macropteric carabids characteristic for their significantly high dispersion potential), rapidly inhabit the clear-cut section. Likely, the parallel furrows should favor their movement while the perpendicularly orientated ones should hamper colonization of new areas by carabids

The occurrence of many forest carabid species depends on the actual habitat microdifferentiation – the composition and thickness of the forest litter layer (Wallin & Ekbom, 1988; Baguette, 1993; Niemelä et al., 1996; Koivula et al., 1999; Koivula, 2001, 2002). The (non-forest) light preferring carabids, in turn, avoid the forest litter, that is additionally enriched with some admixture of leaves (Koivula et al., 1999). It can be assumed that the ploughed soil (eg. following the stripes arrangement) should affect carabids in a different way.

The strips are ploughed so that 25-30 cm deep layer of soil is being removed and the sandy soil lies in the bottom of the furrow. The upper 25-30 cm soil layer is put aside, up side down disturbing thus the natural structure of soil profile. For forest species, which need the presence of forest litter, this new artificial sandy environment creates homogenous and unfavorable conditions. For the open area species and eurytopic species (further in text referred to as non-forest), however, this new environment will not only be heterogeneous but it will also favor their occurrence.

The following exploratory questions were tested in the present study:

Q1 – Large-sized forest species, while moving within the study stand, should mostly utilize the retention groups and seldom enter the clear-cut sections. On the other hand, the catch rates of non-forest species should be lower in retention groups as compared to the clear-cut, open sections.

Q2 – Ploughing results in deep furrows on the top soil of clear-cuts. If these furrows are perpendicular with the clear-cut/mature-forest edge, they may stimulate forest species to enter the clear-cut. If the furrows run parallel to the edge, however, the non-forest species may be more attracted to the clear-cut sections.

Q3 – Both forest and non-forest carabids recognize the micro-variation in a stand and are, consequently, associated with certain types of micro-habitat (e.g., Koivula 2001, 2002). One of the expected results of this should be different catch rates over the clear-cut space.

STUDY AREA

The study site, a 95-years old Scots pine stand, was situated in compartment 114a in Forest range Pustowo, Niedźwiady Forest District (Człuchów Forest) and partly clear-cut in 1995. The undergrowth, occupying as much as 20% area, consisted of beech, spruce and juniper. In the forest floor vegetation, mosses *Entodon Schreberi, Hyllocomium splendes, Lycopodium clavatum* and herbaceous plants (*Vaccinium myrłillus, Vaccinium vitis-idaea, Melampyrum pratense, Chimaphila umbellata, Luzula pilosa. Deschampsia flexuosa)* predominated.

Small patches of ground were covered by bilberry. In the ploughed clear-cut area, no herbaceous vegetation was present during the first two years. Subsequently, *Deschampsia* entered into that area.

Two groups of pine trees were retained in the harvesting operation, sized at 400 and 700 m^2. The soil (proper rusty soil type) was ploughed in the autumn of 1996, and in the spring of 1997. Scots pine seedlings were planted in the clear-cut sections of the study site. Clear-cut section was ploughed with furrows parallel to the stand edge, and part of it was ploughed with perpendicular furrows (Fig.1). The furrows were ploughed at depth of 25-30 cm, their width, however, exceeded 70 cm. Between each two adjacent furrows there was 80 cm not ploughed strip that was fully covered by the soil coming from furrow and put up side down.

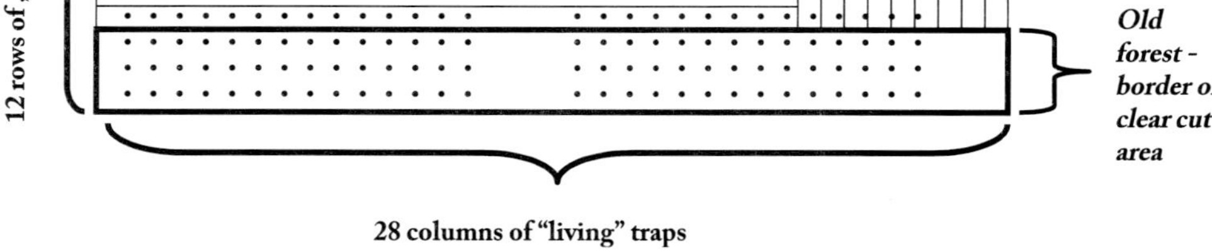

Fig. 1. The study site with clear-cut sections, forest section and two retention groups. Trap locations and directions of ploughing furrows are shown.

The northern (upper longer edge on the map) edge of the study site was adjacent to forest road, and southern edge to the remnants of 95-yr old Scots pine forest (Fig. 1). In the present paper I will use the term "clear-cut edge" to indicate southern border between the clear-cut section and the old pine forest. The western and eastern edges of the map were adjacent to other parts of clear-cut area. The western and eastern edges of these not studied sections were bordered by 95-years old pine forest (distance 250 m). The CMR study was carried out during 1996-1999, during the four first post-harvest summers.

METHODS

Field study

As pitfall traps I used 0.5 l plastic boxes, with the upper diameter 12 cm. The bottom of each box was bored with five holes (diameter 1.0 mm) to enable out flow of rain water. The traps were placed in 28, 12-trap grids of 6 m x 6 m (Fig.1) and inspected daily during summer months (1996: 26th of June – 18th of September; 1997: 21st of June – 16th of October; 1998: 24th of June – 2nd of September; 1999: 29th of June – 1st of September). During inspections, the individual trap number and caught species were recorded. All individuals longer than 12 mm were individually marked with a unique number, and released at a distance of 1.5 m from the trap from where they were caught. The time intervals between subsequent observations were uneven, therefore in order to unify the data sets only observations done between July 1st and August 31st were used.

All captured beetles were marked following Grüm (1959). For this purpose, beetle elytra were imaginarily divided into four sections: the bottom-right section of elytrum (ones, "1"), the upper-right section (threes, "3"), bottom-left section (tens, "10") and upper-left section (fifties, "50"). Different numbers of punctures, made with a needle, in each section would then define the numerical code of a specimen, For instance, number "188" would be two punctures in the section 1, two in the section 3, three in the section 10, and three in section 50.

Data analysis

Exploratory question Q1 explores the movement and catch rates of selected carabid in tree retention group and in the clear-cut section.

This question was subjected to a two-step verification procedure. In step 1, the routes of carabid movement were roughly determined. A necessary assumption was made that between subsequent captures carabids had moved along straight lines (as there was no information on their "true" routes). The majority of the routes of forest carabids should cross, or end in, the retention groups. Movement patterns in the clear-cut sections are shown for *Carabus arcensis* and *C. violaceus*.

Step 2 involved a comparison of catch rates of the forest and non-forest carabids both in the clear-cut and in the unharvested sections. As the empirical data distributions did not follow the normal distribution (Shapiro-Wilk test), the hypothesis was verified using the Mann-Whitney *U* test in Statistica software (StatSoft Inc., 1997).

Exploratory question Q2 concerns the presence of forest species and non-forest ones in the furrows directed either perpendicularly or parallel to the stand edge.

This question predicted that ploughing furrows parallel or perpendicular to the longer side of the clear-cut section should result in different catch rates of forest and non-forest carabid species. In this case, the non-normal data were subject to the Mann -Whitney *U* test. To obtain a simple graphic presentation of carabid occurrence in the clear cut area, catch-rate maps were produced and plotted using the spatial relationships estimated by semi-variances in ESRI ArcMap 9.2 software.

Exploratory question Q3 refers to different reactions of forest and non-forest carabids to the microhabitat variation within a stand

This question was tested by assuming that the variation in the catch rates of a particular species over the study site is presented by the semivariogram (VarioVin 2.21; Pannatier 1996). Semivariance gives information of mean catch rate differences between traps with inter-trap distance h (the vertical axis, i.e. the separating distance see Fig. 2). While using the square grid of traps, their distances are described as being the h parameter (or its multiplied value).

Three parameters are important when interpreting a semivariogram: Sill, Range & Nugget (Kapusta, 2004). Sill denotes the level of maximum stable (steady) variability, which means a reduction in spatial autocorrelation between captures in distant traps. The h distance at which the autocorrelation reduction takes place is called Range. This index bears information on the range of autocorrelation. Large values indicate that the catch rates are spatially homogeneous, even when comparing very distant traps.

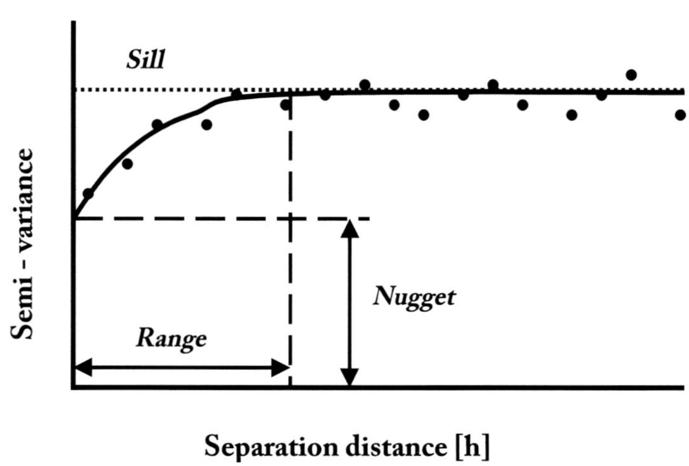

Fig. 2. An example of semivariogram with the parameters: Sill, Range and Nugget.

The Nugget index gives evidence for the role of autocorrelation in the general variation. The larger its value, the smaller the role of autocorrelation in the total/general variation. Small values of the index suggest large catch rate similarity between neighboring traps.

The differences between mean values of the parameters describing the forest and the non-forest species were tested with the Mann-Whitney *U* test.

RESULTS

Exploratory question Q1 – regarding the movement and catch rates of selected carabid in tree retention group and clear-cut section.

The analysis of movements of *C. arcensis* in the course of the fourth year of study revealed that this species moves across tree retention groups (Fig. 3). Analysis was conducted following the situation as it was in the fourth year of study, when *C. arcensis* was most frequently observed (1621, including 729 returns) *vs.* 75 (27 returns) in the first year of study, 57 (57) in the second year, and 709 (399) in the third year of observations. In order to properly present the movement routes of *C. arcensis*, the movements of males and females are shown separately on Fig. 3.

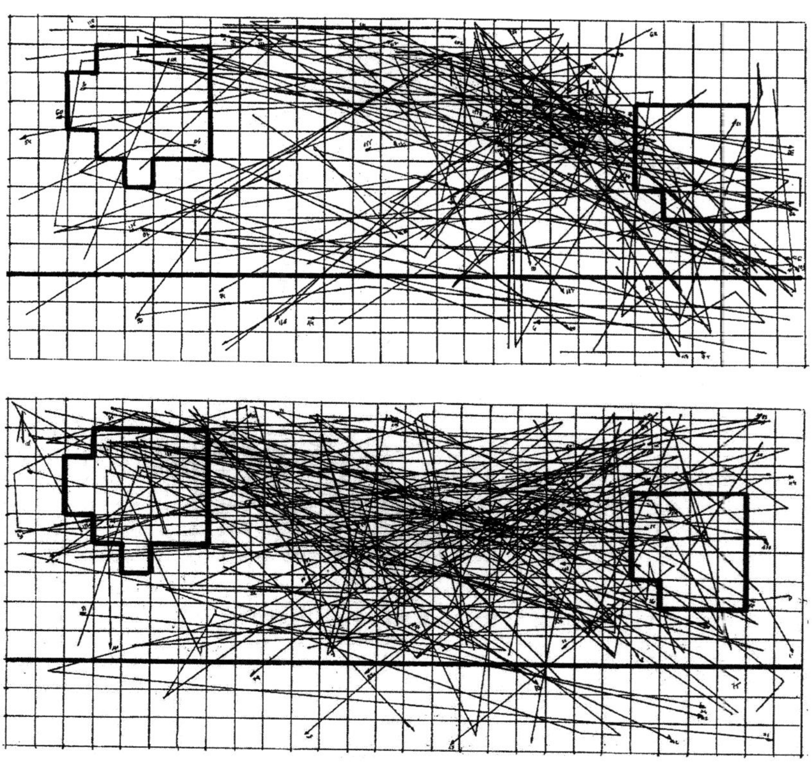

Fig. 3. The movement routes of *C. arcensis* males and females in the clear-cut section, forest section and retention tree groups.

Worth noticing is the concentration of movement routes close to the right-hand side of the clear-cut section, near to or at the smaller retention group. In this particular section of the study site, the furrows were perpendicular to the longer side of the clear-cut. Hence the result may suggest that the furrows force carabids to move along them in the clear-cut. On the other hand, the movement routes of *C. violaceus* were concentrated at the forest edge, rarely reaching the retention groups (Fig. 4).

The average catch rates of the forest species *C. glabratus* and *C. violaceus* were higher in the retention groups as compared with the clear-cut sections, however the difference was not not statistically significant (Table 1). The analysis showed that the catch rates of *P. niger* and *C. arcensis* significantly differed between the clear-cut sections and the retention groups (Table 1). As expected, the non-forest species *P. caerulescens*, *P. virens*, *C. erratus*, *Amara lunicollis* and *Harpalus rufipalpus* were more frequently captured in the clear-cut than in the retention groups (Table 1).

Table 1. Mean catch rates and standard deviations of non-forest and forest carabid species in the Człuchów Forest study site. Catches from clear-cut sections, and from large and small retention tree groups are shown separately. The "F" letter denotes forest species, the "N" non-forest ones. In the U-column, tests of catch rates differences are presented for each species in clear-cut section and large retention group, as well as in clear-cut section and small retention group, respectively.

Carabid taxon	**Clear-cut section**	**Large retention patch**	**Small retention patch**	***U* test**
Pterostichus caerulescens N	5.58 ± 6.38*°	5.14 ± 6.28*	4.73 ± 6.13°	*n.s. °Z = 1.96; p = 0.480
Pterostichus virens N	3.40 ± 3.16*°	1.95 ± 2.31*	2.87 ± 3.01°	*Z = 4.25; p = 0.001 °n.s.
Calathus erratus N	7.94 ± 8.30*°	3.41 ± 3.62*	7.71 ± 14.07°	*Z = 5.14; p = 0.001 °Z = 3.21; p = 0.010
Amara lunicollis N	5.94 ± 8.29*°	3.84 ± 6.77*	2.16 ± 2.83°	*Z = 2.05; p = 0.040 °Z = 3.84; p = 0.001
Harpalus rufitarsis N	10.09 ± 7.99*°	7.03 ± 7.08*	5.20 ± 4.12°	*Z = 3,61; p = 0.001 °Z = 4.50; p = 0.001
Pteroctichus niger F	1.47 ± 1.86*°	0.75 ± 1.06*	0.53 ± 0.81°	*Z = 3.22; p = 0.001 °Z = 3.79; p = 0.001
Carabus violaceus F	0.60 ± 1.04*°	0.49 ± 0.76*	0.67 ± 0.98°	*n.s. °n.s.
Carabus glabratus F	0.10 ± 0.36*°	0.19 ± 0.43*	0.18 ± 0.38°	*n.s. °n.s.
Carabus arcensis F	4.22 ± 3.91*°	2.81 ± 3.71*	3.04 ± 3.18°	*Z = 3.84, p = 0.001 °Z = 2.25, p =0.024

Exploratory question Q2 concerns the presence of forest species and non-forest ones in the furrows directed perpendicularly or parallel to the stand edge.

The analysis demonstrated that the catch rate difference between parallel and perpendicular (relative to the southern edge of the clear-cut section) furrows was significant for *Pterostichus caerulescens, Amara lunicollis, Harpalus rufipalpus* and *Carabus arcensis*. The non-forest species *P. caerulescens, A. lunicollis* and *H. rufipalpus* were captured more frequently in parallel than in perpendicular furrows. Such an orientation enables fast movements along the clear-cut section edge which favors an effective penetration into the clear-cut. On the other hand, *C. arcensis* was more frequently captured in the furrows oriented perpendicularly to the long edge of the clear-cut section.

The catch rate maps of *Carabus arcensis, Calathus erratus* and *Amara lunicollis* provided insights of micro-habitat use differences between forest and non-forest species (Figs 5-7). The forest species *C. arcensis* showed higher catch rates in the vicinity of the right-hand retention group - in this part of the clear–cut, the furrows had been ploughed perpendicular to the clear cut section edge (Fig. 5). As the retention groups themselves had not been ploughed, individuals reaching their edges should be able to easily change the direction of their movement.

The occurrence of the non-forest carabids *Calathus erratus* and *Amara lunicollis* (Figs 6-7) increased in those parts of the clear-cut section that had been ploughed parallel

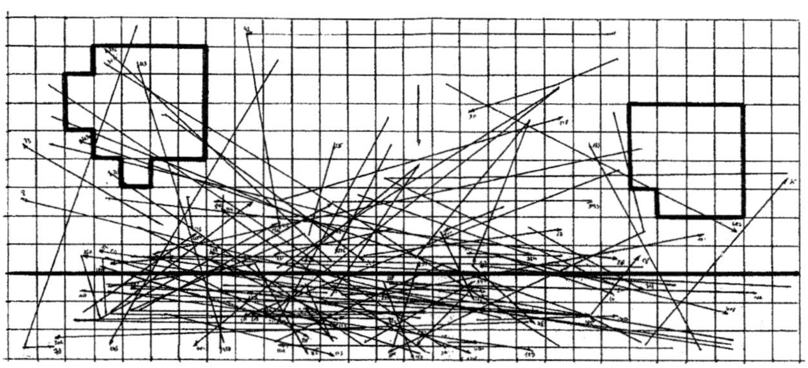

Fig. 4. The movement routes of *C.violaceus* in the clear-cut section, forest section and the tree retention groups.

Fig. 5. Semivariogram map of *C. arcensis* occurrence and catch rate in the clear-cut sections.

Fig. 6. Semivariogram map of *C. erratus* occurrence and catch rate in the clear-cut sections.

Fig. 7. Semivariogram map of *A. lunicollis* occurrenceand catch rate in the clear-cut sections.

to the clear-cut section edge. On other hand, catch rates of these species were lower in perpendicular section than in the parallel one.

Table 2. Carabid mean catch rates with standard deviations (individuals/trap*season) in furrows parallel and perpendicular to the longer edge of the clear-cut sections of the study site.

Carabid taxon	Parallel	Perpendicular	*U* test
P. caerulescens	5.72 ± 6.21	5.00 ± 7.08	Z = 2.43; p = 0.015
P. virens	3.66 ± 2.89	3.56 ± 4.08	n.s
C. erratus	7.64 ± 7.17	9.20 ± 11.85	n.s
A. lunicollis	6.60 ± 8.85	3.20 ± 4.51	Z =4.03; p = 0.001
H. rufipalpus	10.66 ± 8.36	7.75 ± 5.64	Z = 3.27; p = 0.001
P. niger	1.54 ± 1.91	1.21 ± 1.65	n.s
C. violaceus	0.57 ± 1.02	0.72 ± 1.14	n.s
C. glabratus	0.11 ± 0.36	0.08 ± 0.32	n.s
C. arcensis	4.09 ± 4.09	4.75 ± 3.38	Z = 2.99; p = 0.003

Exploratory question Q3 refers to different recognition to the microhabitat variation in a stand by forest and non-forest carabids.

The analysis of semivariance of the catch rates of the non-forest and forest species revealed significant differences in the Nugget index values of these two groups. The differences between these groups were only marginally significant ($p < 0.10$) as the Sill parameter was concerned. Using only the Nugget parameter, one may come to the conclusion that the catch rates of the non-forest species is characterized by a more heterogeneous spatial distribution than that of the forest species. This may in turn suggest that the non-forest species experience the habitat of the open clear-cut area as being more heterogeneous, as compared with the forest species. The shape of the selected semivariograms, as presented for *P. caerulescens* and *C. arcensis* (Fig. 8), reveals that when between-trap distances rise above 6 h (36 m), the autocorrelation of *C. arcensis* catch reached its maximum reduction. In the case of *P. caerulescens*, the maximum reduction in autocorrelation was obtained between traps placed at least 10 h (60 m) apart.

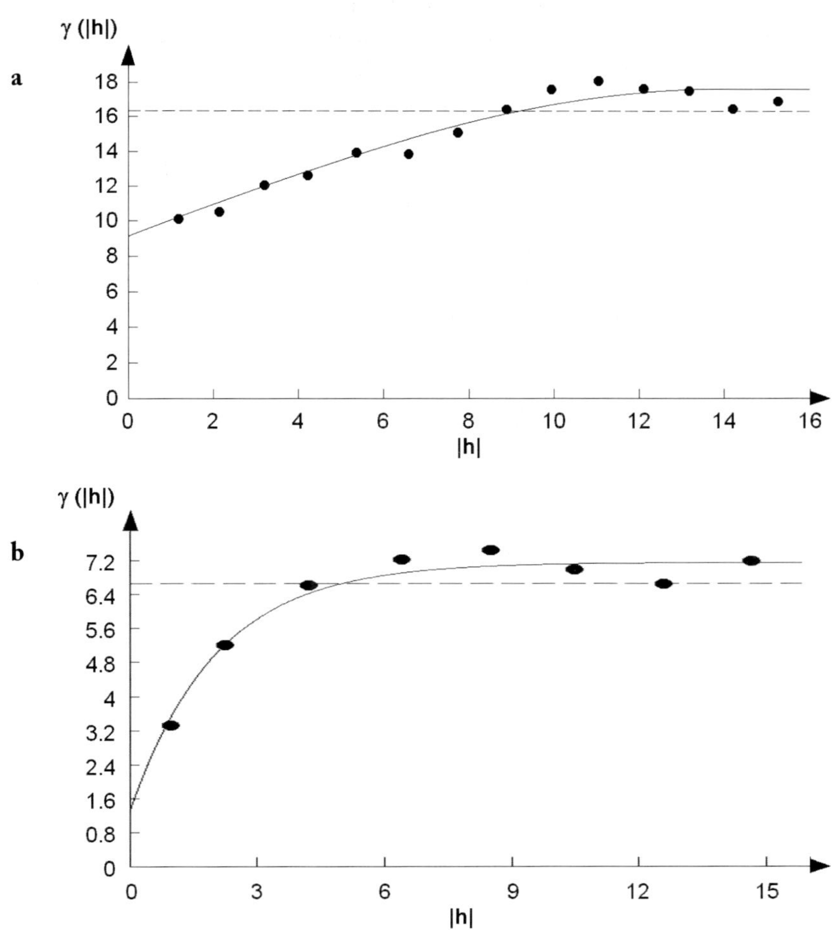

Fig. 8. Semivariograms for (a) *P. caerulescens* and (b) *C. arcensis*.

Table 3. Semivariance analysis of occurrence of non-forest and forest species in clear-cut section: mean values of Nugget, Range and Sill comparison. For parameter definitions, see Methods.

Parameter	Non-forest species	Forest species	Mann-Whitney *U* test
Nugget	11.6	2.5	Z = 2.3735; p = 0.017
Range	10.5	8.7	Z = 0.7302; p = 0.465
Sill	21.9	7.7	Z = 1.8257; p = 0.067

DISCUSSION

Exploratory question Q1 – regarding the movement and catch rates of selected carabid in tree retention group and clear-cut section.

This question predicted that large forest species should spend more time in the retention groups as compared with the clear-cut sections of the study site, while the catch rates of non-forest species should show the opposite trend. This hypothesis got support for latter carabid group, with the exception of *Calathus erratus. Pterostichus caerulescens*, *P. virens*, partly *C. erratus, A. lunicollis* and *H. rufipalpus* used the retention groups to a significantly lesser extent, as compared with the clear-cut site that had been ploughed. As the study lasted 4 years after harvest, the non-forest species should have had enough time to colonize the clear-cut (Szyszko, 1983; Niemelä et al., 1993; Skłodowski, 1995, 2006a; Koivula & Niemelä, 2003). Open-habitat carabids are efficient in colonizing clear-cuts: the first individuals may colonize small (<0.5 ha) clear-cut openings within months after logging (Koivula & Niemelä, 2003).

The catch rates of non-forest species, however, were lower in the (small) retention groups than in the adjacent clear-cut sections. These retention groups were 700 m^2 and 400 m^2 in size, thus many times smaller than the 5000 m^2 groups studied by Matveinen-Huju et al. (2006, 2007) but larger than the smallest ones (ca. 200-300 m^2) studied by Koivula (2002). Considering such small retention groups, the tree crowns shade over only one quarter of the area of the retention group: its northern part is shadowed. If so, it is not the shadow but the remaining intact forest floor that seems to be the most probable factor responsible for decreased catch rates of the non-forest carabids over the area of a tree retention group. A very similar trend of non-forest species catch rate reduction was observed by Skłodowski (2002) in a study on 27 small area islands by using the ordinary glycol traps.

Catch rates of forest carabid species were quite similar in the retention tree groups as compared with the clear-cut sites, though possibly *C. violaceus* and *C. glabratus* may have responded negatively to logging. The low catch rates of these species could have resulted from their reluctance to enter the open area. Previously Assmann & Günther (2000) had made a similar conclusion based on their study of *C. glabratus.*

Surprisingly, the mesophilous *P. niger* and the xerophilous *C. arcensis* were captured statistically more frequently in the clear-cut sites than in the retention groups. Both species are considered forest species in Poland (Burakowski, 1973, 1974; Szujecki et al., 1983; Szyszko, 1983; Skłodowski, 2006). They are both relatively common across forest-farmland ecotones (Skłodowski, 1999, 2001). In the study of the anthropogenic disturbance in Białowieża Primeval Forest, *P. niger* was absent in the open area, being present only in stands (Skłodowski, 2006b). This conclusion, as well as the above cited observations, suggests that the species should be considered forest species in Poland.

Despite its mesophilous life style, *P. niger* may penetrate open areas up to a distance of 40 m from the forest edge (Skłodowski, 1999), sometimes even 250 m (Wallin, 1985). It also dominated the carabid assemblages in five years old plantations in the Białowieża Forest (Skłodowski, 2006b). Both species move widely over the clear-cut section as well as forest section and the retention groups. *Carabus arcensis* may move between the two retention groups (Fig. 5). The retention groups examined were apparently too small to attract forest specialist beetles. Similarly, Matveinen-Huju et al. (2007) concluded that small retention patches may not create conditions suitable for the persistence of forest specialists.

Exploratory question Q2 – concerning the presence of forest species and non-forest ones in the furrows directed perpendicularly or parallel to the stand edge.

The second question predicted that soil preparation by ploughing perpendicularly toward the edge of the clear-cut section would stipulate the movements of some forest species in the clear-cut sites; parallel furrows, on the other hand, should favour non-forest carabids. As expected, the non-forest species *P. caerulescens, A. lunicollis* and *H. rufipalpis* were more frequently captured in the furrows oriented parallel than in those oriented perpendicular toward the clear cut edge (and forest section). Apparently, these rather small species were able to move along the furrows. As the catch rates were lower in the perpendicular furrows, one may suspect that non-forest species might respond to perpendicular furrows as barriers. Such a conclusion may also be drawn from the observations of catch rate maps, with differences between those parts of the clear-cut that had been ploughed perpendicularly as compared with those having parallel furrows.

The capature/recapture rates of *C. arcensis* were high in perpendicular furrows. The catch-rate map for the clear-cut area showed a distinct diagonal cluster pattern of *C. arcensis,* starting on the right side close to the clear-cut area (the perpendicular furrows zone) and ending in the central part of the close-cut area – close to places in which the perpendicular and parallel furrows are crossed. An analysis of movement routes (direct lines between capture and recapture events) for this species led to a similar conclusion. I suggest that *C. arcensis* uses perpendicular furrows to enter the clear-cut sections from the forest. If an individual reaches a retention group (not ploughed but bordering these parallel furrows), it may change its initial movement direction. Indeed, based on the observed routes, *C. arcensis* may use the retention patches as "stepping stones". Jopp & Reuter (2005) have shown, however, that large species, such as *C. hortensis*, may not always use stepping stones while moving in suboptimal areas. On the other hand, the smaller *Abax parallelepipedus*, may utilize retention groups more efficiently, as did *C. arcensis.*

Exploratory question Q3 – refers to different responses of forest and non-forest carabids to the microhabitat variation within a stand.

The last exploratory question predicted that in the clear-cut sections of the study site, both forest and non-forest carabids would recognize micro-scale variation of the habitat, which in turn would lead to different catch and recapture rates of these carabid groups over the clear-cut sections. According to Niemelä et al. (1993), one of the responses of carabids to a clear-cut and associated habitat change is an increase in abundance of species characteristic of dry and open conditions. Clearly, clear-cutting leads to the alteration of forest micro-habitats (e.g. Niemelä et al., 1996). Under such conditions, wing-dimorphic species may survive better than brachypterous species do (Kotze & O'Hara, 2003). Indeed, in the present study brachypterous species - many of which are also forest species - find suboptimal conditions in the clear-cut sections. Hence, forest and non-forest species may detect and colonize the new micro-sites in different ways.

The semivariance analysis of the catch rates of forest and non-forest species indicated differences in the use of the clear-cut sections by these two groups. The low value of the Nugget parameter (equal to approx. 33% of the Sill value) indicates a more homogeneous catch-rate distribution between adjacent traps for forest species. On the other hand, the rather high Nugget value for non-forest species (exceeding 50% of the Sill value) gives evidence for large catch-rate variation between adjacent traps. Thus, the non-forest species, as compared with the forest species, may experience the clear-cut sections structurally more heterogeneous.

Perhaps the forest species experience the clear-cut habitat suboptimal, though it may still provide some shelter against, for example, direct sunlight (i.e., the ploughing furrows). The non-forest hemizoophages, on the other hand, may find the sun-exposed, relatively dry patches provide abundant food, such as grass seeds for *Amara* species. Another possible explanation for this difference in micro-habitat use comes from body size differences between forest and non-forest species, and related "environmental grain" size. The latter are often even 2-4 times smaller than the forest species; hence, the structural elements may be too fine for the forest carabids, while the non-forest carabids might still efficiently find shelter and breeding micro-sites in the clear-cut.

CONCLUSIONS

1. The catch rates of non-forest carabids were lower in the retention groups than in the clear-cut sections, which suggests that these carabids generally avoid the unharvested area independent of whether there are retention trees or not.
2. Unexpectedly, the forest species were almost equally often captured in both the clear-cut sections and the retention groups. This finding does mean, however, that the forest carabid species do not use tree retention groups; some of them may use them as stepping stones while moving across otherwise hostile habitat matrix.
3. The furrows oriented perpendicular toward the long edge of the clear-cut section may favor movements of some forest species to the clear-cut, but they may hamper movements of non-forest species that find the clear-cut sections useful habitat. Thus,

in order to aid the recovery of forest fauna in a recently clear-cut area, the furrows should be ploughed perpendicular not parallel toward the clear-cut edge.

5. I suggest that the forest species experience the clear-cut area structurally more homogeneous than the non-forest species do.

ACKNOWLEDGEMENTS

I would like to thank Matti Koivula and an anonymous reviewer for their help in improving the manuscript. My heartiest thanks go to Matti especially for his useful comments and linguistic revision of the text.

REFERENCES

Assmann, T. & Günther, J.M. (2000). Relict populations in ancient woodlands: genetic differentiation, variability and power of dispersal of *Carabus glabratus* (Coleoptera, Carabidae) in north-western Germany. – In: Brandmayr, P. et al. (eds) Natural history and applied ecology of carabid beetles: 197-206. Pensoft. 304 pp.

Atlegrim, O., Sjöberg, K. & Ball, J.P. (1997). Forestry effects on a boreal ground beetle community in spring: selective logging and clear-cutting compared. Entomologica Fennica 8: 19-26.

Baars, M.A. (1979). Pattern of movement of radioactive carabid beetles. Oecologia, 44: 125-140.

Baguette, M. (1993). Habitat selection of carabids beetles in deciduous woodland of southern Belgium. Pedobiologia 37: 365-378.

Burakowski, B., Mroczkowski, M. & Stefanska, J. (1973-1974). Katalog Fauny Polski, Chrzaszcze Coleoptera, Biegaczowate Carabidae, Vol. XXIII, t. 3 & t. 4. PWN. [in Polish.] 232 + 430 pp.

du Bus de Warnaffe, G. & Lebrun, P. (2004). Effects of forest management on carabid beetles in Belgium: implications for biodiversity conservation. Biological Conservation 118: 219-234.

Charrier, S., Petit, B. & Burel, F. (1997). Movements of *Abax parallelepipedus* (Coleoptera, Carabidae) in woody habitats of a hedgerow network landscape: a radio-tracing study. Agriculture, Ecosystems and Environment 61: 133-144.

ESRI ArcMap 9.2 software (Computer program).

Grüm, L. (1959). Sezonowe zmiany aktywności biegaczowatych (Carabidae). Ekol. Pol. A, 9: 255-267.

Jopp F. & Reuter, H. (2005). Dispersal of carabid beetles – emergence o distribution patterns. Ecological Modelling 186: 389-405.

Joyce, K.A., Holland, J.M. & Doncaster, C.P. (1999). Influences of hedgerow intersections and gaps on the movement of carabid beetles. Bulletin of Entomological Research 89: 523-531.

Kapusta, P. (2004). Metody geostatyczne w ekologii. Geostatistical method in ecology. Wiadomości Ekologiczne 50, No.1: 171-194. [in Polish with English summary.]

Koivula, M. (2001). Carabid beetles (Coleoptera, Carabidae) in boreal managed forests – meso-scale ecological patterns in relation to modern forestry. PhD thesis, University of Helsinki. 120 pp.

Koivula, M. (2002). Alternative harvesting methods and boreal carabid beetles (Coleoptera, Carabidae). Forest Ecology and Management 167: 103-121.

Koivula, M., Puntilla, P., Haila, Y. & Niemelä, J. (1999). Leaf litter and the small-scale distribution of carabids beetles (Coleoptera, Carabidae) in the boreal forest. Ecography 22: 424-435.

Koivula, M. & Niemelä, J. (2003). Gap felling as a forest harvesting method in boreal forests: responses of carabid beetles (Coleoptera, Carabidae). Ecography 26: 179-187.

Kotze, D.J. & O'Hara, R.B. (2003). Species decline – but why? Explanations of carabid beetle (Coleoptera, Carabidae) declines in Europe. Oecologia 135: 138-148.

Lindenmayer, D. & Franklin, J.F. (2006). Conserving Forest Biodiversity: A Comprehensive Multiscaled Approach. Island Press, 352 pp.

Lövei, G., Stringer, I.A.N., Devine, C.D. & Cartellieri, M. (1997). Harmonic radar – a method using inexpensive tags to study invertebrate movement on land. New Zealand Journal of Ecology 21 (2): 187-193.

Matveinen-Huju, K., Niemelä, J., Rita, H. & O'Hara, R.B. (2006). Retention-tree groups in clear-cuts: Do they constitute 'life-boats' for spiders and carabids? Forest Ecology and Management 230: 119-135.

Matveinen-Huju, K., Niemelä, J. & Rauha, A.M. (2007). Spider and carabid assemblages in retention-felled stands – short-term effects in Eastern Finland. Manuscript. – In: Matveinen-Huju, K. Short-term effects o variable retention on epigeic spiders and carabid beetles in Finland. Academic dissertation. Yliopistopaino. Helsinki 2007. 51 pp.

Mools, P.J.M. (1979). Motivation and walking behaviour of the carabid beetle *Pterostichus caerulescens* L. at different densities and distributions of the prey. A preliminary report. Miscallaneous papers LH Wageningen, 18: 185-198.

Mools, P.J.M. (1987). Hanger in relation to searching behaviour, predation and egg production of the carabid beetle *Pterostichus caerulescens* L. results of simulation. Acta. Phytopath. Entom. Hung. Vol. 22 (1-4): 187-205.

Niemelä, J., Langor, D. & Spence, J.R. (1993). Effects of clear-cut harvesting on boreal ground-beetle assemblages (Coleoptera: Carabidae) in Western Canada. Conservation Biology 7: 551-561.

Niemelä, J., Haila, Y. & Punttila, P. (1996). The importance of small-scale heterogeneity in boreal forests: variation in diversity in forest-floor invertebrates across the succession gradient. Ecography 19: 352-368.

Petit, S. & Bruel, F. (1993). Movement of *Abax ater* (Col. Carabidae): Do forest species survive in hedgerow networks? Vie Milieu 43 (2-3): 119-124.

Pontégnie, M., du Bus de Warnaffe, G. & Lebrun, P. (2005). Impacts of silvicultural practices on the structure of hemi-edaphic macrofauna community. Pedobiologia 49: 199-210.

Pannatier, Y. (1996). Variowin: software for spatial data analysis in 2D. Springer Verlag, 91 pp.

Petit, S. & Bruel, F. (1993). Movement of *Abax ater* (Col. Carabidae): Do forest species survive in hedgerow networks? Vie Milieu 42 (2-3): 119-124.

Pihlaja, M., Koivula, M. & Niemelä, J. (2006). Responses of boreal carabid beetle assemblages (Coleoptera, Carabidae) to clear-cutting and top-soil preparation. Forest Ecology and Management 222 (1-3), 182-190.

Riecken, U. & Raths, U. (1996). Use of radio telemetry for studying dispersal and habitat use of *Carabus coriaceus* L. Ann. Zool. Fenn. 33 (1) 1996: 109-116.

Rijnsdorp, A.D. (1980). Pattern of movement and dispersal from a Dutch forest of *Carabus problematicus* Hbst. (Coleoptera Carabidae). Oecologia (Berl.) 45: 274-281.

Sklodowski, J. (1995). Antropogenne przeobrażenia zespołów biegaczowatych (Col. Carabidae) w ekosystemach borów sosnowych Polski. [Anthropogenic changes in carabid assemblages (Col. Carabidae) in pine forest ecosystems In Poland.] – In: A. Szujecki et al. (eds.), Antropogeniczne przeobrażenia epigeicznej i glebowej entomofauny borów sosnowych. Katedra Ochrony Lasu i Ekologii. Fundacja "Rozwój SGGW", Warszawa. [in Polish.] pp: 17-174.

Skłodowski, J. (1999). Movement of selected carabid species (Col. Carabidae) through a pine forest-fallow ecotone. Folia Forestalia Polska Ser. A. 41: 5-23.

Skłodowski, J. (2001). The structure of Carabid communities in some field-ecotones. Baltic Journal of Coleopterology 1 (1-2): 41-53.

Skłodowski, J. (2002). System kolonizacji zrębu zupełnego przez biegaczowate oraz możliwości jego doskonalenia. System of colonization clear-cut area by carabid beetles and possibility of its improving – Wydawnictwo SGGW, Warsaw Agricultural University Press. 134 pp. [in Polish with English summary.]

Skłodowski, J. (2006a). Anthropogenic transformation of ground beetle assemblages (Coleoptera: Carabidae) in Białowieża Forest, Poland: from primeval forests to managed woodlands of various ages. Entomologica Fennica 11: 296-314.

Skłodowski (2006b). Monitoring of anthropogenic changes in Białowieża Primeval Forest. Carabidae. – In: A. Szujecki (Eds), Zooindication-based monitoring of anthropogenic transformations in Białowieża Primeval Forest. Warsaw Agricultural Press: 109-148.

StatSoft Inc. (1997). Statistica for Windows (Computer program). – Tulsa, OK, USA.

Szyszko, J. (1983). State of Carabidae (Col.) fauna in fresh pine forest and tentative valorisation of this environment. –Warsaw Agricultural University Press. 80 pp.

Szyszko, J. (2002). Carabids as an efficient indicator of the quality and functioning of forest ecosystems useful in forest management. – In: Szyszko, J. et al. (Eds), How to protect or What we know about Carabid Beetles. From knowledge to application – from Wijster (1969) to Tuczno (2001): 301-318. Warsaw Agricultural University Press. 378 pp.

Szujecki, A., Mazur, S., Perliński, S. & Szyszko, J. (1983). The process of forest soil macrofauna formation after afforestation of farmland. – Warsaw Agricultural University Press. 196 pp.

Wallin, H. & Ekbom, B.C. (1988). Movements of carabid beetles (Coleoptera: Carabidae) inhabiting cereal fields: a field tracing study. Oecologia 77: 39-43.

Wallin, H. (1985). Spatial and temporal distribution of some abundant carabid beetles (Coleoptera, Carabidae) in cereal fields and adjacent habitats. Pedobiologia 28: 19-34.

L. Penev, T. Erwin & T. Assmann (Eds) 2008
BACK TO THE ROOTS AND BACK TO THE FUTURE. TOWARDS A NEW SYNTHESIS AMONGST TAXONOMIC, ECOLOGICAL AND BIOGEOGRAPHICAL APPROACHES IN CARABIDOLOGY
Proceedings of the XIII European Carabidologists Meeting, Blagoevgrad, August 20-24, 2007, pp. 469-481.

Patterns of succession and conservation value of post-industrial areas in central Poland based on carabid fauna (Coleoptera, Carabidae)

Axel Schwerk & Jan Szyszko

Laboratory of Evaluation and Assessment of Natural Resources,
Warsaw University of Life Sciences

SUMMARY

An inventory of carabid fauna was carried out over the 3year period 2004-2006 on two different post-industrial areas located in Central Poland. The first area was an ash heap produced by a power station, whereas the second area was a heap of stony material produced during brown coal mining process. Three study sites of different age on the ash heap and four study sites of different age on the brown coal mining heap were studied. A pine forest on undisturbed soil located close to the ash heap was chosen as reference study site.

During the three years of study 2324 individuals from 70 species were collected, some of them detected for the first time on the regional level or on the Polish Red List of threatened species. Mean Individual Biomass (MIB) of Carabidae was calculated as indicator of stage of succession. With ongoing succession species characterised by small size are replaced by large bodied species, so that MIB increases as succession progresses. Both MIB as well as multivariate Correspondence Analysis (CA) suggest that succession is delayed on the study sites. However, after about 15 years the carabid fauna changed on the mining heap. Succession seems to be more delayed on the ash heap, despite indications of a similar change in carabid fauna. Many species were restricted to particular habitat types and even more species were found only in particular successional stages.

The observed patterns of succession seem to be characteristic for many post-industrial areas, particularly those, which are subject to the process of primary succession. The conservation importance of landscape diversity with respect to habitats and stages of succession is corroborated by the study.

Keywords: Carabidae, post-industrial areas, succession, conservation

INTRODUCTION

Since the last decades of the 20th century there has been rising awareness that post-industrial areas offer numerous possibilities for nature conservation. Many post-industrial areas have high numbers of rare or endangered species (e.g., Kelcey, 1975; Johnson et al., 1978; Gillham & Smith, 1983; Gemmell & Connell, 1984; Rebele & Dettmar, 1996). On the other hand, revegetation may proceed slowly on these areas delaying their conservation function. In this context successional processes are of special importance, because careful management of successional stages is central to faunal and floral recovery (e.g., Bradshaw, 1984; Jochimsen, 2001). Therefore, study of successional processes on post-industrial areas could be useful.

In this paper we assess the processes of succession in relation to conservation value of two post-industrial areas near the city of Bełchatów in central Poland. In both areas carabid beetles were used to analyse study sites of different age. Prior studies have shown differences between these areas with respect to total carabid fauna as well as single species (Schwerk & Szyszko, 2006). Since carabid beetles are sensitive to changes in environmental conditions (Rainio & Niemelä, 1993) they may be useful as indicators of the successional process on these areas. Mean Individual Biomass (MIB) of Carabidae has been described as a simple and effective indicator of the stage of succession (Szyszko, 1990; Szyszko et al., 2000). With ongoing succession species characterised by small size are replaced by large bodied species so that MIB increases as succession progresses. Because many species are restricted to particular stages of succession, successional diversity of ecosystems is important for biodiversity conservation (e.g., Szyszko, 2004; Lindenmayer et al., 2006). Special value may be assigned to areas, which host rare or endangered species.

The aim of the paper is to answer the following research questions:

(1) What patterns of succession occur at the study areas, using MIB values to describe succession on the study sites?

In addition, unconstrained ordination was conducted to describe patterns of faunal change. Distribution of the study sites along the ordination axes may give insight into successional processes.

(2) How is landscape diversity (i.e., habitat types, successional stages) associated with biodiversity?

This aspect was analysed by calculating the percentage share of species restricted to special habitat types and stages of succession respectively. The higher the percentage of restricted species, the more the respective habitats or successional stages contribute to overall biodiversity.

(3) Do any rare and/or endangered species occur in these post-industrial areas?

The existence of species new to regional checklists or recorded among Red List data books indicates value of the areas with respect to biodiversity conservation.

MATERIAL AND METHODS

Study areas and collecting methods

The study areas are located in central Poland close to the city of Bełchatów (Fig. 1). The industrial activity at Bełchatów consists of brown coal mining and a power station. These activities have generated two heaps of waste material, one of stony material produced during the mining process and a second of ashes produced by the power station. Eight study sites were chosen for investigation, comprising different old parts of the ash heap (3 sites) and pine stands of different age on the brown coal mining heap (4 sites). A pine forest on undisturbed soil was chosen as reference site (Table 1).

Carabids were collected at each site using pitfall traps (Barber, 1931). Three pitfall traps were installed at each site. Traps were jar glasses with a funnel installed flush with the soil surface. A roof was suspended a few cm above the funnel and ethylene glycol was used as a killing agent and preservative. Carabids were sampled from mid-April to mid-October in 2004 and late April to mid-October in 2005 and 2006. Carabids were identified to the species level and named according to Freude et al. (2004).

Data analyses

Mean individual biomass (MIB) of Carabidae was calculated to assess the stage of succession (Szyszko, 1990; Szyszko et al., 2000). MIB is calculated by dividing the biomass of all sampled carabids by the number of specimens caught. Biomass values were obtained using the formula of Szyszko (1983) that describes the relationship between the body length of a single carabid individual (x) and its biomass (y):

$$\ln y = -8.92804283 + 2.55549621 \times \ln x \quad (1)$$

After DCA to select the appropriate statistical model, Correspondence Analysis was carried out using the CANOCO for Windows software package, v. 4.5 (ter Braak, 1987; ter Braak & Šmilauer, 2002). Dominance values (percentage share of the respective species in a sample) of carabid species on the sampling plots were used.

The CA analyses were performed using scaling on inter-sample distances and Hill's scaling, using unweighted and untransformed data about each of all species. Biplots were cre-

Fig. 1. Location of the city of Bełchatów in Poland.

Table 1. Description of the study sites. 1B – 3B, study sites on the ash heap; 4B, forest stand; 5B – 8B, study sites on the brown coal mining heap (data on age relate to the year 2004).

Study site	Type	Description
1B	Ash heap	About 7-8 years old vegetation on about 10 years old ashes, insignificant cover with plants
2B	Ash heap	About 10 years old pioneer vegetation with dense cover of plants on about 12 years old ashes, planting of oak had been carried out on this area, but failed because of game bite
3B	Ash heap	About 12 years old shrub vegetation on about 15 years old ashes, dense cover of plants
4B	Forest stand	Pine stand of about 65 years on undisturbed soil, used as reference plot
5B	Brown coal site	Pine plantation, 3 years old
6B	Brown coal site	Pine plantation, 10 years old
7B	Brown coal site	Pine plantation, 14 years old
8B	Brown coal site	Pine plantation, 21 years old

ated for species and samples with species weight range adjusted so that 15 species with the largest impact on the results of the analysis are displayed (ter Braak & Šmilauer, 2002).

Shares of species restricted to special habitat types as well as stages of succession were calculated, treating the mining heap, the ash heap and the reference forest stand as different habitat types. With respect to stages of succession the study sites were divided into young or old classes according to MIB values. MIB values range from about 40 mg to about 400 mg (e.g., Szyszko et al., 1996, 2000), with some exceptional values falling beyond these extremes. Therefore, MIB values below 100 mg were assumed to indicate young stages of succession, whereas MIB values exceeding 100 mg indicate advanced stages of succession.

Species new for the checklist of the Łódź Uplands (Jaskuła et al., 2002; Jaskuła, 2003) and species recorded at the Red List Data Book of Poland (Głowaciński, 2002) were identified.

RESULTS

Altogether 2324 individuals from 70 species were collected (Table 2, Appendix). Total species numbers differ only slightly between the years, varying between 49 species (2004, 2006) and 53 species (2005). However, on some study sites the species numbers varied more dramatically (Table 2). Total catches of individuals varied between 511 individual (2006) and 942 individual (2005). MIB values ranged from 39.2 mg (study site 5B in 2005) up to 260.2 mg (study Site 7B in 2004). In general, study sites 1B, 2B, 3B, 5B and 6B exhibited low MIB values (far below 100 mg), with the exception of a much higher MIB value (135.5 mg) at study site 3B in 2006. At study sites 4B, 7B and 8B MIB values were constantly above 100 mg.

Table 2. Numbers of species, numbers of individuals and MIB values for the study sites as well as the total sampling results in the years 2004 – 2006.

Study site	Species			Individuals			MIB (mg)		
	2004	2005	2006	2004	2005	2006	2004	2005	2006
1B	18	14	7	198	423	144	49.2	63.9	65.6
2B	15	16	15	149	132	76	55.4	67.0	50.6
3B	15	20	10	82	92	24	46.7	49.0	123.5
4B	12	8	11	198	109	84	182.0	214.4	194.6
5B	10	8	18	105	60	96	44.0	39.2	39.7
6B	3	7	5	13	37	10	59.4	47.5	59.5
7B	8	10	8	69	29	22	260.2	106.1	147.9
8B	12	16	13	57	60	55	237.0	209.4	153.1
Total	49	53	49	871	942	511	108.7	88.6	95.3

Fig. 2 shows the MIB values in relation to age of the study sites. All MIB values on the ash heap ranged 50-75 mg, with exception of the MIB value of the 14-yr old study site B3 in 2006, which was calculated from a small number of individuals. On the mining heap the MIB values varied around 50 mg on the younger study sites, but MIB increased at an age of about 15 years.

The Correspondence Analysis identifies clear separation of the study sites with respect to stage of succession (MIB values) along the first ordination axis (Fig. 3). The study sites characterized by low MIB values are located on the left side and the study

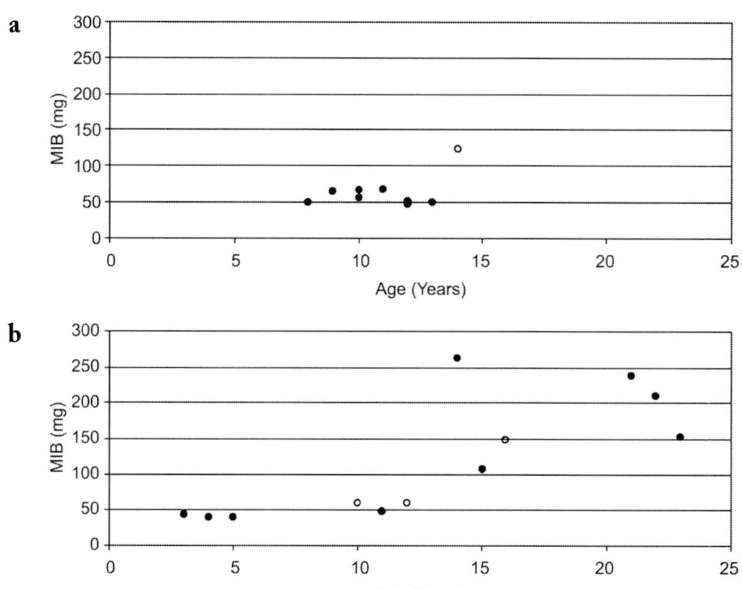

Fig. 2. MIB values on the ash heap (a) and the mining heap (b) in relation to the age of the study sites (Open circles indicate that the MIB value was calculated from less than 25 individuals).

sites characterized by high MIB values located on the right side of the diagram. The pine stands on the mining heap (7B, 8B) are separated from the stand on undisturbed soil (4B) along the second ordination axis. Among the young successional stages site 3B is located most closely to the advanced stages of succession. Species located close to the young successional stages in the diagram are common to open habitats (e.g., *Amara aenea*, *Calathus ambiguus*, *Calathus erratus* and *Harpalus flavescens*). Typical forest species are located closer to advanced stages of succession in the ordination (Fig. 3). However, some of them (e.g., *Pterostichus niger*, *Carabus auronitens*) are centred rather closely to the sites on the mining heap, whereas other species (e.g., *Carabus arvensis*) appear more closely to the reference forest stand.

Almost 60 % of the species collected were restricted to a special habitat type (Fig. 4a). More than 70 % of the species were restricted to either young or old stages of succession, with more than 50 % of these restricted to the young stages of succession (Fig. 4b).

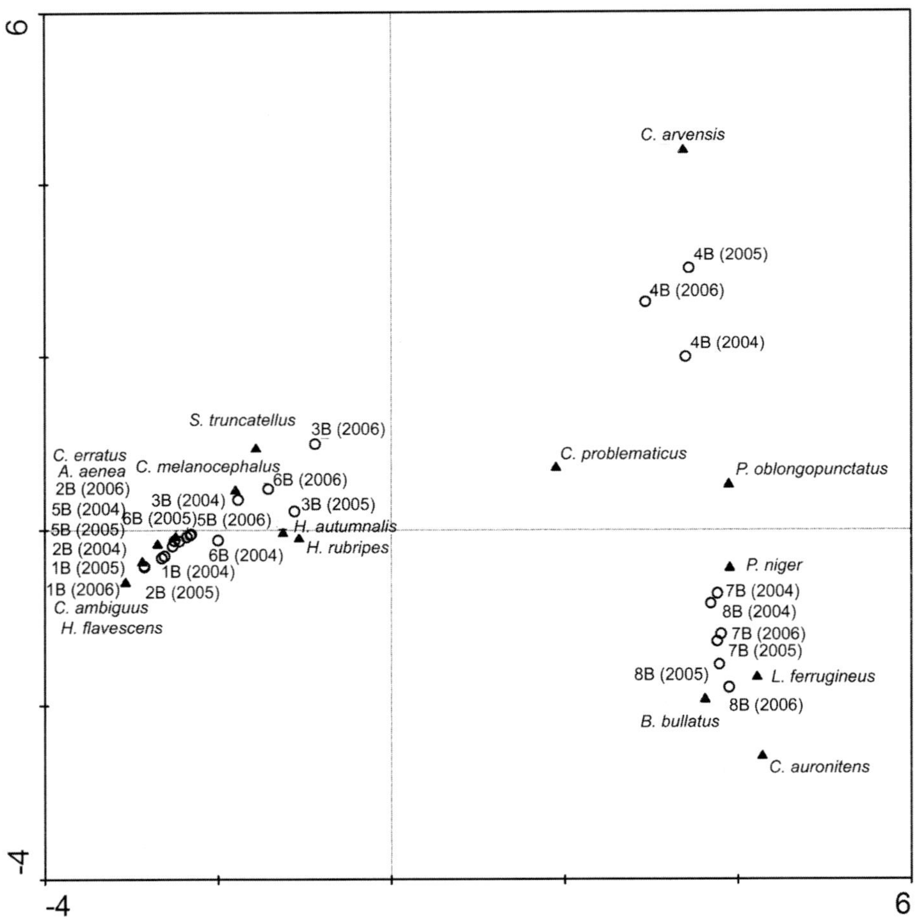

Fig. 3. Correspondence Analysis (CA) carried out with the dataset (year of sampling in brackets behind label of study site).

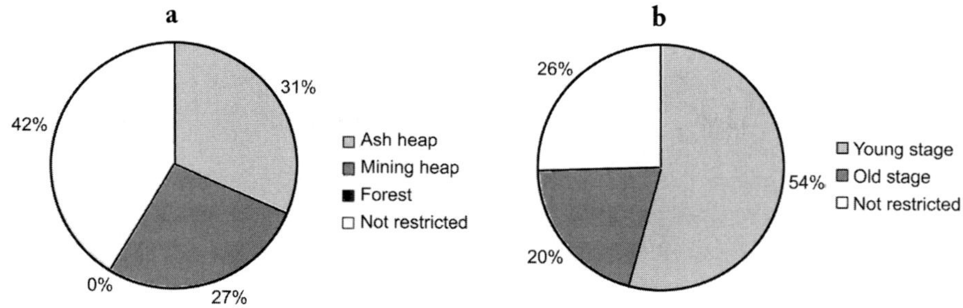

Fig. 4. Percentage share of species restricted to habitat types (a) and stages of succession (b).

Five species were new for the checklist of Carabidae in the Łódź Uplands, namely *Badister lacertosus*, *Calathus cinctus*, *Harpalus calceatus*, *Harpalus hirtipes* and *Licinus depressus*. Two additional species, *Masoreus wetterhallii* (vulnerable) and *Broscus cephalotes* (data deficient), were included in the Red List Data Book of Poland (Głowaciński 2002).

DISCUSSION

In general, rather low numbers of individual carabids were collected. However, there are significant changes in numbers of individuals as well as species on particular study sites across the different years of our study. Such variability is characteristic of carabid populations, a feature explained in the context of metapopulation models (e.g., Hanski & Gilpin, 1991; den Boer & Reddingius, 1996).

The MIB values, however, were generally rather constant on the study sites. The exceptional value observed at study site 3B in 2006 may be assessed as an outlier based on a low number of individuals, or perhaps this suggests that succession is moving forward at this site. Future studies will test, which interpretation is true. Assessment of the state of succession of the study sites based on MIB values is confirmed by the distribution of the study sites and species in the ordination diagram. Here, the location of study site 3B corroborates to some extent the hypothesis that succession is advancing at this site.

In reference to our objectives, characteristic patterns of succession were detected on the post-industrial areas (research question 1). The study shows, that succession is delayed on both disturbed sites. This phenomenon is known from many post-industrial areas (Rebele & Dettmar, 1996; Schwerk et al., 2006; Jochimsen, 2001). This pattern of succession seems to be particularly characteristic for areas underlying the process of primary succession. Majer (1989) reports a strikingly similar pattern with respect to data about recolonisation of Illinois surface-mine spoils by thrips (Thrysanoptera). In this study rapid increase in number of species takes place 10-20 years after succession begins. Schwerk (1996) found that sites younger than 15 years showed MIB values of about 50 mg, whereas a study site older than 35 years showed MIB values clearly exceeding 100 mg at a colliery spoil heap in the Ruhr Valley.

The importance of successional processes in promoting landscape habitat diversity is corroborated by the study (research question 2). Diversity in successional stages is at least as valuable for biodiversity conservation as diversity of habitats, as has been previously stressed (e.g., Szyszko, 2004; Lindenmayer et al., 2006). In particular young stages of succession on post-industrial areas may host high numbers of species, especially after reclamation measures have been applied (e.g., Neumann, 1971; Parmenter & MacMahon, 1987; Vogel & Dunger, 1991).

Post-industrial areas may, in fact, include habitats, which have become rare on modern landscapes, where natural disturbances have been minimised and, therefore, provide habitats for certain species (e.g., Rebele & Dettmar, 1996). The studied areas provided the first regional records for five species, but only two species were represented in the Polish Red List data book (Głowaciński, 2002) (research question 3). With exception of *Calathus cinctus*, all these species are included in the Catalogue of Polish Fauna (Burakowski et al., 1973, 1974). However, since there have been several nomenclature changes in recent years, *Calathus cinctus* might have already been detected in the region, but the record could be hidden in outdated nomenclature. In general, the results suggest conservation value for these reclaimed habitats on the regional level. Since most of the regionally rare species prefer dry and sandy habitats exposed to sun, it would be suitable to keep some parts of such lands in early stages of succession.

ACKNOWLEDGEMENTS

The authors thank mgr inż. Ryszard Bijak and mgr inż. Kazimiersz Grochulski for help with the fieldwork, Karsten Hannig for confirming the identifications of difficult individuals and John Spence for valuable comments on the manuscript and linguistic revision of the text. This paper is communication number 206 of the Laboratory of Evaluation and Assessment of Natural Resources.

REFERENCES

Barber, H.S. (1931). Traps for cave inhabiting insects. – J. Mitchel. Soc. 46: 259-266.

Bradshaw, A.D. (1984). Ecological principles and land reclamation practice. – Landscape Planning 11: 35-48.

Burakowski B., Mroczkowski M. & Stefańska J. (1973). Katalog fauny Polski (Catalogus faunae Poloniae). Część XXIII, tom 2. Chrząsccze (Coloeptera). Biegaczowate (Carabidae), część 1. – Państwowe Wydawnictwo Naukowe, Warszawa.

Burakowski B., Mroczkowski M. & Stefańska J. (1974). Katalog fauny Polski (Catalogus faunae Poloniae). Część XXIII, tom 3. Chrząsccze (Coloeptera). Biegaczowate (Carabidae), część 2. – Państwowe Wydawnictwo Naukowe, Warszawa.

den Boer, P. J. & Reddingius, J. (1996). Regulation and stabilization paradigms in population ecology. – Chapman & Hall, London, Weinheim, New York, Tokyo, Melbourne, Madras.

Freude H., Harde K.-W., Lohse G.A. & Klausnitzer B. (2004). Die Käfer Mitteleuropas. Bd. 2, Adephaga 1, Carabidae (Laufkäfer), 2. (erweiterte) Aufl. – Spektrum-Verlag, Heidelberg/Berlin.

Gemmell, R.P. & Connell, R.K. (1984). Conservation and creation of wildlife habitats on industrial land in Greater Manchester. – Landscape Planning 11: 175-186.

Gillham, M.E. & Smith, J.K. (1983). Industry and wildlife: compromise and coexistence. – Endeavour, N. F. 7: 162-172.

Głowaciński, Z. (Ed.) (2002). Czerwona lista zwierząt ginących i zagrożonych w Polsce. – Instytut Ochrony Przyrody PAN, Kraków.

Hanski, I. & Gilpin, M. (1991). Metapopulation dynamics: brief history and conceptual design. – Biological Journal of the Linnean Society 42: 3-16.

Jaskuła, R. (2003). Biegaczowate (Coleoptera: Carabidae) w wybranych rezerwatach okolic Lodzi. – Parki nar. Rez. Przyr. 22: 549-560.

Jaskuła, R., Kowalczyk, J.K. & Watala, C. (2002). Ground beetles (Coleoptera: Carabidae) of Lodz Upland, Central Poland. – Baltic J. Coleopterol. 2: 117-125.

Jochimsen, M.E. (2001). Vegetation development and species assemblages in a long-term reclamation project on mine spoil. – Ecological Engineering 17: 187-198.

Johnson, M.S., Putwain, P.D. & Holliday, R.J. (1978). Wildlife conservation value of derelict metalliferous mine workings in Wales. – Biol. Conserv. 14: 131-148.

Kelcey, J.G. (1975). Industrial development and wildlife conservation. – Environmental Conservation 2: 99-108.

Lindenmayer, D.B., Franklin, J.F. & Fischer, J. (2006). General management principles and a checklist of strategies to guide forest biodiversity conservation. – Biological Conservation 131: 433-445.

Majer, J.D. (1989). Long-term colonization of fauna in reclaimed land. – In: Animals in primary succession. The role of fauna in reclaimed lands (Majer, J. D., ed.). Cambridge University Press, Cambridge, New York, Port Chester, Melbourne, Sydney, p. 143-174.

Neumann, U. (1971). Die Sukzession der Bodenfauna (Carabidae (Coleoptera), Diplopoda und Isopoda) in den forstlich rekultivierten Gebieten des Rheinischen Braunkohlenreviers. – Pedobiologia 11: 193-226.

Parmenter, R.R. & MacMahon, J. A. (1987). Early successional patterns of arthropod recolonization on reclaimed strip mines in southwestern Wyoming: the ground-dwelling beetle fauna (Coleoptera). – Environ. Entomol. 16: 168-175.

Rebele, F. & Dettmar, J. (1996). Industriebrachen: Ökologie und Management. – Ulmer, Stuttgart (Hohenheim).

Schwerk, A. (1996). Charakterisierung von Laufkäferzönosen (Coleoptera: Carabidae) auf städtischen Industriebrachen anhand freilandökologischer und populationsgenetischer Untersuchungen. – Dissertation. Ruhr-University Bochum.

Schwerk, A., Sałek, P., Duszczyk, M., Abs, M. & Szyszko, J. (2006). Variability of Carabidae in time and space in open areas. – Entomol. Fennica 17: 258-268.

Schwerk, A. & Szyszko, J. (2006). Succession of carabid fauna (Coleoptera: Carabidae) on post-industrial areas near Bełchatów (Central Poland). – Wiad. entomol. 25, Supl. 1: 71-85.

Szyszko, J. (1983). Methods of macrofauna investigations. – In: The process of forest soil macrofauna formation after afforestation of farmland (Szujecki, A., Szyzsko, J., Mazur, S. & Perliński, S. eds). Warsaw Agricultural University Press, Warsaw, p. 10-16.

Szyszko, J. (1990). Planning of prophylaxis in threatened pine forest biocoenoses based on an analysis of the fauna of epigeic Carabidae. – Warsaw Agricultural University Press, Warsaw.

Szyszko, J. (2004). Foundations of Poland's cultural landscape protection – conservation policy. – In: Cultural landscapes and land use (Dieterich M., Van der Straaten J., eds). Kluwer Academic Publishers, The Netherlands, p. 95-110.

Szyszko, J., Vermeulen, H.J.W. & den Boer, P. J. (1996). Survival and reproduction in relation to habitat quality and food availability for *Pterostichus oblongopunctatus* F. (Carabidae, Col.). – Acta Jutlandica 71: 25-40.

Szyszko, J., Vermeulen, H.J.W., Klimaszewski, K., Abs, M. & Schwerk, A. (2000). Mean individual biomass (MIB) of ground beetles (Carabidae) as an indicator of the state of the environment. – In: Natural history and applied ecology of carabid beetles (Brandmayr, P., Lövei, G., Zetto Brandmayr, T., Casale A. & Vigna Taglianti, A., eds). Pensoft Publishers, Sofia-Moscow, p. 289-294.

ter Braak, C.J.F. (1987). CANOCO – A FORTRAN program for canonical community ordination by [partial][detrended][canonical] correspondence analysis, principal components analysis and redundancy analysis (version 2.1). – DLO-Agricultural Mathematics Group, Wageningen.

ter Braak, C.J.F. & Šmilauer, P. (2002). CANOCO reference manual and CanoDraw for Windows User's guide: Software for Canonical Community Ordination (version 4.5). – Microcomputer Power, Ithaca.

Vogel, J. & Dunger, W. (1991). Carabiden und Staphyliniden als Besiedler rekultivierter Tagebau-Halden in Ostdeutschland. – Abhandlungen und Berichte des Naturkundemuseums Görlitz 65: 1-31.

Appendix. Numbers of individuals of Carabidae collected at the study sites (1B – 8B) in the years 2004 – 2006 (list of species in alphabetical order).

Species	1B			2B			3B			4B			5B			6B			7B			8B			Sum
	'04	'05	'06	'04	'05	'06	'04	'05	'06	'04	'05	'06	'04	'05	'06	'04	'05	'06	'04	'05	'06	'04	'05	'06	
Amara aenea							1						3	12	11										27
Amara bifrons								2							2										4
Amara brunnea																							1	2	3
Amara communis							2	4														3	4	2	15
Amara consularis																							1		1
Amara equestris				3	4	4																			11
Amara eurynota													2												2
Amara familiaris						1																			1
Amara fulva		1			2																				3
Amara lunicollis				2			2																		4
Amara plebeja															1										1
Amara spreta	34	6	1	1																					42
Badister bullatus							1	4				1								8	1		2	4	21
Badister lacertosus											1									1			4	5	11
Bembidion lampros															1										1
Bembidion pygmaeum														1	1										2
Bradycellus caucasicus								1																	1
Bradycellus csikii							1																		1
Broscus cephalotes			1																						1
Calathus ambiguus	14	10	1	23	5								10		3										66
Calathus cinctus	4	5			1																				10
Calathus erratus	66	114	51	70	37	39	33	26	1			3	78	25	49	11	28	5							636
Calathus fuscipes					2	2							1	4	1		1								11
Calathus melanocephalus	6	6	1		1	10	5	15	2	1		1			1		2								51
Calathus micropterus										3		4										2			9
Carabus arvensis										88	77	47						1							213
Carabus auronitens																					2	6	12	14	34

Species	1B			2B			3B			4B			5B			6B			7B			8B			Sum
	'04	'05	'06	'04	'05	'06	'04	'05	'06	'04	'05	'06	'04	'05	'06	'04	'05	'06	'04	'05	'06	'04	'05	'06	
Carabus granulatus																			1						1
Carabus hortensis							3	2	2	2	1														10
Carabus problematicus	1							1	3		1	5										4	3	1	19
Carabus violaceus																			9	1		4	3		17
Cicindela hybrida															1										1
Harpalus affinis	1			1		1																			3
Harpalus anxius								1					5	10	7										23
Harpalus autumnalis	1					1							3	3	2		3	2			1			1	17
Harpalus calceatus	1			1	1																				3
Harpalus flavescens	33	265	88	28	59	11																			484
Harpalus griseus		1																							1
Harpalus hirtipes	1	1	1																						3
Harpalus laevipes								1														2	4	8	15
Harpalus latus				1			2	1	1																5
Harpalus rubripes				5	1	1	3	6					1	3	8			1	3	1					33
Harpalus rufipalpis	2									1							1								4
Harpalus rufipes	7	5		6	12	1		1	1	2		1					1					1	1		39
Harpalus servus													1												1
Harpalus smaragdinus													1	2	3	1	1								8
Harpalus solitaris				1																					1
Harpalus tardus	1	1		4			3	6	3						2										20
Harpalus xanthopus																						2	2	3	7
Leistus ferrugineus										4	1								4	5	5				19
Leistus rufomarginatus																								1	1
Leistus terminatus																				1					1
Licinus depressus						1		4	1	1															7
Loricera pilicornis					1																				1
Masoreus wetterhallii							3	2																	5

Species	1B			2B			3B			4B			5B			6B			7B			8B			Sum
	'04	'05	'06	'04	'05	'06	'04	'05	'06	'04	'05	'06	'04	'05	'06	'04	'05	'06	'04	'05	'06	'04	'05	'06	
Microlestes minutulus															1										1
Notiophilus biguttatus																					1				1
Panagaeus bipustulatus		2				1	1	5	4	2	1	3													19
Poecilus lepidus	1			2	3	1																			7
Poecilus versicolor				1	1					4		1													7
Pterostichus diligens							1																		1
Pterostichus melanarius	1																		1	1	1	2	3	1	10
Pterostichus niger						1		1		60	24	14				1			48	8	10	26	16	8	217
Pterostichus nigrita																							1		1
Pterostichus oblongopunctatus								1		30	3	4							1	2		3	2	5	51
Pterostichus strenuus																			2	1	1	2	1		7
Syntomus foveatus		1																							1
Syntomus truncatellus	3				1	1	21	8	6						1			1							42
Synuchus vivalis	21	5																							26
Trechus quadristriatus					1										1										2
Sum	198	423	144	149	132	76	82	92	24	198	109	84	105	60	96	13	37	10	69	29	22	57	60	55	2324

L. Penev, T. Erwin & T. Assmann (Eds) 2008
Back to the Roots and Back to the Future. Towards a New Synthesis amongst Taxonomic, Ecological and Biogeographical Approaches in Carabidology
Proceedings of the XIII European Carabidologists Meeting, Blagoevgrad, August 20-24, 2007, pp. 483-509.

Patterns of urbanisation in the City of Sofia as shown by carabid beetles (Coleoptera, Carabidae), ants (Hymenoptera, Formicidae), and terrestrial gastropods (Mollusca, Gastropoda Terrestria)

Lyubomir Penev, Ivailo Stoyanov, Ivailo Dedov & Vera Antonova

Central Laboratory for General Ecology, Yuri Gagarin Street 2, 1113 Sofia, Bulgaria.
E-mail: info@pensoft.net

SUMMARY

The GLOBENET project in Sofia was expanded from studies on carabid beetles to assess how different soil invertebrates reflect urbanisation. Carabid beetles, terrestrial gastropods and ants were chosen for being well-studied, widely distributed and functionally important groups in ecosystems. At the same time, they are characterised by different life strategies, habitat requirements and trophic relationships.

The urban faunas of all three groups were formed by (1) filtering out several native species from the regional species pools of Sofia Kettle, (2) gaining new autochtonous species, and (3) gaining new allochtonous species. The latter is exhibited mainly by gastropods and less by ants (invasive ant species are found only in houses or greeneries). There were also no clearly invasive carabid beetle species. A trend of colonisation by invasive gastropods species from Sofia invading the surroundings was observed.

In all three groups, there were clearly distinguishable "urban" assemblages that differed from those of the rural and suburban zones. Specific suburban assemblages were formed by ants, but not by gastropods or carabid beetles.

The main factors affecting spatial variation of ground-beetle, ant and terrestrial gastropod assemblages respectively were: distance from the city center; perimeter/area ratio of the fragment, area of the homogenous habitat, distance to the nearest fragment; and altitude, area and level of urbanisation of the sampling sites.

The patterns of urbanisation had much in common amongst the three groups studied, but at the same time they displayed group-specific peculiarities. Hence, bioindication and monitoring activities should be carefully planned to include various taxonomic groups with different life-strategies.

Keywords: terrestrial gastropods, ants, ground-beetles, Sofia, Bulgaria, urbanisation, biodiversity, habitat preference, zoogeographical structure

INTRODUCTION

The spatial variation of biotic communities in urban environments enjoys a prime interest of biologists, landscape ecologists, and specialists in urban planning. In virtually all regions of the Globe, urbanisation is easily detectable and affects the local floras and faunas in many ways – e.g. altering their composition and diversity. Hence, the idea of studying the impact of "a common factor onto different floras/faunas in different ecological/historical conditions" or the so called "intercontinental community convergence" (Schoener, 1986) finds its appropriate application in on times of ongoing global urbanisation.

One of the first steps in this direction became the establishment of the GLOBENET project, aimed at surveying the common anthropogenic impacts on biodiversity on a global scale using carabid beetles (Niemelä et al., 2000; Niemelä & Kotze, 2000). The basic concept utilized within the framework of the GLOBENET initiative is the paradigm of urban-rural gradients as primary interfaces between heavy urban development and the relatively less disturbed natural landscapes (Klausnitzer, 1982; McDonnell & Pickett, 1990). The GLOBENET project engendered immediate interest among the carabidologists. Impact of urbanisation on ground beetle assemblages was studied using the GLOBENET unified methodology and sampling protocols in several cities of the world (e.g. Niemelä et al., 2002; Ishitani et al., 2003; Sadler et al., 2006; Elek & Lövei, 2007; Magura et al., 2008) followed by similar studies on other animal groups (Hornung et al., 2007; Vilisics et al., 2007; Magura et al., 2008).

Increased international interest in the project encouraged the Sofia team to extend its work toward a more detailed examination of the soil properties and the floristic components of the landscape, as well as to include some additional soil mesofauna groups, such as gastropods, spiders, harvestmen, myriapods, ants and nematodes among others. The results of the collaboration between several specialists in abiotic and biotic components of the urban environment of Sofia were summarized in a collection of studies (Penev et al., 2004).

The present paper aims at comparing the effect of urbanisation on different soil invertebrates characterized by different life strategies and habitat requirements. We compare here the general patterns of formation of urban ground beetle, ant and terrestrial gastropod faunas. Spatial variations of some assemblage parameters, such as alpha-diversity, zoogeographical and ecological structure, were observed in field sites where these groups co-occurred.

All three animal groups studied are highly diverse and abundant in almost every terrestrial habitat. They have been chosen for being taxonomically well known, locally abundant and easy to collect by various sampling methods, including quantitative ones. All three groups are "key species" in ecosystems as predators, prey, detritivores, mutualists and herbivores. Moreover, their habitat requirements range broadly from higrophily to xerophily. Carabids, in particular, are known as excellent model objects for monitoring purposes mainly because they are widely distributed, abundant, sensitive to landscape alteration and other environmental factors (e.g. Stork, 1990). Ants are appropriate for inventory and monitoring programs because most of the species have stationary, perennial nests with restricted foraging ranges (from less than one meter to few hundred meters). The low mobility of terrestrial gastropods makes them sensitive to environmental changes. Terrestrial gastropods are known to be one of the dominant groups in city parks (Pisarski et al., 1989). Similarly, ground beetles and ants are also abundant at these locations.

MATERIAL AND METHODS

Site description

The investigated area is situated in West Bulgaria and includes the City of Sofia, Sofia Kettle and the adjacent foothills of the surrounding mountains – Stara Planina, Vitosha, Sredna Gora, and Lyulin, ranging up to about 1000 m above sea level.

One of the main criteria, besides distance from the city centre, used to select the sampling sites, was the prevalence of more or less similar forest types – in our case, broad-leaved forests dominated by oak (*Quercus* spp.) (Penev et al., 2004). The ant study sites were also located in open grassy habitats chosen on the basis of prevalence of common dominant grass and herb species (e.g. *Poa* spp., *Festuca* spp., *Phleum pratense* L., *Avena fatua* L., *Setaria viridis* (L.), *Cynodon dactylon* Pers., *Dactylis glomerata* L., *Trifolium* spp., *Taraxacum officinale* Weber., *Bellis perennis* L., *Capsella bursa-pastoris* Med., and *Plantago* spp.), as well as in forested habitats dominated by oak trees (Antonova & Penev, 2006).

Carabid beetles and gastropods were sampled at eleven sites along an urban-rural gradient (with its urban end in the city parks of Sofia, spreading towards the rural surroundings) (Fig. 1). More detailed site descriptions are provided in Penev et al. (2004). Ants were collected at 56 localities situated in the urban greenery – i.e. wooded and open areas in the parks, green yards, along the transport corridors and streets – specifically 10 "forest" and 46 "meadow" sites. A further 8 localities situated in the vicinity of Sofia (at the foothills of Vitosha, Lyulin and Lozen Mountains, not higher than 150 m above the average altitude of the city with distance up to 10 km from the city ring road), were sampled, too. (Fig. 2).

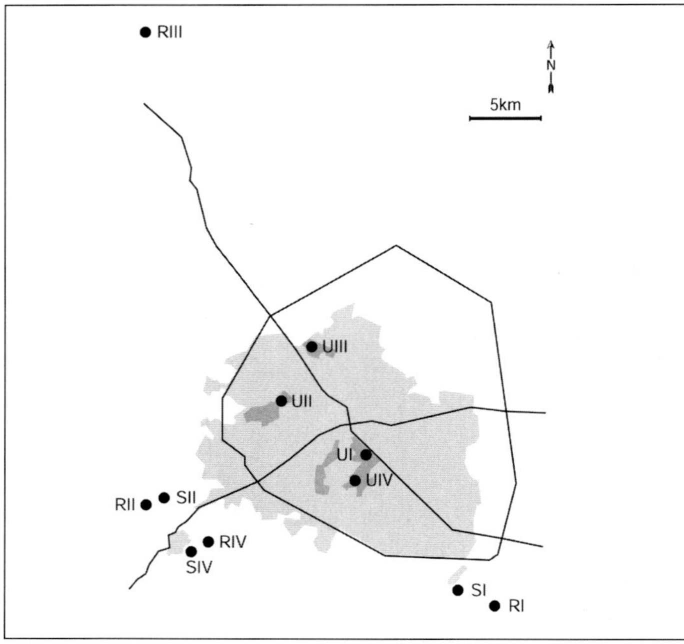

Fig. 1. The GLOBENET sampling sites for carabid beetles and gastropods (UI-UIV – urban sites; SI-SIII – suburban sites; RI-RIV – rural sites).

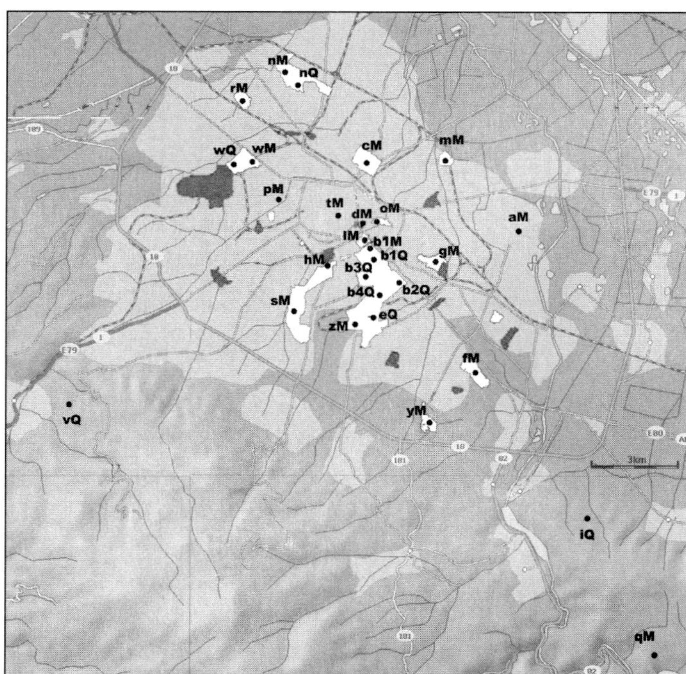

Fig. 2. The sampling sites for ants in Sofia and its surroundings.

Sampling methods

Ground beetles

At each site we collected carabid beetles using pitfall traps which were active during the vegetation seasons of 1998 (May 07-October 22) and 1999 (April 01-October 26) and were emptied monthly (for details of sites and sampling design see Niemelä et al., 2002, Penev et al., 2004). All carabid beetles caught in the pitfall traps were identified to species using standard keys (Freude et al., 1976; Hůrka, 1996). The Balkan and Bulgarian endemic species were identified using an unpublished manuscript by the late O. L. Kryzhanovskij. The systematic nomenclature follows Kryzhanovskij et al. (1995).

Ants

Quantitative quadrate samples were taken: ten 3 m x 3 m (9 m^2) plots in forest habitats and ten 1 m x 1 m (1m^2) plots in meadow sites (Czechowski et al., 1995). Additionally, a "direct sampling" method (Bestelmeyer et al., 2000) was used in the urban green areas of Sofia. In the forest sites the colonies were sampled from the litter, the ground (up to 25 cm depth), under bark and stones, in logs and nuts. In the meadow sites samples were taken from the ground, grass clumps, and under stones. Ants were identified to species after Atanassov & Dlussky (1992), Seifert (1996) and Czechowski et al. (2002). The ant material for the present study was collected in 2003-2005 and is kept in the collection of V. Antonova at CLGE-BAS.

Gastropods

Terrestrial gastropods were sampled quantitatively by a quadrat method (50 x 50 cm) proposed by Oekland (1930). Ten samples were taken at each site from April to October in 1999 and 2000. Leaf litter and underlying soil horizon of each sample were sifted at depth of 10 cm. Hand collecting from the soil surface, under tree bark and stones, in rotten wood and on plants supplemented these plots.

Species characteristics

Ground beetles

The Siberian, Holomediterranean (expansive), and the Ponto-Mediterranean (expansive) species were considered being widely distributed (WD in Table 1), while the Balkan (sub)endemics, the Holomediterranean, and the Ponto-Mediterranean elements (sensu DeLattin, 1967) were classified as species with a more restricted (local) distribution (RD in Table 1). The polytopic/stenotopic dichotomy was defined in the following way:

Table 1. Species list, distribution, and ecological characteristics of the ground beetles.

No	Species	Zone	Zoog.	Ecol.
1	*Ophonus melleti* (Heer, 1837)	U	RD	P
2	*Ophonus diffinis* (Dejean, 1829)	U	RD	P
3	*Harpalus caspius* (Steven, 1806)	U	RD	P
4	*Ophonus rufibarbis* (Fabricius, 1792)	U	WD	P
5	*Panagaeus bipustulatus* (Fabricius, 1775)	U	RD	P
6	*Loricera pilicornis* (Fabricius, 1775)	U	WD	S
7	*Pterostichus anthracinus* (Illiger, 1798)	U	WD	S
8	*Lebia humeralis* Dejean, 1825	U	RD	P
9	*Patrobus atrorufus* (Ström, 1768)	U	WD	S
10	*Leistus ferrugineus* (Linnaeus, 1758)	U	WD	P
11	*Scybalicus oblongiusculus* (Dejean, 1829)	U	RD	P
12	*Badister bullatus* (Schrank, 1798)	U	WD	P
13	*Asaphidion flavipes* (Linnaeus, 1761)	U	WD	P
14	*Brachinus explodens* Duftschmid, 1812	U	WD	P
15	*Amara familiaris* (Duftschmid, 1812)	U	WD	P
16	*Agonum viduum* (Panzer, 1797)	U	WD	P
17	*Syntomus obscuroguttatus* (Duftschmid, 1812)	U	RD	P
18	*Acupalpus flavicollis* (Sturm, 1825)	U	WD	P
19	*Amara majuscula* (Chaudoir, 1850)	U	WD	P
20	*Clivina fossor* (Linnaeus, 1758)	U	WD	P
21	*Pterostichus ovoideus* (Sturm, 1824)	U	WD	S
22	*Dromius linearis* (Olivier, 1795)	U	RD	S
23	*Zabrus tenebrioides* (Goeze, 1777)	UR	WD	P
24	*Pterostichus niger* (Schaller, 1783)	UR	WD	S
25	*Pterostichus oblongopunctatus* (Fabricius, 1787)	UR	WD	S
26	*Ophonus nitidulus* Stephens, 1828	UR	WD	P
27	*Pterostichus melanarius* (Illiger, 1798)	UR	RD	P
28	*Trechus quadristriatus* (Schrank, 1781)	UR	WD	P
29	*Stomis pumicatus* (Panzer, 1796)	UR	RD	P
30	*Platynus assimilis* (Paykull, 1790)	UR	WD	S
31	*Pterostichus strenuus* (Panzer, 1797)	UR	WD	S
32	*Trechus obtusus* (Erichson, 1837)	UR	RD	P
33	*Pterostichus nigrita* (Paykull, 1790)	UR	WD	S
34	*Poecilus cupreus* (Linnaeus, 1758)	UR	WD	P
35	*Synuchus vivalis* (Illiger, 1798)	UR	WD	P
36	*Carabus coriaceus* Linnaeus, 1758	UR	WD	S
37	*Calathus fuscipes* (Goeze, 1777)	UR	RD	P

38	*Brachinus crepitans* (Linnaeus, 1758)	UR	WD	P
39	*Bembidion subcostatum javurkovae* Fassati, 1943	UR	RD	P
40	*Carabus convexus* Fabricius, 1775	UR	WD	P
41	*Calosoma inquisitor* (Linnaeus, 1758)	UR	WD	S
42	*Calathus melanocephalus* (Linnaeus, 1758)	UR	WD	P
43	*Amara convexior* Stephens, 1828	UR	WD	P
44	*Amara aenea* (DeGeer, 1774)	UR	WD	P
45	*Abax carinatus* (Duftschmid, 1812)	UR	RD	S
46	*Bembidion lampros* (Herbst, 1784)	UR	WD	P
47	*Anchomenus dorsalis* (Pontoppidan, 1763)	UR	WD	P
48	*Amara ovata* (Fabricius, 1792)	UR	WD	P
49	*Notiophilus rufipes* Curtis, 1829	UR	WD	P
50	*Laemostenus terricola* (Herbst, 1783)	UR	RD	P
51	*Harpalus rufipes* (DeGeer, 1774)	UR	WD	P
52	*Harpalus rubripes* (Duftschmid, 1812)	UR	WD	P
53	*Notiophilus palustris* (Duftshmid, 1812)	UR	WD	P
54	*Nebria brevicollis* (Fabricius, 1792)	UR	WD	P
55	*Leistus rufomarginatus* (Duftshmid, 1812)	UR	WD	S
56	*Harpalus atratus* Latreille, 1804	UR	RD	P
57	*Diachromus germanus* (Linnaeus, 1758)	UR	RD	P
58	Carabus violaceus Linnaeus, 1758	UR	RD	P
59	*Harpalus luteicornis* (Duftschmid, 1812)	UR	WD	P
60	*Harpalus latus* (Linnaeus, 1758)	UR	WD	P
61	*Harpalus cupreus* fastuosus Faldermann, 1836	UR	RD	P
62	*Calosoma sycophanta* (Linnaeus, 1758)	R	WD	S
63	*Harpalus quadripunctatus* Dejean, 1829	R	WD	S
64	*Carabus hortensis* Linnaeus, 1758	R	WD	S
65	*Cicindela campestris* Linnaeus, 1758	R	WD	P
66	*Molops dilatatus dilatatus* Chaudoir, 1868	R	RD	S
67	*Molops alpestris centralis* Mlynar, 1977	R	RD	S
68	*Platyderus rufus* (Duftschmid, 1812)	R	WD	S
69	*Platynus scrobiculatus* (Fabricius, 1801)	R	RD	S
70	*Cychrus semigranosus* Pallairdi, 1825	R	RD	S
71	*Pterostichus melas* (Creutzer, 1799)	R	WD	P
72	*Gynandromorphus etruscus* (Quensel, 1806)	R	RD	P
73	*Carabus intricatus* Linnaeus, 1761	R	WD	S
74	*Carabus montivagus* Pallairdi, 1825	R	RD	S
75	*Carabus ullrichi* Germar, 1824	R	RD	P
76	*Molops robustus parallelus* Mlynar, 1976	R	RD	S

77	*Amara saphyrea* Dejean, 1828	R	RD	P
78	*Amara similata* Gyllenhal, 1810	R	WD	P
79	*Tapinopterus kaufmanni* (Ganglbauer, 1896)	R	RD	S
80	*Xenion ignitum* (Kraatz, 1875)	R	RD	S
81	*Amara eurynota* (Panzer, 1797)	R	WD	P
82	*Myas chalybaeus* (Palliardi, 1825)	R	RD	S
83	*Pterostichus brucki* Schaum, 1859	R	RD	S
84	*Molops piceus bulgaricus* Maran, 1938	R	RD	S
85	*Aptinus bombarda* (Illiger, 1800)	R	RD	S
86	*Molops rufipes golobardensis* Mlynar, 1976	R	RD	S

Abbreviations: P – polytopic, S – stenotopic; RD – species with a restricted distribution, WD – species with a wide distribution; U – found collected in the urban zone, UR – found collected in both urban and rural (=non-urban) zones, R – found collected in the rural zone.

polytopic species are defined as ecologically tolerant species that occur in both forested and open habitats, while the forest specialist species were classified as stenotopic.

Ants

Stenotopic species were considered those that occurred exclusively in one habitat, while the polytopic ones occurred in multiple habitats. The Holarctic, Palearctic, European and Euro-Siberian species were considered widely distributed (WD), while Euro-Caucasian, Central-European, South-Transpalaearctic, Eurasian steppe, Pontic, Mediterranean, and Boreo-Montane were classified as species with restricted distribution (RD) (Table 2).

Table 2. Species list, distribution, and ecological characteristics of the ants.

No	Species	Zone	Zoog.	Ecol.
1	*Camponotus truncates* (Spinola, 1808)	U	RD	P
2	*Lasius balcanicus* Seifert, 1988	U	RD	P
3	*Lasius psammophilus* Seifert, 1992	U	WD	P
4	*Myrmica rugulososcabrinodis* Karawajew, 1929	U	RD	P
5	*Myrmica salina* Finzi, 1926	U	WD	P
6	*Myrmica rubra* (Linnaeus, 1758)	U	WD	P
7	*Ponera coarctata* (Latreille, 1802)	U	RD	P
8	*Prenolepis nitens* (Mayr, 1852)	U	RD	P
9	*Stenamma debile* (Förster, 1850)	U	RD	P
10	*Temnothorax affinis* Mayr, 1855	U	RD	P
11	*Tetramorium hungaricum* Röszler, 1935	U	RD	P
12	*Camponotus fallax* (Nylander, 1856)	UR	WD	P

13	*Camponotus piceus* (Leach, 1825)	UR	RD	P
14	*Camponotus vagus* (Scopoli, 1763)	UR	WD	P
15	*Dolichoderus quadripunctatus* (Linnaeus, 1771)	UR	WD	P
16	*Formica cinerea* Mayr, 1853	UR	RD	P
17	*Formica cunicularia* Latreille, 1798	UR	RD	P
18	*Formica pratensis* Retzius, 1783	UR	RD	P
19	*Formica rufibarbis* Fabricius, 1793	UR	WD	P
20	*Formica sanguinea* Latreille, 1798	UR	RD	P
21	*Formica fusca* Linnaeus, 1758	UR	WD	P
22	*Lasius alienus* (Förster, 1850)	UR	WD	P
23	*Lasius brunneus* (Latreille, 1798)	UR	RD	P
24	*Lasius citrinus* Emery, 1922	UR	RD	P
25	*Lasius flavus* (Fabricius, 1781)	UR	WD	P
26	*Lasius fuliginosus* (Latreille, 1798)	UR	WD	P
27	*Lasius paralienus* Seifert, 1992	UR	WD	P
28	*Lasius platythorax* Seifert, 1991	UR	WD	P
29	*Lasius niger* (Linnaeus, 1758)	UR	WD	P
30	*Myrmica lonae* Finzi, 1926	UR	WD	S
31	*Myrmica ruginodis* Nylander, 1846	UR	WD	P
32	*Myrmica rugulosa* Nylander, 1846	UR	WD	P
33	*Myrmica sabuleti* Meinert, 1860	UR	WD	P
34	*Myrmica scabrinodis* Nylander, 1846	UR	WD	P
35	*Myrmica schencki* Emery, 1894	UR	WD	P
36	*Myrmica specioides* Bondroit, 1918	UR	RD	P
37	*Myrmecina graminicola* (Latreille, 1802)	UR	WD	P
38	*Plagiolepis pygmaea* (Latreille, 1798)	UR	WD	S
39	*Polyergus rufescens* Latreille, 1798	UR	RD	P
40	*Solenopsis fugax* (Linnaeus, 1758)	UR	RD	P
41	*Tetramorium forte* Forel, 1903	UR	RD	P
42	*Temnothorax tuberum* (Fabricius, 1775)	UR	WD	P
43	*Temnothorax unifasciatus* (Latreille, 1798)	UR	RD	P
44	*Tapinoma erraticum* (Latreille, 1798)	UR	WD	P
45	*Temnothorax nylanderi* (Förster, 1850)	UR	RD	P
46	*Tetramorium caespitum* (Linnaeus, 1758)	UR	WD	P
47	*Camponotus ligniperda* (Latreille, 1802)	R	WD	P
48	*Camponotus aethiops* (Latreille, 1798)	R	RD	P
49	*Cataglyphis nodus* (Brullé, 1832)	R	RD	S
50	*Formica gagates* Latreille, 1798	R	RD	P
51	*Formica rufa* Linnaeus, 1761	R	WD	P

52	*Myrmica sulcinodis* Nylander, 1846	R	RD	P
53	*Messor structor* (Latreille, 1798)	R	RD	P
54	*Plagiolepis taurica* Santschi, 1920	R	RD	S

Abbreviations: P – polytopic, S – stenotopic; RD – species with a restricted distribution, WD – species with a wide distribution; U – found collected in the urban zone, UR – found collected in both urban and rural (=non-urban) zones, R – found collected in the rural zone.

Gastropods

Locally abundant snail species, strongly dependent on one environmental factor (e.g. hygrophilous, calciphilous species) were considered stenotopic (S), while the more globally abundant in Sofia mesophilous and xerophilous gastropod species were classified as polytopic (P). The geographically widely distributed (WD) species included Latomediterranean, European, Palaearctic and Holarctic species. The Bulgarian endemics, Balkan endemics, Ponto-Mediterranean and Holo-Mediterranean gastropods were classified as species having a restricted distribution (RD) (Table 3).

Table 3. Species list, distribution, and ecological characteristics of the terrestrial gastropods.

No	Species	Zone	Zoog.	Ecol.
1	*Lehmania* sp.	U	?	P
2	*Chondrula microtraga* (Rossmässler, 1839)	U	RD	P
3	*Oxychilus translucidus* (Mortillet, 1854)	U	RD	P
4	*Oxyloma elegans* (Risso, 1826)	U	WD	S
5	*Succinea oblonga* Draparnaud, 1801	U	WD	S
6	*Vallonia excentica* Sterki, 1892	U	WD	P
7	*Arion distinctus* Mabille, 1868	U	WD	P
8	*Aegopinella nitens* (Michaud, 1831)	U	WD	P
9	*Oxychilus draparnaudi* (Beck, 1837)	U	WD	P
10	*Tandonia kusceri* (Wagner, 1931)	UR	RD	P
11	*Tandonia cristata* (Kaleniczenko, 1851)	UR	RD	P
12	*Limax conemenosi* Boettger, 1882	UR	RD	P
13	*Lindholmiola girva* (Frivaldsky, 1835)	UR	RD	P
14	*Euomphalia strigella* (Draparnaud, 1801)	UR	RD	P
15	*Helix lucorum* Linnaeus, 1758	UR	RD	P
16	*Vallonia pulchella* (Müller, 1774)	UR	WD	P
17	*Cochlicopa lubrica* (Müller, 1774)	UR	WD	P
18	*Cochlicopa lubricella* (Porro, 1838)	UR	WD	P
19	*Pupilla muscorum* (Linnaeus, 1758)	UR	WD	P

20	*Ena obcura* (Müller, 1774)	UR	WD	P
21	*Chondrula tridens* (Müller, 1774)	UR	WD	P
22	*Zebrina detrita* (Müller, 1774)	UR	WD	P
23	*Arion lusitanicus* Mabille 1868	UR	WD	P
24	*Arion fasciatus* (Nilsson 1823)	UR	WD	P
25	*Arion subfuscus* (Draparnaud, 1801)	UR	WD	P
26	*Arion silvaticus* Lohmander, 1937	UR	WD	P
27	*Vitrina pellucida* (Müller, 1774)	UR	WD	P
28	*Aegopinella minor* (Stabile, 1864)	UR	WD	P
29	*Oxychilus glaber* (Rossmässler, 1835)	UR	WD	P
30	*Daudebardia rufa* (Draparnaud, 1805)	UR	WD	P
31	*Daudebardia brevipes* (Draparnaud, 1805)	UR	WD	P
32	*Tandonia budapestensis* (Hazay, 1881)	UR	WD	P
33	*Laciniaria plicata* (Draparnaud, 1801)	UR	WD	P
34	*Balea biplicata* (Montagu, 1803)	UR	WD	P
35	*Limax maximus* Linnaeus, 1758	UR	WD	P
36	*Limax flavus* Linnaeus, 1758	UR	WD	P
37	*Deroceras reticulatum* (Müller, 1774)	UR	WD	P
38	*Deroceras turcicum* (Simroth, 1894)	UR	WD	P
39	*Deroceras sturanyi* (Simroth, 1894)	UR	WD	P
40	*Deroceras agreste* (Linnaeus, 1758)	UR	WD	P
41	*Perforatella incarnata* (Müller, 1774)	UR	WD	P
42	*Monacha cartusiana* (Müller, 1774)	UR	WD	P
43	*Xerolenta obvia* (Menke, 1828)	UR	WD	P
44	*Cttania balcanica* (Kobelt, 1876)	UR	RD	S
45	*Argna truncatella* (Pfeiffer, 1841)	UR	WD	S
46	*Carychium tridentatum* (Risso, 1826)	R	WD	P
47	*Merdigera obscura* (Müller, 1774)	R	WD	P
48	*Cochlodina laminata* (Montagu, 1803)	R	WD	P
49	*Oxychilus hydatinus* (Rossmässler, 1838)	R	RD	P
50	*Milax parvulus* Wiktor, 1868	R	RD	P
51	*Tandonia serbica* Wagner, 1931	R	RD	P
52	*Deroceras bureshi* (Wagner, 1934)	R	RD	P
53	*Vestia ranojevici* (Pavlović, 1912)	R	RD	P
54	*Candidula rhabdotoides* (Wagner, 1927)	R	RD	P
55	*Sphyradium doliolum* (Draparnaud, 1801)	R	WD	P
56	*Punctum pygmaeum* (Draparnaud, 1801)	R	WD	P
57	*Discus perspectivus* (Mühlfeld, 1818)	R	WD	P
58	*Eucobresia diaphana* (Draparnaud, 1805)	R	WD	P

59	*Vitrea diaphana* (Studer, 1820)	R	WD	P
60	*Vitrea subrimata* (Reinhardt, 1871)	R	WD	P
61	*Vitrea contracta* (Westerlund, 1871)	R	WD	P
62	*Aegopinella pura* (Alder, 1830)	R	WD	P
63	*Oxychilus inopinatus* (Uličny, 1887)	R	WD	P
64	*Oxychilus depressus* (Sterki, 1880)	R	WD	P
65	*Carpathica stussineri* (Wagner, 1805)	R	WD	P
66	*Nesovitrea hammonis* (Ström, 1765)	R	WD	P
67	*Limax cinereoniger* Wolf, 1803	R	WD	P
68	*Lehmania nyctelia* Bourguignat, 1861	R	WD	P
69	*Deroceras laeve* (Müller, 1774)	R	WD	P
70	*Euconulus fulvus* (Müller, 1774)	R	WD	P
71	*Bradybaena fruticum* (Müller, 1774)	R	WD	P
72	*Soosia diodonta* (Ferussac, 1821)	R	WD	P
73	*Pseudotrichia rubiginosa* (Schmidt, 1853)	R	WD	P
74	*Cepaea vindobonensis* (Ferussac, 1821)	R	WD	P
75	*Helix pomatia* Linnaeus, 1758	R	WD	P
76	*Orcula bulgarica* Hesse, 1915	R	RD	S
77	*Macedonica frauenfeldi* (Rossmässler, 1856)	R	RD	S
78	*Arianta pelia* (Hesse, 1912)	R	RD	S
79	*Truncatellina claustralis* (Gredler, 1856)	R	WD	S
80	*Truncatellina cylindrica* (Ferussac, 1807)	R	WD	S
81	*Vertigo antivertigo* (Draparnaud, 1801)	R	WD	S
82	*Vertigo alpestris* Alder, 1838	R	WD	S
83	*Acanthinula aculeata* (Müller, 1774)	R	WD	S
84	*Balea perversa* (Linnaeus, 1758)	R	WD	S
85	*Bulgarica vetusta* (Rossmässler, 1836)	R	WD	S

Abbreviations: P – polytopic, S – stenotopic; RD – species with a restricted distribution, WD – species with a wide distribution; U – collected in the urban zone, UR – collected in both urban and rural (=non-urban) zones, R – collected in the rural zone.

RESULTS AND DISCUSSION

Composition and origin of urban faunas

Ground beetles

The total carabid species diversity recorded from the territory of the City of Sofia and its vicinity (255 species) represents about 35% of the 720 ground-beetle species in the Bulgarian fauna (Stoyanov, 2004). Of them, 237 species were recorded in the city itself (33% of the Bulgarian list). The fauna of the City of Sofia is an almost full subset (222 species, or 94%, shared) of the fauna of the West Bulgarian region (sensu Gueorguiev & Gueorguiev, 1995) and shows the highest similarity with that of the Vitosha Mt. Both regions (Sofia+West Bulgaria and Vitosha Mt.) share 114 carabid species – about 59% of the total 192 species recorded from both regions so far.

Altogether, 86 carabid species (12% of the Bulgarian fauna) were collected at the GLOBENET sites, 63 (73%) of them were recorded at the non-urban sites and 61 (71%) at the urban sites. While there was a considerable overlap between the faunas, 39 (45%) shared species, 25 (40% of the species collected from non-urban sites, or 29% of total species richness) of these were lost in the urban compared to the non-urban zone. However, 22 (36% of species collected from urban sites, or 25% of total species richness) ground-beetle species not found in non-urban sites were recorded in the urban zone (see Table 4).

Table 4. General urbanisation patterns in soil mesofauna groups, studied in the framework of Sofia GLOBENET project.

	Carabidae (12% of the Bulgarian fauna)			Formicidae (41% of the Bulgarian fauna)			Gastropoda (35% of the Bulgarian fauna)		
	U	trend	Non-U	U	trend	Non-U	U	trend	Non-U
Species richness (% of zone total)	71	↑	73	85	↓	80	53	↑	89
Unique species (% of species richness)	36	↑	40	24	↓	19	20	↑	53
Widely distributed species (% of zone total)	69	↓	57	52	↓	48	80	↑	79
Species of restricted distribution (% of zone total)	31	↑	43	54	↓	46	20	↑	21
Polytopic species (% of zone total)	77	↓	57	96	↓	91	91	↓	84
Stenotopic species (% of zone total)	23	↑	43	4	↑	9	9	↑	16

When taking into account the zoogeographical characteristics of the carabid beetles in both contrasting zones of the landscape, it becomes clear that the widely distributed species are more common in the urban zone (42 species, or 69%) than in the non-urban one (36 species, 57%). On the other hand, species of a more restricted geographical distribution (i.e. Balkan (sub)endemics + Holomediterranean + Ponto-Mediterranean) are more numerous in the non-urban zone – 27 species (43%) vs. 19 (31%) in the urbanised area of the city of Sofia.

According to the habitat preferences of the ground-beetle species recorded from each investigated region, the number of polytopic species (i.e., ecologically tolerant species that occur in both forested and open habitats) is quite similar across both of them – 47 species (77%) in the urban and 36 (57%) in the non-urban zone. The situation is more contrasting when considering the stenotopic (forest specialist) species – the non-urban region is inhabited by nearly 2 times more forest-specialist species (27 spp., 43%) than the urban one (14 spp., 23%) (see Table 4).

Ants

Fifty-four species, belonging to 17 genera, of ants were recorded in the studied region, representing 41% of the list of all species known from Bulgaria (131) (see Table 4). Within the City of Sofia, 46 species, comprising 85% of all species in the Sofia region and 35% of the Bulgarian list, were recorded.

All outdoor ant species found during the current study were autochthonous. Three allochthonous species were of tropical origin and occurred in greenhouses and buildings (*Monomorium pharaonis*, *Linepithema humile* and *Hypoponera punctatissima*). Until now, the invasive species *Lasius neglectus* has not been found in Sofia, although it was recorded in other Bulgarian cities – i.e. Varna (Seifert, 2000) and Silistra (unpublished data).

Some of the common, locally dominant species in the vicinity of Sofia and the Sofia region (i.e., *Formica pratensis*, *F. rufa* and *F. sanguinea*), exist in low numbers within the city of Sofia. On the contrary, the sub-dominant species in non-urban communities (i.e. *Lasius platythorax* and several *Myrmica* species) represented a larger component of the ant community in the urban areas. These species occupy the ecological niches vacated by the species that dominate in the native assemblages (Klausnitzer, 1987). Analogously, in urban and semi-urban grasslands, the sub-dominant *Lasius niger*, *Tetramorium caespitum*, *Formica rufibarbis* and *F. cunicularia* take over the leading role in ant assemblages, because the true dominant species of the neighbouring natural communities (i.e., *Formica pratensis*) are rare in the city.

Most of the species were polytopic and occured in more than one habitat – 50 species (93 % of the urban and rural fauna). These are either forest species (i.e., *Myrmica rubra*, *M. ruginodis*, *Temnothorax nylanderi*, *Myrmecina graminicola* and *Lasius platythorax*), or meadow species (*Solenopsis fugax*, *Tetramorium caespitum*, *Lasius niger* and *Formica cinerea*). There were also polytopic species with narrow ecological tolerance, such as *Ponera coarctata*, *Myrmica rugulosa*, *M. scabrinodis*, *M. schenki*, *M. specioides*. There were two stenotopic species in the urban zone (*Plagiolepis pygmaea* and *Myrmica lonae*) (4%) and two in the rural zones (*Plagiolepis taurica* and *Cataglyphis nodus*) (9%) (see Table 4).

Within the Euro-Siberian complex, the highest percentage (by number of species) is represented by the Euro-Siberian s. str. (16%), Euro-Caucasian (14%) and Holarctic (13%) elements in the studied region (Antonova & Penev, 2006). The other categories of the same complex are represented by less than four species (7%). The percentage of Mediterranean species (sensu stricto) is remarkably low (eight species, 15 %). Hence, the percentage of widely distributed (Holarctic, Palearctic, Euro-Siberian and European) species is around 54 % of the whole urban fauna (Table 4).

Thus, the urban ant fauna of Sofia can be regarded as a slightly impoverished version of the native fauna of the Sofia Kettle. There were no truly invasive ant species in the natural and semi-natural habitats of Sofia, because the three alien ant species recorded were entirely synanthropic.

Gastropods

The total number of terrestrial gastropod species recorded in Sofia and its vicinity was 85, that is 35% of all species known from Bulgaria (Table 4). The species richness within the city was almost half (45 species – 53% of the sampled species, or 19% of the Bulgarian list that holds 240 species so far) of the 76 species (89% of the sampled gastropod species) occurring in the areas surrounding Sofia. Thirty-six of the species known from the vicinity of Sofia (80% of all species in Sofia) have been found in the urban environment as well.

There were 10 species of synanthropic and/or introduced gastropod species, alien to the local fauna (24% of the fauna of the city). These can be divided into 2 sub-groups: species established as synanthropic in Sofia, but occurring either in natural or anthropogenic habitats in other regions of Bulgaria (five species), and species introduced by man into the urban fauna of Sofia (five species). In most cases, these were invasive species (except for *Aegopinella nitens*) that were well adapted to urban conditions and often capable of penetrating from their suburban and rural habitats into the City of Sofia (i.e. *Arion lusitanicus*, *Arion fasciatus*).

Polytopic species dominated the urban assemblages (41 species, 91% of all urban fauna). There are a few exceptions of abundant stenotopic species – the hygrophilous *Oxyloma elegans* and *Succinea oblonga*, and the calciphilous species *Helicigona balcanica* and *Argna truncatella* (altogether totalling 9%). The proportion of stenotopic species decreased in the urban environment – four species (9% of all species found in Sofia) versus non-urban environment – 12 stenotopic species (16% of all species in the non-urban zone).

From a zoogeographic point of view, widely distributed gastropod species were the most prevalent element of the fauna of Sofia (80%). These were from the Lato-Mediterranean – eight species (23%), European – 18 (51%), Palaearctic – 3 (9%) and Holarctic – 6 (17%) ranges. Species of restricted distribution (20%) were the Bulgarian endemics – 1 (11%), Balkan endemics – 3 (33%), Ponto-Mediterranean – 4 (44%), Holomediterranean – 1 (11%); 1 species is currently of unclear zoogeographical status.

The origin of the terrestrial gastropod fauna of Sofia can be considered as a result of several parallel processes: (a) fragmentation of the natural ranges of the autochthonous species that occurred in the region prior to urbanisation, (b) introduction of allochthonous species, (c) expansion of xerophilic snails from the environs of Sofia into free ecological spaces “opened” due to the fragmentation of forested habitats, and (d) immigration of mesophillous species from the broad-leaved forests situated around Sofia into the city parks. These processes cannot be distinguished clearly from each other and their action goes on at present with varying intensity. Moreover, not only does immigration of species from the vicinity of Sofia into the urban environment occur, but also the opposite process of invaders (*Arion lusitanicus*, *Arion fasciatus*, and *Deroceras reticulatum*) expanding from the city to the suburbs seems to take place. The resulting patterns are both complex and dynamic (Dedov & Penev, 2000).

Do “urban” assemblages exist?

Ground beetles

As pointed out by Niemelä et al. (2002), the sites in the urban zone are clearly clustered together and separately from the non-urban sites that also form a compact group. Spatially adjacent sites in the city are faunistically more similar to each other than the sites in the non-urban zone (see also Stoyanov & Penev, 2004). The most isolated and heavily transformed sites on the Sofia plain (i.e. sites SI and RIII; Fig. 1) exhibit some weak similarity, while the more or less rural sites at the foothills around Sofia form a distinct group. The observed pattern clearly confirms the “urbanisation” of the carabid fauna of Sofia.

Ants

The cluster analysis of the assemblages based on presence/absence data resulted in grouping into two main clusters: woody and open habitats (Antonova & Penev, in press). The assemblages of the open areas were further subdivided into two groups: grassy habitats and artificial biotopes (asphalt coverings and agricultural land). Most of the assemblages of open areas clearly demonstrated the impact of urbanisation by grouping into three main zones depending on the distance from the city centre: central, peripheral and rural. The central urban zone encompasses sites situated concentrically at about 3,5 km distance from the city centre (“St. Nedelja” square) (Antonova & Penev, in press). The peripheral zone looks like a stripe of fragments within the city, situated between the central urban zone and the ring road.

Gastropods

Due to the low mobility of the terrestrial gastropods (Sverlova, 1998), it could be expected that the urban environment will provoke quick changes and degradation of the urban malacocoenoses. Dedov & Penev (2004) found that in the parks of Sofia distinguishable “urban” malacocoenoses existed. The analysis of the presence/absence data showed that three main groups of assemblages can be distinguished at 35 % similarity level – locality

RIII, all urban sites together with the SI locality and the rest of the rural localities. It can be concluded that there are clearly distinctive malacocoenoses of urban type (UI, UII, UIII, UIV and SI). At the same time, the urban snail communities demonstrate greater differences among each other than the rural ones do. A clearly distinct suburban zone could not be observed, and the suburban sites SII and SIV were grouped to the rural and SI to the urban localities, respectively.

Which are the main factors in spatial variation of assemblages in urban environment?

Ground beetles

The multivariate analysis of the carabid-beetle assemblages performed by Stoyanov & Penev (2004) clearly demonstrated the presence of one prominent ordination axis that summarizes the essential part of the dataset variation. The urban sites form a relatively compact cluster that clearly stands apart from the non-urban sites. Some of the non-urban sites (RIV) separate from an adjacent group of sites mainly on a geographical basis, while others (RIII, SI) cluster together on basis of their very depauperated ground-beetle fauna, probably resulting from the strong isolation of the respective forest patches within the landscape. The rest of the non-urban sites, while being widely scattered along the second ordination axis, still form a cluster on a geographical proximity basis (RII, SII, SIV) leaving the more geographically distant site (RI) relatively isolated from that cluster. The observed pattern allows us to confidently identify a major "latent" urbanisation gradient that explains most of the observed variation in the carabid beetle data, and a second one illustrating the influence of geographical isolation and landscape configuration of the sampling sites.

Ants

According to a preliminary DCA analysis (Antonova & Penev, in press), the ant assemblages were separated along two hypothetical gradients. One of the gradients explained the variation in ant assemblages from forest to open areas. The second one reflected the effect of the distance from the city centre. The factors of urbanisation most significant for the meadow dwelling ant assemblages appeared to be distance from the city center, area of the homogenous habitat, and distance of the fragment to the nearest one.

Gastropods

According to a CCA ordination analysis performed by Dedov & Penev (2004), a rural-urban gradient was observed along the first axis. Sites UI, UII, UIII, UIV and SI (Fig. 1) were isolated as localities of the urban type, while sites RI, RII, RIII, RIV, SII and SIV were distinguished as localities of non-urban type. The locality near the village of

Drenovo (RIII) is clearly distinguished along the second axis. Factors partly explaining the spatial variation in gastropod assemblages appeared to be altitude, area, and level of urban development (measured as a percentage of built-up area of the sampling sites). The first two factors seem to have a stronger effect in the localities of non-urban type, while the percentage of built-up areas was correlated with assemblages from the urban biotopes. Another major factor was the "origin of forest" – semi-natural for the localities of non-urban type and anthropogenic for urban localities. Other important factors related to those described above were soil type, slope, level of development of grasses, soil pH, soil nitrogen, leaf litter thickness, moisture of the leaf litter and the mechanical composition of the soil (Dedov & Penev, 2004).

GENERAL DISCUSSION

The fauna of a city results from an interaction of several complex factors, such as geographical position, relief in and around the city, the number, area size and quality of the parks, presence/absence of connections with natural habitats around the city, distribution patterns of pollution in the city, stage of fragmentation of the landscape and the ability of a certain animal group to adapt to urban conditions. Also of primary importance seems to be the present-day intensity of construction, de- and aforestation trends and other town-planning activities. All these factors act together in great complexity. This makes any attempt to reconstruct the pathways of urban faunas formation largely speculative (Dedov & Penev, 2000).

The increase of the anthropogenic pressure towards the city centre leads to reduction and extinction of stenotopic species and results in impoverishment of invertebrate faunas (Vepsäläinen & Wuorenrinne, 1978; Klausnitzer, 1987; Chudzicka & Skibinska, 1994, 1998a,b; Sverlova, 1997; Vepsäläinen et al., 2008). In many invertebrate groups, biodiversity is lost in urban habitats (Vepsäläinen & Wuorenrinne, 1978; Czechowski, 1982; Kasprzak, 1981; Niedbala et al., 1982; Pisarski, 1982; Skibinska, 1978, 1982, 1986c; Sawoniewicz, 1982, 1986; Winiarska, 1986). Therefore, the ecological conditions in cities could be considered a hostile barrier for many invertebrate animal groups. This is valid to a great extent for terrestrial gastropods in Sofia, where the diversity loss was nearly 40% (76 species in non-urban versus 45 species in the urban area). A similar abrupt decrease in terrestrial gastropod species number has also been reported in Lvov, Ukraine (Sverlova, 1999). This might be a result of the lower mobility and the corresponding higher sensitivity of snails to deforestation and habitat fragmentation. It may also be a result of anthropogenic pressure of visitors in the city parks, loss of unique habitats, (micro) climatic changes, and pollution of air, waters and soil (Klausnitzer, 1987; Whiteley, 1994). In ground-beetles, the decrease in species richness is negligible – about 8 %, while in ants we observe even the opposite tendency (Table 4) which, is most probably due to the incomplete sampling in the non-urban zone.

In similar studies carried out in Helsinki, it was found that species diversity decreased towards the urban zone (Niemelä et al., 2002; Venn et al., 2003; but see also Alaruikka et al., 2002, who established no trend in species richness along urban-rural gradient). The same decrease of carabid species richness in the urban areas was observed by Niemelä et al. (2002) and Hartley et al. (2007) in the city of Edmonton (Canada) and by Ishitani et al. (2003) in the city if Hiroshima (Japan). In the city of Debrecen (Hungary), however, Magura et al. (2004) demonstrated that the ground beetle species richness was higher in the rural and urban areas, as compared to the suburban one. In ants, changes in species diversity and abundance towards the city centre have been recorded by Kondoh (1978), Pisarski & Czechowski (1978), Vepsäläinen & Wuorenrinne (1978), Pisarski (1982), Czechowski (1991), Behr et al. (1996), Dauber (1997), Dauber & Eisenbeis, (1997), Schlick-Steiner & Steiner (1999), and Vepsäläinen et al. (2008). The low rates of biodiversity loss in ants and ground beetles in our study may be explained by the great mobility of these two groups on one hand, and by the large-sized and relatively well preserved city parks of Sofia that resemble natural and semi-natural forests, on the other.

Despite the general tendency of impoverishment of the fauna in the urban zone, under certain conditions (micro-habitats with slight anthropological influence) an increase of species diversity may be observed. For example, in some parks of Sofia (West Park, North Park, Loven Park), the diversity of snails is increased mainly at the expense of introduced and/or synanthropic species (as *A. nitens*, *Ar. subfuscus*, *Ar. hortensis*, etc.). According to some authors (see Trojan, 1981), the low anthropogenic pressure can increase the species richness of some social insects, including Formicidae. The reason for this may be the establishment of new habitats, their mosaic patterns and the great number of ecotones created by urbanisation (Frankie & Ehler, 1978), as well as the organic refuse (Trojan et al., 1982). Twenty-two species of carabids were gained in the urban sites, compensating for the loss of other species towards the city centre, which leads to a more or less balanced diversity of urban and non-urban carabid assemblages.

It may be considered as a rule that in urban habitats widely distributed species prevail over those with restricted distributions (Klausnitzer, 1987). This holds for a number of taxa studied in urban conditions (for example Vepsäläinen & Wuorenrinne, 1978; Vepsäläinen et al., 2008 – Formicidae; Burakowski & Nowakowski, 1981a,b – Cerambycidae and Elateridae; Cholewicka, 1981 – Curculiondae; Czechowski, 1981 – Carabidae; Czechowski et al., 1981 – Opiliones; Jedryczkowski, 1981 – Isopoda; Kasprzak, 1981 – Oligochaeta; Krzyzanowska et al., 1981 – Aranei; Kubicka, 1981 – Scarabaeidae; Pilipiuk, 1981 – Lumbricidae; Czechowski, 1982 – Carabidae; Jedryczkowski, 1982 – Diplopoda; Pisarski, 1982 – Formicoidea; Sterzynska, 1982 – Collembola; Langourov, 2004 – Phoridae).

In the City of Sofia the same trend of increased percentage of widely distributed species was confirmed for carabid beetles and ants (Stoyanov, 2004; Antonova & Penev, 2006). The share of Holarctic ant species was twice as large in Sofia than in the vicinity. The percentage of widespread Euro-Siberian species also increases in the city. In general, the bulk of the myrmecofauna of Sofia and its suburbs is composed by widespread species. Similar trends have been reported for other urban myrmecofaunas (Pisarski & Czechowski, 1987; Vepsäläinen & Pisarski, 1982; Pisarski & Kulesza, 1982). For the carabid beetles, a similar trend has been previously observed by Magura et al. (2004), who found that in urban areas opportunistic species dominated.

The terrestrial gastropods exhibited a trend that seemed to be opposite to the ants and carabid beetles, although the percentage of widely and locally distributed species were very close to each other. This slight difference between most soil mesofauna groups and terrestrial gastropods might be a result of the low mobility of gastropods on one hand, and the specific microhabitat requirements and the peculiarities in their biology (Barker, 2001), on the other. As slowly moving organisms, their distribution across wide areas is also relatively slow and hence, many species adapt to local conditions and manifest narrower ecological tolerance. This can be seen in the formation of many endemic species with different taxonomic lineages (multiple genera and families). This leads to a relatively lower number of terrestrial gastropod species with wide ranges in comparison to more mobile invertebrate taxa. Consequently, in the cities, the share of widely distributed species is lower while several stenotopic species co-occur in the relatively undisturbed parts of city parks (Bajdashnikov, 1992). Another common pattern observed along urbanisation gradients in most invertebrate groups is that the percentage of ecologically tolerant species increases with anthropogenic pressure (see Klausnither, 1987 and references therein). The higher percentage of polytopic species may also be an indicator for anthropogenic influence (Chudzicka & Skibinska, 1998a). According to Whiteley (1994), in urban environment, the specialist species disappear, while the number of generalist species increases. It is also known that higher levels of anthropogenic pressure result in depauperation of faunas (Vepsäläinen & Wuorenrinne, 1978; Bankowska et al., 1984; Pisarski et al., 1989; Chudzicka & Skibinska, 1998a,b), and eventually numerous extinctions (Skibinska, 1986b; Chudzicka & Skibinska, 1998a). The effect of faunal homogenization in the big cities (Whiteley, 1994) brought some researchers to the pessimistic conclusion that most of the species in urban parks do not have a conservation value and thus do not need to be protected (Bankowska et al., 1984). Considering cities as a whole, Luniak (1996) notes that the biodiversity they hold is clearly lower in comparison to the adjacent territories, but points out that urbanocoenoses in many cases are richer in species and more abundant than anticipated – the same isn't valid for the most heavily influenced assemblages. If the multitude of urban habitats is compared, the view of Whiteley (1994), that diversity is dramatically influenced in different urban habitats, finds its logical extension in the view of Trojan (1994), that urbanisation leads in some cases to an increase in biological diversity and a decrease in others.

The analysis of gastropod assemblages showed that in the city of Sofia the percentage of stenotopic species decreased (9% of all species found in the area of Sofia) in comparison to the non-urban territories (16% of all species found in the non-urban zone). The carabids and ants shared a similar trend: carabids (23% and 42% respectively) and ants (4% and 9% respectively) (Table 4). This may be interpreted as a result of the degradation of habitats within the city, as well as a result of destruction and disappearance of some habitats and the rare species inhabiting them (Vepsäläinen & Wuorenrinne, 1978; Chudzicka & Skibinska, 1994, 1998a,b; Sverlova, 1997; Vepsäläinen et al., 2008).

The city parks (including those in Sofia) are among the relatively less influenced urban habitats, and as such function as refuges for many animal groups (Czechowska & Bielawski, 1981; Czechowski, 1982; Czechowska, 1986; Skibinska, 1986a,b; Sterzynska, 1987). They are also the biggest green "islands" within cities that maintaining a higher diversity through their greater area, as found by numerous studies (see e.g. Klausnitzer, 1987; Whiteley, 1994).

Our results confirmed the thesis of Klausnitzer (1987) that the number of animal species in the city increases proportionally with the age and area of a homogeneous park biotope. The pattern observed in the ants was similar to those described by Suarez et al. (1998) and Yamaguchi (2004), who used a multiple regression to find that the fragment area and time of its isolation from the rural habitat (age) were the best predictors for the number of native ants that will be found in that fragment. A significant positive correlation was also found between the area of the habitat and ant species richness as well as between area and nest density in humid habitats in Belgium (Maes et al., 2003). According to these authors, this is due to higher plant diversity and presence of proper nesting places, respectively. Similar correlations between age and size of the parks and diversity of snails have been observed in Lvov, Ukraine by Sverlova (1997, 2000) and Bajdashnikov (1985). The bigger and older the forested areas are, the higher is the diversity of snails. Moreover, several rare stenotopic species may occur in the oldest parts of the parks (Bajdashnikov, 1992).

It has been shown that the urban assemblages are quite distinct (Czechowski, 1982; Nowakowski, 1986; Winiarska, 1986; Pisarski et al., 1989; Chudzicka & Skibinska, 1994, 1998a), and may even be termed "urbanocoenoses" (Luniak, 1996). Our results clearly support the specificity of the urban assemblages which differ from those in suburban and rural areas. However, although being clearly distinguished in species composition and abundancies, urban assemblages of the different animal groups also appear quite different in structure and diversity (Whiteley, 1994). This reinforces the need to use various taxonomic groups when designing bioindication and monitoring schemes in cities.

CONCLUSIONS

1. In all three groups, the urban faunas can generally be regarded as impoverished variants of the regional species pools. For gastropods, however, the urban fauna is enriched by invasive species and may even serve as a source of distribution of such species into the neighbouring suburban and rural territories. For ants, such an enrichment is not recorded, with the exception of three fully synanthropic species, not occurring in natural or semi-natural habitats. There are no invasive species of ground beetles and the urban fauna can be described as a modified, "urbanised" version of the native fauna. Both faunas of ground beetles – the native and the urban one – are of comparable diversity but of slightly different species composition.
2. The diversity of urban faunas decreased relative to neighbouring territories. The degree of decrease differed with taxa and ranged from 2% (ground beetles) to 36%

(gastropods). Species composition was also observed to have changed between these two zones.

3. Urbanisation leads to an increase in the proportion of widely distributed and eurytopic species. This trend was quite prominent in gastropods, but less so in ants and ground beetles. In both latter groups, the proportion of eurytopic and stenotopic species in the city was relatively equal.
4. "Urban" assemblages could be delimited in all three groups. As compared to the native assemblages from the surroundings of the city, they can be characterised by changes in species composition, dominant structure, and the presence of alien species (gastropods). Suburban assemblages were not clearly distinguished in all three groups.
5. The main influence on the spatial variation in assemblages seemed to be a complex urbanisation factor that led to a clear separation of "urban" assemblages from suburban and rural ones. Some peculiar factors revealed by ordination analyses in the different groups were altitude, type of habitat (woody/open), distance from the city centre, area and percentage of built-up area of the sampling sites.
6. Patterns of urbanisation shown by the different groups were similar in many respects, but their qualitative and quantitative assessment differed. This reinforces the necessity of a broad approach, based on different taxonomic groups, when planning bioindication and monitoring activities in cities.

ACKNOWLEDEMENTS

This project was supported by Grants TKB-1616/2006 and INI 03/2005 (BioCore) of the National Science Fund, Ministry of Education and Science, Bulgaria. Our special thanks are due to Terry Erwin (Smithsonian Institution, Washington DC, USA) and Evan Esch (University of Alberta, Edmonton, Canada) for critical reading and linguistic editing of the manuscript. We also thank Tibor Magura (Hortobagy National Directorate, Hungary) for the many useful suggestions that helped us to improve the manuscript.

REFERENCES

Alaruikka D., Kotze D.J., Matveinen K. & Niemelä J. (2002). Carabid beetle and spider assemblages along a forested urban–rural gradient in southern Finland. Journal of Insect Conservation 6(4): 195-206.

Antonova, V. & Penev, L. (2006). Change in the zoogeographical structure of ants (Hymenoptera: Formicidae) caused by urban pressure in the Sofia region (Bulgaria). – Myrmecologische Nachrichten 8: 271-276.

Antonova, V. & Penev, L. (in press). Classification of assemblages of ants in the green areas in Sofia City. – Acta Zoologica Bulgarica 00: 00-00.

Atanassov, N. & Dlussky, G. (1992). Fauna Bulgarica. 22. Hymenoptera, Formicidae. Bulgarian Academy of Sciences, Sofia, 310 pp. (in Bulgarian).

Bajdashnikov, A.A. (1985) Land molluscs of the transcarpathian region of the USSR and their distribution in connection with main landscaps and plant asociations. Proc. Zool. Ins., AN. SSSR 135: 44-66. (in Russian).

Bajdashnikov, A.A. (1992). Terrestrial mollusk fauna of the Ukrainian Polesie Area. Communication 1. Specific composition and connection with vegetative cover. – Vestnik Zoologii 4: 13-19. (in Russian).

Bankowska, R., Czechowski, W., Garbarczyk, H. & Trojan, P. (1984). Present and prognosticated fauna of the housing estate Bialoleka. – Memorabilia Zool. 40: 157-163.

Barker, G.M. (Ed.) (2001). The biology of terrestrial molluscs. CABI Publishing.

Behr, D., Lippke, S. & Cölln, K. (1996). Zur Kenntnis der Ameisen von Köln (Hymenoptera, Formicidae). – Decheniana-Beihefte 35: 215-232.

Bestelmeyer, B.T., Agosti, D., Alonso, L.E., Brandão, R.F., BrownJr.,W.L., Delabie, H.C.J. & Silvestre, R. (2000). Field Techniques for Study of Ground-Dwelling Ants: An overview, Description and Evaluation. – In: Ants: standard methods for measuring and monitoring biodiversity. (Agosti, D., Majer, J. D., Alonso, L. E.& T. Schultz., eds). 280 pp. Washington and London, Biological diversity handbook series, Smithsonian Institution Press, p. 122-144.

Burakowski, B. & Nowakowski, E. (1981a). Longicorns (Coleoptera, Cerambycidae) of Warsaw and Mazovia. – Memorabilia Zool. 34: 199-218.

Burakowski, B. & Nowakowski,E. (1981b). Click beetles (Coleoptera, Elateridae) of Warsaw and Mazovia. – Memorabilia Zool. 34: 165-180.

Cholewicka, K. (1981). Curculionids (Coleoptera, Curculionidae) of Warsaw and Mazovia. – Memorabilia Zool. 34: 235-260.

Chudzicka, E. & Skibinska, E. (1994). An evaluation of an urban environment on the basis of faunistic data. Memorabilia Zool. 49: 175-185.

Chudzicka, E. & Skibinska, E. (1998a). Diversity of reactions insect communities as a response to anthropogenic pressure. – Memorabilia Zool. 51: 13-30.

Chudzicka, E. & Skibinska, E. (1998b). Monitoring and role of terrestrial invertebrates in biondicatory evaluation of environment and changes. – Memorabilia Zool. 51: 3-12.

Czechowska,W. (1986). Structure of Neuropterian (Neuropteroidea) communities in urban greenareas of Warsaw. – Memorabilia Zool. 41: 187-214.

Czechowski, W. (1981). Carabids beetles (Coleoptera, Carabidae) of Warsaw and Mazovia. – Memorabilia Zool. 34: 135-144.

Czechowski, W. (1982). Occurence of Carabids (Coleoptera, Carabidae) in the urban green of Warsaw according to the land utilization and cultivation. – Memorabilia Zool. 39: 3-108.

Czechowski, W. (1991). Comparison of the myrmecofaunas (Hymenoptera, Formicoidea) of tree stands and lawns in Warsaw parks. – Fragmenta Faunistica 35(12): 179-183.

Czechowska, W. & Bielawski, R. (1981). Coccinelids (Coleoptera, Coccinellidae) of Warsaw and Mazovia. – Memorabilia Zool. 34: 181-197.

Czechowski, W., Kubicka, A. & Starega, W. (1981). Harvestmen (Arachnoidea, Opiliones) of Warsaw and Mazovia. – Memorabilia Zool. 34: 111-118.

Czechowski, W., Pisarski, B. & Yamauchi, K. (1995). Succession of ant communities (Hymenoptera, Formicidae) in moist pine forests. – Fragmenta Faunistica 38(24): 447-487.

Czechowski, W., Radchenko, A. & Czechowska, W. (2002). The ants of Poland. Museum and Institute of Zoology PAS, Warszawa.

Dauber, J. (1997). Ameisenfauna einer urbanen Landschaft. – Naturschutz und Landschaftsplanung 29: 303-309.

Dauber, J. & Eisenbeis, G. (1997). Untersuchungen zur Ameisenfauna einer urbanen Landschaft am Beispiel der Stadt Mainz. – Abh. Ber. Naturkundemus. Görlitz 69(2): 237-244.

Dedov, I. & Penev, L. (2000). Species composition and origins of the terrestrial gastropod fauna of Sofia City, Bulgaria. – Ruthenica 10: 121-131.

Dedov, I. & Penev, L. (2004). Spatial variation in terrestrial gastropod communities (Gastropoda, Pulmonata) along urban-rural gradients in Sofia City, Bulgaria. – In: Ecology of the City of Sofia. Species and Communities in an Urban Environment (Penev, L., Niemelä, J., Kotze J. & Chipev, N., eds). Pensoft Publishers, p. 307–318.

De Lattin, G. (1967). Grundriss der Zoogeographie. Gustav Fischer, Jena.

Elek, Z. & Lövei, G. L. (2007). Patterns in ground beetle (Coleoptera: Carabidae) assemblages along an urbanisation gradient in Denmark. – Acta Oecologica 32: 104-111.

Frankie, G.W. & Ehler, L.E. (1978). Ecology of insects in urban environments. – Ann Rev. Entomol. 23: 367-387.

Freude, H., Harde, K.W. & Lohse, G.A. (1976). Die Käfer Mitteleuropas. Goecke und Evers, Krefeld.

Gueorguiev, V.B. & Gueorguiev, B.V. (1995). Catalogue of the ground-beetles of Bulgaria (Coleoptera: Carabidae). Pensoft, Sofia-Moscow.

Hartley, D.J. Koivula, M.J., Spence, J.R., Pelletier, R. & Ball, G.E. (2007). Effects of urbanization on ground beetle assemblages (Coleoptera, Carabidae) of grassland habitats in western Canada. – Ecography 30: 673-684.

Hornung, E., Tóthmérész, B., Magura, T. & Vilisics, F. (2007). Changes of isopod assemblages along an urban-suburban-rural gradient in Hungary. – European Journal of Soil Biology 43, 158-165.

Hůrka, K. (1996). Carabidae of the Czech and Slovak Republics. Kabourek, Zlin.

Ishitani, M., Kotze, D.J. & Niemelä, J. (2003). Changes in carabid beetle assemblages across an urban-rural gradient in Japan. – Ecography 26: 481-489.

Jedryczkowski, W. (1981). Isopods (Isopoda) of Warsaw and Mazovia. – Memorabilia Zool. 34: 79-86.

Jedryczkowski, W.J. (1982). Millipedes (Diplopoda) of Warsaw and Mazovia. – Memorabilia Zool. 36: 253-261.

Kasprzak, K. (1981). Enchytraeids (Oligochaeta, Enchytraeidae) of Warsaw and Mazovia. – Memorabilia Zool. 34: 59-67.

Klausnitzer, B. (1982). Zur Kenntniss urbaner Gradienten. – Tagungsber. 1. Leipziger Symp. urb. Ök. 1981: 13-20.

Klausnitzer, B. (1987). Ökologie der Großstadfauna. Gustav Fischer Verlag, Jena.

Kondoh, M. (1978). A comparison among ant communities in the antropogenic environment. – Memorabilia Zool. 29: 79-92

Krzyzanowska, E., Dziabaszewski, A., Jackowska, B. & Starega, W. (1981). Spiders (Arachnoidea, Aranei) of Warsaw and Mazovia. – Memorabilia Zool. 34: 87-110.

Kryzhanovskij, O.L., Belousov, I.A., Kabak, I.I., Kataev, B.M., Makarov, V.G. & Shilenkov, V.G. (1995). A checklist of the Ground-beetles of Russia and Adjacent Lands (Insecta, Coleoptera, Carabidae). Pensoft Publishers, Sofia-Moscow.

Kubicka, A. (1981). Scarabaeids (Coleoptera, Scarabaeidae) of Warsaw and Mazovia. – Memorabilia Zool. 34: 145-164.

Langourov, M. (2004). Scuttle flies (Diptera: Phoridae) from urban and suburban areas in the Sofia Plain. – In: Ecology of the City of Sofia. Species and Communities in an Urban Environment (Penev, L., Niemelä, J., Kotze, D.J. & Chipev, N., eds), Pensoft Publishers, Sofia-Moscow, p. 429-436.

Luniak, M. (1996). Synurbization of Animals as a Factor Increasing Diversity of Urban Fauna. – Biodiversity, Science and Development 50: 566-575.

Maes, D., Van Dyck, H., Vanreusel, W. & Cortens, J. (2003). Ant communities (Hymenoptera, Formicidae) of Flemish (northern Belgium) wet heathlands, a declining habitat in Europe. – Eur. J. Entomology 100: 545-555.

Magura, T., Tóthmérész, B. & Molnár, T. (2004). Changes in carabid beetle assemblages along an urbanisation gradient in the city of Debrecen, Hungary. – Landscape Ecology 19: 747-759.

Magura, T., Tóthmérész, B., Hornung, E. & Horváth, R. (2008). Urbanisation and ground-dwelling invertebrates. – In: Urbanization: 21st Century Issues and Challenges (Wagner, L.N., ed.). New York: Nova Science Publishers Inc. pp. 213-225.

Magura, T., Tóthmérész, B. & Molnár, T. (2008). A species-level comparison of occurrence patterns in carabids along an urbanisation gradient. – Landscape and Urban Planning 86: 134-140.

McDonnell, M.J.& Pickett, S.T.A. (1990). Ecosystem structure and function along urban-rural gradients: an unexploited opportunity for ecology. – Ecology 71(4): 1232-1237.

Niedbala, W., Blaszak, C., Bloszyk, J., Kaliszewski, M. & Kazmierski, A. (1982). Soils mites (Acari) of Warsaw and Mazovia. – Memorabilia Zool. 36: 235-252.

Niemelä, J., Kotze, J. (2000). GLOBENET: the search for common anthropogenic impact on biodiversity. – In: Natural History and Applied Ecology of Carabid Beetles, Proceedings of the IX European Carabidologists Meeting (Brandmayr, P., Lövei, G., Brandmayr, T.Z., Casale, A. & Vigna Taglianti, A., eds). Pensoft, Sofia-Moscow, p. 187-196.

Niemelä, J., Kotze, J., Ashworth, A., Brandmayr, P., Desender, K., New, T., Penev, L., Samways, M. & Spence, J. (2000). The search for conmmon anthropogenic impacts on biodiversity: a global network. – Journal of Insect Conservation 4: 3-9.

Niemelä, J., Kotze, D.J., Venn, S., Penev, L., Stoyanov, I., Spence, J., Hartley, D., MontesDeOca, E. (2002). Carabid beetle assemblages (Coleoptera, Carabidae) across urban-rural gradients: an international comparison. – Landscape Ecology 17: 387-401.

Nowakowski, E. (1986). Structure of soil click beetle (Coleoptera, Elateridae) communities in the urban green areas of Warsaw. – Memorabilia Zool. 41: 81-102.

Ökland, F. (1930). Quantitative Untersuchungen der Landschneckenfauna. – Z. f. Morphol. u. Ökol. d. Tiere. 16: 748-803.

Penev, L., Stoyanov, I., Dedov, I., Dimitrov, D., Grozeva, M. & Doichinova, V. (2004). The Sofia (Bulgaria) GLOBENET sites: Description and spatial variation of the landscape mosaic. – In: Ecology of the City of Sofia: Species and communities in an urban environment (Penev, L., Niemelä, J., Kotze, D.J. & Chipev, N., eds). Pensoft, Sofia-Moscow, p. 3-24.

Pilipiuk, I. (1981). Earthworms (Oligochaeta, Lumbricidae) of Warsaw and Mazovia. – Memorabilia Zool. 34: 69-77.

Pisarski, B. (1982). Ants (Hymenoptera, Formicoidea) of Warsaw and Mazovia. – Memorabilia Zool. 36: 73-90.

Pisarski, B. & Czechowski, W. (1978). Influence de la pression urbaine sur la myrmécofaune. – Memorabilia Zool. 29: 109-128.

Pisarski, B. & Czechowski, W. (1987). Structure and origin of ant communities of Warsaw. – In: Chemistry and biology of social insects (Eder, J. & Rembold, H., eds). Verlag J. Peperny, Munich, p. 605.

Pisarski, B. & Kulesza, M. (1982). Characteristics of animal colonizing urban habitats. – Memorabilia Zool. 37: 71-77.

Pisarski, B., Pilipiuk, I. & Sterzynska, M. (1989). Structural changes of communities of the soil fauna in an urban environment – the example of Warsaw. – Unesco Program "Der Mensch und Die Biosphäre", Berlin, 30: 71-84.

Sadler, J.P., E.C. Small, H. Fiszpan, M.G. Telfer & Niemelä, J. (2006). Investigating environmental variation and landscape characteristics of an urban–rural gradient using woodland carabid assemblages. – J. Biogeogr. 33: 1126-1138.

Sawoniewicz, J. (1982). Ichneumonidae (Hymenoptera) of Warsaw and Mazovia. – Memorabilia Zool. 36: 5-39.

Sawoniewicz, J. (1986). Structure of Ichneumonidae (Hymenoptera) communities in the urban green areas of Warsaw. – Memorabilia Zool. 41: 103-124.

Schlick-Steiner, B.C. & Steiner, F.M. (1999). Faunistisch-ökologische Untersuchungen an den freilebenden Ameisen (Hymenoptera: Formicidae) Wiens. – Myrmecologische Nachrichten 3: 9-53.

Schoener, T. W. (1986). Mechanistic approaches to community ecology: a new reductionism? – Am. Zool. 26: 81-106.

Seifert, B. (1996). Ameisen, beobachten, bestimmen. Naturbuch Verlag, Augsburg.

Seifert, B. (2000). Rapid range expansion in *Lasius neglectus* (Hymenoptera, Formicidae) – an Asian invader swamps Europe. – Mitt. Mus. Nat.kd. Berl., Dtsch. entomol. Z. 47(2): 173-179.

Skibinska, E. (1978). Influence de la pression urbaine sur les Groupements de Vespidae. – Memorabilia Zool. 29: 173-181.

Skibinska, E. (1982). Sphecidae (Hymenoptera) of Warsaw and Mazovia. – Memorabilia Zool. 36: 103-127.

Skibinska, E. (1986a). Structure of Sphecidae (Hymenoptera) communities in urban green areas of Warsaw. – Memorabilia Zool. 41: 125-186.

Skibinska, E. (1986b). Structure of wasp (Hymenoptera, Vespoidea) communities in the urban green of Warsaw. – Memorabilia Zool. 42: 37-54.

Skibinska, E. (1986c). Effect of anthropogenic pressure on Vespoidea and Sphecidae communities. – Memorabilia Zool. 42: 55-66.

Sterzynska, M. (1982). Springtails (Collembola) of Warsaw and Mazovia. – Memorabilia Zool. 36: 217-234.

Sterzynska, M. (1987). Structure of springtail (Collembola) communities in the urban green of Warsaw. – Memorabilia Zool. 42 (2): 3-18.

Stork, N. (1990). The role of ground beetles in ecological and environmental studies. Intercept, Andover.

Stoyanov, I. (2004). The Ground-Beetle (Coleoptera: Carabidae) Fauna of Sofia, Bulgaria: a Checklist. – In: Ecology of the City of Sofia. Species and Communities in an Urban Environment (Penev, L., Niemelä, J., Kotze, D. J. & Chipev, N., eds). Pensoft, Sofia-Moscow, p. 401-415.

Stoyanov, I. & Penev, L. (2004). Spatial structure of carabid beetle assemblages along an urban-rural gradient. – In: Ecology of the City of Sofia. Species and Communities in an Urban

Environment (Penev, L., Niemelä, J., Kotze, D. J. & Chipev, N., eds). Pensoft, Sofia-Moscow, p. 371-400.

Suarez, A.V., Bolger, D.T.& Case, T.J. (1998). Effects of fragmentation and invasion on native ant communities in coastal Southern California. – Ecology 79(6): 2041-2056.

Sverlova, N.V. (1997). Options for conserving land molluscs (Gastropoda, Pulmonata). – Biodiversity in urban habitats. Conservation and Biodiversity in Ukraine, p. 50-50 (in Ukrainian).

Sverlova, N.V. (1998). Terrestrial molluscs as a indicator of the Parks. – Naukovii visnik 9: 63-64 (in Ukrainian).

Sverlova, N.V. (1999). Terrestrial malacocoenoses in Lvov and relation with the vegetation belts in the city. – Naukovo tovaristva imeni Shevchenka 3: 249-253 (in Ukrainian).

Sverlova, N.V. (2000). Zur städtischen Landschneckenfauna der Ukraine (Gastropoda: Pulmonata). – Malac. Abhandl. Mus. Tierek (Dresden) 20: 111-117.

Trojan, P. (1981). Urban fauna: faunistic, zoogeographical and ecological problems. – Memorabilia Zool. 34: 3-12.

Trojan, P., Górska, D. & Wegner, E. (1982). Process of synanthropization of competitive animal associations. - Memorabilia zool. 37: 125-135.

Trojan, P. (1994). The shaping of the diversity of invertebrate species in the urban green spaces of Warsaw. - Memorabilia Zool. 49: 167-173.

Venn, S.J., Kotze, D.J. & Niemelä, J. (2003). Urbanization effects on carabid diversity in boreal forests. - European Journal of Entomology 100: 73-80.

Vepsäläinen, K., Ikonen, H. & Koivula, M. (2008). The structure of ant assemblages in an urban area of Helsinki, southern Finland. - Ann. Zool. Fennici 45: 109-127.

Vepsäläinen, K. & Pisarski, B. (1982). The structure of urban ant communities along a geographical gradient from North Finland to Poland. – *Ossolineum* (Animals in urban environment. Symposium in Warszawa Labłonna), p. 155-168.

Vepsäläinen, K. & Wuorenrinne, H. (1978). Ecological effects of urbanization on the mound-building Formica L. species. - Memorabilia Zool. 29: 191-202.

Vilisics, F., Elek, Z, Lövei, G.L. & Hornung, E. (2007). Composition of terrestrial isopod assemblages along an urbanisation gradient in Denmark. - Pedobiologia 51: 45-53.

Whiteley, D. (1994). The state of knowledge of the invertebrates in urban areas in Britain with examples taken from the city of Sheffield. - Memorabilia Zool. 49: 207-220.

Winiarska, G. (1986). Noctuid moth (Lepidoptera, Noctuidae) communities in urban parks of Warsaw. - Memorabilia Zool. 42: 125-148.

Yamaguchi, T. (2004). Influence of urbanization on ant distribution in parks of Tokyo and Chiba City, Japan. - Ecol. Research 19: 209-216.